Advances in Engineering Education in the Middle East and North Africa

Mahmoud Abdulwahed • Mazen O. Hasna
Jeffrey E. Froyd
Editors

Advances in Engineering Education in the Middle East and North Africa

Current Status, and Future Insights

Springer

Editors
Mahmoud Abdulwahed
Qatar University
Doha, Qatar

Mazen O. Hasna
Qatar University
Doha, Qatar

Jeffrey E. Froyd
Texas A&M University
College Station, TX, USA

ISBN 978-3-319-15322-3 ISBN 978-3-319-15323-0 (eBook)
DOI 10.1007/978-3-319-15323-0

Library of Congress Control Number: 2015954174

Springer Cham Heidelberg New York Dordrecht London

Printed on acid-free paper

Springer International Publishing AG Switzerland is part of Springer Science+Business Media (www.springer.com)

Foreword

Academia: By Dr Sheikha Al-Misnad, President of Qatar University With ambitious plans to transition to a knowledge-based economy, Qatar is witnessing tremendous growth in the industrial and information sectors. Strategically, the country has made significant investments in establishing a modern education system, drawing on international best practices while firmly rooted in local needs. The Qatar National Vision 2030 presents a model national vision that articulates the focus on building human capacity. With a national mandate to support and strengthen science, technology, engineering, and mathematics (STEM) disciplines, the focus on engineering is now stronger than ever.

The College of Engineering at Qatar University has been one of the leading colleges in the region in terms of its investment in the engineering education curriculum, practice, and scholarship. The College of Engineering first attained Accreditation Board for Engineering and Technology (ABET) accreditation in 2005 as a testament to its quality of teaching, research, and scholarly activity. Since then, there has been significant growth in engineering teaching and learning innovations, K–12 outreach projects, education research funding, and scholarly publications in the area of engineering education. This book is the college's contribution to raising awareness in the Middle East and North Africa (MENA) of the significance of engineering education. It stands as one of the very few available scholarly compilations of advances in engineering education in the MENA region, and Qatar University is tremendously proud of this seeding contribution.

I want to thank the authors and editors of the book for their efforts. I am also grateful to Qatar Petrochemical Company (QAPCO) for supporting this milestone initiative, trusting this book makes a valuable addition to the advancement of engineering education in the MENA region and beyond.

Industry: By Dr Mohammed Al-Mulla, Vice Chairman and Chief Executive Officer, Qatar Petrochemical Company (QAPCO) The economy of Qatar as well as many other neighboring countries depends to a larger extent on engineering. Regional engineering colleges provide a main source of engineering graduates, and hence the quality and outcomes of engineering education in MENA engineering colleges are of significant importance to the MENA industrial sector. From an industrial perspective, engineering graduates are needed with both technical and soft skills. Competencies such as teamwork, communication, leadership, and project management are as important as sound knowledge in engineering and sciences fundamentals. Furthermore, the industry sector urges engineering colleges to embed higher levels of experiential and industrially linked practices into the engineering curriculum. Industry increasingly demands for engineers with higher levels of inter- and multidisciplinary knowledge and experience; in particular, in the Middle East and North Africa, the latter assertion is of significant importance. As industry, we believe that a systematic innovation in engineering education driven by contextual needs, scholarly research, and highly linked with industrial stakeholders could drive the quality of engineering graduates to significantly higher levels. This book contribution is a significant initial step in sharing best practices and innovations in engineering education in the MENA region, and is an eye-opener on the emerging field of engineering education research in a region where it is not that popular. We are pleased in QAPCO to have collaborated with a national engineering college, the College of Engineering at Qatar University, to bring various regional innovations in engineering education into a scholarly book. We hope this could be a starting point where both industry and academia in the region work together on systematic and sustainable scholarship and research in engineering education for meeting contextual and global needs.

Acknowledgments

This work has been developed under a grant funded by Qatar Petrochemical Company (QAPCO).

List of International Reviewers

The editors would like to acknowledge the efforts of the international experts reviewers listed below who kindly peer-reviewed the contributed chapters in this book.

Name	Affiliation	Country
Ramzi Nasser	Chief Reviewer, Qatar University	Qatar
Ali Elkamel	University of Waterloo	Canada
Sandra Courter	University of Wisconsin–Madison	USA
Demetrios G. Sampson	University of Piraeus	Greece
John M. Long	Deakin University	Australia
Brad Stappenbelt	University of Wollongong	Australia
John W. Prados	The University of Tennessee	USA
Clint Steele	Swinburne University of Technology	Australia
Julia Williams	Rose-Hulman Institute of Technology	USA
Brett Shelton	Boise State University	USA
David Soldan	Kansas State University	USA
John E. Mitchell	University College London	UK
Mark B. Yeary	University of Oklahoma	USA
Lahiru Silva	University of Calgary	Canada
Denis Bédard	L'Université de Sherbrooke	Canada
Michele Rosano	Curtin University	Australia
Ahmed Samer Ezeldin	The American University in Cairo	Egypt
Johannes Strobel	Texas A&M University	USA
Shannon M. Chance	Hampton University	USA
Edwin C. Jones	Iowa State University	USA
Omar Al Zoubi	Carnegie Mellon University in Qatar	USA
Linda Vanasupa	California Polytechnic State University	USA
Thomas J. Joyce	Newcastle University	UK
Corey Schimpf	Purdue University	USA
Arto Vihavainen	University of Helsinki	Finland
Kim Blackmore	Australian National University	Australia
Ivica Boticki	University of Zagreb	Croatia

Name	Affiliation	Country
Jonathan Hilpert	Georgia Southern University	USA
Eugenio Pellicer	Polytechnic University of Valencia	Spain
Ezequiel Scott	The National University of the Center of the Buenos Aires	Argentina
Jørgen Erik Christensen	Technical University of Denmark	Denmark
Michael Buckley	University at Buffalo	USA
Gerald Burnham	University of Texas at Dallas	USA
David Cropley	University of South Australia	Australia
Niall Holmes	Dublin Institute of Technology	Ireland
Leonhard Emil Bernold	University of New South Wales	Australia
Kali P. Nepal	Deakin University	Australia
K. W. Chau	Hong Kong Polytechnic University	Hong Kong
Rick L. Homkes	Purdue University	USA
Sivachandran Chandrasekaran	Deakin University	Australia
Sally A. Male	University of Western Australia	Australia
S. K. Ramesh	California State University, Northridge	USA
Richard Burns	West Chester University	USA
Keith Sheppard	Stevens Institute of Technology	USA
S. Mohd Said	University of Malaya	Malaysia
Rui M. Lima	University of Minho	Portugal
James Trevelyan	University of Western Australia	Australia
Kinda Khalaf	Khalifa University of Science, Technology and Research	UAE
Matthew Scott Stanford	United States Air Force Academy	USA
Patricia Brackin	Rose-Hulman Institute of Technology	USA
Lars G. Hanson	Technical University of Denmark	Denmark
Aleksandra Klašnja-Milicevica	University of Novi Sad	Serbia
Mahmoud El-Halwagi	Texas A&M University	USA

Contents

Contributors

Mahmoud Abdulwahed Qatar University, Doha, Qatar

Manal A. Al-Bishawi Faculty of Engineering and Technology, Architectural Engineering Department, Birzeit University-Palestine, Birzeit, Palestine

Saleh Alhazbi Department of Computer Science and Engineering, Qatar University, Doha, Qatar

Mohamed Ally Centre for Distance Education, Athabasca University, Athabasca, AB, Canada

Yacob Astatke Morgan State University, Baltimore, MD, USA

Shadi Balawi Department of Aerospace Engineering, Khalifa University of Science, Technology, and Research, Abu Dhabi, United Arab Emirates

Rezaul K. Chowdhury Department of Civil and Environmental Engineering, United Arab Emirates University, Al Ain, United Arab Emirates

Kenneth Connor Rensselaer Polytechnic Institute, Troy, NY, USA

Maurice Danaher Zayed University, Dubai and Abu Dhabi, United Arab Emirates

Jennifer DeBoer Purdue University, West Lafayette, IN, USA

Mohamed Essaaidi National Higher School of Information Technology (ENSIAS), Mohammed V University, Rabat, Morocco

Jeffrey E. Froyd Engineering Academic and Student Affairs, Texas A&M University, College Station, TX, USA

Erik de Graaff UCPBL, Aalborg University, Aalborg, Denmark

Abdel Magid Hamouda Mechanical and Industrial Engineering Department, College of Engineering, Qatar University, Doha, Qatar

Farook Hamzeh Department of Civil and Environmental Engineering, American University of Beirut, Beirut, Lebanon

Mazen O. Hasna Qatar University, Doha, Qatar

G. Wesley Hitt Department of Applied Mathematics and Sciences, Khalifa University of Science, Technology, and Research, Abu Dhabi, United Arab Emirates

Petronella James Morgan State University, Baltimore, MD, USA

Leon Jololian Zayed University, Dubai and Abu Dhabi, United Arab Emirates

Kinda Khalaf Department of Biomedical Engineering, Khalifa University of Science, Technology, and Research, Abu Dhabi, United Arab Emirates

Jumoke O. Ladeji-Osias Morgan State University, Baltimore, MD, USA

Munjed A. Maraqa Civil and Environmental Engineering Department, United Arab Emirates University, Al-Ain, United Arab Emirates

Farzad Moazzami Morgan State University, Baltimore, MD, USA

Mohamed M. Mohamed Civil and Environmental Engineering Department, United Arab Emirates University, Al-Ain, United Arab Emirates

Shahrin Mohammad Centre for Engineering Education, Universiti Teknologi Malaysia, Johor Bahru, Malaysia

Ramzi N. Nasser Dhofar University, Salalah, Oman

Mabrouk Ouederni Qatar Petrochemical Company (QAPCO), Doha, Qatar

Fatin Aliah Phang Centre for Engineering Education, Universiti Teknologi Malaysia, Johor Bahru, Malaysia

Michael H. Romanowski Qatar University College of Education, Qatar University, Doha, Qatar

Carel Rouhana Department of Civil and Environmental Engineering, American University of Beirut, Beirut, Lebanon

Abdurrahim Saka Yeni şafak, Istanbul, Turkey

Mohammed Samaka Department of Computer Science & Engineering, Qatar University, Doha, Qatar

Kevin Schoepp Zayed University, Dubai and Abu Dhabi, United Arab Emirates

Craig Scott Morgan State University, Baltimore, MD, USA

Amr M. I. Sweedan Civil and Environmental Engineering Department, UAE University, Al-Ain, United Arab Emirates

Faris Tarlochan Mechanical and Industrial Engineering Department, College of Engineering, Qatar University, Doha, Qatar

Christina Teokari Department of Civil and Environmental Engineering, American University of Beirut, Beirut, Lebanon

Khairiyah Mohd Yusof Centre for Engineering Education, Universiti Teknologi Malaysia, Johor Bahru, Malaysia

Essam Zaneldin Civil and Environmental Engineering Department, UAE University, Al-Ain, United Arab Emirates

About the Editors

Mahmoud Abdulwahed has completed his BSc, MSc, and PhD in electrical, control, and systems engineering. He continued his postgraduate studies in Germany, Sweden, France, and the UK where he worked before joining the College of Engineering at Qatar University in fall 2011. His main expertise is in innovation and education; he has published in technology, enhanced- & e-learning, engineering & STEM education, entrepreneurship, and leadership. Mahmoud has authored 60+ peer-reviewed conference and journal articles, and has attained a number of recognitions, awards, and fellowships alongside his studies and career. Mahmoud is currently the adviser of strategic initiatives for the vice president and chief academic officer at Qatar University, where his role focuses on inception, conceptualization, and/or management of large-scale strategic initiatives; he is also a faculty member with the College of Engineering.

Mazen O. Hasna received a BS degree from Qatar University, Doha, Qatar in 1994, an MS degree from the University of Southern California (USC), Los Angeles in 1998, and a PhD degree from the University of Minnesota, Twin Cities in 2003, all in electrical engineering. In 2003, Dr. Hasna joined the department of electrical engineering at Qatar University as an assistant professor. Currently, he serves as the vice president and chief academic officer of Qatar University. Dr. Hasna's research interests span the general area of digital communication theory and its application to performance evaluation of wireless communication systems over fading channels. Current specific research interests include cooperative communications, ad hoc networks, cognitive radio, and network coding.

Jeffrey E. Froyd is a TEES research professor in the Office of Engineering Academic and Student Affairs at Texas A&M University, College Station. He received a BS degree in mathematics from Rose-Hulman Institute of Technology and MS and PhD degrees in electrical engineering from the University of Minnesota, Minneapolis. He was an assistant professor, associate professor, and professor of electrical and computer engineering at Rose-Hulman Institute of Technology. At Rose-Hulman, he co-created the integrated first-year curriculum in science, engineering and mathematics, which was recognized in 1997 with a Hesburgh Award Certificate of Excellence. He served as project director at the National Science Foundation (NSF)

Engineering Education Coalition in which six institutions systematically renewed, assessed, and institutionalized innovative undergraduate engineering curricula. He has authored over 70 journal articles and conference papers and offered over 30 workshops on faculty development, curricular change processes, curriculum redesign, and assessment. He has served as a program co-chair for three Frontiers in Education conferences and the general chair for the 2009 conference. He also serves on the IEEE Curricula and Pedagogy Committee, which is part of the University Resources Committee, which is part of the Educational Activities Board. Prof. Froyd is a fellow of the IEEE, a fellow of the American Society for Engineering Education (ASEE), an ABET program evaluator, the editor-in-chief for the *IEEE Transactions on Education*, a senior associate editor for the *Journal of Engineering Education*, and an associate editor for the *International Journal of STEM Education*

Part I
Background, Editorship, and Global Perspectives

Chapter 1
Purpose of the Book and Editorial Perspectives

Mahmoud Abdulwahed, Mazen O. Hasna and Jeffrey E. Froyd

1.1 Purpose of the Book

Engineering and engineering education play an essential role in the Middle East. Global debates regarding roles of engineering and required transformations in engineering education have been extensive during the past few years. The past decade has witnessed the emergence of engineering education research as an independent field, led by tremendous investment in the USA. Some institutions have started embracing radical and transformative approaches to engineering education and curriculum design driven by recent research findings and pressing national needs, for example, Georgia Tech., Olin College, and Singapore University of Technology and Design. Parallel to this, engineering education in the Middle East has witnessed significant shifts, both quantitatively and qualitatively. For instance, many national institutions have standardized their curricula through Accreditation Board for Engineering and Technology (ABET) accreditation, many international universities have opened branch campuses in various Middle Eastern countries, and the number of new universities is rapidly increasing in many countries (e.g., Saudi Arabia, Turkey, and United Arab Emirates). Furthermore, engineering continues to be a preferred higher education choice for a significant portion of Middle Eastern students compared to the trends currently witnessed in the West. Nevertheless, apart

M. Abdulwahed (✉) · M. O. Hasna
Qatar University, Doha, Qatar
e-mail: m.abdulwahed@qu.edu.qa

M. O. Hasna
e-mail: hasna@qu.edu.qa

J. E. Froyd
Engineering Academic and Student Affairs, Texas A&M University, College Station, TX, USA
e-mail: froyd@tamu.edu

M. Abdulwahed et al. (eds.), *Advances in Engineering Education in the Middle East and North Africa,* DOI 10.1007/978-3-319-15323-0_1

from seeking ABET accreditation, efforts to innovate in engineering curricula and/or adopt/adapt novel approaches in the Middle East seem to be isolated, and depend, to a large extent, on instructors' choice.

Lack of systematic, comprehensive documentation on undergraduate engineering curricula, practices for teaching and curriculum development, and the status of engineering and engineering education in the Middle East hampers constructive conversations about how to improve engineering education in the Middle East. Without clearer pictures of the current state, progress to envisioned, future states is more difficult. This book aims to help address this gap in the literature on engineering education, and to stimulate intellectual and critical discourse on the next wave of engineering innovation and education in the Middle East.

The book provides a set of recent innovations in engineering education in the Middle East spanning four dimensions: (1) learning and teaching, (2) curriculum development, (3) assessment and accreditation, and (4) challenges and sustainability. This has been preceded by introductory and concluding shorter sections. At the end of the book, a set of recommendations for implementations in engineering education and engineering colleges is provided.

1.2 Perspectives on Engineering Education in the Middle East

Engineering education in the Middle East and North Africa (MENA) region can be rooted to the eleventh and twelfth centuries when the advancement in several sciences was booming in this part of the world (Al-Hassani 2012). At that time, differentiation between disciplines was not clear, and it was common to find scholars who studied medicine, astronomy, chemistry, etc. Engineering was not known as a discipline with such a title, and formal education did not exist at that time. Rather, what is called today as "applied sciences" was studied and practiced as a discipline. Scientists at that time managed to discover many scientific theories and put them into practice. Formal documentation of modern engineering education in the MENA region is also missing and scattered, and this can be a mega project on its own. Most engineering colleges in the region started in the middle of the twentieth century and have been evolving ever since. In most cases, the colleges followed the engineering education models of those countries that occupied their states in the past.

In the Gulf region, attention to engineering sciences started with the discovery of oil in the beginning of the twenty-first century. Initially, countries relied on foreign international companies to take care of the business. In Qatar, for example, it was till 1980 when the first engineering school started. Prior to that, Qatar Petroleum (or Qatar General Petroleum Corporation as it was called) relied on sending very few students to pursue engineering studies abroad, mainly in the USA. In 1980, a male-only engineering college started as part of the Qatar

University (QU), comprising four disciplines: electrical, mechanical, chemical, and civil engineering. It comes as a surprise that petroleum engineering is missing from this college, but at the point of time that the college was created, this discipline was suffering from an overflow in the market need worldwide. The college followed an American model of studies (credit hours based), and so did most, if not all, engineering colleges in the Gulf region. On the contrary, most engineering schools in the rest of the MENA region followed a yearly system of studies with predefined modules for each year.

The College of Engineering at QU started with less than 30 students. In parallel, the scholarship program of the government continued to send students to pursue engineering outside Qatar. Engineering education in Qatar did not witness any major change in both quantity and quality until the late 1990s when there was a decision to pursue ABET accreditation following the newly developed criterion promoting outcome-based assessment (EC2000). In few years, another major change happened in Qatar when the Qatar Foundation (QF) signed an agreement with Texas A&M in 2003 to open a branch in Qatar. The agreement included establishing four coed engineering programs in electrical, mechanical, chemical, and petroleum disciplines. Parallel to that, QU has been going through a uniform process that affected the engineering college, and computer science was moved from the science college to join engineering. More interestingly, two new engineering programs were established targeting female students only, namely, computer engineering and industrial and systems engineering. Following that, both Texas A&M University at Qatar (TAMUQ) and College of Engineering at QU have been growing while adding programs on electrical and chemical engineering for females at QU.

Up to almost 2010, engineering education in Qatar focused on "education" only with little research component and no graduate studies. In 2011, the first PhD program in Qatar as a whole was introduced in QU and in the engineering discipline. Following that, many master's programs were introduced in both QU and TAMUQ. The research component has grown substantially also with the establishment of the Qatar National Research Fund (QNRF) in 2008 through which most of the funding has gone to engineering topics so far.

With the release of Qatar National Vision 2030, the role of engineering has come to the focus again due to the core role engineering plays in transforming and maintaining a knowledge-based economy (KBE; Abdulwahed et al. 2013). A new university (Hamad Bin Khalifa University) was established with engineering, and design is one of the focus areas. It is expected that this new college will focus more on graduate studies in engineering than undergraduate degrees.

Looking ahead, the future of engineering education in the MENA region looks promising. ABET accreditation (or equivalent) is no longer thought of as a prestige, and most reputable engineering schools in the region have that seal now. The expansion now is twofold: first, an increased focus on research and graduate studies to serve the KBE transformation and, second, a relook at the effectiveness of undergraduate degrees with a trial to have a deeper professional component.

1.3 Perspectives on Engineering Education in North America

Since 1900, five significant shifts in engineering education that have occurred or are occurring have been documented (Froyd et al. 2012): (1) dominance of engineering science over engineering practice (Grinter 1955; Seely 1999), (2) accreditation of engineering programs based on student achievement of learning outcomes (Prados et al. 2005), (3) escalating resurgence of engineering design (Seely 1999), (4) increasing application of research on learning and teaching (Froyd and Lohmann 2014; Singer et al. 2012), and (5) growing importance of computing, information, and communication technologies in engineering education (e.g., massive open online courses (MOOCs) and personal response systems (Kay and LeSage 2009)). The first two shifts have been accomplished and the implications for engineering education are well understood. However, influences of the last three shifts, in both North America and the entire world, are less clear because these shifts are still in progress. Implications of the last three shifts will be considered in the remainder of this introductory chapter.

The first of the three shifts still in progress, increasing emphasis on engineering design (Dym et al. 2005) has encouraged engineering educators to explore ways to assess and develop engineering design skills. The most common way that engineering students encounter design is the capstone engineering course(Dutson et al. 1997; Howe 2010; Todd et al. 1995). Another common curriculum feature that emphasizes engineering design is one or more design courses in the first year (Brannan and Wankat 2005; Sheppard and Jennison 1997). However, few 4-year engineering curricula feature an emphasis on engineering design in the second or third year (Sheppard et al. 2008). However, instructional strategies, such as problem-based learning, project-based learning, inquiry-based learning, service learning, and other approaches that have been described as inductive teaching strategies (Prince and Felder 2006), are being explored to increase emphasis on developing engineering design knowledge and skills, especially where there are not explicit courses on engineering design in the second and third years. Litzinger et al. have synthesized "the literature on development of expertise, students' approaches to learning, students' responses to instructional practices, and the role of motivation in learning [to formulate] a set of instructional practices that are most likely to elicit learning that supports the development of expert professional practice" (Litzinger et al. 2011), including design. Complimenting broadening implementation of inductive instructional strategies is the development of approaches to assessing and evaluating development of design knowledge and skills (Davis et al. 2002). Formative assessment and evaluation can provide feedback to students (Black and Wiliam 1998) that they can use to improve the way they approach engineering design problems.

The second shift, employing research on learning and teaching, has developed multiple instructional strategies that promote student engagement with content

(Smith et al. 2005), such as various approaches to active learning (Bonwell and Eison 1991), peer instruction (Mazur 1997), cooperative learning (Johnson et al. 2006), peer-led team learning, formative assessment (Nicol and Macfarlane-Dick 2006), and team-based learning (Michaelsen and Sweet 2011). Efficacy of these approaches has been demonstrated in multiple studies conducted in diverse learning environments over the last 20 years (Freeman et al. 2014; Hake 1998; Hoffman et al. 2006; Olson and Riordan 2012; Prince 2004; Springer et al. 1999; Tien et al. 2002; Woods et al. 1997). The major challenge over the next decade will be broadening implementation of instructional approaches that have demonstrated results with respect to learning that are superior to the results obtained via lectures (Froyd et al. 2008). Some studies have provided information on the extent to which adoption of research-based instructional strategies has occurred (Borrego et al. 2010; Froyd et al. 2013; Henderson et al. 2012; Prince et al. 2013).

Continuing developments in computing, information, and communication technologies have been used to develop instructional approaches such as blended learning (Derntl and Motschnig-Pitrik 2005; Garrison and Kanuka 2004), hybrid learning, just-in-time teaching (Novak and Patterson 1998), and flipped classrooms in which students are expected to have engaged with course content, often prior to coming to physical classrooms. The same technologies have also been applied to offer courses and degree programs completely online. The current culmination of online courses has been MOOCs, which provide widespread access to course content. All of these instructional and assessment approaches now offer engineering educators diverse, and sometimes overwhelming, options for course design. One of the challenges for engineering educators in using instructional approaches and applying developments in computing, information, and communication technologies is that many of the examples focus on delivery of content to students. However, a major task for engineering educators is evaluation of the extent to which students have achieved course learning outcomes. Classroom response systems provide one excellent example of the use of instructional technologies for formative assessment. Another example is the use of learning analytics, systematic analysis of student learning activities (e.g., attendance, submission of homework problems, quiz scores, and examination scores) to identify patterns (e.g., what students are more likely at risk to do poorly in a course) that can guide interventions. One example of products that support learning analytics is Civitas (Civitas Learning 2013). More examples of technology-enabled assessment methods are needed.

Implications of the three shifts still in progress—(1) increased emphasis on engineering design, (2) increasing application of research on learning and teaching, and (3) increasing application of technological developments in communications, computing, and information—have broadened the scope of possibilities that engineering educators around the world can use while designing or redesigning their courses. One of the motivations for this book is to examine how engineering educators in the Middle East and North Africa have been responding to the increasing range of instructional and assessment approaches.

References

Abdulwahed, M., Pokharel, S., Balid, W., & Hasna, M. O. (2013). Systemic framework of competencies of 21st century technical and engineering human resource: A development case for a knowledge based economy in Qatar. In GCC Conference and Exhibition (GCC), November 7th IEEE (pp. 413–418). IEEE.

Al-Hassani, S. T. S. (Ed.). (2012). *1001 inventions: The enduring legacy of Muslim civilization.* Washington, DC: National Geographic.

Black, P., & Wiliam, D. (1998). Assessment and classroom learning. *Assessment in Education: Principles, Policy & Practice, 5*(1), 7–74. doi:10.1080/0969595980050102.

Bonwell, C. C., & Eison, J. A. (1991). *Active learning: Creating excitement in the classroom.* Washington, DC: George Washington University Press.

Borrego, M., Froyd, J. E., & Hall, T. S. (2010). Diffusion of engineering education innovations: A survey of awareness and adoption rates in U.S. engineering departments. *Journal of Engineering Education, 99*(3), 185–207. doi:10.1002/j.2168-9830.2010.tb01056.x.

Brannan, K. P., & Wankat, P. C. (2005). *Survey of first-year programs.* Paper presented at the ASEE Annual Conference & Exposition, Portland, OR, USA. http://search.asee.org/search/fetch?url=file%3A%2F%2Flocalhost%2FE%3A%2Fsearch%2Fconference%2F29%2FAC%25202005Paper855.pdf&index=conference_papers&space=129746797203605791716676178&type=application%2Fpdf&charset=.

Civitas Learning. (2013). Civitas Learning. http://www.civitaslearning.com/. Accessed 15 Oct 2014.

Davis, D. C., Gentili, K. L., Trevisan, M. S., & Calkins, D. E. (2002). Engineering design assessment processes and scoring scales for program improvement and accountability. *Journal of Engineering Education, 91*(2), 211–221. doi:10.1002/j.2168-9830.2002.tb00694.x.

Derntl, M., & Motschnig-Pitrik, R. (2005). The role of structure, patterns, and people in blended learning. *The Internet and Higher Education, 8*(2), 111–130. doi:10.1016/j.iheduc.2005.03.002.

Dutson, A. J., Todd, R. H., Magleby, S. P., & Sorensen, C. D. (1997). Engineering design through project-oriented capstone courses. *Journal of Engineering Education, 86*(1), 17–28.

Dym, C. M., Agogino, A. M., Eris, O., Frey, D., & Leifer, L. J. (2005). Engineering design thinking, teaching, and learning. *Journal of Engineering Education, 94*(1), 103–120.

Freeman, S., Eddy, S. L., McDonough, M., Smith, M. K., Okoroafor, N., Jordt, H., & Wenderoth, M. P. (2014). Active learning increases student performance in science, engineering, and mathematics. *Proceedings of the National Academy of Sciences of the United States of America.* doi:10.1073/pnas.1319030111.

Froyd, J. E., & Lohmann, J. R. (2014). Chronological and ontological development of engineering education as a field of scientific inquiry. In A. Johri & B. M. Olds (Eds.), *Cambridge handbook of engineering education research.* New York: Cambridge University Press.

Froyd, J. E., Beach, A., Henderson, C., & Finkelstein, N. (2008). Improving educational change agents' efficacy in science, engineering, and mathematics education. In H. Hartman (Ed.), *Integrating the sciences and society: Challenges, practices, and potentials* (Vol. 16, pp. 227–255). London: Emerald Group Publishing Limited.

Froyd, J. E., Wankat, P. C., & Smith, K. A. (2012). Five major shifts in 100 years of engineering education. *Proceedings of the IEEE, 100*(13), 1344–1360. doi:10.1109/JPROC.2012.2190167.

Froyd, J. E., Borrego, M., Cutler, S., Henderson, C., & Prince, M. J. (2013). Estimates of use of research-based instructional strategies in core electrical or computer engineering courses. *IEEE Transactions on Education, 56*(4), 393–399. doi:10.1109/TE.2013.2244602.

Garrison, D. R., & Kanuka, H. (2004). Blended learning: Uncovering its transformative potential in higher education. *The Internet and Higher Education, 7*(2), 95–105. doi:10.1016/j.iheduc.2004.02.001.

Grinter, L. E. (1955). Report on evaluation of engineering education. *Journal of Engineering Education, 46*(1), 25–63.

Hake, R. R. (1998). Interactive-engagement versus traditional methods: A six-thousand-student survey of mechanics test data for introductory physics courses. *American Journal of Physics, 66*(1), 64–74. doi:10.1119/1.18809.

Henderson, C., Dancy, M. H., & Niewiadomska-Bugaj, M. (2012). Use of research-based instructional strategies in introductory physics: Where do faculty leave the innovation-decision process? *Physical Review Special Topics—Physics Education Research, 8*(2), 020104–020115. doi:10.1103/PhysRevSTPER.8.020104.

Hoffman, K., Hosokawa, M., Blake, R., Headrick, L., & Johnson, G. (2006). Problem-based learning outcomes: Ten years of experience at the University of Missouri-Columbia School of Medicine. *Academic Medicine, 81*(7), 617–625. doi:10.1097/01.ACM.0000232411.97399.c6.

Howe, S. (2010). Where are we now? Statistics on capstone courses nationwide. *Advances in Engineering Education, 2*(1), 1–27. http://advances.asee.org/vol02/issue01/03.cfm.

Johnson, D. W., Johnson, R. T., & Smith, K. A. (2006). *Cooperation in the college classroom* (3rd ed.). Edina: Interaction Book Company.

Kay, R. H., & LeSage, A. (2009). Examining the benefits and challenges of using audience response systems: A review of the literature. *Computers & Education, 53*(3), 819–827. doi:10.1016/j.compedu.2009.05.001.

Litzinger, T. A., Lattuca, L. R., Hadgraft, R. G., & Newstetter, W. C. (2011). Engineering education and the development of expertise. *Journal of Engineering Education, 100*(1), 123–150.

Mazur, E. (1997). *Peer instruction: A user's manual*. Englewood Cliffs: Prentice Hall.

Michaelsen, L. K., & Sweet, M. (2011). Team-based learning. *New Directions for Teaching and Learning, 128*, 41–51. doi:10.1002/tl.467.

Nicol, D. J., & Macfarlane-Dick, D. (2006). Formative assessment and self-regulated learning: a model and seven principles of good feedback practice. *Studies in Higher Education, 31*(2), 199–218. doi:10.1080/03075070600572090.

Novak, G. M., & Patterson, E. T. (1998). *Just-in-time teaching: Active learning pedagogy with WWW*. Paper presented at the IASTED International Conference on Computers and Advanced Technology in Education, Cancun, Mexico. http://webphysics.iupui.edu/JITT/ccjitt.html.

Olson, S., & Riordan, D. G. (2012). Engage to excel: Producing one million additional college graduates with degrees in science, technology, engineering, and mathematics. Report to the President. Washington, DC, USA: Executive Office of the President, President's Council of Advisors on Science and Technology.

Prados, J. W., Peterson, G. D., & Lattuca, L. R. (2005). Quality assurance of engineering education through accreditation: The impact of Engineering Criteria 2000 and its global influence. *Journal of Engineering Education, 94*(1), 165–184. doi:10.1002/j.2168-9830.2005.tb00836.x.

Prince, M. J. (2004). Does active learning work? A review of the research. *Journal of Engineering Education, 93*(3), 223–231. doi:10.1002/j.2168-9830.2004.tb00809.x.

Prince, M. J., & Felder, R. M. (2006). Inductive teaching and learning methods: Definitions, comparisons, and research bases. *Journal of Engineering Education, 95*(2), 123–128. doi:10.1002/j.2168-9830.2006.tb00884.x.

Prince, M. J., Borrego, M., Cutler, S., Henderson, C., & Froyd, J. E. (2013). Use of research-based instructional strategies in core chemical engineering courses. *Chemical Engineering Education, 47*(1), 27–37.

Seely, B. E. (1999). The other re-engineering of engineering education, 1900–1965. *Journal of Engineering Education, 88*(3), 285–294. doi:10.1002/j.2168-9830.1999.tb00449.x.

Sheppard, S., & Jennison, R. (1997). Freshman engineering design experiences and organizational framework. *International Journal of Engineering Education, 13*(3), 190–197.

Sheppard, S., Macatangay, K., & Colby, A. (2008). *Educating engineers: Designing for the future of the field*. San Francisco: Jossey-Bass.

Singer, S. R., Nielsen, N. R., & Schweingruber, H. A. (Eds.). (2012). *Discipline-based education research: Understanding and improving learning in undergraduate science and engineering*. Washington, DC: National Academies Press.

Smith, K. A., Sheppard, S. D., Johnson, D. W., & Johnson, R. T. (2005). Pedagogies of engagement: Classroom-based practices. *Journal of Engineering Education, 94*(1), 87–101.

Springer, L., Stanne, M. E., & Donovan, S. S. (1999). Effects of small-group learning on undergraduates in science, mathematics, engineering, and technology: A meta-analysis. *Review of Educational Research, 69*(1), 21–51. doi:10.3102/00346543069001021.

Tien, L. T., Roth, V., & Kampmeier, J. A. (2002). Implementation of a peer-led team learning instructional approach in an undergraduate organic chemistry course. *Journal of Research in Science Teaching, 39*(7), 606–632. doi:10.1002/tea.10038.

Todd, R. H., Magleby, S. P., Sorensen, C. D., Swan, B. R., & Anthony, D. K. (1995). A survey of capstone engineering courses in North America. *Journal of Engineering Education, 84*(2), 165–174.

Woods, D. R., Hrymak, A. N., Marshall, R. R., Wood, P. E., Crowe, C. M., Hoffman, T. W., Wright, J. D., Taylor, P. A., Woodhouse K. A., & Bouchard, C. G. K. (1997). Developing problem solving skills: The McMaster problem solving program. *Journal of Engineering Education, 86*(2), 75–91.

Mahmoud Abdulwahed completed his BSc, MSc, and PhD in electrical, control, and systems engineering. He continued his postgraduate studies in Germany, Sweden, France, and the UK where he worked before joining the College of Engineering at the Qatar University in fall 2011. His main expertise is in innovation and education; he has published in technology-enhanced and e-learning, engineering and science, technology, engineering, and mathematics (STEM) education, entrepreneurship, and leadership. Mahmoud has written more than 60 peer-reviewed conference and journal articles, and has attained a number of recognitions, awards, and fellowships alongside his studies and career. He is currently the adviser of strategic initiatives for the vice president and chief academic officer at the Qatar University, where his role is focused on inception, conceptualization, and/or management of large-scale strategic initiatives; he is also a faculty member at the College of Engineering.

Mazen O. Hasna received his BS degree from the Qatar University, Doha, Qatar, in 1994; MS degree from the University of Southern California (USC), Los Angeles, in 1998; and PhD degree from the University of Minnesota, Twin Cities, in 2003—all in electrical engineering. In 2003, Dr. Hasna joined the Department of Electrical Engineering at the Qatar University as an assistant professor. Currently, he serves as the vice president and chief academic officer of the Qatar University. Dr. Hasna's research interests span the general area of digital communication theory and its application to performance evaluation of wireless communication systems over fading channels. His current specific research interests include cooperative communications, ad hoc networks, cognitive radio, and network coding.

Jeffrey E. Froyd is a Texas Engineering Experiment Station (TEES) research professor in the Office of Engineering Academic and Student Affairs at Texas A&M University, College Station. He received his BS degree in mathematics from the Rose–Hulman Institute of Technology and MS and PhD degrees in electrical engineering from the University of Minnesota, Minneapolis. He was an assistant professor, associate professor, and professor of electrical and computer engineering at Rose–Hulman Institute of Technology. At Rose–Hulman, he cocreated the Integrated, First-Year Curriculum in Science, Engineering and Mathematics, which was recognized in 1997 with a Hesburgh Award Certificate of Excellence. He served as project director of a National Science Foundation (NSF) Engineering Education Coalition in which six institutions systematically renewed, assessed, and institutionalized innovative undergraduate engineering curricula. He has authored more than 70 journal articles and conference papers and has offered more than 30 workshops on faculty development, curricular change processes, curriculum redesign, and assessment. He has served as a program cochair for three Frontiers in Education conferences and the general chair for the 2009 conference. He also serves on the Institute of Electrical and Electronics Engineers (IEEE) Curricula and Pedagogy Committee, which is part of the University Resources Committee, which is part of the Educational Activities Board. Prof. Froyd is a fellow of the IEEE and the American Society for Engineering Education (ASEE), an Accreditation Board for Engineering and Technology (ABET) program evaluator, the editor in chief for the *IEEE Transactions on Education, a senior associate editor for the Journal of Engineering Education, and an associate editor for the International Journal of STEM Education.*

Chapter 2
Developments in Engineering Education and Engineering Education Research in Europe

Erik de Graaff

2.1 Introduction

Our civilisation is highly dependent on engineering. We need engineers to design and manage the tools that our lifestyle requires, from mechanised agriculture to electronic gadgets. Without exaggeration, we can say that engineering education is the key to future success. This chapter aims to present an overview of the development of engineering education in Europe. The cultural diversity of so many different nations results in a fragmented picture of the historical developments with examples from just a few countries. The original economic-driven initiative of the European Union (EU; The European Coal and Steel Community, ECSC) established in 1951 paved the way for a political union and a desire to extend the collaboration to other areas including higher education.

An important step in this development was the signing of the Bologna declaration by the ministers of education of 29 European countries on 19 June 1999. The Bologna declaration propositioned a European Higher Education Area in which students and graduates could move freely between countries, using prior qualifications in one country as acceptable entry requirements for further study in another. In the wake of the Bologna declaration, an ambitious tuning project was started aiming to equate the different systems of higher education while maintaining the rich diversity of European education and the independence of academic and subject specialisms. As a result, Europe is more and more able to set uniform goals and to develop a common policy in promoting Europe as a knowledge society.

Furthermore, this chapter presents an overview of the present situation of engineering education and the development of engineering education research (EER), which is subsequently analysed to provide a foundation for a vision of future developments.

E. de Graaff (✉)
UCPBL, Aalborg University, Aalborg, Denmark
e-mail: degraaff@plan.aau.dk

M. Abdulwahed et al. (eds.), *Advances in Engineering Education in the Middle East and North Africa,* DOI 10.1007/978-3-319-15323-0_2

2.2 A Historical Perspective on Engineering in Europe

2.2.1 Introduction

Engineers have built Europe. All over Europe, man-made structures dominate the landscape. Without engineers, half of the country of the Netherlands would disappear below the waves of the North Sea. Europe also played a major part in the development of engineering. After a few early engineering feats, like the pyramids in Egypt, the hanging gardens in Babylon or the great wall in China, the initiative moved to Greece and later Rome. The Greek contributions to mathematics greatly helped the development of engineering. In the sixth century B.C., with the application of geometry principles, the Greek engineer Eupalinos managed to dig a tunnel through Mount Kastro at Samos, starting from the two ends and meeting in the middle.

Engineering accomplishments are a hallmark of civilisation. However, besides grave monuments and temples, most practical engineering throughout the ages has been carried out in a military setting. For instance, the Roman army employed engineers to construct roads, bridges, fortified campsites and war machines. To intimidate the Germans, Julius Caesar made it a point to build a bridge across the river Rhine, just to show he could.

2.2.2 Emergence of Higher Engineering Education

In many Western countries today, engineering education still has a background in the military. For instance, in Holland the first engineering school was established in 1600 at Leiden University on the initiative of the military leader of the young Dutch republic Prins Maurits. Under the name *'Duytsche Mathematicque'*, the school aimed to train land surveyors and builders of fortifications.

In France, the Ecole Polytechnique was established in 1794, named *'Ecole Centrale des Travaux Publics'*. When Napoleon came into power, he adapted the regime at this engineering school, introducing military discipline. Since then, the Ecole Polytechnique developed as a place where the French military and policy elite received their schooling. Years later, the French engineer and mathematician Poincaré analysed the functionality of the French system of engineering education in order to explain the loss of the French in the war against Germany (Galison 2003). At the other side of the Atlantic, the third president of the USA, Thomas Jefferson, established the first engineering program in America at the military academy West Point in 1802.

The transfer to the civil environment and the establishment of an academic status took quite some time. The development of Delft University of Technology (TU Delft) in the Netherlands can serve as an example. On 8 January 1842, King Willem II founded the 'Royal Academy for the education of civilian engineers, for serving both nation and industry, and of apprentices for trade'. The academy also educated civil servants for the colonies and revenue officers of the Dutch East Indies. Just over 20 years later, the Royal Academy was transformed into a polytechnic school, bringing the school under the influence of the rules applying to secondary education. This school went on to educate architects and engineers in the fields of civil works, shipbuilding, mechanical engineering and mining during the rest of the nineteenth century. It was not before 22 May 1905 that an act was passed, acknowledging the academic level of the school's technical education—it became a Technische Hogeschool, or an institute of technology. The institute was granted corporate rights by an act passed on 7 June 1956. Recognition of the institute as a university of technology had to wait until 1986.

Engineering is foremost a practical profession. In the old days, the way to learn a craft was to start working as apprentice to an established craftsman. The master–apprentice system was formalised in medieval times. A student had to stay in the service of the master for a fixed period before he could prove his skills with a master-test and become a full member of the guild. This way, the guild protected the quality of their profession. Institutionalisation of engineering education in Europe dates back to the eighteenth century. Research and science were closely connected in the formation of technology and were the basis in engineering education. In the nineteenth century, when the most engineering schools were established, there was a clear pedagogical idea: to teach science and to present research.

2.3 The Current Status and Impact of Engineering Education Research

2.3.1 Introduction

Since the end of World War II, participation in higher education—including engineering education—increased substantially. As a consequence, higher education changed dramatically, from highly individualised personal study plans to programmes for mass education and lecture theatres with several hundred seats. Presently, engineering education in Europe is developing towards a common market. A specific goal of the Bologna declaration is to promote mobility among (engineering) students in Europe. As a consequence, universities are also engaging in an international competition to attract students. It is not surprising that presently we observe an increasing interest for improvement and innovation in engineering education. All over Europe, 'Centres of Expertise in Learning and Teaching' within universities are being established or, in case of older existing institutes, reinstated. The position of such institutes in the university organisation is by no means clear, and also the tasks and responsibilities vary widely from one university to the next.

In most cases, the position within the organisation is not stable and varies with every change in local university politics. For instance, at TU Delft in the Netherlands, originally an educational service centre was created in the central university staff bureau. In the mid-1980s, a scientific department was established, headed by a full professor on university teaching and educational development in higher education. When the professor left, just before the turn of the century, the department started to dwindle. Finally, the section was split up in a research group and a teacher-training centre (De Graaff and Sjoer 2006). A few years later, the research group was discontinued, and the last member of the original group left TU Delft at the end of 2011.

The conclusion of the cycle of events depicted above is twofold: First, a university of technology needs expertise on educational innovation in higher education, and second, it is not clear where this educational expertise should be positioned in the organisation. Below, different organisational positions of educational expertise centres at Dutch universities will be explored from a historical perspective (De Graaff and Sjoer 2006).

2.3.2 Research and Development Centres in the 1960s

Europe has a respectable tradition of social scientists researching engineering education (de Graaff and Borri 2006). During the 1960s, the number of students enrolling in higher education increased dramatically. As a consequence, universities had to adjust their teaching methods in order to deal with mass higher education (Wiegersma 1989). Innovation and improvement soon became keywords in dealing with this issue. In a scientific environment, it seems natural that research plays a major part in order to establish a solid foundation for quality improvement. However, in order to achieve that, a major gap needs to be bridged. As it was put by a reporter at the end of the first national convention on Research of Higher Education in the Netherlands, Eindhoven, 27–28 April 1966: ‘If one wants to improve the quality of teaching in higher education, then, first of all it is necessary to establish contact between the following two groups of people: those who are concerned with the teaching of science and those who are engaged with the science of teaching’ (Vroeijensteijn 1981).

At the beginning of the increase in interest for research on higher education, the Dutch schools of technology played a central part. The third national convention on Research of Higher Education was organised in Delft, 15–16 January 1976 (Vroeijensteijn and van Woerden 1976). At the opening of this conference, the minister of higher education addresses the position of the Research of Higher Education (in Dutch, Research van het Wetenschappelijke Onderwijs, RWO) centres. The minister points out that the position of the RWO institutes differs markedly from that of other research institutes because their finances are drawn directly from the university funds. Although the financial position of the RWO institutes is sound, the positioning of these institutes into the university organisation is difficult. The gap between research and application of the outcomes is one of the problems singled out by the minister. Since the average university professor does not have enough time

Table 2.1 Educational units in Dutch higher education during the mid-1980s

Scientific department in one of the faculties
Rotterdam
Maastricht
Scientific department with professional teacher training
Utrecht
Leiden
Research institute with an affiliation to the local university
Amsterdam
Nijmegen
Groningen
Educational centres as part of the administrative university staff
Amsterdam (free university)
Delft (technical university)
Enschede (technical university)
Eindhoven (technical university)
Tilburg
Wageningen

to study educational science along with his own profession, the minister states: 'it is not surprising that it is hard to implement new educational insights in the practice of higher education'.

One might say that Research on Higher Education became a booming business during the 1980s. The respective institutes were allowed to hire more and more researchers. The various educational centres joined forces in an informal network of Centres for Research on Higher Education (CRWO). The network includes both general universities and universities of technology. The CRWO network continued to organise study conferences (Bartelds et al. 1987). Also, a section Higher Education was established as a thematic group operating within the national framework of educational researchers. The CRWO has played a major part in the start of educational research in the Netherlands. More recently, the CRWO transformed into a society of professional educational consultants in higher education (EHON).

By the time of the fourth national convention on Research in Higher Education in 1981, research institutes were established at or in close relation to most universities. Although some of these institutes were strictly service oriented (notably the educational service at TU Delft), most were research departments with their own scientific staff, fitting to the trend in industry of establishing 'Research and Development Centres'. Table 2.1 lists the organisational positions of educational units in the mid-1980s.

The classification above is not beyond debate. In some cases, staff units style themselves as research group. The presence of a professor with an assignment of research in higher education is taken as primary criterion. Such professors were appointed at the scientific departments and at some of the research institutes, but not at a staff unit. However, starting with this grouping, in the next section, we can have a look at the changes taking place in the following years.

2.3.3 Emerging of Consultancy and Training Centres

Despite the initial optimism, the problem of a deficient connection between research outcomes and educational practice in higher education only became bigger. In part because the researchers tended to become more and more involved in their own theories and in part because the teachers did not apply the same sound scientific methods they used for their own professional field to their teaching tasks (van der Vleuten 1997). The logical consequence of this gap was a growing pressure on the RWO institutes from their financers to spend less time on scientific research and invest more in innovation projects with a concrete and measurable outcome. Gradually, the institutes drifted apart. Some institutes even adapted to the style of commercial consultancy firms and managed to continue the growth process. Others turned into research institutes drawing on research funds; the ones who were more dependent on the primary stream of university funding tended to lean more heavily on consultancy and training. In some cases, the research-oriented departments were curtailed and repositioned within the universities' administrative staff. An example of this development is Rotterdam where the Erasmus University first closed down its department of educational research completely. A few years later, a new institute was founded, focusing on consultancy rather than on research. As related in the introduction, something to the same effect happened in Delft. Also at Twente University, a successful educational unit was downgraded in size as well as in organisational position. In Eindhoven, the Educational Service Centre was finally closed down. It is interesting to note that a recent study could not find evidence supporting a relationship between satisfaction with the operation of the institute and particular positioning in the organisation. Apparently, each solution has its own benefits. Table 2.2 summarises the situation in 2006.

Educational research institutes that acquire their main funding out of research grants tend to operate rather independently of the host university. University management appears to be unhappy with this independent position; consequently, they prefer to organise the educational consultants as a staff unit with limited or no research tasks. As a consequence, scientific research in higher education is cut short. What started out as research-based institutes aiming to sustain innovation in higher education through scientific research ended up as consultancy units supporting existing practice. Yet, if you aim for the highest quality, educational innovation based on applied research of higher education is needed (de Graaff and Kolmos 2013).

2.3.4 Revival of Educational Research

In the past years, EER has been on the rise once more all around the world (Borrego and Bernhard 2011). Engineers more and more often carry out their own educational research. In the USA, the book *Scholarship Reconsidered* (Boyer 1990) had a big impact. Ernest Boyer argued that the intellectual rigor of scholarly activities should

Table 2.2 Educational units in Dutch higher education in 2006

Scientific department or institute
Utrecht
Leiden
Maastricht
Amsterdam (free university)
Groningen
Heerlen (open university)
Consultancy or research institute with affiliation to the local university
Amsterdam (UvA)
Nijmegen
Rotterdam
Enschede (technical university)
Educational centres as part of the administrative university staff
Tilburg
Eindhoven (technical university)
Wageningen
Delft (technical university)

be extended to encompass four interrelated activities: discovery, integration, application and *teaching*. The Scholarship of Teaching and Learning aims to improve student learning and is usually motivated by a teacher's interest in how students in their own classrooms are learning (Huber and Hutchings 2005).

In the USA, the National Science Foundation (NSF) makes substantial funds available for EER. Universities like Stanford, Purdue and Virginia Tech have fast-growing departments for research on engineering education. Also in Australia and Asia, the trend to strengthen engineering education through research and innovation is manifest. The contributions of the USA (and Australia) dominate the worldwide network on Research in Engineering Education (REES).

As we have seen, many universities in Europe have their own locally funded educational support centres. However, as most of these centres focus on consultancy and training, there is not much room for research. Altogether, it appears to be harder to raise funds for educational research in Europe than it is in the USA. Apparently, the financing system of the European projects does not favour the topic of EER. An explanation can be found in the EU system of applying for funding. Brussels provides ample funding for European projects aiming to improve research and education in Europe each year. However, research and education are administered by two different directorates: the Directorate General (DG) for Education and Culture and the DG for Research and Innovation. The topic of EER has the misfortune to fall between these two directorates. In the long run, this could mean Europe runs the risk of falling behind compared with the rest of the world in a research area where it once had the lead.

2.4 Methods of Engineering Education Research

2.4.1 Methodological Diversification

In the beginning, research on higher education was conducted by social scientists, employing their own set of research methods (see Sect. 3.1). As engineering education develops as a field of applied research in its own right, more and more teachers with a background in technology and the natural sciences are involved. A characteristic of a fully developed branch of science is an all-embracing research methodology. The following criteria define a scientific discipline:

- Well-established chairs
- A public debate on research methodology
- A coherent theory, distinct from other domains

Presently, engineering education fulfils the first criterion. There are chairs in engineering education established around the world. The other two criteria appear as yet to be more difficult. The debate on research methodology is starting both in the international engineering education journals and in conferences and seminars.

Presently, practitioners of EER either have a background in natural sciences or in social sciences, or in both disciplines. As a consequence, EER uses methods both from the natural sciences and from the social sciences. Researchers will have to motivate their choice between quantitative and qualitative, as well as mixed methods research designs. The choice of methods depends on the worldview and of course on the research questions. In general, quantitative research implies that there are some well-developed theories that need further confirmation. Qualitative and mixed methods research aims at building theoretical understanding from a more diffuse starting point (Creswell 2009).

2.4.2 A Taxonomy Proposal for Engineering Education Research

EER is a rich field of investigation. It covers research on learning and teaching in all engineering disciplines and also in supporting sciences, like physics, chemistry, computing and mathematics, which form the scientific base of engineering research (Malmi et al. 2012). Furthermore, EER also applies theories and research methodologies from social sciences, like education, psychology and sociology, to investigate many-faced aspects of learning and teaching engineering. In order to get an overview of the whole field, there is a need to look at both: what is being researched and how the research is carried out.

Different types of studies require a different selection of methods. The literature reports an abundance of research strategies, suitable for various purposes. As a consequence, it can be hard to compare research outcomes and relate different studies to each other. In order to gain an overview of the research in a particular field—and to take notice of gaps in the topics covered—taxonomy is needed. In a joint initiative of the Société Européenne pour la Formation des Ingénieurs (SEFI) working group EER and Line B of the EU project EUGENE, a series of workshops were organised aiming to construct a taxonomy for EER from a European perspective (de Graaff and Kolmos 2010). The resulting proposal for a taxonomy was tested on conference and journal contributions (Malmi et al. 2010, 2012). Based on these papers, the section below summarises the outline of the European taxonomy for EER.

The taxonomy could be used as a measurement tool to reveal differences between various publication forums, thus giving suggestions for authors where to submit certain types of papers. Furthermore, it could make visible hidden trends or emerging research paradigms in the field. By clarifying the difference between case reports and research papers, we can also point out how the scientific level of papers should be increased when we aim at more generalisable results and deeper insights. The taxonomy could also be used as a reference when the publication forums are defining review criteria for different types of papers. At the moment, we see it clearly problematic that the review criteria in many conferences and also in some journals do not give clear enough guidelines both for authors and for reviewers concerning what is expected for the papers.

Finally, as EER is gradually gathering recognition as an emergent field moving from the margins to the mainstream (Streveler and Smith 2010, NEERC 2006), a taxonomy can help to provide a map of the terrain for new scholars entering the field. We expect the results of this analysis to provide us with data supporting the goal of building recommendations to improve scientific writing in the EER community (Table 2.3).

2.4.3 Publication on Engineering Education Research

An active research field requires a platform for publication of research findings. In the beginning, the social-science-based researchers used to publish their findings primarily in their own professional journals and conferences. This habit emphasised the gap between the researchers and the field of practice. Only on special occasion—for instance, with an invitation as guest speaker at an engineering education conference—would they address the community of engineering educators directly.

In Europe in the 1970s, this community started to organise itself. The first international society aiming at uniting engineering teachers was Internationale Gesellschaft für Ingenieurpädagogik (IGIP), founded in 1972 at the University of Klagenfurt (Austria). In the German language, IGIP stands for Internationale Gesellschaft für Ingenieurpädagogik. A year later, the European Society for Engineering Education, known as SEFI (the French acronym for Société Européenne pour la Formation des Ingénieurs), was established in Belgium. Both societies offer the opportunity to present experiences, ideas and research on teaching and learning in engineering at annual conferences. A few times, these conferences have even been organised jointly, allowing members of both societies to mingle. Shortly after its

Table 2.3 Taxonomy for engineering education research (EER). (Malmi et al. 2012)

After processing all comments, the proposed taxonomy categorises research in six dimensions. In principle, we consider the dimensions independent of each other, but we recognise that certain research paradigms prefer certain types of collected data and analysis methods
1. Explanatory framework (EF) dimension
Neither research nor practical development takes place in isolation. We are building on previous work by other researchers and practitioners. This is the basic premise of all academic work. Scholarly work should also always recognise the premises of one's work and methods. Equally important is to give credit to others' work on which we are building our own work by mentioning this in the text and properly referencing their publications and works. The EF dimension aims at making visible how the target publication is linked to previous work (Jolly et al. 2011). We limit our investigation to such conceptual constructs that we expect to be known in a wider community of EER researchers. These constructs, which we call EFs, can be, for example, the following
Theory can refer to well-established theories, such as constructivism, situated learning or cognitive load theory
Model/framework/taxonomy/formal construct refer to established conceptual constructions, which are not generally called theories. Some examples could be Bloom's or Solo taxonomies, concept maps, IEEE curriculum definitions and pedagogical patterns
Very often, papers build on previous research, which does not have an established widely known status, in various ways. These could be the following
Using previous work as motivation. For example, addressing open questions, which were identified in other publications
Extending previous research to a new data set or reanalysing previous data sets in a new way
Using previous results as a starting point for new research
Applying a methodology, which was developed in another paper
To simplify the analysis, we will not classify the latter types of references, but consider only such EFs that we consider well-known within the EER community. Though, at the same time, we recognise that 'well-known' is an ambiguous concept and thus needs to be negotiated
We also do not list links to technical tools or frameworks. There is a multitude of such applied in EER, as engineering is about designing, implementing and applying technologies. Neither do we classify methodological references here, such as phenomenography, content analysis or various statistical tests or analysis methods. The methodologies are captured by other dimensions
Each EF is accepted on face validity. If the authors claim they are using it, we do not question this, as we can analyse in reasonable time only the publication we have. We will not report EFs, which are not explicitly mentioned in the paper, that is, we will not try to interpret from the paper whether the work is based on some EF
2. Research strategy (RS) dimension

Table 2.3 (continued)

There are many different ways how research is carried out. Here, we differentiate the general research design from more detailed-level data analysis methods. The former captures the choice of research questions and how they are generally approached, while the latter concern the concrete analysis methods used in processing collected data. Here, we face a problem which terminology we should use for the wider design of the research. In some contexts, we could use here either term research paradigm or research approach, but these are not used in all cases we cover, and especially the term paradigm is a too wide concept for us. On the other hand, the term research design typically refers to a rather detailed description of the research setting. Malmi et al. (2010) used the term research framework: Research framework '…is an overall orientation or approach that guides or describes the research, as opposed to a specific method or technique. A research framework may have associated theoretical, epistemological and/or ontological assumptions (e.g. phenomenography); may prescribe or suggest the use of particular methods (e.g. grounded theory); or may simply be a descriptive term for a kind of research activity that has certain characteristics (e.g. action research (AR), case study). Not all papers will have a research framework'
A similar dimension has also been proposed by several other researchers, though with different names: emerging methodologies (Case and Light 2011) or research strategies (Chism 2010). Also Merriam (2002), Creswell (2007) and Denzin and Lincoln (2005) present a similar type of classification for methodologies. We will adopt the term research strategy, instead of research framework, to avoid confusion with explanatory framework
We propose the following set of research strategies, though we recognise that the list can be expanded (definitions from Malmi et al. 2010)
Action research (AR)
A self-reflective systematic inquiry undertaken by participants to improve practice. Typically conducted as an iterative cycle of planning, action, change and reflection. A large number of EER papers have these characteristics—as their goal is to improve EER practice, and in many cases the teacher is participating in the development project. However, we do not classify reports on single research/development activities as AR (but classify them based on how they have been implemented). Instead, AR should be restricted to papers reporting on a longer term or sequential development activities where the authors have been participating themselves. AR often also involves people who are part of the activity that is under the study. The role of the researcher can be anything from an external expert who tells what to do differently (technical AR) to the catalyst who gets the things moving and gets people to participate and act themselves to improve the situation they are in (critical AR). Action research may be found under other names, like participatory research, collaborative inquiry, emancipatory research, action learning, and contextual action research.
Case study (CS)

Table 2.3 (continued)

A case study is an in-depth, descriptive examination conducted in situ, usually of a small number of cases/examples. Typically, a case study includes using several different data sources, which provide a rich description of the investigated case and possibly allows triangulating the investigated topic. We note that case study is not a clearly defined concept, and there are different interpretations of it. If the authors claim they have used the case study approach, we accept it by face validity
Constructive research (CR)
Research that aims to demonstrate and/or evaluate the feasibility of a proposed idea (concept implementation, proof-of-concept research). It revolves around the development of, for example, software, technology, a teaching approach or an evaluation instrument. Papers concerning educational technology fall within CR if the paper focuses on presenting the novel tool/learning environment and its functionality. In many cases, the paper also includes some kind of evaluation study, which adds another research strategy, like experimental research (Exp). Also, papers that present a novel teaching method and focus on how it is implemented and why it is used, with possible evaluation, fall within CR
Delphi
A specific method of seeking consensus by showing a group of raters a summary of their ratings, with justifications, then iteratively inviting them to reconsider their ratings in the light of what the others have said. Papers that apply this method usually identify it explicitly
Ethnography (Eth)
A branch of anthropology that deals with the scientific description of individual cultures
Experimental research (Exp)
Quantitative research based on manipulating some variables while varying and measuring others. This requires formation of control and experimental groups of participants with random assignment of participants or use of naturally formed groups. All papers reporting on a treatment-control group setting include this strategy. The focus is on testing the effect of some treatment. Survey research papers may have similar outlook, but there the focus is on investigating the variables without explicit treatment
Grounded theory (GT)
Qualitative, data-driven research in the tradition of Glaser and/or Strauss that aims to formulate theories or hypotheses based on data. A qualitative study that analyses data to build a model to explain that the working of the investigated topic fits well here
Phenomenography (PhG)
Investigation of (other) people's ways of experiencing a phenomenon (2nd-order perspective). Phenomenographic research is well established and all papers applying it say it explicitly
Phenomenology (PhL)
Investigation of one's own experience of a phenomenon (1st-order perspective)

Table 2.3 (continued)

Survey research (Survey)
Quantitative research based on exploring the incidence, distribution and/or relationships of variables in nonexperimental settings. Survey research focuses on an investigated topic as it is, not with a treatment. For example, the effect of gender or age group on study success
A paper may have more than one research strategy. On the other hand, a paper may not have any research strategy that we can identify. Each explicitly mentioned research strategy will be accepted on validity. If the authors claim they are using it, we will not question this
Contrary to EFs, we will distinguish between cases where the authors explicitly mention a research strategy (though probably they do not use this term but say, for example, that they were carrying out AR), and cases where we conclude from the report that some research framework was used. The reason for this is that mentioning a research strategy/framework/paradigm/approach is not a common convention. For instance, we record case study as one research strategy, but many times this is not explicitly written in a paper. In summary, we give research strategy an additional tag 'explicit', 'implicit' or 'none' if no research framework can be identified in the paper
3. Data source (DS) dimension
Data collection implies what kind of data has been used in the empirical part of the work. Most papers have at least one data source, but very often include several. Examples of categories in this dimension include the following (the list is not exhaustive)
Students' submitted work (essays, project reports, learning diaries…)
Examinations and tests
Questionnaires
Instruments
Interviews
Observation data
Databases (e.g. study register data)
Software log data
Researchers' own experiences (e.g. 'lessons learned')
Literature (e.g. literature reviews, meta-studies)

Table 2.3 (continued)

An instrument is a special case of a questionnaire. It is a psychometrically validated questionnaire, which is used to measure some aspect of human behaviour, such as the Myers–Briggs personality test. We list such instruments, if encountered, as they are useful tools in many aspects of EER, and we wish to promote their wider application, instead of building similar tools ad hoc. Typically, most papers report some of researchers'/teachers' own experiences and reflections. However, we will report them as data source only if their share of the paper is significant compared to other collected data. We will not any more identify data sources as intrusive or natural data. We just mention that certain data sources like interviews are always intrusive, whereas some are natural, like study register data, and some can be used in both ways
Finally, we will register what is the scope of data collection (as a whole), that is, has it been carried out in the individual level, group level (like classroom, student group in one course/unit, whole course/unit), institution level (curriculum, program, university, …) or multi-institutional level (many universities, whole country, …)
4. Data analysis (DA) dimension
The data analysis section will describe how empirical research data are being analysed or what other means are used to draw conclusions in the paper. Most papers feature at least one kind of analysis method. If a paper has an RF, the framework often directs the analysis methods that are used. However, the same analysis method can, of course, be found in a paper that is applying some other framework or has no specified RF at all. The number of possible analysis methods is extensively large, and we need to gather them into rough categories. We will apply the following categories to cover the data analysis method dimension
Quantitative complex
Any form of statistical methods exceeding simple descriptive statistics, such as statistical tests, correlations, regression analysis, factorial or cluster analysis, data mining techniques, …
Quantitative simple
Descriptive statistics including cross tabulation and graphics
Qualitative enhanced
Any qualitative methods which have a clearly specified analysis process (is it reported to such aspect that it could be repeated?)
Qualitative simple or not specified
Qualitative analysis which only includes identifying important themes/topics/items of interest, like interviews without specifying a method or structure. This category will also be used in cases where it seems that some method has been applied, but due to missing description we cannot know what it was
Other
Other methods, to be described
None
Plain common sense and using one's own reflections as a kind of analysis method are counted here as well

Table 2.3 (continued)

A paper can include several of the above methods, and we will list all that we recognise regardless of whether it is explicitly stated or implicitly included. Based on that, we can later on derive whether the paper is a quantitative/qualitative/mixed methods paper
5. Reporting (RP) dimension
Reporting the research setting and process is a central part of good scientific communication. We will make observations in the text on the following aspects
1. Research questions
a. Research questions/research goals are explicitly emphasised in the text. This means that they can be found easily with some visual cue, such as within bullet or number lists or italicised within other text
b. Research questions/goals can be found in other text, typically in the introduction or in the beginning of the section which describes the paper contribution (implicit). This means that we can identify some text in the paper, which presents the goals of the research and/or questions to which the research seeks answers. We do not require that these are presented as questions, but may be written as ordinary text
c. No research question or goal can be identified in the text (none). It is not possible to identify any text presenting the goals of the paper, either in the abstract, introduction or before the method/results section
2. Methodology section
a. Methodology has clearly its own section in the paper (explicit). This can be identified based on paragraph titles
b. Methodology is implicitly described within other text, typically in the results section (implicit). It is possible to identify that there is text that presents the methodology used, regardless of where it is in the paper. Some educational technology papers, for instance, simply describe the design, implementation or functionality of a novel software or tool without any specific method section. We categorise these papers as technology papers
c. No clear methodology description can be identified in the text (none)
3. Validity/generalisability discussion
a. The paper has a separate section/subsection discussing the validity/generalisability/trustworthiness/limitations of the research (explicit)
b. The paper has some critical discussion of some of these issues in the text, typically in the results section or conclusion (implicit). It is possible to identify a discussion of limitations of the paper regardless of where it is and whether it is in one or several places in the text
c. No critical discussion can be found or it is very vague (none)
6. Nature (NT) dimension
Finally, the NT tries to capture the general character of the paper as a whole. We categorise the NT as follows
Empirical paper: It is a paper, which has the basic elements of empirical research, including clear data collection, analysis and reporting results. The paper may or may not have hypotheses. Data analysis may be based on quantitative or qualitative or mixed methods

Table 2.3 (continued)

Case report: It describes a novel educational setting, such as a new teaching method, assessment method, learning resource, learning specific software, etc. The focus of the paper is on describing the new contribution. There is no evaluation, or the evaluation is very shallow, typically reporting some student results, student feedback and/or teachers' experiences with no clear research setting (such as comparison to previous year). A case report typically has a limited scope, related to some specific course, and the research setting and method aspect of the paper is vague—the focus is on the novel thing, whether it be teaching method, software or something else. Usually, the focus of the paper is to improve practice
Position paper/proposal: This is a paper where the authors want to raise some issue for discussion among the EER community or propose something new to be considered in engineering education practice or EER
Theory paper: It discusses theoretical aspects of teaching and learning; for example, it compares some learning theories in some context. The paper is based on theoretical discussion and argumentation and has little or no empirical data to support its claims
We will not differentiate whether the nature of the paper is explicitly or implicitly stated as this is basically our interpretation of this issue. All papers have some nature

foundation, in 1975, SEFI initiated the publication of a quarterly publication. At the start, the journal was a mixture between a society newsletter and a platform for scholarly articles on engineering education. Since 1975, *European Journal of Engineering Education* (EJEE) gradually became more focused on research papers and review articles with a theoretical foundation.

In the USA, the American Society for Engineering Education (ASEE) *Journal of Engineering Education* (JEE) has a much longer history, publishing on engineering education for over 100 years (Wankat 2004). A third journal, with no ties to a specific organisation, is the *International Journal of Engineering Education*. Presently, there are also several other national journals on engineering education, for instance in Australia and Brazil.

International recognition indicates that JEE and EJEE have developed into leading journals in the field of engineering education (Osorio and Osorio 2004).

Recently, this was confirmed by the keynote presented by Phil Wankat at the 40th Annual SEFI conference in Leuven. Together with a few colleagues, he made a bibliometric analysis comparing the US-based JEE with the EJEE. The paper based on the text of the keynote is published in the EJEE (Wankat et al. 2014).

The analysis clearly demonstrates that both journals gradually developed a more pronounced research orientation: 'Initially there were little or no research papers in the 70's and 80's and then there was gradually increasing research content' (Wankat et al. 2014). The conclusion of the study states that both journals became more frequently cited and that there was a qualitative evolution in breath of scholarship with increasingly interdisciplinary research articles with more collaboration and more references. While the international character of both journals also evolved over the 40 years, geographical spread is notably broader in EJEE.

During the past years, several new conferences and symposia created new opportunities for presenting studies on engineering education, like the international Research in Engineering Education Network (REEN) and the International Research Symposium on Problem-Based Learning (IRSPBL). Also a number of online journals emerged, further enriching the possibilities of engineering education researchers to publish their work. In any case, nowadays there are ample opportunities for publishing.

2.5 The Future of Engineering Education and Research in Europe

2.5.1 Introduction

To achieve the political ambitions of Europe as a knowledge society, we need many engineers with leadership skills and creativity who are capable of coping with the increasing complexity of the technological foundations of our society. To train engineers who can meet these challenges, we need to innovate our approaches to teaching and learning at the universities of technology and polytechnic schools.

Participation in higher education is expected to remain at a high level. A lot of efforts are made trying to attract more young people to choose a career in engineering. Engineering could possibly attract more female students, and most importantly, as some people say that the present day half-life of technical knowledge is said to be less than 5 years, modern engineering requires a lifelong study. As a consequence of this development, the ability to access and manage information becomes more important than knowledge per se.

A fundamental change of the process of teaching and learning is the result. In fact, this may be one of the most striking factors of change in the present era. Traditionally, the primary task of teachers is transfer of knowledge. A teacher possesses expert knowledge or skills in a particular field, which consists of the basis of teaching. For a long period of time, lecturing has been the dominant educational format in higher education. As a dispenser of knowledge, the lecturer used to be responsible for both the content and the structure of teaching.

Due to the rapid developments of information and communication technology (ICT) over the past decades, the old-fashioned method of lecturing is now all but obsolete. The increased use of ICT leads to an increase in the dynamism of organisations. Consequently, people in those organisations will have to learn to deal with new and unexpected situations. Teachers in higher education for instance will have to face the fact that they do not hold the monopoly on knowledge anymore. Information is freely available and can be accessed from anywhere at all times. Moreover, in many areas, the state of the art is changing so rapidly that it is hard for teachers even to keep up with their students.

All over the world, institutes in higher education have seen this change coming, and they have tried to adjust to it in various ways. Internet and e-mail are widely used as means for communication between teachers and students. Learning content, full text as well as supporting slides, is available online.

Following the example of the open courseware project of Massachusetts Institute of Technology (MIT), several universities make learning content of their study programmes available for free on the Web. Materials may comprise streaming video lectures, interactive Web demonstrations, homework problems and exams. A recent trend is to offer massive online open courses (MOOCs).

Nowadays, students can exercise their skills in many ways supported by software packages rather than by a teacher. However, several people already recognised that it does not suffice simply to transfer the old content to the new media. To take full advantage of the opportunities offered by the ICT technology, the learning environment needs to be redesigned completely.

2.5.2 Scenarios for Future Developments

The challenge for the field of higher education is to take full advantage of the wealth of opportunities that are opened through the new teaching and learning tools. In order to prepare a policy for dealing with higher education/research relations in

Europe, a Strategic Analysis of Policy Issues - European Technology Assessment Network (STRATA–ETAN) expert group was set up by the EU in December 2001 with the mission to prepare a Foresight report. The 21 members of the expert group, chaired by Maurice Godelier and rapporteur Etienne Bourgeois, come from a wide variety of countries (EU states members, Accession countries and Canada) and disciplinary background. Each one of them wrote 'issue papers'. In 2002, the final report of the STRATA–ETAN Expert Group on Foresight for the Development of Higher Education/Research Relations was presented (Bourgeois 2002). The following section draws upon the issue paper I prepared for the STRATA–ETAN Expert Group (De Graaff 2002).

In this issue paper, I propose two scenarios depicting projected developments in higher education. The focus of the first scenario is on the potential of individual freedom and self-direction in learning. The second scenario highlights the collective aspect of collaboration and communication in teams. A development that is deemed unavoidable is a further differentiation in the tasks of a teacher. In both scenarios, three different teacher roles will be identified: the facilitator, the educational designer and the assessor.

Scenario 1: The Individualist

As a result of the rapid technological development, an increasing number of tools are available that reduce our dependency on other people. In order to prepare a text for printing, an author used to need at least a dozen people, skilled in typesetting, layout, correction, etc. Presently, anyone with a PC can easily deliver a camera-ready print proof, corrected for spelling errors and grammar. An example even closer to the school can be found in the widely available software packages that support learning to speak a foreign language. Until recently you needed an expert, preferably a native speaker, to correct your pronunciation—a tiring exercise, boring both for the students and for the teacher. Nowadays, a student can do these exercises whenever and as many times he wants. The advantage of not needing anyone will tempt some people to isolate themselves. They can get whatever they need and they can do whatever they want without having to see anyone.

The Individualist as a Facilitator As a facilitator, the individualist has a hard time. He will have to focus his attention on the interaction process among learners, and, in a sense, act as an expert in group dynamics. The individualist tends to feel uncomfortable in this position, and he will easily fall back in the old role of the expert providing the answers. It is not so much that the individualist is incapable of fulfilling this task; he will probably not enjoy doing it.

The Individualist as an Educational Designer In the role of a designer, the individualist probably feels more at home. However, the present-day educational design projects usually involve teamwork. Even if the creative spark of the individual constitutes an essential element in the design process, the filial of each member's contribution depends on the ability to communicate and co-operate. The individualist will have to learn to compromise in this respect in order to become an effective team member.

The Individualist as an Assessor The role of the assessor also involves a strong element of communication. However, in this position the individualist retains the authority of the content expert. In order to be successful in this task, it is important that the individualist learns to be specific with respect to the criteria that he applies. Such criteria specifying the core characteristics of expertise within a particular domain are at the same time concrete learning objectives for the students.

Scenario 2: The Team Player

As a result of the rapid technological development, things get so complicated that you cannot virtually do anything on your own anymore. A single person never will possess all relevant information. Therefore, in order to achieve something worthwhile, you need the participation of others. Each individual tends to focus on his own specialty. Sharing the own expert information with other team members becomes a crucial skill. Presently, organisational structures are becoming more flexible, and teams within organisations are only temporary structures. The ideal team member is adapted to the volatile interpersonal relationships within a network structure.

The Team Player as a Facilitator The team player usually feels quite comfortable in the role of a facilitator. The main risk with this type of person is that he becomes too much involved in achieving the objectives. When there are problems in the group, he may tend to give practical solutions rather than helping the group to find their own way out. When the group is doing fine without his advice, such a person may feel obsolete and frustrated. However, a natural team player has the ability to enjoy the sharing of experiences in the team, including his own.

The Team Player as an Educational Designer Present-day educational design is a typical team task. This teamwork calls for specialist expertise in particular domains, creative ideas and critical reflection on the input of others. As a member of such a team, a teacher who is a natural team player can fully blossom.

The Team Player as an Assessor The role of the assessor will be less easy for a team player. Judging others is usually not the best way to inspire collaboration within a team. A team player will tend to emphasise the feedback value of an assessment and hesitate before taking pass/fail decisions. A common error in project work, for instance, is the tendency to focus on the collective group product and to neglect the individual contributions. Another problem in this respect is that the assessor may feel insufficiently equipped to judge the overall performance, as this will go beyond his own area of expertise. However, in general, team players can operate satisfactory as assessors, provided they feel the assessment does justice to the individual team member.

Maybe at first glance, the two scenarios depicted above appear to be mutually exclusive. However, it is possible to regard the scenarios as the two extreme ends of a bipolar scale. Training teachers to operate effectively within the emerging new structure of higher education involves getting them to know themselves. The scenarios could be used as a point of reference.

2.6 Concluding Remarks

For a long time, Europe dominated the world in political and economic respect as well as in culture and science. France, Germany, England, Italy and even a small country as the Netherlands competed mostly among themselves. The Industrial Revolution started in Europe and spread around the world from there.

It was a bit of shock to recognise the increasing impact of former colonies like the USA at the beginning of the last century. For some, it may have lasted until after the World War II before they could accept the fact that Europe was no longer at the centre of world power.

With the political and economic power, the centre of gravity of science and technology also shifted. For years now, the five highest-ranking universities in the category Engineering and Technology of the Times Higher Education World University Ranking are situated in the USA. Different ranking systems show a slightly different picture. In the Shanghai ranking, the top 11 positions are taken by US institutes. However, the Quacquarelli Symonds (QS) ranking allows three universities from the UK, two from Switzerland and one from Singapore in their top ten. The QS ranking is also the one that most clearly shows the on-going worldwide development, with increasingly high rankings for universities from China and other countries in Asia. The power balance is shifting again, and science and technology seem to follow.

In the meantime, these past years we have experienced a period of economic crisis. In particular, Europe appears to have been hit hard, spreading from banks to governments. As a result, now all over Europe, budgets for higher education are being cut. When things get difficult, people tend to play safe. Universities, like commercial companies in trouble, tend to choose to layoff tenured staff. As the Delft professor of economics of innovation Alfred Kleinknecht (2013) points out in his farewell lecture, letting go people too easily is detrimental for innovation.

Unsurprisingly, the funding of research on higher education has become problematic. Yet, in order to increase the efficiency of our universities, we still need research to find out what really determines the effectiveness of learning in higher engineering education. Economising by sending away the experts that can direct this research may prove highly inefficient in the long run. As far as engineering education is concerned, Europe is truly becoming the land of the setting sun, unless we manage to build a machine to reverse the tide.

References

Bartelds, J., Brants, K., van der Hoop Gerritsen, W., Hulshof, M., & van der Weiden, M. (Eds.) (1987). *Studiedag Contactgroep Research Wetenschappelijk Onderwijs [Study Conference Contact Group Research Higher Education]*. Nijmegen/Utrecht: IOWO/VSNU.

Borrego, M., & Bernhard, J. (2011). The emergence of engineering education research as an internationally connected field of inquiry. *Journal of Engineering Education, 100*(1), 14–47 (January 2011).

Bourgeois, E. (2002). *Higher education and research for the era: Current trends and challenges for the near future; final report of the Strata-Etan Expert Group on foresight for the development of higher education/research relations*. Brussels: European Commission, Directorate-General for Research (October 2002).
Boyer, E. L. (1990). *Scholarship reconsidered: Priorities of the professoriate*. San Francisco: Jossey-Bass.
Case, J. M., & Light, G. (2011). Emerging Methodologies in Engineering Education Research. *Journal of Engineering Education, 100*(1), 186–210.
Chism, N. (2010). Qualitative research basics: A guide for engineering educators (pp. 9–10). http://cleerhub.org/resources/8.
Creswell, J. W. (2007). Qualitative inquiry and research design: Choosing among five approaches (2nd edn.). Thousand Oaks: Sage.
Creswell, J. W. (2009). *Research design: Qualitative, quantitative and mixed methods approach* (3nd edn.). Thousand Oaks: Sage.
de Graaff, E. (2002). Educating independent learners in a digital learning environment; 'Past performance is no guarantee for future success'. Issue paper Stratan Etan Group. Delft: 2002.
de Graaff, E., & Borri, C. (2006). Research and innovation of engineering education in Europe the contribution of SEFI. Proceedings of the 2006 JSEE annual conference (pp. 12–17).
de Graaff, E., & Kolmos, A. (2010). Research methods in engineering education research. Proceedings of Joint International IGIP-SEFI annual conference 2010.
de Graaff E., & Kolmos, A. (2013). Innovation and research on engineering education. In J. Michael Spector, M. David Merril, J. Elen, & M. J. Bishop (Eds.), *Handbook of research on educational communications and technology* (pp. 565–571). New York: Springer.
de Graaff, E., & Sjoer, E. (2006). Positioning educational consultancy and research in engineering education. Proceedings of the 34th Annual SEFI Conference (pp. 98–101). Uppsala: Uppsala University.
Denzin, N. K., & Lincoln, Y. S. (2005). Introduction: The discipline and practice of qualitative research. In N. K. Denzin & Y. S. Lincoln (Eds.), *The Sage handbook of qualitative research* (3rd edn., pp. 1–32). Thousand Oaks: Sage.
Galison, P. (Ed.). (2003). *Einstein's clocks, Poincarė's maps*. London: Hodder and Stoughton.
Huber, M. T., & Hutchings, P. (2005). *The advancement of learning: Building the teaching commons*. San Francisco: Jossey-Bass.
Jolly, L., Willey, K., Tibbits, G., & Gardner, A. (2011). Conferences, reviews and conversations about improving engineering education, Proceedings of the Research in Engineering Education Symposium (REES 2011). Madrid, Spain.
Kleinknecht, A. (2013) *'Een Erfenis van Schumpeter'* Afscheidsrede. ('An inheritance from Schumpeter', farewell lecture). Delft: TU Delft.
Malmi, L., Sheard, J., Simon, Bednarik, R., Helminen, J., Korhonen, A., Myller, N., Sorva, J., & Taherkhani, A. (2010). Characterizing research in computing education: A preliminary analysis of the literature. Proceedings of the sixth international workshop on Computing Education Research (ICER'10). New York: ACM. pp. 3–12. doi:10.1145/1839594.1839597.
Malmi, L., De Graaff, E., Adawi, T., Curmi, R., Duffy, G., Kauz, C., Kinnunen, P., & Williams, B. (2012). *Developing a taxonomical classification of EER papers*. Paper presented at the 34th Annual SEFI Conference Engineering Education 2020: Meet the Future, Thessaloniki, Greece.
Merriam, S. B. (2002). Qualitative research in practice: Examples for discussion and analysis. San Francisco: Jossey-Bass.
NEERC. (2006). National Engineering Education Research Colloquies' special report "The Research Agenda for the New Discipline of Engineering Education,". *Journal of Engineering Education, 94*(4), 257–261.
Osorio, Nestor L., & Mary A. Osorio. (2004). Engineering Education in Europe and the USA. *Science & Technology Libraries, 23*(1), 49–70.
Streveler, R., & Smith, K. (2010). Guest editorial: From the margins to the mainstream: The emerging landscape of engineering education research. *Journal of Engineering Education, 99*(4), 285–287.

van der Vleuten, C. P. M. (1997). De intuïtie voorbij [Beyond intuition] Tijdschrift voor Hoger Onderwijs, 15.1. pp. 34–46.

Vroeijensteijn, A. (red.). (1981). Kwaliteitsverbetering Hoger Onderwijs [Improvement of quality in Higher Education]. Proceedings of the fourth national convention Research in Higher Education, December 1-180-1981. Eindhoven: Stichting Nationaal Congres/1981.

Vroeijensteijn, A., & van Woerden, V. (red.). (1976). *Onderwijsresearch & Praktijk deel 1 en 2 [Educational Research & Practice]*. Delft: Delftse Universitaire pers.

Wankat, P. (2004). Analysis of the first ten years of the Journal of Engineering Education. *Journal of Engineering Education, 93*(1), 13–21.

Wankat, P., Williams, B., & Neto, P. (2014). Engineering education research in EJEE and JEE: Citation and reference discipline analysis. *European Journal of Engineering Education, 39*(1), 7–17.

Wiegersma, S. (1989). *Innovatie van het hoger onderwijs [Innovation in Higher education]. Hoger onderwijsreeks*. Groningen: Wolters-Noordhoff.

Erik de Graaff is trained as psychologist and holds a PhD in social sciences. He started his academic career at the University of Maastricht in 1979, working with Problem Based learning (PBL. In 1994 he was appointed as associate professor in the field of educational innovation at the Faculty of Technology Policy and Management of Delft University of Technology. Dr. de Graaff has been a visiting professor at the University of Newcastle, Australia in 1995 and a guest professor at Aalborg University in Denmark in 1997. The collaboration with Aalborg University led to an appointment as full professor at the department of Development and Planning in 2011. Dr. de Graaff is recognized as an international expert on PBL. He contributed to the promotion of knowledge and understanding of higher engineering education with numerous publications and through active participation in professional organizations like SEFI, IGIP, IFEES and ALE. He has published over 200 articles and papers and he has presented more than 70 keynotes and invited lectures on various topics related to PBL in higher education, like: Working with PBL, Management of change, Assessment and evaluation, Methods of applied research and Collaboration between university and industry. Since January 2008 he is Editor-in-Chief of the European Journal of Engineering Education.

Chapter 3
Engineering Education in Southeast Asia: Practice and Research

Khairiyah Mohd Yusof, Fatin Aliah Phang and Shahrin Mohammad

3.1 Introduction

Engineering education is viewed as a crucial field throughout the world, with efforts in improving and dissemination of knowledge through various means, which includes conferences and journal publications. These efforts have started even as early as the beginning of the twentieth century; the *Journal of Engineering Education,* for example, was first published in 1911. By the end of the twentieth century, there were intensive efforts on engineering education in North America, Europe, Australia, and South Africa.

In Asia, particularly in the southeast, efforts to create interest in engineering education started off at the turn of the twenty-first century. While initial efforts were created through conferences in collaboration with associations that include Australia, such as the United Nations Educational, Scientific and Cultural Organization (UNESCO) Centre for Engineering Education (CEE) based in the University of Melbourne, Australia, the interest started to take hold through local efforts, especially with the advent of the Washington Accord (WA) membership and challenges of the changing requirements in the twenty-first century.

This chapter focuses on the rapid emergence of engineering education practice and research in Southeast Asia, with particular emphasis on the experience of Malaysia, as one of the early adopters in the region to embrace and integrate engineering education practice with research. Activities that promote engineering education in East Asia, such as Korea, Hong Kong, and Japan, will also be briefly discussed to reflect the extent of interest that has been growing rapidly around the region.

K. M. Yusof (✉) · F. A. Phang · S. Mohammad
Centre for Engineering Education, Universiti Teknologi Malaysia, Johor Bahru, Malaysia
e-mail: khairiyah@cheme.utm.my

M. Abdulwahed et al. (eds.), *Advances in Engineering Education in the Middle East and North Africa,* DOI 10.1007/978-3-319-15323-0_3

3.2 Creating Interest in Engineering Education

Interest in engineering education had been slow to start in Asia and likewise in Southeast Asia. While there were sporadic, isolated interested parties, their network would mostly come from those in North America, Europe, or Australia. In the twentieth century, and even now at times, many engineering academics, and sometimes even associations or bodies in Southeast and East Asia, define engineering education simply as the technical engineering content in itself that are taught in universities at the undergraduate or graduate levels. Thus, when engineering education research is mentioned, some would refer to it as research in the engineering field, rather than using educational research methods for improving how we educate engineers. This confusion was manifested in the early emergence of engineering education seminars and conferences, where purely technical papers from various engineering fields were submitted and at times accepted for presentation.

The twenty-first century brought about major challenges through numerous changes in how industries and businesses function throughout the world. The current challenges are marked by rapid development in technology, explosion in information, borderless economy, sustainability and security concerns, and many other complex, novel problems that have never been seen before, causing various sectors, such as businesses, industries, governments, and various entities, to change their modus operandi (NAE 2005; Duderstadt 2008). The need to change to meet these challenges, and improve the quality of graduates, are also strongly felt, even in Southeast and East Asia, resulting in several nations from the region to become members of the WA in the first decade of the twenty-first century, such as Japan, Korea, Singapore, and Malaysia. This move, as a consequence, affects the accreditation of engineering programs, transforming the programs into the outcome-based approach (OBA) to abide by the accord. The impact of accreditation in enhancing the quality of engineering education in Malaysia is discussed further in one of the following sections.

3.2.1 Engineering Education Societies and Organizations

Engineering education societies and organizations are spreading the interest and are building a community of practice in engineering education. These societies are normally able to consolidate efforts, at least at the national level, to plan and organize bigger events to garner the support of stakeholders. More importantly, the societies would have more resources and harness the necessary support from the international community to enhance engineering education at the national level.

There are a several engineering education societies in Southeast and East Asia, such as in South Korea, Japan, and Malaysia. Japan established its Japanese Society for Engineering Education as early as 1952 (JSEE 2014). The Korean Society for Engineering Education (KSEE) was established in 1993 (IFEES 2013) and holds annual conferences that gathered academics and industries in engineering education. The KSEE is a platform for the industries and academics in engineering to meet and discuss the direction and future of engineering education in South Korea. The KSEE and the JSEE are members of the Association for Engineering Education in Southeast and East Asia and the Pacific (AEESEAP), which was established in 1970 (AEESEAP 2014). In other countries, however, members of the AEESEAP consist of only certain member universities, usually just one university from one country. For example, in Singapore, the member university is the National University of Singapore, while in Malaysia, the member country is the University of Malaya. China had also established the Chinese Society for Engineering Education (CSEE; IFEES 2014).

In Malaysia, the Society for Engineering Education Malaysia (SEEM) was established in 2007 (CEE 2012e) to promote engineering education in Malaysia. The SEEM was initiated by the Malaysian Council of Engineering Deans through a pro-tem committee in 2005. Activities under the SEEM are done independently, or in collaboration with other groups or centers of engineering education in Malaysian universities. These activities consist of training programs on engineering education for academic staff, engineering education conferences, seminars, and exhibitions. There are also other engineering associations that have engineering education sections or divisions, such as the Institution of Engineers Malaysia (IEM), the Malaysian Society for Engineering and Technology (MySET), the Institution of Chemical Engineers (IChemE) Malaysia, and the Institution of Electrical and Electronic Engineers (IEEE) Malaysia.

Many of the engineering education societies form international linkages and co-operation. These are normally established in the form of Memorandum of Understanding (MoU). One such example in the region is the MoU between the KSEE and the SEEM, which was signed in 2012 (CEE 2012d). Through this MoU, representatives from the SEEM are invited to attend the KSEE annual conference and speak at the plenary session while representatives from the KSEE are invited to conferences organized by the SEEM. On a larger platform, engineering education societies from various countries can join and network through the International Federation of Engineering Education Societies (IFEES). The SEEM, KSEE, JSEE, and CSEE are also members of the IFEES.

3.2.2 Conferences, Seminars, and Symposiums in Engineering Education

Conferences in engineering education play an important role in giving awareness of the importance of engineering education. This can be seen through the annual conferences organized by the KSEE and the JSEE, which gathered big crowds of participants from the industries and academics every year. In 2014, the JSEE is organizing its 62nd annual conference (JSEE 2014), while the KSEE its 21st. International participants were invited to speak or attend the conferences (CEE 2012d). The AEESEAP which was established in 1970 (AEESEAP 2014) also held conferences in several member countries, such as Japan, Malaysia, Korea, Australia, etc.

However, sometimes, the papers of the AEESEAP conferences are mixed between technical research and engineering education sessions. The KSEE conference, though, is solely for engineering education papers. Lately, the number of papers submitted has been increasing, showing evidence of growing interest in engineering education among Korean academics.

In Malaysia, among the first conferences devoted solely to engineering education was the Conference on Engineering Education organized by Universiti Teknologi Malaysia (UTM) on 14–15 December in 2004. Although other engineering conferences had papers on engineering education, most were mixed with engineering papers. Realizing the importance of engineering education in the region, the following conference in 2005 was upgraded to become the Regional Conference on Engineering Education (RCEE). To attract audience to the conference as well as provide expert input that was very much needed in outcome-based education, Prof. Richard Felder and Dr. Rebecca Brent were invited to hold pre- and post-conference workshops as well as deliver a keynote address during RCEE 2005. From then on, it became a requirement for the conference to have internationally renowned speakers in engineering education to instill awareness to practitioners in engineering education on the importance of scholarship as well as rigorous research in engineering education (RREE). RCEE 2007 had Prof. Karl Smith from the University of Minnesota and Purdue University, USA, as the keynote and workshop speaker, where the focus was on pedagogies of engagement, especially on cooperative learning. RCEE 2010 had Profs. Karl Smith and Ruth Streveler from Purdue University as the joint keynote and workshop speakers, where the focus was on systematic curriculum design and RREE. To introduce the rigorous research aspect of engineering education, RCEE was renamed as the Regional Conference on Engineering Education and Research in Higher Education (RCEE and RHEd). The aim is for engineering those who do rigorous research in higher education to network and learn from one another, which could hopefully lead to collaboration that involves RREE. RCEE and RHEd 2012 had several invited speakers including Prof. Dr. David Radcliffe, head of the School of Engineering Education, Purdue University, Prof. Dr. Duncan Fraser, from the University of Cape Town in South Africa and also the chair of the Research in Engineering Education Network Governing Board, and Prof. Dr. Susan Oh, from the Korean Society of Engineering Education. Pre- and post-conference workshops were given by experienced researchers in RREE, such as Assoc. Prof. Dr. Johannes Stroble and Assoc. Prof. Dr. Heidi Diefes-Dux, both from Purdue University, and Assoc. Prof. Dr. Elliot Douglas from the University of Florida (CEE 2012g). Thus far, up until 2014, six conferences had been organized in the series, which was held once every 2 years. Since 2012, the conference was held in collaboration with the SEEM and the Higher Education Leadership Academy under the Ministry of Education. The quality of papers has also steadily increased, from experience sharing to scholarly based implementations and research papers in engineering education.

These activities led to higher-quality research-based conferences hosted by UTM and the SEEM, such as the 2013 International Research Symposium on Problem-based Learning (IRSPBL 2013) and the 2013 Research in Engineering Education

Symposium (REES 2013), which was held back-to-back in July in Putrajaya, Malaysia. Both these conferences were held for the first time in Asia, after being hosted in the USA, Europe, and Australia. The stringent vetting of papers at the conference ensured high-quality research papers. Though there were many papers from all over the world, 40–50 % of the papers from both conferences came from Malaysia and the Southeast Asian region. This enabled local participants to share and get valuable feedback from other presenters as well as learn and network with those in the international engineering education research community. The REES drew a big community of experts in engineering education research to come to the conference, and it is an eye-opening experience for many new researchers in engineering education in Asia, especially in Malaysia. Further discussion on the impact of REES will be included in Sect. 3.4.

Lately, with interest in engineering education on the rise in the region, there are other conferences hosted solely by local universities and organizations as well as conferences hosted by universities in collaboration with organizations and universities from other countries. Since 2009, the International Conference on Engineering Education (ICEED) was organized annually in Malaysia by Universiti Teknologi MARA under the IEEE Malaysia Engineering Education Chapter. In 2012, Universiti Kebangsaan Malaysia (UKM) collaborated with Sharjah University, United Arab Emirates, to host the 6th International Forum on Engineering Education in Kuala Lumpur. Another newly started IEEE conference in the region is the IEEE Teaching, Assessment, and Learning in Engineering (TALE), which had its first conference in Hong Kong in 2012, and the second in Bali, Indonesia, in 2013. In 2014, UTM and Uppsala University, Sweden, collaborated to host the RCEE and RHEd in conjunction with the Learning and Teaching in Computing and Engineering (LaTiCE), which is a conference affiliated with the IEEE Computing Education Society, in Kuching, Sarawak. In 2012, LaTiCE was held in Macau, and in 2016, it will be held in Taiwan. From the number of conferences happening around this region, there is clearly rapidly rising interest among engineering educators on engineering education.

3.3 Engineering Education Practice

Engineering education practice is of utmost importance among Southeast and East Asian Nations. The meteoric rise of the economy of nations in this part of the world can be mostly attributed to advancements in industries which require engineering knowledge and expertise. While many of the nations in this region provided scholarships for their qualified and selected citizens to study engineering in countries, such as the United States of America, the United Kingdom, and other European nations, they also developed their engineering programs in parallel while strategically increasing the number of students, and thus graduates, as the capacity to educate engineers increase.

With the increasing importance of engineering education, demand on quality also increased. There is also a need to uphold the sanctity of the engineering profession. This brought in the need for accreditation of engineering programs to preserve and ensure that graduates have the required foundation to practice as engineers.

In the twenty-first century, the shift is now from creating awareness to the need for increasing and ensuring quality of engineering education to ensure that this region has a pool of capable engineers. Many see the need to shift the practice of engineering education from the traditional mode to those that are more fitting in preparing graduates to face up to twenty-first-century challenges. From the formation of accrediting bodies, signing up for WA membership to providing the required training of auditors and engineering educators in the OBA, nations in this region can be seen to actively invest efforts to upgrade their practices to be scholarly and of high academic quality. This section discusses engineering education practice, with particular emphasis on ensuring academic quality and accreditation.

3.3.1 Accreditation and Accrediting Bodies

Various accrediting bodies were formed at the national level in Southeast and East Asian nations in the first decade of the twenty-first century, with many of them becoming signatories of the WA. This is in contrast to other nations in Asia that take up accreditation by foreign accrediting bodies, such as the American Accreditation Board of Engineering and Technology (ABET). For example, the Accreditation Board for Engineering Education of Korea (ABEEK) started accrediting engineering programs in 2001 and became a member of the WA in 2007 (ABEEK 2014). The Japanese Accreditation Board (JABEE) was established in 1999 and became a member of the WA in 2005 (JABEE 2014). In Singapore, the Engineering Accreditation Board was established in May 2002 under the Institution of Engineers Singapore (IES). They became a member of the WA in June 2006 (IES 2014).

Engineering accrediting bodies became signatories of the WA to ensure that accredited programs from the country are at par with those of member countries. Japan, Korea, Singapore, and Malaysia are WA signatories in the Asian region. The graduates of accredited programs are recognized in member countries, such as the USA, Australia, Germany, South Africa, Canada, etc., and therefore can practice as engineers globally. A very important requirement under the WA is that the curriculum of accredited programs must embrace outcome-based education, which results in major changes of how programs are designed, implemented, and assessed. While accreditation is voluntary in Japan and Korea, in Malaysia it is compulsory. This is because in Malaysia, graduates from engineering programs that are not accredited are not legally allowed to practice as an engineer. Further details of the history of academic quality assurance and accreditation of engineering programs in Malaysia are elaborated in the next section.

3.3.2 Accreditation and Quality Assurance: The Malaysian Experience

In Malaysia, the engineering education quality assurance initiatives at the public institutions of higher learning can be traced to as early as in 1957 when the Public Services Department (JPA), a body that was given the mandate to carry out the accreditation exercise, audited engineering programs to ensure that graduates are acceptable for entry into the government service. In 1959, the IEM also carried out a similar exercise following the UK and Australia model. But after the Engineering Act was passed by the parliament in 1967, the Board of Engineers took over the responsibility and carried out joint accreditation with the IEM.

The increase in the number of private institutions of higher learning in the 1990s has led to the formation of the National Accreditation Board (LAN), which was responsible to accredit programs at private institutions. Then in 1997, the Malaysian Qualifications Agency (MQA) was established to accredit all programs both at the public and private higher learning institutions. However for the engineering programs, there is a provision that the professional bodies will continue to accredit the engineering programs. Consequently, in 2000, the Engineering Accreditation Council (EAC) comprising of the Board of Engineers Malaysia (BEM), IEM, the MQA, and JPA was established. The BEM stipulated clearly the attributes that are necessary in preparing for engineering practice such as the ability to apply mathematics, science, and engineering in solving engineering tasks; the ability to understand environmental, economic, and community impacts on development; and the ability to communicate effectively and ethically in discharging duties. However, the main portion of the accreditation exercise was carried out quantitatively such as looking at numbers of credits, number of courses, and students to staff ratio. These are clearly evident in the September 1999 version of the BEM guidelines.

However, as Malaysia aspires to be an educational hub in Asia, international recognition of engineering programs is deemed necessary. Therefore, in 2003, the EAC, which is under the BEM, was admitted as a provisional member of the WA. By being a member, all signatory countries will recognize substantial equivalency of an accreditation system within a country. This assures that graduates of accredited programs in their countries are prepared to practice engineering at the entry of the profession. For the EAC, the route to be a WA signatory was a long and arduous 6-year journey. This is mainly due to the major paradigm shift from the teacher-centered, traditional approach to the student-centered OBA for the institutions of higher learning as well as the shift from quantitative to a more qualitative approach of accreditation exercise that emphasizes on continuous quality improvements. Consequently, major changes had to be made not only at the higher learning institutions, but even the EAC was expected to improve and benchmark with international practice.

This drastic jump in the requirement and framework for academic quality exerted by the EAC made the OBA to prevail as a basis for accreditation for engineering programs in Malaysia. Well-established engineering programs in Malaysian institutions of higher learning had to comply strictly to the 2007 EAC Manual that have since evolved several times. These efforts bore the desired fruit, when Malaysia, in 2009, was accepted to be a full signatory member of the WA. Nevertheless, the transition had not been easy for engineering programs in Malaysia. During that time frame, none of the engineering programs was accredited for 5 years, the most was 3 years. As an example, of all the electrical and electronic engineering programs audited between 2009 and 2012, 69 % were only granted a 2-year accreditation, while 19.6 % were granted a 3-year accreditation, out of a maximum of 5 years accreditation (Liew and Puteh 2014). The accreditation evaluation of the engineering programs defined that a program can get 5 years accreditation if it satisfies the minimum accreditation requirement with no shortcomings and has addressed all the requirements effectively. However, in the transition to fulfill the new accreditation requirements, there was a period where the program owners, as well as the auditors from the side of the EAC, were uncertain about the definition of the standards. This confusion results in some accreditors taking an overly strict definition of the minimum standard, causing inconsistencies during the accreditation process, which led to inconsistencies in the outcome of the accreditation. Not surprisingly, this caused an uproar among program owners and within the Ministry of Education. Nonetheless, these are transitional problems that were brought to the attention of the Engineering Accreditation Department under the Board of Engineers and were resolved with the guidance from the various WA mentors. For example, to tackle the problem of inconsistencies, a newly required moderation process is now instituted, resulting in a more moderate and facilitative approach by the EAC auditors in the recent accreditation process.

The current changes in the EAC manual in 2012 to comply the latest WA requirements show that the EAC is in-line with the latest development and practiced continuous quality improvement. These new requirements reflect the needs and challenges of the twenty-first century which engineering programs need to prepare graduates for in the university, which is actually a push towards increasing the quality of engineering programs. While engineering program owners are now well versed in the OBA, the latest requirements include challenging outcomes, such as complex problem-solving skills and complex engineering activities (EAC 2012), and pose another challenge in increasing the quality of teaching and learning for developing students. Consequently, Malaysian institutions of higher learning, which are now wiser after going through the initial changes, are making a lot of effort to comply with the EAC 2012 manual. With the continuous efforts to improve the quality of accreditation required under the WA, the EAC is directly responsible to ensure compliance by engineering programs. Therefore, in Malaysia, membership in the WA, which shaped the current accreditation process, has also managed to exert pressure on institutions of higher learning to change and improve engineering programs to meet the new requirements for accreditation and thus pushed program owners to think, discuss, and implement required measures to enhance the quality of engineering education in the country.

3.3.3 Engineering Education Centers

During the past decade, centers for engineering education were established across several Asian countries. Some of these countries include Korea, Hong Kong (China), and Malaysia. Although these centers focus on different areas, the main aim is to enhance engineering education, especially in their own respective countries and institutions. Several areas of focus include innovation in teaching and learning in engineering, research in engineering education, engineering service learning, women in engineering, developing the talent pipeline, and engineering faculty development through training and mentoring.

In South Korea, the commitment of the government towards engineering education can be seen in the funding of 65 Centers for Innovative Engineering Education (CIEE) in different universities across the country, which are affiliated to five hub centers. Each hub center has different focus, such as globalization, multidisciplinary design, accreditation, industry driven programs, etc. (Song 2012). In 2014, however, the number of CIEE in Korea has risen to more than 70. For example, Pusan National University under its Innovation Center for Engineering Education (PICEE) has been organizing international service learning to produce global engineers by collaborating the program with Politeknik Elektronika Negeri Subaraya (PENS), Indonesia, and UTM (CEE 2013a). The hub centers are also active in organizing workshops, seminars, and competitions. Sung kyun kwan University (SKKU) Hub Centre for Innovative Engineering Education organized the International Workshop on Innovative Engineering Education, inviting international speakers and experts with participation among the member centers in November 2011 and January 2013 (CEE 2012a). In November 2013, the Pusan National University Hub Centre organized the Capstone Design Fair (E2Festa), which is a capstone design competition among engineering students in Korea and several partner institutions from other countries. In conjunction with E2Festa, there was also a seminar on engineering design education with a section for invited speakers from all over the world.

In Hong Kong, the Hong Kong University for Science and Technology (HKUST) under the Centre for Engineering Education Innovation (E^2I) plays an important role in improving the teaching and learning of engineering for HKUST. Workshops and training are given to engineering lecturers, and resources for engineering education are readily accessible on the center's website (Centre for Engineering Education Innovation 2014). E^2I also carries out research among the students and lecturers of HKUST on some specific areas that can impact the engineering education of HKUST such as team working skills and assessment in engineering education. The center also organizes Workshop on International Innovative Engineering Education, where participation was by invitation only.

In Malaysia, there are two centers that focus on different aspect of engineering education: the UKM's Centre for Engineering Education Research (CEER) and the UTM Centre for Engineering Education (CEE). The UKM CEER, which was established in 2009, focuses on research in engineering education in various topics since its formation was initially from the engineering education research group (CEER

2014). CEER emphasized their activities on research, publication, and conferences in engineering education. Some research areas of CEER members are action research, quantitative research using the Rasch model, project learning, and topics related to teaching, learning, assessment, industrial training, and curriculum.

The UTM CEE was established in 2011 to consolidate the various efforts in engineering education that had been going on in UTM since the early 2000. CEE gives emphasis on both engineering education research and improving engineering education among engineering lecturers and students through training and educational programs. The tagline for CEE, "Training and Research in Engineering Education (TREE)," aptly summarizes the thrust of the center. To ensure high-quality research as well as train others in conducting research in engineering education, CEE offers postgraduate studies in engineering education that can further strengthen the research in engineering education. CEE promotes the virtuous cycle of research (VCR; Phang and Mohd-Yusof 2013) to ensure that the research is grounded in educational theories and carried out through rigorous educational research methodology. By doing so, the research results can be used to improve the teaching and learning of engineering. These findings are disseminated not only through seminars and publications but also through training sessions with engineering academics. This is in-line with the recommendation by Jamieson and Lohman in the 2012 ASEE report, Innovation with Impact (Jamieson and Lohman 2012). Elaboration on research in engineering education will be elaborated in the following section.

In looking at all the engineering education centers in the various countries in this region, most are promoting good scholarly efforts in engineering education through seminars, workshops, sharing sessions, and systematic training. Many conferences are also organized by the centers, which aims to disseminate efforts in engineering education to a large audience. However, concerted efforts to offer regular workshops and training, which normally cater to a smaller number of audiences compared to conferences, are necessary to systematically train and upgrade the skills of engineering academics throughout the region to encourage the implementation of scholarly good practices in engineering classrooms.

Currently, training is very much localized at the institutional or national level. As mentioned earlier, there are centers in Korea, as well as in Hong Kong, which conduct regular workshops and training. Similarly, the UTM CEE conducts regular training that can also be offered in-house at the request of specific institutions. In Malaysia, training at the national level is also offered by the Higher Education Leadership Academy, under the Malaysian Ministry of Education, although they are not solely conducted for engineering academics.

Undoubtedly, having dedicated centers for engineering education expedites the promotion and enhancement of quality in engineering education practice. Many of these centers assist their own institutions as well as form part of national level initiatives to provide support for fulfilling accreditation requirements, such as training and multidisciplinary project offerings. The numerous varied activities, plus the linkages among some of these centers, encourage the sharing and dissemination of scholarly practices among practitioners as well as create interest among new engineering academics.

3.4 Engineering Education Research

The aim of research in engineering education is to improve the practice of engineering education for the betterment of teaching and learning in engineering. Hence, research and practice in engineering education go hand in hand, and it is easier to use the term engineering education to refer to both the research in engineering education (including those that are rigorous) and practices in teaching engineering. Engineering education is now a knowledge discipline of its own where rigorous and meaningful research is conducted through VCR (Phang and Mohd-Yusof 2013). The VCR is essentially a study on ideas or issues where the research output can be used to inform and improve current practice. This will in turn generate new questions and ideas, which can then be further researched on to further improve the implementation. More elaboration on the VCR is in the following section.

In the Asian region, engineering education research was introduced through various means, namely postgraduate studies in engineering education, research and innovation centers for engineering education, societies for engineering education, conferences in engineering education, and improvement activities related to accreditation exercise. However, the most effective approach to ensuring rigorous engineering education research is through the postgraduate programs in engineering education, such as the master of philosophy and the doctor of philosophy in engineering education. This is because of the academic rigor of such programs would require expert input and reviews as well as the assessment of the research conducted. In addition, most countries, such as Malaysia, have its own quality assurance requirements. Having the postgraduate program is also a productive way of having experts in engineering education research, who can understand and be the middle people in explaining engineering as well as rigorous educational research requirements to those from engineering and also those from education. Further elaboration will be provided in the following section.

3.4.1 Virtuous Cycle of Research in Engineering Education

Publications in engineering education basically can be divided into two: experience sharing (scholarly and non-scholarly) and research papers. Most of the papers are experience sharing because RREE introduced by Streveler and Smith (2006) is not a common practice among some active participants of the engineering education community in this region. The ASEAN Journal of Engineering Education (AJEE) tried to accommodate the community by accepting both scholarly experience sharing and research papers as a means to encourage engineering educators to publish in engineering education while simultaneously introducing and exposing readers to elements of RREE.

The idea of RREE was first introduced to the Southeast Asian community during the 2007 Regional Conference in Engineering Education in Johor Bahru, Malaysia, by Prof. Karl Smith from Purdue University, USA, during a preconference workshop session. This idea was further expanded to include the VCR during the 2010 Regional Conference on Engineering Education and Research in Higher Education (RCEE and RHEd 2010) in Kuching, Sarawak, Malaysia, during the keynote by Smith and Streveler (2010). Phang and Mohd-Yusof (2013) then further promoted the VCR in engineering education to the Asian community, arguing that innovations and new ideas implemented should be studied to determine their impact and/or other questions that may arise and can lead to improvements. Conducting engineering education research that becomes VCR ensures that the research done is of significant importance in engineering education. This idea of VCR has been encouraged through trainings courses held by the UTM CEE to help enhance the quality of engineering education in the region. The CEE also provides training workshops on conducting RREE, which leads to VCR.

As most of the participants in the engineering education research community in the region are trained in the engineering background, VCR relates the cycle of research using the similar steps in most of the scientific research, which are (Phang and Mohd-Yusof 2013):

a. Identify the problem
b. Review literature and established theories to connect the problem with grounding principles
c. Design the methodology
d. Analyze data
e. Find meaning to conclude and generalize

Following the initial idea given by Phang and Mohd-Yusof (2013), the VCR can be simplified as shown in Fig. 3.1. The research problems must come from the current practice such as problems in learning certain engineering topics and innovative teaching and learning approaches. Some examples are:

1. Chemical process control (Mohd-Yusof et al. 2011, 2013b; Helmi et al. 2013)
2. Engineering education for sustainable development (Abdul Aziz et al. 2013a, 2013b; Mohd-Yusof et al. 2013c)
3. Problem-based learning, problem solving, and cooperative learning (Mohd-Yusof et al. 2011, 2013a; Mohammad-Zamry et al. 2011; Phang et al. 2012; Helmi et al. 2011, Helmi et al. 2013)
4. First-year experience (Abdul Aziz et al. 2013c; Mohd-Yusof et al. 2014)
5. Thermodynamics (Mulop et al. 2013; Mulop et al. 2014)
6. 3-D CAD (Adnan et al. 2013)

After the problem is identified, it needs to be clearly formulated into a researchable framework. This has been explained by Phang and Mohd-Yusof (2013). The next step is to review the literature of the existing theories from the past and current research or practice reports. This is to ensure that the research is well grounded in strong educational principles or theories (Streveler and Smith 2006).

The steps of research methodology and data analysis must draw from sound research methodology. This has been explained in Phang and Mohd-Yusof (2013)

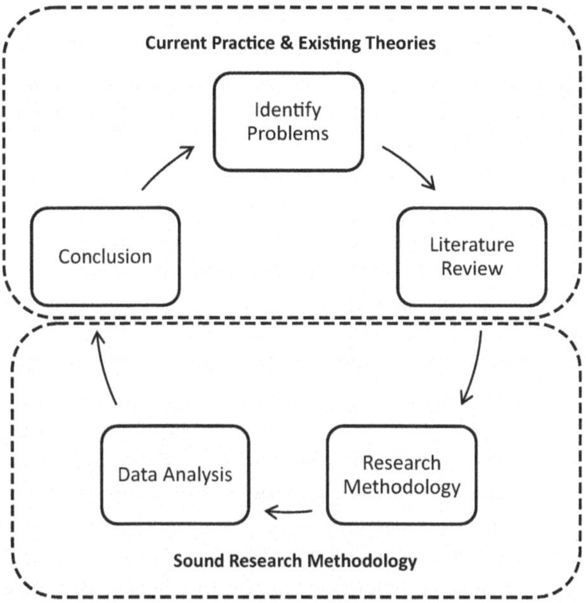

Fig. 3.1 Virtuous cycle of research (VCR) in engineering education

with a few examples of research methods guidebooks. Research methodology ranged from the objectivist paradigms (e.g., postpositivism, empericism) to the subjectivist paradigms (e.g., interpretivism, constructivism; Crotty 1998). In the objectivist paradigms, researchers tend to answer research questions through deductive approach and employ research strategies that aim at testing and confirming existing theories, such as survey and experimental research. Usually, statistical analysis is used to analyze the data to accept or reject certain hypotheses, and generalization is made as a conclusion of the research. Sometimes, it is more convenient to label this as quantitative research approach (Creswell 1994).

On the other hand, the subjectivists favor inductive approach to generate theories that can explain phenomena in an educational setting by using research methodology like grounded theory (Glaser and Strauss 1967), ethnography (Hammersley and Atkinson 1995), case study (Stake et al. 2005), action research (McNiff 1988), and so on. Usually, in-depth analysis is employed to understand and explore the questions of how and why in order to produce rich description or explanation of a phenomenon. This approach is usually coined as the qualitative research approach.

Despite the debate and competition between the two major paradigms, the selection of a sound research methodology also depends on the research questions and the nature of the research problems. As the research questions stem from the research problems (Phang and Mohd-Yusof 2013), a researcher should not plan how to do the research (and select the research methodology) prior to the formulation of research problems and research questions. Research questions that require answers using statistical data should be answered using the quantitative research methods, while research questions that require rich and in-depth explanation should opt for the qualitative research methods.

The quantitative research methods collect quantitative data (numbers), while the qualitative research methods collect qualitative data (texts). The quantitative research methods may include close-ended questionnaire survey, test, interview, and observation using checklists that collect quantitative data. The qualitative research methods may include open-ended questionnaire survey, interview, observation using field note, artifact analysis, and so on that collect qualitative data.

Research data collection should follow as close as possible to the research method procedure to ensure the validity and reliability of the data. This should be followed by the correct and appropriate data analysis method. Quantitative data analysis methods usually involve calculation and the use of statistics, either descriptive or inferential statistics. It may be aided with computer software, such as Microsoft Excel, Statistical Package for Social Sciences (SPSS), R Project for Statistical Computing (R-Statistics), Winsteps, AMOS, and so on. In the qualitative data analysis methods, there are constant comparative method (from grounded theory methodology; Strauss and Corbin 2008), Miles and Huberman (1994) qualitative data analysis, content analysis (Krippendorff and Bock 2008), thematic analysis (Braun and Clarke 2006), and so on. Some computer software may assist the qualitative data analysis in terms of managing the large quantity of data, such as NVivo, Atlas, Computer-Assisted Qualitative Data Analysis Software (CAQDAS), MAXQDA, and so on.

A good engineering education researcher should have some basic knowledge of a wide variety of research methodologies and methods in engineering education research and master a few of them. Most importantly, a researcher should know how to secure the validity and reliability (Cohen et al. 2007) of the data collection and analysis. Validity refers to the accuracy of measurement, and interpretation of the data represents what the researcher intends to measure. It is the appropriateness, meaningfulness, and usefulness of the specific inferences a researcher makes based on the data collected. For example, if we intend to measure conceptual understanding of dynamics among engineering students, we should ensure that the questions asked in the conceptual test are actually testing their conceptual understanding, not problem solving or factual memorization.

On the other hand, reliability is the consistency of the data obtained. This basically means that every time the data are collected from the respondents, same or similar data should be obtained. For example, a test is given to a group of students today, and x mean score is obtained after the analysis. If the test is repeated to the same group of students tomorrow, the same or similar x mean score should be obtained as well. This shows that the measurement and instrument are reliable and can be used to make generalized conclusion. It should be noted that the research results should yield a certain level of confidence to the practitioners to be able to accept the conclusion of the research. This is because, ultimately, the research conclusion will be used to make educational decisions that may change how engineering should be taught.

Furthermore, in making a conclusion from the data, it should be able to match with the current practice as the problem initially came from the current practice. Sound educational decisions can be made through recommendations that are drawn from the valid interpretation of the data. The conclusion may provide new insights into the existing theories.

3.4.2 Postgraduate Studies in Engineering Education

UTM is the first university in Asia that offered a doctoral study in engineering education in the year 2008 (Mohd-Yusof et al. 2012). Some of the other universities that offer a PhD in the engineering education program in the world are Purdue University (USA), Virginia Tech University (USA), Utah State University (USA), and Aalborg University (AAU; Denmark). Later, UKM joined the bandwagon to offer a full research doctoral program in the year 2013.

Establishing PhD in the engineering education program is not a simple matter because it is a multidisciplinary program that requires expertise in engineering as well as educational principles and research methods. In addition, there are not that many engineering education programs to learn from and benchmark against. Since the candidates for PhD in engineering education programs are from an engineering- and technology-related background, proper support in learning and understanding the fundamental theories of education and educational research methods must be given. Otherwise, most of them will be overwhelmed in their effort to unravel the mountains of information that they have to digest if no guidance were given. Therefore, offering PhD in the engineering education program requires careful thought and planning; simply offering the program without much thought will make it very difficult for the students, as they try to do their research.

In Malaysia (and Asia), there are two universities offering PhD in the engineering education program: UTM and UKM. UTM have produced more than 12 PhD graduates since its inception. In addition, UTM offers the MPhil study that focuses on engineering education research. UTM's PhD program in engineering education is conducted in full research mode with some compulsory supplementary courses that help engineers or engineering educators who enroll into the program to cope with the transformation from technical sciences to interdisciplinary sciences (technical and social sciences). Thus far, from the students' feedback the courses were useful in providing a broad overview that support students with good starting points for understanding at a deeper level. For the program to be truly interdisciplinary, it is governed by the CEE, not by a faculty or school that is usually limited to single disciplinary. Finding the examiners for the students are not that easy because in Malaysia, and even in Asia, there were not that many experts in engineering education research. Having experts in the field as the external examiners is crucial because they reflect the trustworthiness of the quality of research and the graduates. For the first cohort of students, their viva sessions were attended by external examiners who are experts in engineering education from universities that offer PhD in engineering education, such as Profs. Karl Smith, Johannes Strobel, and Heidi Deifus-Dux from Purdue University and Prof. Anette Kolmos from AAU (Mohd-Yusof et al. 2012).

UTM also offers a joint PhD in the engineering education program with AAU, Denmark. Students have to be in both universities for at least 6 months within the normal duration of 3 years and enroll in specified courses. They will also have at least one supervisor from each university. Those who graduate from the program will receive a certificate with the seal of both universities. The strength of this joint program is the cross-cultural experience that students can experience and the support of supervisors and access to facilities from UTM and AAU. In addition, this program can also provide expertise in the development of problem-based learning implementation at various levels in engineering programs, especially in the Southeast Asian region.

The PhD in the engineering education program in UTM has a diverse student body consisting of students not only from Malaysia but also from the USA, Pakistan, Iran, and Nigeria. The students come from a variety of backgrounds, such as engineering academics, engineers, technical education teachers, matriculation teachers, mathematics lecturers, computer scientists, and so on. This PhD research in engineering education contributes to the body of knowledge and serves to enhance implementation in UTM and Malaysia as well as in Asia in general.

The theses are based on RREE that covers a wide range of educational areas, such as self-regulation in learning statics, conceptual knowledge in 3-D CAD, practicing engineers' training needs, accreditation standards in engineering programs, courseware design for the teaching and learning of thermodynamics, the impact of cooperative problem-based learning towards engineering students' motivation, problem solving and learning, electrical engineering students' conceptual understanding of electrical circuits, and so on. The research conducted by the students were also published in international academic journals and presented at engineering education conferences worldwide.

3.4.3 Publication and Dissemination of Research Findings

The advancement of RREE in Malaysia really took off with the advent of PhD in engineering education programs. The increasing number of research activities as well as funding and the positive outcomes from the research conducted managed to attract engineering academics to not only participate in the research teams but also properly learn educational research methods and the related grounding theories. Nevertheless, there is still a lot to be done, especially in terms of journal publications.

Contribution in the form of research publications is still very low from the region, although the number of experience sharing papers is on the rise. A lot more high-quality research on significant areas of study must be conducted and published to bring out the Asian experience and perspective of engineering education in scholarly, research-based journals. This will disseminate research findings based on

Asian context that will further provide suitable models and frameworks for impactful innovations and advancement in Asian engineering education.

The development of the engineering education research community in Southeast Asia received support from the international community in 2013, when both the International Research Symposium on Problem-Based Learning (IRSPBL) and the REES were held back-to-back in Malaysia. This is the first time both conferences were held in Asia hosted by the UTM CEE in collaboration with the UNESCO Chair on PBL in engineering education, which is based in AAU in Denmark, and the Research in Engineering Education Network (REEN), which is an international network of communities that promote research in engineering education. Known for the close-knit community of practitioners and researchers, both symposiums and the workshops held in conjunction with the events gave a chance for the Asian, and especially the Malaysian, community to learn and network with participants who are passionate in engineering education research and promoting good practices in the area. By having both symposiums in the region, the local community was able to gauge the standard of their own practice and research and thus enhance the quality to be at par with those at the international level. The PhD students who were still learning and developing their research were also able to get their work critiqued and verified; those who participated felt that the symposiums were an immense opportunity and learning experience.

Having the IRSPBL and REES in Malaysia in 2013 was indeed an eye-opening experience for most of the Asian participants. Many of them did not have the opportunity to join the active engineering education research communities in North America, Europe, and Australia. The extent of rigor and the conducive and supportive environment during the paper discussion sessions relayed the passion and commitment of the community on the importance of engineering education research. While other conferences in the region were mostly at the experience-sharing level, the requirements and examples given through both symposiums illustrated the kind of effort, planning, and rigor needed in conducting research in engineering education that can inform and improve practice. Thus, the determination and all the energy invested to host the IRSPBL 2013 and REES 2013 by the UTM CEE were indeed worth the effort because the aim of providing interest and enhancing the quality of research and publication in the region was indeed fulfilled.

3.5 Bridging Research and Practice

There is a very strong movement within the engineering education community to bridge theory, research, and practice. This is also the aim of the VCR in engineering education. Unfortunately, many of the researches conducted were not translated into practice. On the other hand, many innovations in engineering education are made without proper study of evidence, or even scholarly foundation. For this reason, the 2012 ASEE report by Jamieson and Lohmann, "Innovation with Impact," highlights the need for a culture of systematic innovation with impact—systematically gathering evidence of impact is highly needed to determine the effectiveness of an innovation in engineering education, and not just simply carrying out the innovation without knowing the impact.

It is clear, therefore, that research and practice should go hand in hand. Implementing the VCR would be ideal within a certain group or institution, but the findings from the research should also be published and disseminated in other manners to have a larger impact, not only on the local community but also at a larger global level.

For this reason, training of engineering academics based on scholarly and research-based knowledge and practices is seen as crucial and is therefore a thrust activity for the UTM CEE. Other than developing research ability of the engineering education students, training is also given as part of continuous professional development for engineering academics who are already teaching in engineering schools or faculties, many with PhD degrees in engineering. Among the variety of training given are those included which impart pedagogical knowledge and skills, problem solving, assessment, course design and research in engineering education. The courses given include sound principles in the area as well as the evidence of impact in engineering education. Participants are also encouraged to share their scholarly experiences through publication and conference presentations, before perhaps moving on to systematic studies. Through writing and sharing of experience, the training workshop participants can also become reflective practitioners, who can continuously improve their practice.

Like other regions in the world, translating research into practice is still elusive in Southeast Asia. With the current approach to research and training embraced in the CEE, it is hoped that this model can facilitate the translation of research into practice in Malaysia and the Southeast Asian region.

3.6 Conclusion

Formal education of engineers had been taking place in Southeast Asia for more than 100 years. However, interest in engineering education as a field throughout the region has only increased in the early twenty-first century. The advent of outcome-based education as required by the WA and the requirement for accreditation of engineering programs drove more institutional and governmental support for deeper involvement among academic staff into the field. The OBA has also driven some institutions to take a more scholarly approach to move beyond the initial documentation phase.

Since the early twenty-first century, numerous events have been regularly held to disseminate and gain interest in engineering education hosted by different organizations and bodies. While most of the current conferences and seminars consist of experience-sharing papers, there are efforts to increase the quality by requiring scholarly implementation. In fact, some conferences had insisted on proper research papers on engineering education. The increased quantity and quality of activities have contributed to further interest in the area, especially towards enhancing the quality of engineering education in the region.

Nevertheless, the Southeast Asian region has still a long way to go in implementing systematic, scholarly, and evidence-based approaches to enhance engineering education. With the numerous challenges that engineering graduates have to face, engineering education leaders in the region realize that meaningful innovations with impact in engineering education is necessary for the region to remain competitive and rise up to the challenges ahead. Therefore, despite the slow start to embrace engineering education as a field, the realization of its importance is pushing a rapid increase in efforts to organize and participate in engineering education events, ensuring that the region is well poised for producing quality future engineering graduates.

References

Abdul Aziz, A., Mohd-Yusof, K., Udin, A., Abdul Latif, A, & Mohamed Yatim, J. (2013a). *Inculcating sustainable development among engineering students, part 2: Assessing the impact on knowledge and behaviour change*. Proceedings Engineering Education for Sustainable Development 2013, University of Cambridge, United Kingdom, Sept 20–22, 2013.

Abdul Aziz, A., Mohd-Yusof, K., Udin, A., Abdul Latif, A., & Mohamed Yatim, J. (2013b). Development of students' knowledge-behavioural changes in relation to sustainability through a case study. *Procedia—Social and Behavioral Sciences, 102,* 568–576. doi:10.1016/j.sbspro.2013.10.773.

Abdul Aziz, A., Mohd-Yusof, K., Udin, A., & Mohamad Yatim, J. (2013c). *Inculcating awareness of sustainable development in first year engineering students—a comparative study of pedagogical approaches*. Proceedings 5th Research in Engineering Education Symposium 2013, Putrajaya, Malaysia, Paper Session 4.1, July 5–6, 2013. ISBN 978–967-0194-35-6.

ABEEK. (2014). President's message. http://www.abeek.or.kr/htmls_kr/en/contents.jsp?menu_l=60. Accessed 30 March 2014.

Adnan, M. F., Daud, M. F., & Saud, M. S. (2013). *Contextual knowledge in 3D CAD modeling*. Proceedings 5th Research in Engineering Education Symposium 2013, Putrajaya, Malaysia, pp. 15–25, July 5–6, 2013. ISBN 978-967-0194-35-6.

AEESEAP. (2014). About us. http://www.aeeseap.com/about/. Accessed 1 March 2014.

Braun, V., & Clarke, V. (2006). Using thematic analysis in psychology. *Qualitative Research in Psychology, 3*(2), 77–101.

Centre for Engineering Education (CEE). (2012a). CEE Director's visit to Korea. *Engineering Education Updates, 2012*(1), 3.

Centre for Engineering Education (CEE). (2012b). List of past lectures and workshops. *Engineering Education Updates, 2012*(1), 6.

Centre for Engineering Education (CEE). (2012d). MoU between KSEE and SEEM. *Engineering Education Updates, 2012*(1), 3.

Centre for Engineering Education (CEE). (2012e). Brief introduction to SEEM. *Engineering Education Updates, 2012*(1), 5.

Centre for Engineering Education (CEE). (2012g). Workshops during RCEE & RHEd 2012. *Engineering Education Updates, 2012*(2), 3.

Centre for Engineering Education (CEE). (2013a). UTM-PNU-PENS international cross-cultural service design & knowledge transfer. *Engineering Education Updates, 2013*(1), 1–2.

Centre for Engineering Education (CEE). (2014a). *ASEAN Journal of Engineering Education*. http://tree.utm.my/ajee/. Accessed 28 Feb 2014.

Centre for Engineering Education Innovation. (2014). Engineering education resources. http://e2i.ust.hk/err/. Accessed 28 Feb 2014.

Centre for Engineering Education Research (CEER). (2014). Background. http://www.ukm.my/p3k/index.php/en/. Accessed 28 Feb 2014.

Cohen, L., Manion, L., & Morrison, K. (2007). *Research methods in education* (6th ed.). Oxon: Routledge.

Creswell, J. W. (1994). *Research design: Qualitative and quantitative approaches*. Thousand Oaks: Sage.

Crotty, M. (1998). *The foundations of social science research: Meaning and perspective in the research process*. New South Wales: Allen and Uwin.

Duderstadt, J. J. (2008). Engineering for a changing world: A roadmap to the future of engineering practice, research and education. The Millennium Project, The University of Michigan.

EAC. (2012). Engineering Programme Accreditation Manual 2012. Board of Engineers, Malaysia.

Glaser, B. G., & Strauss, A. L. (1967). *The discovery of grounded theory—strategies for qualitative research*. London: Weidenfeld and Nicolson.

Hammersley, M., & Atkinson, P. A. (1995). *Ethnography: Principles and practice* (2nd ed.). London: Routledge.

Helmi, S. A., Mohd-Yusof, K., Abu, M. S., & Mohammad, S. (2011). An instrument to assess students' engineering problem solving ability in cooperative problem-based learning (CPBL). 2011 ASEE Annual Conference, Vancouver, Canada, paper AC 2011–2720.

Helmi, S. A., Mohd-Yusof, K., Phang, F. A., Mohammad S., & Abu, M. S. (2013). Inculcating team-based problem solving skills, part 1: enhancing problem solving elements. Research in Engineering Education Symposium (REES 2013), Putrajaya, Malaysia on 4–6 July 2013.

IES. (2014). Engineering accreditation board. http://www.ies.org.sg/pageview.php?page_id=110. Accessed 30 March 2014.

IFEES. (2014). IFEES members. http://www.ifees.net/members/ifees-members. Accessed 30 March 2014.

International Federation of Engineering Education Societies (IFEES). (2013). The 19th KSEE Annual Conference and Engineering Education Festa 2012. http://www.sefi.be/ifees/?p=896. Accessed 28 Feb 2014.

JABEE. (2014). JABEE & accreditation. http://www.jabee.org/english/jabee_accreditation/. Accessed 30 March 2014.

Jamieson, L. H., & Lohmann, J. R. (2012). *Innovation with impact: Creating a culture for scholarly and systematic innovation in engineering education*. Washington: ASEE.

Japanese Society for Engineering Education (JSEE). (2014). About us. https://www.jsee.or.jp/English/. Accessed 1 March 2014.

Krippendorff, K., & Bock, M. A. (2008). *The content analysis reader*. Thousand Oaks: Sage.

Liew, C. P., Puteh, M., & Mohammad, S. (2014). Best practices in Washington Accord signatories: With reference to the accreditation criteria, systems and procedures. 2014 International Conference on Teaching and Learning in Computing and Engineering, 9–12 April 2014, Kuching, Sarawak, Malaysia.

McNiff, J. (1988). *Action research: Principles and practice*. Basingstoke: Macmillan.

Miles, M. B., & Huberman, A. M. (1994). *Qualitative data analysis: An expanded sourcebook*. Thousand Oaks: Sage.

Mohammad-Zamry, J., Mohd-Yusof, K., Harun, N. F., Helmi, S. A. (2012). A guide to the art of crafting engineering problems for problem based learning (PBL). In K. Mohd-Yusof, N. A. Azli, A. M. Kosnin, S. K. Yusof, & Y. M. Yusof (Eds.), *Outcome-based science, technology, engineering, and mathematics education: Innovative practices* (pp. 62–84), Hershey: IGI Global.

Mohd-Yusof, K, Helmi, S. A., Mohammad-Zamry, J., & Nor-Farida, H. (2011). Cooperative problem based learning: A practical model for a typical course. *International Journal Emerging Technologies, 6*(3),12–20.

Mohd-Yusof, K., Phang, F. A., Helmi, S. A., & Alias, R. (2012). Collaborative efforts in developing the PhD in engineering education program in Universiti Teknologi Malaysia. *ASEAN Journal of Engineering Education, 1*(1), 54–57.

Mohd-Yusof, K., Phang, F. A., & Helmi, S. A. (2013a). *How to develop engineering students' problem solving skills using cooperative problem based learning (CPBL)*. World Congress of Engineering Education (WCEE), Qatar on 7–9 January 2013.

Mohd-Yusof, K., Helmi, S. A., & Phang F. A. (2013b). *Inculcating team-based problem solving skills, part 2: enhancing team working skills*. Research in Engineering Education Symposium (REES 2013), Putrajaya, Malaysia on 4–6 July 2013.

Mohd-Yusof, K., Phang, F. A., Kamaruddin, M.J., Hassim, M.H., Hashim, H., Sadikin, A. N., Jamaluddin, J., Othman, N., Hassan, H., Helmi, S. A., Abdul Aziz, A., & Ujang, Z. (2013c). Inculcating sustainable development among engineering students, part 1: designing problems and learning environments with impact. Proceedings Engineering Education for Sustainable Development 2013, University of Cambridge, United Kingdom, Sept 20–22, 2013.

Mohd-Yusof, K., Phang, F. A., Sadikin, A. N., & Helmi, S. A. (2014). *Determining the effect of an engineering overview assignment on students*. Proceedings for the 2014 ASEE Annual Conference and Exposition on Engineering Education, Indianapolis, USA, June 15–18, 2014.

Mulop, N., Mohd-Yusof, K., & Tasir, Z. (2013). *The performance of students learning thermodynamics through an online courseware*. Proceedings 5th Research in Engineering Education Symposium 2013, Putrajaya, Malaysia, Paper Session 2.3, July 5–6, 2013. ISBN 978-967-0194-35-6.

Mulop, N., Mohd-Yusof, K., & Tasir, Z. (2014). The improvement of confidence level of students learning thermodynamics through a multimedia courseware. 2014 IEEE Educon, Istanbul, Turkey, April 2014.

National Academy of Engineering. (2005). *The engineer of 2020: Visions of engineering in the new century*. Washington: Academic Press.

Phang, F. A., & Mohd-Yusof, K. (2013). Taking the "guess-work" out of engineering education: Establishing the virtuous cycle of research. *Procedia—Social and Behavioral Sciences, 102,* 212–220.

Phang, F. A., Mohd Yusof, K., & Syed, A. H. S. H. (2012). Engineering students' perception on learning through CPBL for the first time. 2012 ASEE Annual Conference & Exposition.

Smith, K. A., & Streveler, R. A. (2010). Challenges and opportunities in engineering education research. Keynote for the 2010 Regional Conference on Engineering Education, 7–8 June 2010, Kuching, Sarawak, Malaysia.

Song, S. J. (2012). SKKU innovative hub centre in engineering education presentation. Univeristi Teknologi Malaysia, Johor Bahru, 2012.

Stake, R. E. (2005). Qualitative case studies. In N. K. Denzin & Y. S. Lincoln (Eds.), *The Sage handbook of qualitative research* (3rd ed., pp. 443–466). Thousand Oaks: Sage

Strauss, A. C., & Corbin, J. M. (2008). *Basics of qualitative research: Second edition: Techniques and procedures for developing grounded theory*. Thousand Oaks: Sage.

Streveler, R. A., & Smith, K. A. (2006). Conducting rigorous research in engineering education. *Journal of Engineering Education, 95*(2), 103–105.

Khairiyah Mohd-Yusof Khairiyah is the Director of the Centre for Engineering Education, Universiti Teknologi Malaysia. She is currently the Vice President for Asia and Oceania for the International Federation of Engineering Education Societies (IFEES), the Secretary for the Society of Engineering Education Malaysia (SEEM), and board member for the International Research in Engineering Education Network (REEN). Her engineering education research focuses on innovative curriculum and teaching and learning practices, faculty development, talent pipeline and Engineering Education for Sustainable Development.

Fatin Aliah Phang Fatin completed her PhD at University of Cambridge, UK in 2009 and appointed as a lecturer at the Faculty of Education, Universiti Teknologi Malaysia (UTM). She was the Academic Manager of the Centre for Engineering Education (CEE), UTM and now one of the Academic Fellows at the centre. She is actively involved in research and training in Engineering Education as well as other research areas such as Science Education and Environmental Education.

Shahrin Mohammad Shahrin is the Senior Director, Centre for Quality and Risk Management, Universiti Teknologi Malaysia and currently the President, Society of Engineering Education Malaysia. He is also the Associate Director at the Engineering Accreditation Division and involved as a one of the refered personnel for Malaysian Qualification Agency. With over 30 years of academic experiences, he is involved in the planning, managing and assuring the quality academic programmes. To date he had written four books and more than 90 journals, seminars and conferences papers. These papers addressed his research findings as well as sharing experiences on the teaching and learning practices. His research interest are in structural steel design and engineering education. He has also conducted more than 100 lectures/workshops related to Outcome Based Education, ISO9000, and Academic Quality Assurance.

Part II
Innovations in Learning, Teaching, and Engineering Education

Chapter 4
Leveraging Pedagogical Innovations for Science, Technology, Engineering, and Mathematics (STEM) Education in the Middle East Context

Shadi Balawi, Kinda Khalaf and G. Wesley Hitt

4.1 Introduction

4.1.1 Twenty-First-Century Engineering Skills: Skills Gaps Throughout the World

Engineering education stakeholders, from academic institutions, professors, and alumni to private sector industries, governmental education agencies, and accreditation bodies universally agree that current engineering graduates lack the critical skills essential for the twenty-first century's interconnected dynamic world that is rapidly being transformed by information explosion and monumental scientific and technological advances. The National Academy of Science in the USA identifies five essential twenty-first century skills: adaptability, complex communication/social skills, nonroutine problem-solving, self-management/self-development, and systems thinking (National Academy of Sciences 2010). These competencies are echoed in the UNESCO report "Learning: The Treasure Within: Education for the Twenty First Century" (UNESCO's Report 1999) and in a recent European Community report, which identifies eight key competences essential in a knowledge-based society (European Communities 2007). The EU report emphasizes that these skills

S. Balawi (✉)
Department of Aerospace Engineering, Khalifa University of Science, Technology, and Research, Abu Dhabi Campus, P.O. Box 127788 Abu Dhabi, United Arab Emirates
e-mail: shadi.balawi@kustar.ac.ae

K. Khalaf
Department of Biomedical Engineering, Khalifa University of Science, Technology, and Research, Abu Dhabi Campus, P.O. Box 127788 Abu Dhabi, United Arab Emirates

G. W. Hitt
Department of Applied Mathematics and Sciences, Khalifa University of Science, Technology, and Research, Abu Dhabi Campus, P.O. Box 127788 Abu Dhabi, United Arab Emirates

M. Abdulwahed et al. (eds.), *Advances in Engineering Education in the Middle East and North Africa,* DOI 10.1007/978-3-319-15323-0_4

are not only critical in providing the labor force flexibility that allow for quick adaptation to dynamic changes but also serve as foundations for innovation, productivity, and competitiveness; proficiencies highly valued in a globalized economy and many economically struggling member countries (European Communities 2007).

Research shows that the inadequate preparation of engineers in key competencies in fact extends internationally. A recent UNESCO report (Skills Gaps Throughout the World: An Analysis for UNESCO Global Monitoring Report 2012) warns that skills gaps are constraining companies' ability to grow, innovate, deliver products and services on time, meet quality standards, and meet environmental and social requirements in countries where they operate. The report identifies the lack of available talent and trained resources in the Middle East as the greatest threat for sustainable development of the region. Gulf leaders are among the least satisfied with the supply of employable graduates including engineers, with only 37 % citing their satisfaction (Maktoum Foundation 2012). Employability skills were classified into four categories (technical, cultural, interpersonal, and intrapersonal) and included 15 specific skills: independent task execution, appropriate approach to problem-solving, ability to monitor and evaluate own activities, ability to relate specific issues to wider contexts, ability to apply knowledge to new situations, ability to devise ways to improve own actions, ability to deal with different cultural practices, openness and flexibility, negotiation and mediation skills, self-motivation and initiative, ability to network, creativity and innovation, ability to relate to a wide range of people, team participation, and sense of identity and self-confidence (UNESCO Report 2012). Misalignment between education and employers' needs was cited as one of the main reasons behind the skills gap.

The current engineering curriculum, delivered by the vast majority of institutions worldwide including the Middle East, continues to follow the traditional science model of engineering education in which the first 2 years are typically devoted to basic sciences and mathematics, with minimal exposure to "real-world" engineering problems (Froyd and Ohland 2005; Dym et al. 2005b; Sheppard et al. 2009; Khalaf et al. 2013a). Furthermore, engineering curricula continue to be mostly delivered by traditional passive lecture mode in which instructors start with theories and mathematical models, and then move to textbook examples, which may or may not ultimately extend to real-world applications (Kardash and Wallace 2001; Prince 2004; Prince and Felder 2006). The combination of the traditional model of engineering education, which clearly delays student exposure to engineering integrative thinking and experience, with deductive passive course delivery leads to the current mismatch between the traditional structure of the engineering disciplines and the emerging complexities of modern engineering systems (Creed et al. 2002; Dutson et al. 2007; Dym 2004; Evans et al. 1990; Litzinger et al. 2011). Research shows that students will not develop the aforementioned competencies by following mostly theoretical, disconnected curricula while sitting passively in lecture halls, taking notes, and memorizing content (Newstetter et al. 2012). Even more interactive methods such as Personal Response Systems or Student-Centered Active Learning Environments for Undergraduate Programs (SCALE-UP; Beichner et al. 2000), both of which promote greater student interaction, are not specifically designed to help students develop these competencies because the nature of the problems given to students in traditional engineering classes, while a first step in becoming a successful engineer, is not sufficiently complex to allow students to practice essential twenty-first century skills (Cheville and Bunting 2011; Newstetter et al. 2012). These challenges in developing countries, such as the United Arab Emirates (UAE), have more severe implications, given that the industrial sector is in its infancy and hence has an even higher need for problem solvers, critical thinkers, and independent learners.

4.1.2 Is Problem-Driven Learning (PDL) Culturally Transferrable?

In response to the need for fostering the critical skills for successful modern engineers mentioned above, various inductive learning models have started to make inroads into engineering education (Prince and Felder 2006), joining a broad reform movement across many disciplines in tertiary education towards outcomes-based education (OBE). For example, many of the principles of OBE are now reflected in criteria-based accreditation standards, such as the famous Accreditation Board for Engineering and Technology (ABET) 2000 "a–k" engineering student outcomes criteria (ABET 1997). A necessary process, inherent to any curriculum or subject reform towards OBE, is *constructive alignment* (Biggs and Tang 2007), where learning tasks and assessments that engage students with constructing their own knowledge are made logically consistent with the chosen learning outcomes. Consequently, traditional, passively attended lecture instruction and multiple-choice summative assessments are nearly impossible to reconcile with the requirements of OBE, regardless of discipline or context. Alternative (and often superior) pedagogical models, however, are quite diverse, often tailor-made to suit a particular program in a particular institution. These models include a wide spectrum of pedagogies ranging from discovery learning and case-based learning to problem- and project-based learning, active and cooperative learning, and just-in-time lectures. The main feature shared by these models is the presentation of a specific challenge or complex problem to the students as the initial point of learning after which they are coached to self-learn upon recognizing the need for theories, facts, skills, and concepts (Prince and Felder 2006). Problem-based learning (PBL), as defined by H. S. Barrows, who was one of the pioneers who developed and implemented PBL in medical education over three decades ago, is the learning method based on using problems as a starting point for the acquisition and integration of new knowledge (Barrows and Tamblyn 1980). As a pedagogy centered around solving of complex problems, PBL inherently aids constructive alignment in engineering curricula and offers engineering educators innovative and effective means to successfully engage students deeply with content to acculturate them into the practices of a particular community. Students in PBL courses practice specific skill sets such as spoken and written communication, assume responsibility to be self-directed and lifelong learners, and develop the necessary analytical skills needed to tackle the multifaceted, twenty-first-century engineering challenges.

For our purposes, we adopt a slightly different term—problem-driven learning (PDL). We take this term from the research we have been doing on trying to understand reasoning, problem solving, and learning in authentic sites of interdisciplinary practice—university research labs (Osbeck et al. 2010). Over the last 10 years, we have investigated a tissue-engineering lab, a neuroengineering lab, and two integrated systems biology labs using ethnographic research methods. We then sought to translate our findings on learning in those sites into new models for engineering education (Newstetter et al. 2010). We found in these sites of authentic engineering activity that learning is powered by the need to solve complex problems. PDL fuels advances in knowledge and lab breakthroughs. However, the laboratory problems look nothing like textbook problems. They are complex, ill structured, and ill constrained. They require the integration of knowledge and skills across the bioscience/engineering divide. Adapting to new and changing conditions both in terms of personnel problem types and the ever-present impasses encountered in frontier science is a fact of life. Researchers need to navigate what, when, and how they learn; they work collaboratively when the intractability of the problem demands a collection of heads and hands. Our investigations of these laboratories illuminated why BME majors need to practice early and often the skills of tackling, defining, constraining, and working through complex, interdisciplinary problems to be able to effectively participate as complex problem solvers in industry or research. Thus, the mantra of an introductory course in biomedical engineering (BME) needs to proclaim: Empower students to be agents of their own learning, who are fearless in the face of a complex problem.

In addition to the skill deficiencies that engineering students suffer from on a global level (see Introduction), students in a particular culture may require promotion/validation of certain skills, equally important for the modern engineer, yet lacking in that culture. One example is the empowerment of women in the UAE and the Arab world. Females in this part of the world typically attend all-girl schools and aside from their male relatives, do not interact socially with men. The PBL course is one of the first experiences in coed institutions and cross-gender professional engagement and hence provides an opportunity to promote women empowerment and leadership, through research on achievements of other women as related to the core problems, as well as particular focus on team and communication skills in a coed environment.

Another important skill that was particularly reinforced at KU is "learning to learn" or autonomous self-directed learning. Inherent to PBL, this skill is critical yet nonstandard to a culture that mostly adheres to passive learning didactic lecture models and in which many students, particularly females, are the first generation in their family to attend college. Student teams were empowered to assume initiative and responsibility for their learning and were engaged in the selection, management, and assessment of their learning activities. The main goal is to train life-long learners and independent thinkers equipped to undertake a leading role in a future knowledge-based economy.

4.1.3 Cultural Transfer of Pedagogical Reform in STEM

Over the past 20 years, educational research across all science, technology, engineering, and mathematics (STEM) disciplines has revealed traits common to many of the most successful pedagogical innovations; those that feature learning contexts that enable rich peer interactions (e.g., Crouch et al. 2007) and learning tasks that present students with real-world, ill-posed problems (e.g., Barrows 1998) produce the most substantial learning gains in their students (e.g., Hake 1998). North American pedagogical reforms that have enabled effective, complex problem-solving and rich peer interactions have done so, for example, by doing away with passive lectures (e.g., Laws 1991; Beichner et al. 2007), assigning students to engage in learning tasks in structured teams (e.g., Heller and Heller 1999), and by using challenges or complex, realistic, ill-posed problems (e.g., Barrows 1998). These innovative reforms are increasingly allowing for the progressive evolution in the traditional student and teacher roles, where the students or "apprentices" are empowered to assume increasing responsibility for their learning, while the instructor assumes the role of a facilitator or "master tradesman" coaching and scaffolding expert problem-solving strategies (Barrows 1998; Newstetter 2006; Litzinger et al. 2011).

In many cases, the chance to "construct one's own knowledge" is well received by students, while reliance on fickle teammates negatively violates the expectations of many (Gaffney et al. 2010). These consequences of pedagogical reform have a clear, causal impact on learning gains (Adams et al. 2006). Further refinements to pedagogies in the West increasingly come from research into related "affective factors": motivations, expectations, and self-efficacy. This state of affairs raises important questions for the developing world. How will pedagogical innovations and educational reforms, designed for Western cultures, function when implemented in a novel context, particularly in the Middle East? To what degree does optimizing pedagogy, designed based on the motivations and expectations of Western students, make it incompatible elsewhere? What is the effect of implementing such pedagogies in the Middle East, in particular? What are the mechanics of the pedagogical innovation transfer across cultures in light of encountered challenges and opportunities?

Factors associated with the classroom context are not the only ones causal in successful pedagogical reform efforts. The large body of data on secondary implementations for reformed pedagogies in STEM courses has revealed that institutional (departmental, faculty, and course) factors are, in fact, as important (Wieman et al. 2010). Important departmental factors include authority and support of pedagogical reform by the chair/head of program; scale and breadth of faculty involvement and support, presence, and integration of STEM education specialists within the department/program; and the consistency of reward structures for improved teaching. Important faculty factors include beliefs about and satisfaction level with student learning, level of prior exposure to educational research findings, and overall importance of teaching in their personal interests. Important course factors include workload structures, staff support, and physical resources (Henderson and Dancy 2007).

The broader national and ethnocultural values of the society a university serves can have a strong influence on both the state of student and institutional factors listed above and over how easily they can be changed in an effort to reform pedagogy. In the framework of *expectancy violation* theory (Burgoon 1978, 1995), this influence stems from students' respective prior experiences and the expectations and value judgments about the learning environment that those prior experiences engender. For example, suppose a student has learned physical science by passive attendance at lectures, rote memorization of formulae, formative assessment and feedback through fill-in-the-blank worksheets, and summative assessment through identically structured worksheets-as-final exams. At university level, what does this student expect of his/her first physics course? A reasonable hypothesis would be that they would expect similar instructional design but one featuring more challenging topical content. But if the student arrives on the first day to a 20×20 m studio with round tables, lab equipment, and no projectors or white boards and they are tasked to spend the semester designing experiments to answer fundamental questions, how would they react? What kind of value judgment would they make about the difference? Their expectations have very likely been violated. It is at this point that culture will play a powerful role as it would most likely shape how the student would cope with the expectancy violation. What does a student expect a learning task to be like? What kind of relationship does he/she expect to have with peers in the course and with the instructor? What do faculty expect of their interactions with students? Should students actively engage in class or challenge the instructor's interpretation of a topic? Many of these questions have been investigated in pedagogical reforms in North American cultural contexts (e.g., Gaffney et al. 2010 in undergraduate physics pedagogy) and expectations in these contexts have been shown to be causal in student learning (e.g., Adams et al. 2006), but much less has been done to characterize possible confounding influences in cross-cultural pedagogical reforms.

These considerations also reveal one of the fundamental operational differences between traditional lecture-centered instruction and that based on alternative pedagogy. Instructional strategies derived from effective alternative pedagogies inherently have much larger design spaces, whereas the instructional strategy for a traditional lecture course is essentially one-dimensional or singular. In essence, there is really only one way to run a course that is lecture-centered; the "talk-and-chalk" instructional strategy (Goodwin 1978). Put differently, while traditional lecture appears to function poorly on metrics, such as pre-/post-gain on a concept inventory (see Adams and Wieman 2010 for review), across a wide variety of contexts, alternative pedagogy that shows superior performance to traditional lecture in one context, may or may not exhibit the same performance in another context. To be successful, course reformers must design instructional strategies that operationalize a chosen alternative pedagogy in a manner that matches the needs and respects the constraints of the context, including student and other stakeholder expectations, into which the reformed course is to be implemented.

Figure 4.1 presents a simple working model of a course reform process, whereby the choice of alternative pedagogy is primarily driven by learning outcomes, consistent with the process of constructive alignment (Biggs and Tang 2007), though this

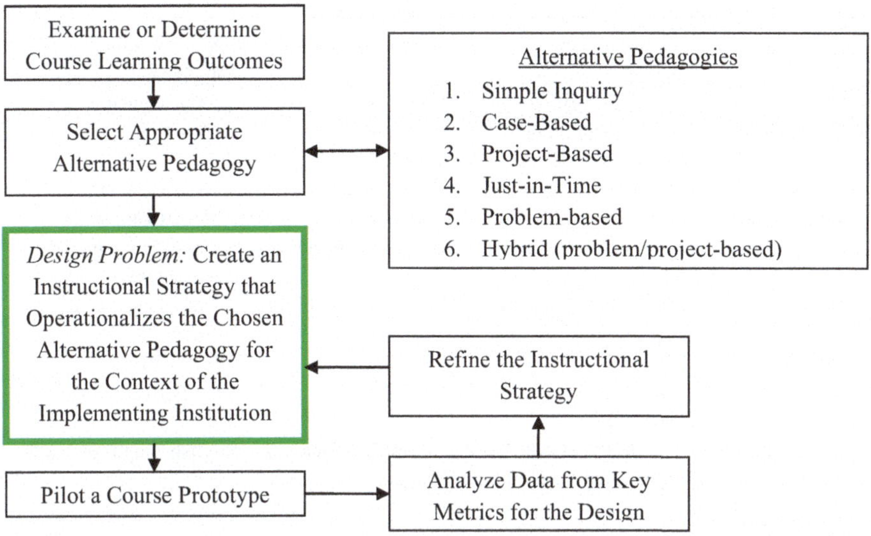

Fig. 4.1 A simple model of the course reform process. The case study for Collaborative workshop physics focuses on the design problem part of the process

process by itself does not determine the optimal instructional strategy. The model is inspired by and, hence, highly reminiscent of the general engineering design process and shows that the emphasis of the present work is on the "second-order" design problem that must be solved when implementing an instructional strategy, developed and aligned to similar outcomes in a US cultural and institutional setting but now being adapted into a different cultural context. Alternative pedagogies are categorized using the scheme of Prince and Felder (2007) and listed in order of overall difficulty, with simple inquiry typically being the least difficult to implement and hybrid problem-/project-based being most difficult. This excludes the additional difficulty imposed by requiring students to navigate learning tasks in teams, as opposed to individually. As Prince and Felder clearly explain, "If instructional objectives are at a low cognitive level…there is no reason to use an inductive method."

A best practice in any kind of design is to begin with a careful study of the design's end-user or client needs, and curriculum is no exception. A deep understanding of the students for whom the course reform should benefit is necessary for determining many objectives and constraints for the design of the instructional strategy. In addition to that, a pre-reform baseline or benchmarking, to quantify and evaluate consequent changes as part of the reform, is extremely valuable.

Starting in the fall of 2009, when KU opened its Abu Dhabi campus, we began gathering a wide variety of data on student background and performance in a traditional, lecture-centered implementation of courses. Table 4.1 shows some of the most salient features of the KU student demographics and course performance. The differences with the typical US engineering major population are remarkable and

Table 4.1 Some salient student demographic and baseline data

Demographic trait	US engineering majors[a]	KU direct admission[b]	KU conditional admission[b]
Gender	19.5 % women	70 % women	35 % women
Ethnic majority	82 % Caucasian or Asian	80 % other MENA nationals	90 % Emirati nationals
Under rep. minority	13 % Black or Hispanic	20 % Asian or SS-African	10 % Asian or SS-African
Family education	1 in 5 first generation college	1 in 5 first generation college	1 in 5 first generation sec. school
Citizenship	1 in 9 not a US citizen	9 in 10 not a UAE citizen	1 in 10 not a UAE citizen
Language	17 % ELLs	90+% ELLs	100 % ELLs
Proficiency	(8.0+ IELTS is "native")	6.5 IELTS average	5.7 IELTS average

KU Khalifa University, *MENA* Middle East and North Africa, *ELL* English language learner, *IELTS* International English Language Testing System

[a] Atman et al. 2010

[b] Hitt et al. 2014

serve to emphasize the importance of a careful conceptual design for any instructional strategy; the instructional strategies published in US educational research literature were designed for very different student populations.

Table 4.1 shows the striking differences between the KU undergraduate engineering student population and their average counterpart population in the USA. We distinguish further between two major components in the KU population, directly and conditionally admitted students, because of the large and significant differences between them. By percentage, conditionally admitted students make up 70–80 % of the total undergraduate student population. Therefore, to design a pedagogy that best serves the largest number of students, it was important for us to focus our efforts on this group's particular learning needs and challenges.

The conditionally admitted student population has several important learning needs, critical for the design of an instructional strategy. While approximately one third are women (KU is a gender-integrated campus), 90 % of conditionally admitted students are Emirati nationals and culturally are not accustomed to working in mixed-gender situations outside of the family. For most of them, KU is the first place where they have to sit alongside students of the opposite gender since age 10. Furthermore, KU policy states that no student can be forced to work in a mixed-gender situation if they have a conscientious objection to doing so. With women making up such a large fraction of the conditionally admitted population, this often combinatorically frustrates instructor-selected team formation. Aside from gender issues, the conditionally admitted student population is also essentially 100 % English language learners (ELLs). The fact that English is the language of instruction places further cognitive burdens on students when tackling learning tasks. Last, conditionally admitted students, being mostly Emirati nationals, are often the first generation in their close family to have attempted advanced study. Many Emirati students live with their families who, without an understanding of the demands and challenges of advanced study, often do not know how to best support and enable their children's university education.

The combination of social anxiety caused by gender-mixing, cognitive overload caused by studying science and mathematics in a foreign language, and potential societal and family pressures associated with being the first university student in the family all make for a very challenging learning experience for KU's conditionally admitted students. It comes as no surprise that as many as 50 % of conditionally admitted students drop/fail/withdraw (DFW) from many core courses in their traditional, lecture-centered form. More worrying, students taking the traditional, lecture-centered version of these courses often show no measurable positive impact on their conceptual understanding of the courses' core concepts.

4.2 KU's Three Course Reform Projects

In this section, we closely examine the case of KU in Abu Dhabi and three course reform projects case studies designed to introduce such pedagogical innovations to its largely Gulf Arab student body. These three are (1) the Engineering Problem Solving Course offered by the Departments of Aerospace and Mechanical Engineering, (2) the University Physics I course offered by the Department of Applied Mathematics and Sciences, and the (3) Introduction to Biomedical Engineering offered by the Biomedical Engineering Department. All three were implemented in experimental formats, making significant to exclusive use of project-based learning, collaborative learning, or problem-based learning, respectively.

4.2.1 Case Study 1: The Engineering Design-and-Build Course

4.2.1.1 Background

The Engineering Design-and-Build course at KU is an innovative, interdisciplinary course created to improve placement, content, and pedagogy for introductory engineering design education. Infused at the freshman level, the course aims to promote expert design thinking by using PBL as the mode of delivery. The course is structured to actively engage the students in the various phases of a prescriptive design-and-build cycle using ill-structured, open-ended problems inspired from industry and is supported by technological tools such as robotics kits and rapid prototyping machines. One of the main contributions is the integration of the prescriptive design cycle with PBL to promote effective inquiry and the systematic, iterative interplay between divergent and convergent questioning in the engineering design process. The inherent alignment of PBL pedagogy and the prescriptive design cycle enhances students' ability to tackle complex challenging problems and reach optimal solutions by following an iterative loop of divergent–convergent processing and decision-making.

The Engineering Design-and-Build course at KU presents to students an innovative incorporation of design pedagogy and classroom technologies, infused at the freshman level of their curriculum, to confront and overcome these challenges and to enable realistic and holistic design education, including a deep emphasis on hands-on conceptual design and engineering design thinking. Three major classroom technologies: (1) LEGO Mindstorms Robotics Kits, (2) a C++ programming interface, and (3) a 3D rapid prototyping printer enable students to quickly and repeatedly "turn the bend" from conceptual design to preliminary design, the most challenging and difficult-to-implement part of the design cycle. The pedagogical mode of delivery is PBL, rather than lecture-centered, and is delivered in a studio format to enhance the engagement of students, the relevance of design in their worldview, and the retention of information.

Traditional engineering design curricula often suffer from a very weak treatment of conceptual design. Often, this is due to several difficult challenges: (1) the fact that there is no standard singly accepted definition of design, (2) that reforms attempting to improve design education often confuse mathematical rigor with design, adding *difficulty* but not necessarily enhancing *design*, and (3) that conceptual design without prototyping, without real fabrication (which is difficult to add to a course), does not have a significant impact on student design thinking. Many curricula are designed to avoid rather than solve this problem of hard-to-define, hard-to-realize conceptual design education by offering computer-aided drafting (CAD) courses or freshman survey courses about engineering under the label of "design." *But students do very little designing themselves in such courses.* Rather, *they read or render* other people's designs, but *they do not design or, more importantly, learn how to think as designers*. Figure 4.2 represents the current multiple hands-on experiences of the complete design cycle that are enabled by replacing lectures with PBL pedagogy and tasking students to tackle design challenges in teams. Emphasis on the complete design cycle at the freshman level enables its repeated and effective use in later stages of a vertically integrated engineering design curriculum.

4.2.1.2 Course Objectives

Tables 4.2 and 4.3 present the goals and the constraints of the design course, respectively. They also present the motivations behind them. Each goal or constraint may have different requirements as presented in the tables.

4.2.1.3 Course Constraints

4.2.1.4 Course Structure

The main components taken into consideration when managing the Engineering Design-and-Build course are team formation, student performance and accountability assessment, instructor supervision, and equipment and assets as shown in Fig. 4.3.

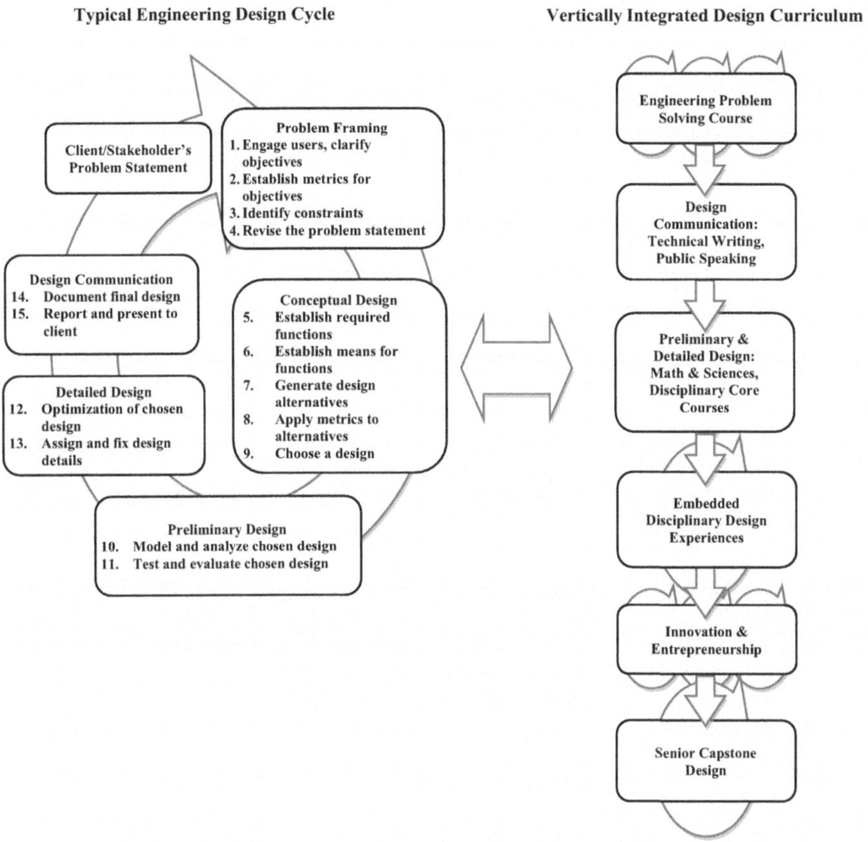

Fig. 4.2 Students taking the freshman Engineering Design-and-Build course experience the iterative design cycle multiple times. Experiencing the complete design cycle is enabled by implementing pedagogical learning practices and the use of advanced classroom technologies, such as LEGO Mindstorms Robotics kits and 3D rapid prototyping printer

Team members are selected by the instructors for all projects except for the final one, where the students are asked to form their own teams. The size of each team is three to four members and in all cases does not exceed four members. The instructors select the teams, balancing gender, ethnicity, and academic level as recommended in problem-based collaborative learning (Newstetter 2006). From instructors' perspective, it is challenging to select students solely based on grade point average (GPA). For example, team dynamics may be different based on their analytical abilities versus their experience in hands-on projects. So, the teams were selected based on a combination of the GPA, team dynamics, and instructors' previous observations of team members' performance. The authors plan to look further into the issue of team formation. Most of the time, the formation is based on internal discussions between instructors reflecting team dynamics in previous courses and projects. The first project is usually assigned a lower-grade weight compared with the following projects.

Table 4.2 Design course goals with their motivations and required means for achievement

No.	Goal	Motivation for setting goal	Requirements
G1	Lower attrition rates	Align to university recruitment and retention priorities	Efficacy with "at-risk" students
G2	Increase awareness of engineer's role	Helping students make the right informed decisions about their future major	Increases familiarity with mechanical and aerospace engineering fields
G3	Provide hands-on lab	Most students have no laboratory experience from secondary school	Includes effective lab projects
G4	Introduce the design cycle	Limit criticism that gains are due to use of "easier" conceptual questions	Increases traditional problem-solving skill
G5	Develop team work skills	Team management as a requirement for modern engineering practice in the industry	Using MS Project to develop WBS and other team management tools
G6	Develop communication tools, oral and written	To be able to express ideas orally using presentations and posters as well as written reports	Using MS Office for editing and generating posters and reports

WBS work breakdown structures

Table 4.3 Design course constraints with their motivations and required means for achievement

No.	Constraints	Motivation for setting constraint	Requirements
C1	Cannot change the contact time model	"2+1" model (2 credit h lecture + 1 h (3 contact h) lab) enshrined in policy, needs greater credibility to change	Adapted to "2+1" contact time
C2	Cannot hire new teaching staff	As a start-up in a developing country, successful hiring searches are typically 18+ months	"Low-effort" methodology[a]
C3	Cannot buy new lab or classroom technology	As in C2, purchasing and customs clearance times are long and unpredictable	No additional lab equipment needed
C4	Must use traditional geometry classroom and lab spaces	The small, start-up campus building is a capacity, expansions and new construction are only in planning phases	Allow for small-scale modifications of the lab space
C5	Must create effectively single-gender learning situations	Avoid severe negative expectancy violation, acceptable compromise	Demonstrated efficacy in single-gender settings
C6	Use of lower-end hand tools and power tools	Not enough time for proper training for higher end tooling	Keep tooling requirement low and done under instructors' supervision

[a] "Low-effort" designation given to pedagogies created by small iterative changes to the traditional lecture model

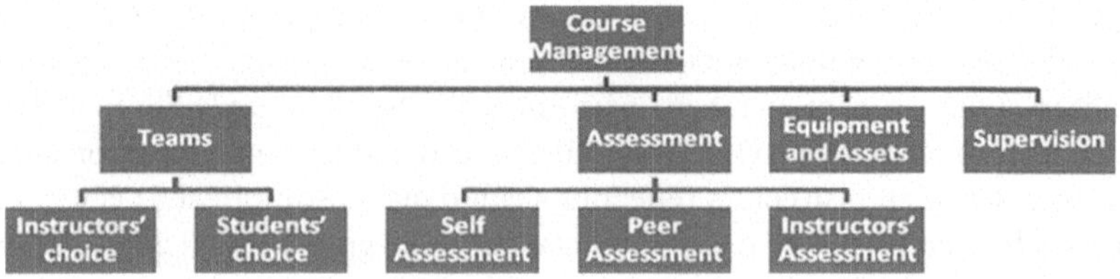

Fig. 4.3 Design course management tree

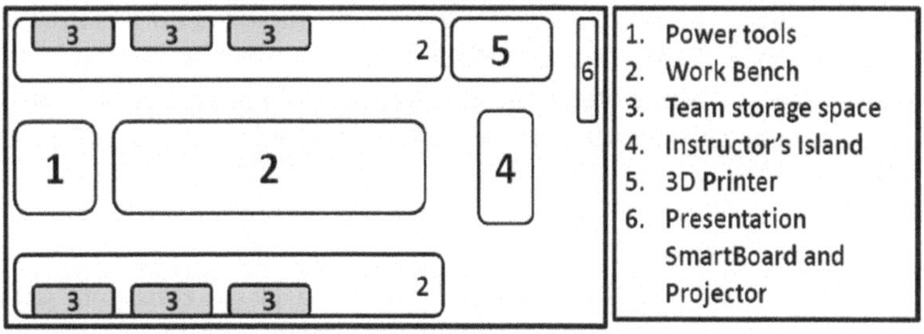

Fig. 4.4 Schematic diagram for the course lab layout

A special lab/studio is set for the course. The layout of the lab is presented in Fig. 4.4. The lab is equipped with hand tools, hardware (LEGO kits and 3D printer), and PCs for the controlling and interfacing with the LEGO kit sensors. Each student group has a storage space for their LEGO kits and designs. The PCs are also equipped with Pro/Engineer solid modeling software. The two main pieces of hardware used in the laboratory are the LEGO Mindstorms kits and a 3D printer, both of which add value in delivering hands-on project experience. Access to the lab equipment is not restricted to class or lab times. Students have the opportunity to continue their work at their convenience. A lab log is kept to keep track of the time spent in the lab. Each team is assigned a LEGO kit, which they sign for and label at the beginning of the course. The students are fully in charge of all their tools and kits and have to do a full inventory at the end of the course. This enhances their self-direction, sense of ownership as well as engagement in the learning process. Before prototyping (3D printing), students are required to get approval of the instructor to ensure that they have systematically undergone the necessary design iterations prior to prototyping.

A team of instructors facilitate and supervise the course (two professors and one lab instructor from three different engineering disciplines). All problems/projects are solved a priori to anticipate any potential problems. The instructors' main role is as facilitators who monitor teams' progress and provide them with the necessary feedback, scaffolding expert problem-solving strategies as needed. This is done to verify that all the members understand and participate in the various design stages of each project (individual accountability) and to ensure that the teams are able to deliver the project's requirements (team accountability). Discussions are triggered mainly through questions posed by instructors in a Socratic style, guiding the team's thinking and analytical process towards achieving the target while allowing them to self-direct and manage. The instructors also motivate the students to be fully engaged by planning peer-evaluated demos, competitive events, and challenges.

4.2.1.5 Assessment

In order to access students' performance, two metrics are introduced: the Problem-Solving Attitudes Survey score and the correlation of design-knowledge with design-performance. The survey was designed and conducted on freshman design students (Khalaf et al. 2010). Due to the lack of well-defined design concept inventory similar to the physics Force Concept Inventory (FCI; Hestenes et al. 1992), the current Problem-Solving Attitudes Survey is designed for the use of the current work. This survey needs to be developed even more to check on its validity and reliability. We also see evidence of good external reliability for the problem-solving survey as successive classes of students are consistent with one another; the average responses of design students surveyed in 2010 and 2011 are consistent with each other, as are those of non-design students surveyed in the same years. Regarding the survey's internal reliability, we have calculated Cronbach's α-scores for the eight statements of the problem-solving portion of the survey. From the professional engineers' perspective, the problem-solving portion of the survey appears to clearly measure a coherent cognitive construct. However, the less initiated into engineering design the subject population (design course students post instruction, then all other engineering students, respectively), the lower the Cronbach's α-scores (see Table 3 in Khalaf et al. 2013a). The same survey is used to compare various modes of delivery (see Table 1 in Khalaf et al. 2013b), and it consistently distinguished between responses of the above subject populations.

To briefly summarize the survey design, the survey is designed to measure the degree to which students possess expert-like attitudes towards engineering design. The motivation for measuring attitudes rather than skills is twofold. First, such an instrument is currently missing from the engineering education literature, but there is a clear need for it (e.g., Reed-Rhodes and Imbrie 2008). Second, measuring student attitudes should not be hindered by a lack of technical engineering knowledge, making it an attractive measure of course effectiveness for freshman students. The measurement target of the survey is conveniently described in terms of three core dimensions (see Table 1 in Khalaf et al. 2010). Each of the dimensions: (1) problem-solving, (2) teamwork, and (3) communication is probed by asking students to register their agreement or disagreement on a five-point Likert scale (strongly agree, agree, neutral, disagree, strongly disagree) with statements that describe an example behavior or belief. This question format for the survey is inspired by similar attitudes instruments in physics education, such as the Maryland Physics Expectations (MPEX) survey (Redish et al. 1998), the Views About Science Survey (VASS; Halloun 1997), the Epistemological Beliefs Assessment for Physical Science (EBAPS; Elby 1998), and the Colorado Learning Attitudes About Science Survey (CLASS; Adams et al. 2006). To maximize the utility of the design course survey in this capacity, statements have been carefully worded to avoid as much professional engineering jargon as possible so as to make it conceptually accessible to freshmen students who may yet have no exposure to engineering concepts. This makes the design course survey very different from a standard end-of-course/instructor evaluation and potentially much more informative.

Student design performance is measured by their grades that are based on three types of performance assessment: self-assessment, peer assessment, and instructor assessment. In self-assessment, the students mark themselves and their other group members in terms of % contribution to the group effort. Peer assessment involves

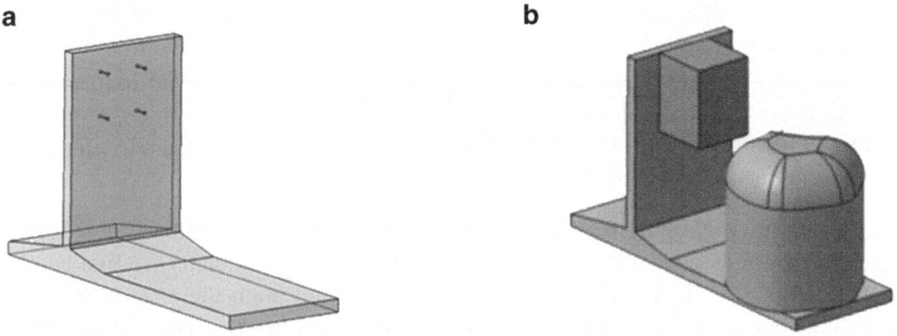

Fig. 4.5 Illustration of client design constraints for the Deployable Cantilever Beam Project case study. The mounting bracket requirement **a** and the volume constraints **b**

the students evaluating other team designs in terms of functionality and creativity on a Likert scale. Students are also asked to challenge other teams' designs, who in turn are asked to defend their designs. The results of the self- and peer evaluations are partly taken into consideration when instructors evaluate team projects. Instructors also follow up on the Moodle discussions when the team is not in the lab. Based on the contributions and the involvement in the team discussions, the highly motivated team members and contributors are rewarded.

4.2.1.6 Example

PBL is best explained by case study, as we now show with the "Deployable Cantilever Beam Project" from the Engineering Design-and-Build course. The methodology of the course is closely aligned with the engineering design cycle shown in Figs. 4.1 and 4.2 and is prescribed for students. First, student teams read and analyze the client's problem statement, after which they proceed to the conceptual design stage, during which various tasks/milestones are used to generate concepts in the form of design alternatives or potential acceptable designs:

Client Statement A deployable beam, cantilevered to a provided wall, is needed to carry loads up to 250 g at the tip. The tip deflection from the horizontal plane should not exceed ±2 mm at any load case. The cantilever beam must be initially packed within a volume of max. 15 cm × 15 cm × 20 cm, then deployed to a length of at least 50 cm, measured from the wall (see Fig. 4.5a, b). The deployment phase must be carried out by a mechanism within the volume (i.e., no external forces by humans). Control system of deployment (such as the NXT brick for on/off, etc.) can be outside the volume (Fig. 4.5b).

For this problem, the students were asked for the following deliverables:

- A report including the brainstorming exercise, objective tree, function mean tree, morphological chart, decision selection matrix as well as any other brainstorming and decision-making exercises.

- Drawings of all alternative designs using CAD drawings, hand sketches, and any other means.
- Code used in C++ that deploys the system, including the closed-loop control system for the tip deflection.
- Drawings of any custom-made parts to be prototyped using the 3D printer in integration with parts from the LEGO kit.

We now show highlights from one student team as they solved this problem and take their work as queues to describe the learning process. Students are first taught to critically read client problem statements and expand upon them through client interview and several conceptual design tools such as "objective trees", "pair-wise comparison charts", and more. The objective tree created by Team #1 is shown in Fig. 4.6.

Objective trees help students brainstorm on specific goals for a design that are gleaned from the broad goals present in the client problem statement. Students then create a pair-wise comparison chart to determine, by team consensus, what the priorities are associated with each objective, as is shown by Team #1 in Table 4.4. Summed ranks in the right column give an estimate of importance.

Students are then tasked to return to and revise the client problem statement based on the results of these and other exercises. This form of intensive critical reading and writing deepens students' concern for client/user perspectives and values and helps establish motive for later coursework in the social sciences. Following these and other exercises, Team #1's revised statement reads:

Revised Problem Statement The client is asking us to design a deployable cantilever beam. This beam should be packed in a box that has 15 cm × 15 cm × 20 cm dimen-

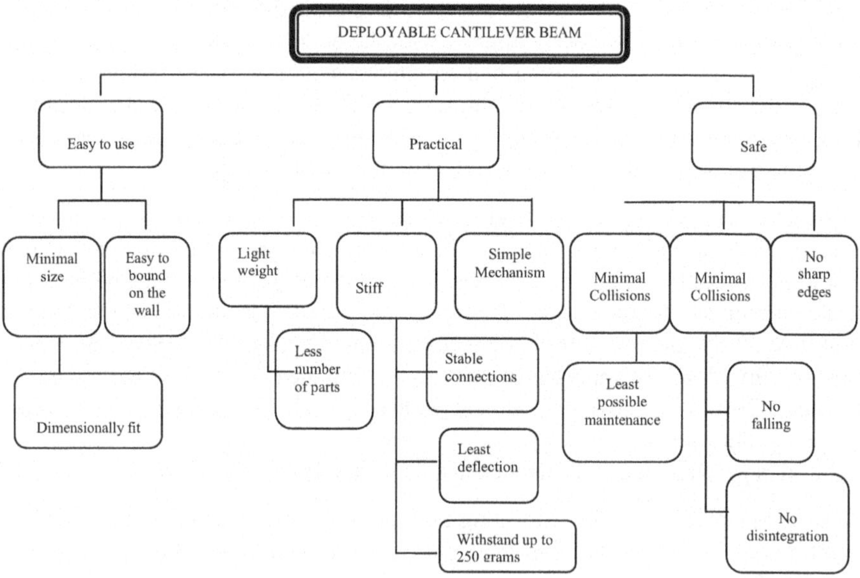

Fig. 4.6 A flowchart representation of the objective tree as presented by Team #1

Table 4.4 Pair-wise comparison chart as presented by Team #1

Objective	Safe	Accurate	Within size limit	Smart	Stable	Tolerates weight	Does not deflect	Total
Safe	XXX	0	1	1	0	0	0	2
Accurate	1	XXX	0	1	0	0	0	2
Within size limit	0	1	XXX	1	0	0	0	2
Smart	0	0	0	XXX	1	0	0	1
Stable	1	1	1	0	XXX	0	0	3
Tolerates weight	1	1	1	1	1	XXX	0	5
Does not deflect	1	1	1	1	1	1	XXX	6

sions. The beam has to be mounted in the wall by four bolts with equal distance between them that are 10 cm. The beam should have a minimum length of 500 mm when it is unpacked from the box. It has to withstand a 250-g load as maximum. The load should be stable while being on the beam, and it should not slide or fall off the beam. If the maximum deflection of the tip of the beam exceeded ± 2 mm, the beam should adjust its situation automatically and alter this deflection. At least two functional parts (not dead loads) have to be designed using PRO/E and printed using the 3D printer. The period of time in which the design should be ready is 4 weeks starting from 12 October 2011 until 16 November 2011. Good results should be obtained from testing more than once.

Student teams next move to the often-neglected conceptual design phase of the engineering design cycle. Teams are coached in a "divergent-convergent" thinking process, whereby means for functions and entire design alternatives are imagined without criticism or constraint. This phase offers many opportunities for instructors to teach sketching and other graphical communication skills as team members vie with one another to communicate their ideas. The culmination of this phase is the conception of at least three substantially different design alternatives and a metrics-based, evidence-based analysis (called a "decision matrix") of which alternative shows the most promise as a candidate for prototyping and preliminary design. Concept drawings and a decision matrix created by Team #1 is shown in Fig. 4.7 and Table 4.5, respectively.

One of the key innovations of the engineering design-and-build course is the integration of important technological enablers for the classroom environment. These are LEGO Mindstorms Robotics Kits, a C++ programming interface, and a 3D printer, all of which allow for teams to rapidly prototype their conceptual designs. Figure 4.8 shows four design alternatives that were physical fabricated by Team #1 using these classroom technologies.

From this stage in the design process, teams are then able to select a preliminary design based on evidence from their own experimentation and testing. This then enables more quantitative modeling and optimization as students prepare for a live, in-class demonstration of their design solutions and compose a design report to report their work and results to the client. Design demonstrations are qualified based on adherence to constraints. Successful designs are distinguished based on the degree to which client objectives are obtained. Design reports also require students to

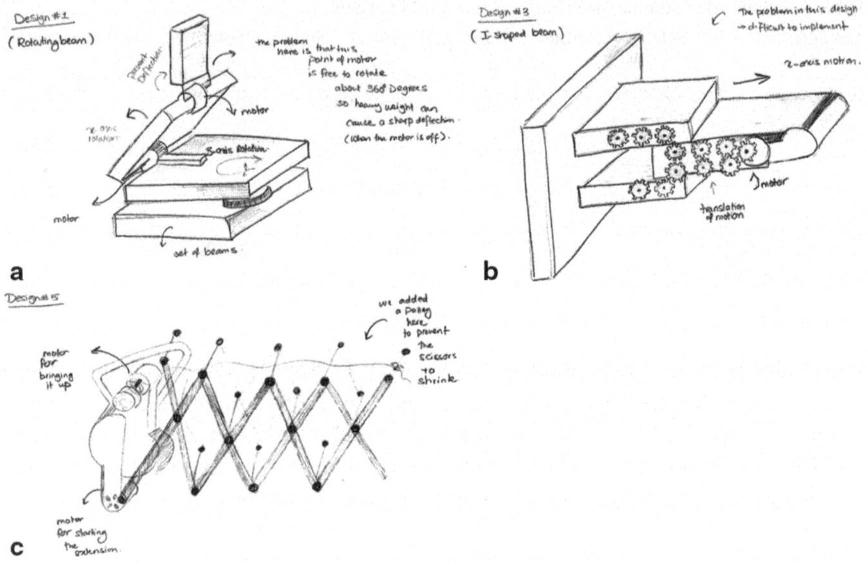

Fig. 4.7 Alternative conceptual designs as presented by Team #1

develop relevant written communication skills. Table 4.6 shows the criteria for a team's design report, in line with the design process they have been guided through.

In summary, the methodology of the course teaches students engineering design thinking and the engineering design process experientially. Students are able to complete the entire design process largely because of the clear definition of design

Table 4.5 Decision matrix for conceptual designs as presented by Team #1

Design constraints	Zigzag design	Wood slider design	Open up bridge design
The size should fit the box (15 × 15 × 20)	Y	Y	Y
It should be simple	N	Y	Y
Design should be 50 cm in length	Y	Y	Y
Design should be stiff to resist deflection	N	Y	Y
Design should be safe with no sharp edges	Y	Y	Y
Design must oppose maximum moment	N	Y	Y
Design objectives	*Score*	*Score*	*Score*
Stability	40	80	90
Safety	100	100	100
Length	100	100	100
Simplicity	50	85	75
Manufacturability	60	40	80
Tolerates *weight*	50	80	80
TOTAL	*400*	*485*	*525*

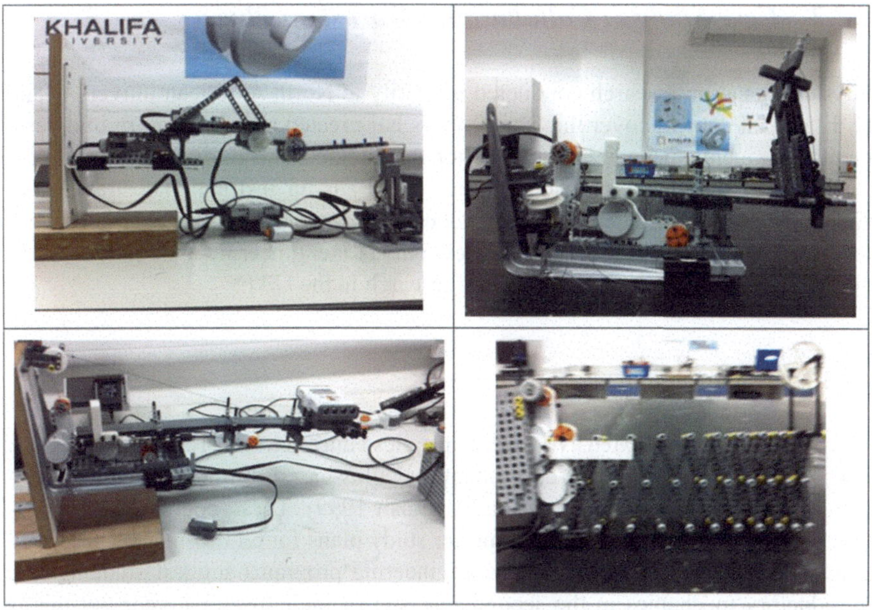

Fig. 4.8 Alternative prototype designs as presented by Team #1

Table 4.6 Example content criteria for student design reports

Design report components	Weight
Going through the prescriptive five-stage design process, identify the inputs, tasks, and output of each stage	25
Clearly present the objective tree similar to Figs. 3.1 and 3.2 (Dym and Little 2003)	10
A pair-wise comparison chart similar to Tables 3.2–3.4 (Dym and Little 2003)	10
Construct your revised project statement	15
Identify the constraints for your design	5
You have to present your alternative using free-hand or computer sketches that show at least three alternative designs	15
Present all of your work in a report format	20

and the explicit effort (through the above milestones/deliverables) to promote design thinking, the project-based pedagogical model, and the technological enables that allow them to quickly prototype their conceptual designs. Student teams are reshuffled at the conclusion of each project and the beginning of the next, allowing all students to experience as interdisciplinary of a mixture of skill sets and perspectives as is possible among freshman students. Students complete five such projects, each lasts 3–5 weeks, during the semester, ensuring that repeated exposure to the complete design cycle maximizes their opportunities to learn and to master it as a way of thinking: systematic design thinking.

4.2.2 Case Study 2: Collaborative Workshop Physics

Physics education research conducted primarily in North American university and K–12 school contexts over the last 20 years has created a rich literature of assessment instruments, pedagogies, instructional strategies, and reform project success stories. Collaborative Workshop Physics (CWP) is an instructional strategy designed to adapt and implement best practices and content from this corpus of work to an introductory calculus-based physics course for engineering undergraduate students at KU of science, technology, and research in the UAE.

4.2.2.1 Background

The earliest and most extensive work in physics education research (PER) in North America is focused on large-enrollment, calculus-based, introductory mechanics courses (Hake 1998; McDermott and Reddish 1999). These courses are foundational in typical second- or third-semester study plans for STEM majors the world over. The majority of undergraduate engineering programs suggest students take introductory mechanics in the second semester of their freshman year but this is not always the case. Topical content of these courses vary considerably, particularly in the last few weeks of the syllabus, but all typically begin with the traditional sequence of kinematics, forces, Newton's Laws for linear motion, work and energy, mechanical conservation laws, rotational kinematics and dynamics, and statics. Past this point in the sequence, different courses diverge on content coverage and are equally likely to focus on fluid mechanics, thermal physics, or waves and oscillations (or even a survey of all of the above and more).

Much contemporary PER on the introductory, calculus-based mechanics course has its origins in two highly cited studies: (1) the publication of the FCI by David Hestenes and colleagues (Hestenes et al. 1992) and (2) the large-scale comparative study by Richard Hake of FCI data gathered from more than 6000 students across dozens of US high schools, colleges, and universities (Hake 1998). In the first paper, Hestenes et al. introduced a 30-question multiple-choice instrument for measuring the respondent's conceptual expertise through simple, everyday classical mechanics problems. The key innovation in the instrument's design was the use of "distracters"; incorrect options created on the basis of extensive open-ended interview of likely respondents (students) trying to answer and explain their reasoning to the same questions. Wrong answers produced in interviews are codified, associated frequencies are determined, and those occurring most frequently are adopted as alternative answers in the multiple-choice version of the same question. The finished multiple-choice version of the instrument is, in a classical testing sense, a highly discriminating instrument, able to reliably detect expertise or diagnose misconceptions in the respondent.

The second paper, by Hake, definitively established the superiority of pedagogical alternatives to traditional, passive-audience lecturing on the basis of a large body of FCI data. R. Hake's data set included class-averaged FCI scores gathered prior to and following calculus-based mechanics instruction from the same respondents, more than 6000 in total, for dozens of high school, college, and university courses in the USA. Comparisons were made using a now widely accepted measure of normalized learning or "gain" g on closed-form tests:

$$g = \frac{\langle S_f \rangle - \langle S_i \rangle}{1 - \langle S_i \rangle},$$

where angle brackets denote class averages taken over the student responses in a given course offering and S_i and S_f refer to the pre-instruction and post-instruction scores on the instrument, respectively. The normalization factor $(1-\langle S_i \rangle)^{-1}$ weighs the improvement $\Delta = \langle S_f \rangle - \langle S_i \rangle$ on the basis of maximum possible improvement, eliminating the advantage low-pretest-scoring populations have for producing large Δ values. This procedure allows for side-by-side comparison of student learning across populations with widely varying levels of prior knowledge. In his study, Hake found that for traditionally taught introductory, calculus-based mechanics courses (i.e., those featuring (1) lecture instruction to passive audiences, (2) algorithmic end-of-chapter style homework and exam questions, and (3) verification or "recipe" labs), the average gain on FCI is $g_{\text{trad.}} = 0.22 \pm 0.02$ (standard error in mean $\sigma / \sqrt{N}$). For courses using any one of a variety of alternative pedagogies (collectively labeled "interactive engagement" (IE) by Hake), the average gain on FCI is $g_{\text{IE}} = 0.48 \pm 0.14$.

The larger deviation in the set of IE course gains is indicative of both the inherently greater variety of instructional techniques among alternative pedagogies and the lower reliability achievable for successful implementations thereof. More recently, these performance variations among pedagogically reformed physics courses have become the subject of close study. While the number of well-known PER-based instructional strategies is just over a dozen, the number of departments and institutions that have attempted secondary implementations of these course designs is likely in the hundreds (Beichner 2007). For the introductory calculus-based mechanics courses, the rich and growing body of data produced from secondary implementations has revealed several traits that the successful cases have in common.

In the context of student learning tasks and classroom norms, successful secondary implementations must establish certain classroom norms that are core functions of the originating institution's design. These norms have broadly been identified under the categories of (1) rich student–student interactions, (2) equality in teacher–student interactions concerning learning tasks, and (3) the superiority of sense-making over answer-making in student work (Dancy and Henderson 2010; Turpen and Finkelstein 2010). In each of these three categories, classroom practice that elevates value (i.e., students' belief that the learning task itself is important) and expectancy (i.e., students' belief that the learning outcome for the task can be achieved) for these categories of norms are strongly associated with, among other measures, large normalized gains on instruments like FCI.

To elaborate and to provide an example, suppose an instructor adopts the well-known instructional strategy of peer instruction (PI; Mazur 1997) for reforming pedagogy in their course. This PER-based instructional strategy uses a relatively easy-to-implement recipe for making traditional lectures more interactive. Reading assignments are given several days before the associated lecture content is delivered. Using a learning management system (LMS), students are quizzed on their understanding of the reading before attending the lecture. The instructor uses that data to organize the topics to be emphasized during coverage. During the lecture, the instructor periodically shows a multiple-choice conceptual question and asks students, using index cards, personal response systems (i.e., clickers or other systems), or a show of hands, to give an individually reasoned answer to the question. The instructor, either visually or electronically, determines the distribution of answers in the room and, if there is a large deviation from the correct response, grants 5 min for everyone to convince their neighbors "that they are correct." Following debate and discussion, students are then prompted to respond again. The instructor then shows his or her own reasoning through the question and reflects upon the key concepts and major principles used in their reasoning. At face value, this approach operationalizes the three necessary types of norms. There is opportunity for student–student interaction several times throughout every lecture; the teachers lay out their reasoning about the learning task just as the students do (implying equality of their perspectives); and the task is conceptual, where the important outcome necessarily is correct *reasoning,* not correct *numbers*.

The underlying mechanism of PI's widespread success in secondary implementations (Lasry et al. 2008) can be undermined by failing to build value and expectancy for these classroom norms (Turpen and Finkelstein 2010). As reported by Turpen and Finkelstein, some secondary implementers of the PI instructional strategy display the correct answer after student discussion but do not consistently present their reasoning for the consideration or critique of students. This omission communicates to students the sense that the instructor has the "final word" and that the correct answer should be believed on authority rather than on correct reasoning. Consequently, the students' sense of value for the task is undermined, and they wait instead for the instructor to tell them the right answer. This further encourages student belief that the answer is more valuable than the reasoning, thereby undermining sense-making over answer-making. Turpen and Finkelstein conclude that the subtle behavioral and psychological shifts caused by classroom practices like this can lead to a course that is superficially reformed pedagogically, but in reality never breaks with or quickly reverts to the values and norms of traditional instruction.

In the present case under study, CWP is the instructional strategy designed to operationalize an alternative pedagogy for calculus-based introductory physics at KU. In what follows below, we will work through the decision-making process followed through each of the steps in the simple course reform model show in Fig. 4.1. The first step is to select an alternative pedagogy as the basis for the course reform that is justified by the course learning outcomes. As is the case in most institutions, freshman-level core math and science courses for STEM majors have rigidly imposed learning outcomes, and such is the case for PHYS-121 (University Physics) at KU. The course's learning outcome is that "At the end of the course, students would be able to":

1. Apply Newton's laws to find equations of motion for simple mechanical systems
2. Understand and use important physical concepts, such as force, work, energy, momentum, and mechanical equilibrium
3. Use energy and momentum concepts to analyze simple mechanical systems
4. Analyze simple harmonic motion
5. Calculate properties of one-dimensional waves

These learning outcomes mostly suggest middle- to high-cognitive-level objectives for the course. Outcome 2 and 3, in particular, requiring the use of abstract concepts such as energy and momentum in analytical problem-solving, easily rank in the top third of Bloom's taxonomy for skills in the cognitive domain (Bloom 1956; Anderson et al. 2000). These learning outcomes also clearly emphasize analytical problem-solving competency. Therefore, given the requirement to demonstrate mid- to high-level cognitive skills in solving analytical problems, just-in-time teaching, problem-based, or hybrid (problem-/project-based) pedagogies are appropriate alternative pedagogies for the course reform. As is well known, just-in-time teaching is the least difficult to implement among the pedagogies in this subset. However, at the time of the conceptual design of CWP, KU did not have the prerequisite LMS to administer pre-lecture quizzes, leaving the latter two: problem-based and hybrid problem-/project-based. The decision between these two pedagogies emerged from stakeholder engagement and user (student) characterization, as discussed below.

4.2.2.2 Course Objectives

Normalized gain on FCI and course DFW rate are two key metrics that emerged from our baseline and benchmarking study of students taking the traditional, lecture-centered version of the course. On these metrics, the performance of Emirati nationals who are conditionally admitted to KU are a particularly important constituency, given that they account for a large fraction of the overall student body and are particularly at risk for course failure. Table 4.7 summarizes the results of similar considerations for goals of the CWP instructional strategy. The first column codifies the goal (G1, G2, etc.), and the second, third, and fourth columns concisely state the goal, give the underlying motivation for adopting it, and give the requirements that the goal places on an acceptable PER-based instructional strategy, respectively.

4.2.2.3 Course Constraints

Similar to Table 4.7, Table 4.8 summarizes constraints placed on the design of a new instructional strategy for the PHYS-121 course at KU. The first column gives a label for the constraint (C1, C2, etc.), and the second, third, and fourth columns

Table 4.7 Collaborative Workshop Physics (CWP) design goals with their motivations and required means for achievement

No.	Goal	Motivation for setting goal	Requirements on a PER method
G1	Lower course failure rate	Align to university recruitment and retention priorities	Efficacy with "at-risk" students
G2	Increase FCI gains	Key performance indicator for the reform project	Increases FCI (or FMCE) gains
G3	Provide hands-on lab	Critical IE norm, also most students have no lab experience from secondary school	Includes effective lab curriculum
G4	Increase "end-of-chapter" problem-solving skill	Limit criticism that gains are due to use of "easier" conceptual questions, exam rules preclude multiple choice (MC) quest	Increases traditional problem-solving skill
G5	Increase engineering course performance	Build credibility with the College of Engineering for future reform efforts	Retains engineers, improves student performance in statics, dynamics, etc.
G6	Enable rich peer interactions	Critical IE norm and assumption that prior-knowledge and language-level diversity enhances learning	Effective for mixed ethnicity teams
G7	Avoid high faculty skepticism	Limit course management intervention that could spoil experimental controls in a reform pilot study	Can leave lecture unchanged (in the event that faculty not part of the reform project are assigned some of the lecture sections to teach)
G8	Minimize student anxiety	Same as R1	Has "traditional" course components
G9	Avoid loose integration of IE elements	Main cause of failure to achieve G2 (Finkelstein and Pollock 2005)	Course components and tasks are tightly integrated by design
G10	Avoid lack of PER-based teaching experience	Limit risks to the reform project pilot, PER literature indicates that most reform efforts go through a period of mediocre performance as faculty adjust to unanticipated complications	KU faculty that are involved in the reform project have some prior experience

PER physics education research, *FCI* Force Concept Inventory, *FMCE* Force-Motion Concept Evaluation, *IE* interactive engagement, *KU* Khalifa University

concisely state the constraint, give the motivation for adopting the constraint, and give the requirements the constraint places on an acceptable PER-based instructional strategy, respectively. Clearly, many of the motivations for these constraints are associated with departmental and institutional factors, which are typically the case in any university. The notable exception is the need to create effective single-gender learning situations (C6), which is directly motivated by student and ethno/national cultural factors. In many US institutions, this would be an unacceptable constraint, given the long documented evidence of performance-based gender gaps

Table 4.8 Collaborative Workshop Physics (CWP) design constraints with their motivations and required means for achievement

No.	Constraints	Motivation for setting constraint	Requirements on a PER method
C1	Cannot change the contact time model	"3+1" model (3 credit h lecture+1 h (3 contact h) lab) enshrined in policy, need greater credibility to change	Adapted to "3+1" contact time
C2	Cannot hire new teaching staff	As a start-up in a developing country, successful hiring searches are typically 18+ months	"Low-effort" methodology[a]
C3	Cannot buy new lab or classroom technology	As in C2, purchasing and customs clearance times are long and unpredictable	No additional lab equipment needed
C4	Must use traditional-geometry classroom and lab spaces	The small, start-up campus building is a capacity, expansions and new construction are only in planning phases	Does not require radically different rooms/physical geometry for classroom spaces
C5	Must deliver same topical content	External accrediting bodies and internal curriculum oversight make topical changes time-consuming and difficult, limit criticism of the reform project that improvements are achieved by narrowing the scope of the course	No coverage reduction in topical contents
C6	Must create effectively single-gender learning situations	Avoid severe negative expectancy violation, acceptable compromise given no evidence of gender gap in FCI pre-test or gain scores	Demonstrated efficacy in single-gender settings

PER physics education research, *FCI* Force Concept Inventory

[a] "Low-effort" designation given to pedagogies created by small iterative changes to the traditional lecture model

in STEM courses and the strong evidence that gender-mixed (and even female-majority) groups outperform and post greater learning gains than single-gender groups working on the same learning tasks. However, as has been demonstrated previously (Hitt et al. 2014), there is little or no significant evidence that such a performance-based gender gap exists for the KU undergraduate student population in PHYS-121. Furthermore, evidence of greater gains among gender-mixed groups in US studies is a perfect example of a performance feature that is strongly context dependent. The negative violation of expectations and social anxiety that mixed-gender teaming creates among KU students is more likely to paralyze learning than to enable it.

4.2.2.4 Course Structure

Given the long history of course reforms for introductory calculus-based mechanics in PER from US institutions, it is reasonable to begin the process of generating design alternatives for a KU instructional strategy with a careful examination of the prior US-based designs. Table 4.9 shows six of the published, well-studied

Table 4.9 Evaluation matrix of prior USA-based instructional strategy designs, possible inspirations for the Collaborative Workshop Physics (CWP) design

Goal	Means to achieve goal	Published evidence that a method has provided the needed means					
		CGPS	MI	SCALE-UP	SDI Labs	UW Ttls	WP
G1	Efficacy with "at-risk" students	✓[b]	✓[h]	✓[n]	✓[o]	?	?
G2	Increases FCI normalized gains	✓[c]	✓[i]	✓[n]	✓[i]	✓[q]	✓[s]
G3	Includes effective lab curriculum	✓[d]	✓[j]	✓[n]	✓[p]	✓[r]	✓[t]
G4	Increases traditional problem-solving skill	✓[e]	✓[k]	✓[n]	?	?	?
G5	Retains engineers, improves student performance in statics, dynamics, etc.	?	✓[l]	✓[n]	✓[o]	?	?
G6	Effective for mixed-ethnicity teams	✓[f]	?	?	?	?	?
G7	Can leave lecture unchanged	✓[e]	✗[j]	✗[n]	✓[o]	✓[r]	✗[u]
G8	Has "traditional" course components	✓[g]	✗[j]	✗[n]	✓[o]	✓[r]	✗[u]
G9	Tight integration of tasks, etc. by design	✗[g]	✓[j]	✓[n]	✗[o]	?	✓[t]
G10	KU faculty have prior experience	✓	✗	✗	✗	✗	✗
Const. requirement on a PER-method							
C1	Adapted to "3+1" contact time	✓[g]	?	✗[n]	✓[o]	✓[r]	✗[t]
C2	"Low-effort" methodology[a]	✓[g]	✗[m]	✗[n]	✗[o]	✓[r]	✗[t]
C3	No additional lab equipment needed	✗[d]	✗[j]	✗[n]	✗[o]	✓[r]	✗[t]
C4	Use traditional geometry class/lab space	✗	✗	✗	✓[p]	✓[r]	✗[u]
C5	No coverage reduction	✓[g]	✓[j]	✓[n]	✗[p]	✓[r]	✓[t]
C6	Effective for single-gender teams	✓[b]	✓[h]	✓[n]	?	?	?
Total positive excess (✓ − (✗ + ?)):		*+8*	*+1*	*0*	*0*	*+2*	*−7*

CGPS Cooperative Group Problem Solving, *MI* Modeling Instruction, *SCALE-UP* Student-Centered Active Learning Environments for Undergraduate Programs, *SDI Labs* Socratic-Dialog-Inducing Labs, *UW Ttls* University of Washington's Tutorials in Introductory Physics, *WP* workshop physics, *FCI* Force Concept Inventory, *KU* Khalifa University, *PER* physics education research

[a] "Low-effort" designation given to pedagogies created by small iterative changes to the traditional lecture model

[b] Etkina et al. 1999

[c] Cummings et al. 1999

[d] Heller and Heller 2007a

[e] Heller et al. 1992

[f] Heller and Hollabaugh 1992

[g] Heller and Heller 2007b

[h] Brewe et al. 2010

[i] Hake 1998

[j] Hestenes 1987

[k] Malone 2008

[l] Brewe et al. 2009

[m] http://modeling.asu.edu

[n] Beichner 2008

[o] Hake 1992

[p] http://www.physics.indiana.edu/_sdi/

[q] Kost et al. 2009

[r] McDermott and Shaffer 2002

[s] Saul and Reddish 1997

[t] Laws 2004

[u] Laws 1991

PER-based instructional strategies that operationalize problem-based or hybrid problem-/project-based pedagogy, mostly in team formats. These six are:

1. Cooperative Group Problem Solving (CGPS; Heller and Heller 2007a, b)
2. Modeling Instruction (MI; Hestenes 1987, and see http://modeling.asu.edu)
3. SCALE-UP (Beichner 2007)
4. Socratic-Dialog-Inducing Labs (SDI Labs; Hake 1992, and see http://www.physics.indiana.edu/_sdi/)
5. University of Washington's *Tutorials in Introductory Physics* (UW Ttls; McDermott and Schaffer 2002)
6. Workshop Physics (WP; Laws 2004)

Rows in Table 4.4 list either the goals or constraints for the CWP design space. The table is an evaluation matrix, used to analyze existing PER-based instructional strategy designs for their usefulness in the KU context, as modeled by these goals and constraints. Entries in the evaluation matrix are populated with literature references where the PER-based method has demonstrated evidence that it provides the needed means for a goal or meets the requirements imposed by a constraint. If the evidence is positive, the citation is preceded by a (✓) symbol. If the evidence is negative or none could be found, the citation is preceded by a (✗) or a (?), respectively. The last row of Table 4.9 shows the total positive excess of evidence calculated as $N_{\checkmark} - (N_{\times} + N_{?})$. Comparing this quantity between the various existing PER-based strategies gives some qualitative indication, which ones, if any, should be examined more closely as possible inspirations for the CWP instructional strategy design.

There are two important conclusions drawn based on the exercise represented in Table 4.9. First, all of the PER-based methods examined either explicitly break constraints or do not evidence that they meet at the requirements imposed by the constraints. This necessarily means that no published PER-based instructional strategy can be used at KU in a "turn-key" or "off-the-shelf" manner. Any one of these methods, if adopted, will require some degree of modification. Alternatively, this means the CWP method must contain new innovations in order to deliver on creating important PER-based classroom norms, measure well on our goals, and conform to our constraints. The encouraging outcome of the exercise represented in Table 4.9 is large lead the CGPS holds over other methods. It is for this reason that the CGPS instructional strategy was more closely examined and used as an inspiration and basis upon which we made our own instructional innovations.

We now briefly review the main features of the CGPS instructional strategy, given its prominence in our survey of prior designs in Table 4.9. CGPS was designed at the University of Minnesota by Heller and Heller (Heller and Heller 2007a, b) and was meant to provide institutions delivering large-enrollment, traditional physics courses a manageable and iterative way to slowly transform to a fully alternative pedagogy. Their reform project began in the problem-solving tutorial sessions, and later the labs, only moving to redesigning the large lecture experience long after improvements produced by the first two changes had been sustainably established and provided improvements necessary to increase the reform project's credibility within the department. A similar approach, to leave the lecture sessions alone at first, seemed very attractive for our case as well, given goals G7 and G8 and constraints C1 and C5. Therefore, we turned our attention exclusively to reforming the 3-h laboratory session only.

The central feature of CGPS is the "context-rich group problem" learning task. A "context-rich" problem bears much in common with good design problems in engineering education; they should be "ill-posed," "under-defined," "realistic," and generally have multiple satisfactory solutions (Dym et al. 2005b). Also, similar to design problems, they must be tackled in teams, as the University of Minnesota group explicitly designed them to operationalize cooperative learning principles in the physics domain. Team members propose solutions and debate with one another, but only one solution can be submitted and it must represent the team consensus. At the University of Minnesota, however, this strategy was first implemented in 2-h problem-solving/tutorial/recitation sessions. This session did not exist in the KU "schedule foot-print" for the PHYS-121 course, which turned our attention to the CGPS labs curriculum (Heller and Heller 2007b). The CGPS labs curriculum also employs the "context-rich" group problem scheme; however, a constraint-breaking drawback is the frequent use of video cameras and video motion analysis systems (Heller and Heller 2007b), equipment that KU did not have at the time of the initial CWP design. This meant that almost none of the laboratory measurements prescribed in the CGPS lab manual could be duplicated without new equipment purchases.

On a pedagogical level, the CGPS labs and context-rich problems also seemed too difficult for KU students, on the basis of instructor experiences at the time, without significant additional scaffolding. The very first CGPS lab exercise requires students to meet in groups prior to class and predict the shape of kinematical graphs for the lab scenario described in the lab problem (Heller and Heller 2007b). However, many KU students could not complete a similar task without one-on-one coaching. It was this point that drew attention to UW Ttls; McDermott and Schaffer 2002) and its reasonable showing in the Table 4.9 evaluation exercise. UW Ttls were specifically designed to demonstrate and coach students in the use of representations, in both conceptual reasoning and quantitative problem-solving (McDermott and Schaffer 2002). This initiated an exploration into a hybrid instructional strategy using UW Ttls and CGPS, but by themselves, neither provided the necessary kinesthetic experience for students that all effective reformed lab curricula feature (G3). Kinesthetic experience is particularly important because of the sensory-motor nature of novices' mechanical misconceptions and the need to queue them and challenge them in the same way in which they were formed (kinesthetic experience). To mitigate this, we took inspiration from the "box-of-probes" philosophy of workshop physics and the equipment already available on-site. The simplest way to replace existing recipe labs and to make them CGPS-like was to narrow the goals of the experiments in our traditional lab manual, take away the given procedures, and pose them as open-ended questions.

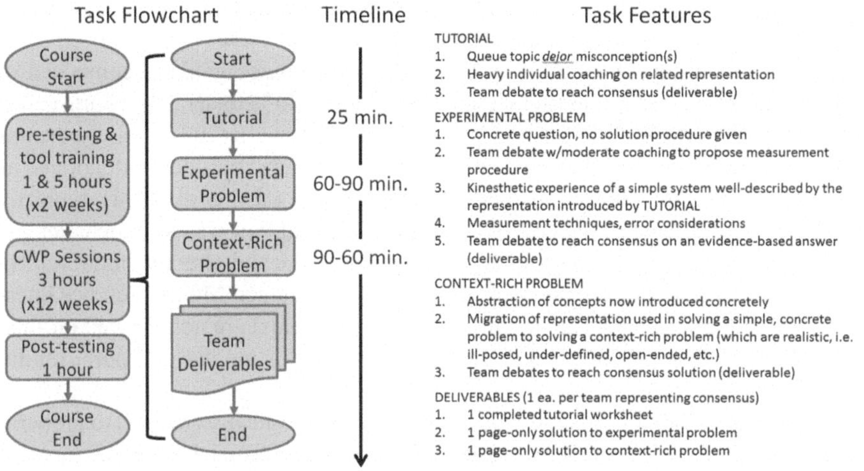

Fig. 4.9 A flowchart representation of the learning tasks used during a typical Collaborative Workshop Physics (CWP) studio session

Figure 4.9 shows a flowchart representation of the CWP instructional strategy prototype that emerged from the above considerations. From Table 4.9, the remaining issue was with G9: tight integration of learning tasks. Tight integration is enabled in CWP instructional strategy by structuring the 3-h CWP session around a small cluster of concepts, then presenting students with progressively more open-ended and less-scaffolded learning tasks, where skills or concepts introduced in early tasks are built upon by later tasks. Students first work through a modified UW Tutorial activity that is heavily scaffolded and intended to introduce a useful physics representation (kinematic graphs, energy bar chart, etc.) Next, student teams are presented with an experiment question where they must design a simple experiment using a "box-of-probes" available from the lab equipment. The phenomenon on display in the experiment problem is chosen such that it is one that is most conveniently analyzed using the representation taught in the preceding tutorial. In a further effort to promote conceptual transfer and abstract understanding of new concepts, the concluding context-rich problem features a question involving the same underlying physics principles as the experiment problem but one having novel situations and/or objects.

4.2.2.5 Assessment

As is suggested by goal G4 and others that are similarly motivated by the need to limit criticisms of the reform project, the assessment scheme was kept unchanged from that of the original, traditional PHYS-121 course. This serves the dual purpose of lowering the difficulty for the instructors to implement the new instructional strategy and allows for a more direct comparison of pre- versus post-reform student performance on course exams. For PHYS-121, the course grade is composed of final exam (40 %), mid-semester exams (1 or 2 tests, 20 % total), lab reports (9–12 experiment reports, 20 % total), and quizzes/other class work (20 % total). In the traditional form of the course, final exam questions were appropriately categorized as moderately difficult "end-of-chapter"-styled problems (Hsu et al. 2004), typically modeled on problems in the course textbook. This tradition was left unchallenged on purpose so that student performance on exams before and after the initial reform could be compared side by side in a straightforward manner. Multiple-choice questions (often the format of choice for conceptual problems) are not allowed on final exams by university policy, and in the initial reform, this too was left purposefully unchallenged.

The nature of laboratory assessment, however, was radically changed to properly measure and incentivize the CWP session. Prior to the reform, lab exercises took the traditional form of a "verification" or "recipe" lab (Domin 1999). The report assessment was little more than a worksheet, requiring students to follow an established procedure, log data into tables and plotting spaces provided, and complete several short-answer questions in conclusion. Reports were not written "from scratch" and were by no means "writing intensive." Introduction and background theory was explained in the given procedure, and the student was not required to demonstrate an understanding of it or to even repeat it in writing in their own words. Consequently, the lab session was widely viewed as the least effective portion of the traditional PHYS-121 instructional strategy.

4.2.2.6 Example

Figures 4.10, 4.11, and 4.12 provide some examples of how learning tasks imported from the three preexisting instructional strategies (UW Ttls, WP-style experiment problems, CGPS context-rich problems) were modified to achieve conceptual coherence and tight integration across the activity sequence. This example is from the course's unit on 1D motion under constant acceleration and the core concepts being taught included: (1) parabolic form of the trajectory, (2) independence of orthogonal components of motion, (3) freedom of choice in selection of axes, (4) relationship between trajectories and the direction of velocity and acceleration vectors, and (5) the ways in which the acceleration vector can be independent from the velocity vector and trajectory. Consequently, as shown in Fig. 4.10, the representations introduced in this CWP session were vector diagrams and kinematic graphs.

The common misconceptions intended for queuing and confrontation in this CWP session included the belief that (1) the acceleration vector is always parallel to the velocity vector, (2) equations of motion are determined by the units of the kinematic quantity (e.g., the units of velocity are [distance]/[time], thus the equation

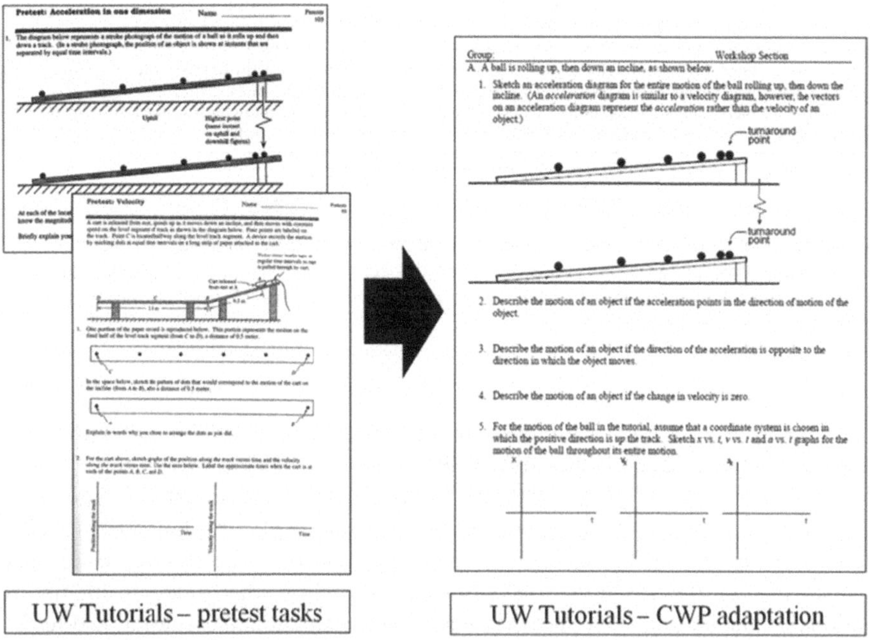

Fig. 4.10 A schematic of how University of Washington Tutorials pretest tasks are adapted for use in Collaborative Workshop Physics (CWP) sessions

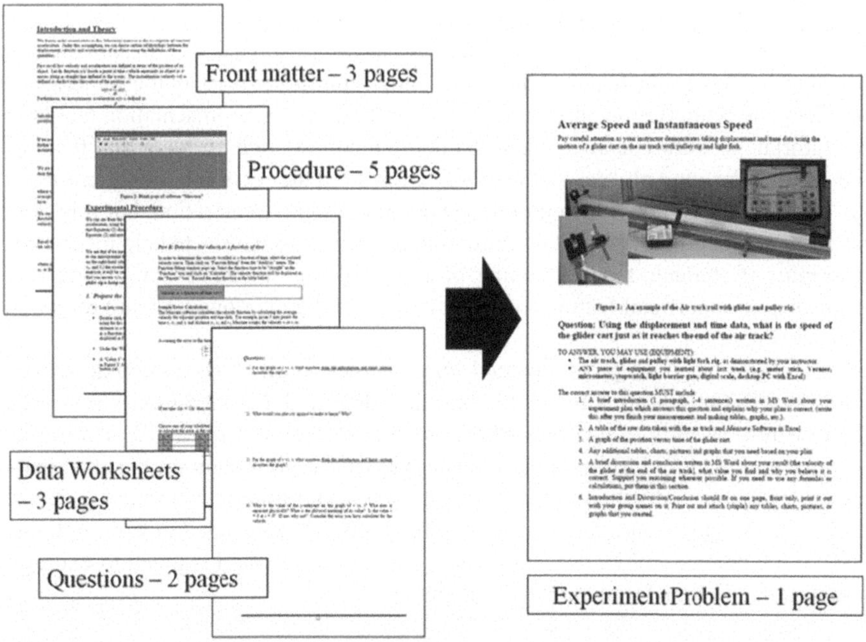

Fig. 4.11 A schematic of how Collaborative Workshop Physics (CWP) experiment problems are crafted using existing laboratory manual content and laboratory equipment

As you are driving to school one day, you pass a construction site for a new building and stop to watch for a few minutes. A crane is lifting a batch of bricks on a pallet to an upper floor of the building. Suddenly, a brick falls off the rising pallet. You clock the time it takes for the brick to hit the ground at 2.5 seconds. The crane, fortunately, has height markings and you see the brick fell off the pallet at a height of 22 meters above the ground. A falling brick can be dangerous, and you wonder how fast the brick was going when it hit the ground. Since you are taking physics, you quickly calculate the answer.

University of Minnesota - Context Rich Problem

Fig. 4.12 An example of a context-rich group problem developed at University of Minnesota that would make a complete Collaborative Workshop Physics (CWP) session with the tasks shown in Figs. 4.3 and 4.4. Context-rich problems were most often used "as-is" from (Heller and Heller 2007a) with little or no modifications

for velocity is "*d/t*"), (3) the closely related misconception that all velocities are average velocities, and (4) that the acceleration of an object is always zero when it is momentarily at rest (Rosenblatt and Heckler 2011). Consequently, as shown in Fig. 4.11, an experimental problem was crafted from the typical kinematics recipe lab that featured (as in the Tutorial) motion under constant acceleration, which then plays havoc on the above misconceptions about motion when students attempt to rationalize the motion. Note that the representation taught in the Tutorial is directly useful for designing an experiment to answer the question posed in the experimental problem and that the source of constant acceleration is altered, from a component of gravity on the incline in the Tutorial to tension in a string from a hanging mass in the experiment problem, to exercise students in migrating the representation across superficial changes in the problem features.

Last, as shown in Fig. 4.12, students are tasked to solve a CGPS context-rich problem, which again features the same underlying principles as the experiment problem and is analyzable most straightforwardly using the representation taught in the tutorial. As before, care is taken to choose a context-rich problem that features the same underlying physics as the other activities in the session but that features a difference context; now the motion is 1D under constant acceleration, but it is free-fall motion, not horizontal motion or motion on an incline. This is to help instructors determine if students are developing problem-solving strategies that are naively connected to superficial features of the particular problem or ones that are rooted in physics principles.

The above basic recipe is repeated for constructing all CWP session activities: (1) a set of core concepts for instruction are identified, (2) literature is reviewed for research-based investigations of known misconceptions, (3) a UW Tutorial is selected and modified for teaching relevant physical representations, (4) an experimental problem is distilled down from an existing traditional laboratory exercise and features a demonstration of motion that confounds associated misconceptions, and (5) a University of Minnesota context-rich group problem is given to exercise abstraction and transfer of the newly taught concepts.

None of the above process is useful without the proper instructor execution and formation of classroom norms. It is critically important that instructors do not lecture during the 3-h session. Student teams are repeatedly reminded that the session is for them to read, comprehend, and do the activities sequence. Instructors are there to *ask* questions, not answer them. Consistent with the University of Minnesota approach and training (Heller and Heller 2007a), instructors are trained to give no opening lecture, but to spend just 5 min making students aware of resources available to them during the session. Thereafter, instructors periodically come around the room to examine teams' work and ask probing questions. Questions are more straightforward and student follow-up is not graded during the tutorial activity to help students learn the representation with additional pressure. However, once the experimental problem and context-rich problem begin, teams are graded on-the-fly by the visiting instructor on the basis of how well they are addressing questions asked in earlier visits. Combined, an instructor typically visits and questions every team two to three times during the experiment problem and twice during the context-rich problem.

4.2.3 Case Study 3: Biomedical Engineering Problem-Driven Course

The PDL course is an introductory course in BME designed and developed at KU in Abu Dhabi in collaboration with Georgia Institute of Technology (GT) in Atlanta. Although the core of the PBL problems and scaffolding approach were adopted from GT, as well as the general course structure, the open-ended, ill-structured problems were specifically designed to "custom-fit" the KU and the UAE culture (Newstetter et al. 2012; Khalaf et al. 2013c).

4.2.3.1 Background

Biomedical Engineering: A Discipline Under Construction
The field of BME lies at the intersection of engineering, medicine, and the biosciences. As such, in addition to the typical challenges encountered in STEM and engineering education as mentioned above, BME education entertains its own unique challenges. Newstetter et al. (2010) summarize the challenges as ones encountered on two main fronts: the educator front and the student front. From the perspective of educators, BME education needs to bridge the gap between engineering and medicine and hence must combine the design and problem-solving skills of engineering with medical and biological sciences knowledge and skills. And yet, to date, almost no textbooks specifically targeting BME exist at the undergraduate level. The learning challenges on the student front are significant. Learners must master

three traditionally distinct intellectual faculties: (1) modeling and quantitative skills required for engineering, (2) qualitative systems analysis skills integral to the life sciences, and (3) clinical sensibilities inherent in medicine. It is, therefore, obvious that biomedical engineering educators need to foster in students the cognitive flexibility inherent in true integrative thinking and system analysis in order to embrace the merging of these distinct practices and historically separated disciplines (Newstetter et al. 2010).

An additional set of challenges in the highly interdisciplinary BME education stems from the dynamic nature and fast pace of evolution of this young discipline. Educators and students alike operate in a discipline with continuously shifting grounds and highly dynamic boundaries and constraints. The typical biomedical engineer of the 1970s and 1980s whose main training was in electrical or mechanical engineering with a few "picked up as needed" courses in biology and physiology did not need the skills crucial for today's tissue engineer who works on designing entire organs from stem cells and hence faces a whole range of engineering, biological, clinical, and ethical complexities. The twenty-first-century set of skills and competencies is critical here not only for innovation, productivity, and competitiveness but more importantly for maintenance and enhancement of the ultimate machine—the human body.

4.2.3.2 Course Objectives

Table 4.10 summarizes the PDL class design goals, motivations, and requirements similar to case study 1. The first column codifies the specific goal (G1, G2, etc.), while the remaining columns concisely state the goals, give the underlying motivation for adopting it, and list the required means for achievement.

4.2.3.3 Course Constraints

Similar to Table 4.10, Table 4.11 summarizes constraints placed on the design of the PDL course in BME (BMED 101). The first column gives a label for the constraint (C1, C2, etc.) and the second, third, and fourth columns concisely state the constraint, give the motivation for adopting the constraint, and give the requirements the constraint places on an acceptable instructional strategy, respectively. Similar to the first case study, noticeably many of the motivations for these constraints are associated with departmental and institutional factors, which is typically the case in any university, except for C6 (the need to create effective single-gender learning situations), which is directly motivated by student and ethno/national cultural factors.

Table 4.10 Problem-driven learning (PDL) design goals with their motivations and required means for achievement

No.	Goal	Motivation for setting goal	Requirements
G1	Tackle a complex real-world problem	Build engineering skills	Define the problem and identify the problem goals
			Explore the problem statement to identify critical problem features
			Develop provisional models and hypotheses that frame problem-solving
			Plan an attack strategy
			Carry out strategy and evaluate it
G2	Conduct self-directed inquiry	Build self-learning skills	Recognize inadequacies of existing knowledge
			Identify learning needs
			Set specific learning objectives
			Make a plan to address these objectives
			Evaluate inquiry
			Assess reliability of sources
			Digest findings and communicate effectively to self and others
			Apply knowledge to problem
G3	Demonstrate effective group skills	Build modern engineering skills	Help group develop team skills
			Willingly forego personal goals for group goals
			Avoid contributing excessive or irrelevant information
			Express disappointment or disagreement directly
			Give emotional support to others
			Demonstrate enthusiasm and involvement
			Complete tasks on time
			Monitor group progress
			Facilitate interaction with other members
			Assess group skills of self and others
G4	Build knowledge in disciplines relevant to BME	Build interdisciplinary BME skills	Digest findings and communicate them effectively to others
			Identify deep principles for organizing knowledge
			Construct an extensive knowledge base in all problem aspects
			Ask probing questions to propel further analysis of problem
G5	Communicate solutions of problems	Build communication skills	Written reports
			Oral presentations

BME Biomedical engineering

Table 4.11 Problem-driven learning (PDL) and institutional design constraints with their motivations and required means for achievement

No.	Constraints	Motivation for setting constraint	Requirements
C1	Cannot hire new teaching staff	As a start-up in a developing country, successful hiring searches are typically 18+ months	Extra workload for BME faculty
C2	Cannot buy new lab or classroom technology	As in C1, purchasing and customs clearance times are long and unpredictable	No additional lab equipment needed
C3	Must deal with language barriers, particularly in medical terminology	Most incoming freshman students struggling with English as a second language and having to additionally deal with medical terms	Specified time for problem comprehension
			Peering weak students with strong ones
C4	Must create effectively single-gender learning situations	Avoid negative expectancy violation	Demonstrated efficacy in single-gender settings

BME biomedical engineering

4.2.3.4 Course Structure

PDL Model at GT—The Development of "Generic" Cross-Cultural Core Problems

The development of a PDL curriculum at Georgia Tech commenced in 2000 as the newly founded Department of Biomedical Engineering was accepting its first PhD students. Faculty began by creating a first-year graduate course that used the white board scaffolding found in medical PBL in the context of six problems representative of the varied branches of BME. Special PBL rooms were commissioned for the new BME building. In the following year, the first undergraduate course titled *Problems in Biomedical Engineering I* was piloted. Over the next 3 years, a number of new problems for this course were developed and run with student teams to determine their appropriateness and relevance for an introductory course in BME. In time, three problems emerged from an iterative process of prototyping, running, analyzing, and redesigning that we now consider as cores over which different skins can be affixed. To illustrate, the first problem focuses on screening or treatment in the context of disease (see example problem in Appendix). The problem brings together probability statistics (sensitivity/specificity/positive predictive value) in health screening, issues of scale and systems in disease, and the development of quantitative methods of analysis for evaluation/decision-making in the face of conflicting and changing information. A significant intended learning outcome for the whole course generally, but for this problem very specifically, is the development of efficient/effective inquiry skills, which are very much needed when sifting through the peer-reviewed journal articles. Each term, a new disease can be explored. Generally, cancer of one kind or another has been used, but more recently, endometriosis and sickle cell disease have been the problem skins.

The second problem has experimental design at its core, and the third has mathematical modeling and computer simulation. These core problems offer enough flexibility that each semester is very different for both students and faculty. For example, through modeling and simulation, students were asked one semester to determine what steps the campus should take to prevent the spread of H1N1, while the next semester, they looked at the potential for experimental viral traps to halt the spread of HIV. This potential to "re-skin" the core problem each term with a different story line, a story line that often comes from current health and science news, keeps the course fresh and current for both faculty and students. Importantly, students really have the sense that they are working alongside other biomedical researchers on significant problems rather than just doing homework sets from textbooks.

In conjunction with problem development, a rubric laying out the intended course learning outcomes and student behaviors was developed for facilitators to use in observing student behaviors on the teams and for students to self- and peer assess. Rubrics for scoring each problem presentation and report as well as writing guidelines for each problem were also developed. This collection of materials scaffold student activities across the problem, making it possible for freshman teams to take on significant authentic problems. Finally, a strategy for a final exam was developed, piloted, evaluated, and redesigned using the same "skin" concept as the problems. Prior to collaboration with Khalifa, the Georgia Tech team had reached a steady state whereby more than 150 students were going through this experience every term facilitated by 12 or more faculty and postdocs a term.

Cross Cultural Globalization—The Development of "Cultural-Specific" Skins The PDL model adopted at the BME Department at KU is based on the system designed by GT in terms of "the core" problems described above. As previously mentioned, other attributes such as the scaffolding approach, the three problems per semester structure as well as the general course structure were also maintained. On the other hand, what we refer to as "skins" or outer shells affixed to these open-ended, ill structured, and poorly constrained core problems were specifically designed to "custom-fit" the KU and the UAE culture.

The following elements were incorporated in the process of the cross-cultural system transfer:

1. Problem Topics—Cultural Relevance: Motivation and Constraints

The topics were carefully selected based on cultural and societal relevance, emphasizing current health challenges in Abu Dhabi and the UAE. For example, as mentioned above, a typical core problem used at GT for the first problem is the identification of optimal methods for disease screening. In alignment with GT, this problem was selected due to the large amount of inquiry involved towards the solution ranging from the disease mechanisms at the molecular level to the physics behind imaging technologies, to the protocols involved in a various screening, and to the highly experimental research that has the potential to create new screening paradigms (Newstetter et al. 2010).

At KU, fresh skins were affixed to the core such that a cultural relevance and benefits were clearly established. For example, the following two health challenges were selected at KU for problem 1:

- Diabetes mellitus type 2: The UAE has the second highest rate of type 2 diabetes prevalence in the world (19.6 %), projected to increase to 63 % by the year 2030.
- Obesity: The UAE has one of the world's highest rates in overweight and obesity (71 % of men and women being either overweight (34 %) or obese (36 %)).

On the other hand, topics such as HIV, drug abuse, or life support were avoided due to cultural constraints.

The main objective of the second core problem, which is typically related to investigating the accuracy of a particular (medical) device, lies in the design of an experiment meant to test a hypothesis. The team has to use the literature to develop a testable hypothesis. Then, they need to develop an experimental protocol for collecting data to either verify or disprove their hypothesis. They must also design and set up an experiment so as to determine whether the results are statistically significant or not. Further, they need to determine what an appropriate sample size will be to achieve significance. And finally, every team member has to individually become institutional review board (IRB)-certified, and the group must get IRB approval beforehand (Newstetter et al. 2010).

An example of a skin affixed to such a problem at KU based on cultural relevance is the design and testing of an intelligent speed control system. Relevance is immediately established when the text of the problem states that Abu Dhabi has one of the highest rates of road deaths in the world, amounting to an alarming 27.4 of 100,000 people as compared to 15.2 in the USA and 11.9 in the EU (HAAD health statistics 2011).

2. Skill-based focus to promote metacognitive learning that is of particular importance, yet nonstandard to culture

4.2.3.5 Assessment

Cultural-Specific Assessment—What Works Best in Line of Cultural Values and Constraints

Assessment in the PBL classes at GT targets four specific areas: self-directed inquiry, knowledge building, collaboration skills, and problem-solving strategies. Various alternative assessment methods are used cumulatively at GT towards assessing these skills through the semester. These include inquiry updates, post-problem self- and peer evaluations, concept maps, written and oral presentations, and written assessment. While all of these are useful tools to monitor and assess the four target areas, cultural constraints may again play a role in the success of these assessment tools. For example, the concepts for peer and self-assessment at KU proved quite challenging as specific cultural values resulted in systematic underestimation of the students of their own performance and overestimation of that of their peers. The solution (affixed skin) was to share the assessment rubric with the students and have them quantify each of the categories by developing "skill lines" as an instrument to gauge the progress. The students were, hence, engaged in the skill assessment and quantification throughout the problem cycle for each of the three problems in a quantitative manner that helped them overcome the cultural assessment constraint. This engagement helped them learn to calibrate and objectively gauge skills (both self and team members). Table 4.12 summarizes the assessment rubric used for the post-problem self- and peer evaluations, while Fig. 4.13 depicts the skill line that the students used in the skill assessment process using an example of inquiry skills.

4.2.3.6 Example

Case Study: BMED 101 Problem 1: Road Safety in Abu Dhabi: The Design of an Intelligent Speed Control System

Client Statement
According to the Health Authority Abu Dhabi (HAAD), Abu Dhabi has one of the highest rates of road deaths in the world amounting to an alarming 27.4 of 100,000 people as compared to 15.2 in the USA and 11.9 in the EU (HAAD health statistics 2011). In fact, the data show that over the past 2 years, fatal injuries related to road accidents were the main cause of death in the Emirate of Abu Dhabi and the main cause of death among children 0–14 years old.

In response to the alarming statistics and recognizing the need for greater awareness about road safety, HAAD has launched a road safety initiative "Drive Safe Save Lives" to educate the public, with Sheikh Khalid Al Qassimi as their spokesperson.

Your team has been selected by the HADD to evaluate the current strategies implemented by HAAD to deal with this enormous health challenge. You are expected to evaluate HADD's existing program and propose an engineered system to decrease the rate of fatalities due to traffic accidents in Abu Dhabi. You will also quantitatively analyze and predict the expected outcomes of your system.

Problem
This problem was chosen as the first problem due to its cultural and societal relevance and in line with the recommended selection criteria for the first problem in an introductory BME PBL class (Newstetter 2006). Clearly, the problem requires a fair amount of inquiry on the part of students. The inquiry spans societal and technical domains. It can range from assessing and analyzing road safety in Abu Dhabi to establishing some fundamental reasons behind the alarming rates of fatal road accidents, to critically assessing the proposed strategies by the HAAD for addressing this challenge, to proposing technical engineering systems and solutions that may

Table 4.12 Assessment rubric used for the post-problem self- and peer evaluations

	Exceptional (A)	Proficient (B)	Fair (C)	Poor (D)
Inquiry skills	*Actively looks for* and recognizes inadequacies of existing knowledge	Recognizes inadequacies of existing knowledge	Occasionally claims areas of inquiry but mostly takes what's left	Takes whatever is left for inquiry
	Consistently seeks and asks probing questions	Generally asks probing questions	Occasionally asks questions	Rarely, if ever, asks questions
	Identifies learning needs and sets learning objectives	Utilizes appropriate search strategies	Uses search engines like Google to find easily available information of questionable reliability/appropriateness	Fails to recognize limits of understanding/knowledge
	Utilizes *advanced* search strategies	Mostly evaluates inquiry by assessing reliability and appropriateness of sources		Fails to assess the reliability or appropriateness of sources
	Always evaluates inquiry by assessing reliability and appropriateness of sources	Utilizes effective search strategies		Demonstrates unsystematic search strategies
Knowledge building	*Thoroughly* digests findings and communicates *effectively* to self and others	Digest findings and communicates to self and others	Reads inquiry results to group without thorough understanding of material	Fails to understand or be able to communicate inquiry findings
	Consistently identifies deep principles for organizing knowledge as evidenced in research notebook	Identifies deep principles for organizing knowledge	Learns own area of inquiry but not those of others	Rarely, if ever, asks questions
	Constructs *an extensive and thorough knowledge* base in all problem aspects	Constructs a thorough knowledge base in most problem aspects	Occasionally asks questions	Fails to use the problem to develop/enhance BME knowledge
	Continually asks probing questions	Asks probing questions		
Problem-solving	*Repeatedly* explores the problem statement to identify critical features	Explores the problem statement to identify critical features	Relies on group to identify critical features	Fails to define problem

Table 4.12 (continued)

	Exceptional (A)	Proficient (B)	Fair (C)	Poor (D)
	Defines/redefines the problem and identifies problem goals	Seeks to understand problem goals	Lets group identify problem goals and then follows along	Articulates no problem goals
	Breaks problem down into appropriate parts	Identifies criteria	Sometimes applies inquiry to problem-solving	Never uses the white boards
	Identifies and defines appropriate criteria	Uses inquiry in problem-solving		Fails to apply inquiry to problem
	Frequently uses white boards to assist in problem-solving	Uses white boards to assist in problem- solving		Never suggests a plan of attack
	Consistently applies inquiry results to problem	Occasionally develops models/hypotheses		Fails to develop analytic framework
	Develops models and hypotheses			
Team skills	*Actively* helps group develop team skills	Supports group goals	Goes along with the group	Does not help in developing team skills
	Willingly foregoes personal goals for group goals	Avoids contributing irrelevant information	Follows but does not lead	Gives no emotional or intellectual support to the team
	Always avoids contributing excessive or irrelevant information	Expresses disagreement directly	Avoids confrontation even when angry or frustrated	Lets group down by failing to complete tasks
	Consistently expresses disappointment or disagreement directly	Gives emotional support to others	Engages in limited interaction with other members	Observes silently contributing little to process
	Consistently gives emotional support to others	Demonstrates enthusiasm and involvement	Occasionally comes unprepared with no explanation	Shows little or no enthusiasm or involvement
	Clearly demonstrates enthusiasm and involvement	Facilitates interaction with other members		
	Monitors group progress and facilitates interaction with other members	Completes tasks on time		
	Always completes tasks on time			

BME biomedical engineering

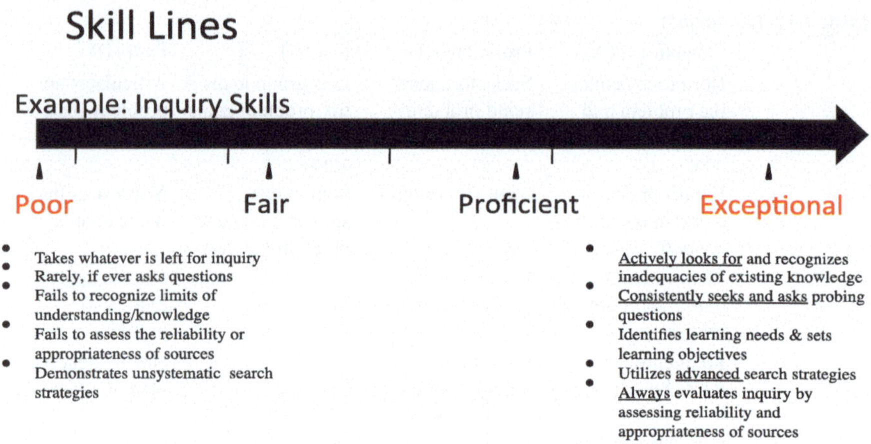

Fig. 4.13 Example of a skill line used to monitor and assess students' skills by instructor, peers, and self

be implemented in order to decrease the rate of fatalities, and to quantitatively assessing the expected outcomes of implementing such as system.

Aside from the provided statistics for motivating the students and establishing the significance of the problem, the problem was designed to be open-ended, ill structured, and poorly constrained. It was deliberately formulated to present minimal information about the situation in order to promote self-directed inquiry and learning as well as systematic knowledge-building skills. The technical engineering component was added to further promote critical thinking and problem-solving abilities. The complexity of the problem and its multifaceted nature required all team members to be involved in the iterative processes of inquiry, knowledge building, and problem-solving.

Team and Action Plan

Team 1 was composed of seven females of four nationalities (4 Emiratis, 1 Jordanian, 1 Pakistani, and 1 Sudanese). They started by establishing the magnitude of the problem in Abu Dhabi and at the university among their professors and peers. Towards this goal, they did interviews at the police department who provided them with information about the scope of the problem, the latest statistics, and some current strategies that they use to deal with this challenge. The students also did a survey on campus with 60 faculty, staff, and students and analyzed the data. Based on police data and their own survey results, they established that speed was the main cause of traffic accidents and fatalities due to these accidents in Abu Dhabi (60 % based on police data and 53 % according to university data). They also concluded that the majority of the accidents are caused by young drivers (95 % of fatal accidents involve drivers between the ages of 18 and 25).

System

Accordingly, the students worked on the design of an intelligent system to control vehicle speeding in Abu Dhabi. Their main design objectives were to: (1) design an intelligent system that has continuous monitoring and feedback of speed and that is capable of communicating with radars and the police department, (2) ensure that the system has a user-friendly high-tech interface, which appeals to young drivers, and (3) establish a reward system as an incentive for safe driving as opposed to the current penalty system, which does not seem to operate very well.

Upon an iterative process of research and inquiry and several meetings with experts (IT professors and engineers), the students decided to base their system design on vehicular ad-hoc network (VANET). VANET is a technology that uses moving cars as nodes in a network to create a mobile network in which participating vehicles are turned into a wireless router or node. The student team built their system using three components: (1) recycled generic tablets, (2) built-in global positioning system (GPS), which detects the speed of each road, and (3) sensors connected to car speedometers.

The system compares the speed of the road with the speed of the car (from the speedometer). If the car is over-speeding, the system allows the driver a cushion of X km/h above the speed limit for 1–2 min. If the driver does not slow down, the system will record a point. This point will be displayed on the screen as well as recorded inside the memory of the car computer. Whenever the car passes by the radar, the radar will be detected by the VANET sensors and the point information will be transferred to the radar as signals. This information will be sent directly to the police headquarters for evaluation (Figs. 4.14 and 4.15).

Implementation: Incentive Versus Penalty System

Based on the number of points, monthly and annual incentives will be given to the drivers. These incentives may include free gas, free toll cards, discount on car registration, discount on insurance, etc. The drivers will be informed about the incentives through short messaging service (SMS) messages or e-mail, for example, "Congratulations, You have no speeding points this month, please collect your ticket online from this website, www.xyz.com."

The students calculated the cost for building the system prototype as approximately AED 700 (about US$200). They anticipate that production in mass quantities would lower this cost substantially. Furthermore, the cost is insignificant compared with the total amount of money spent by the Abu Dhabi Police Department

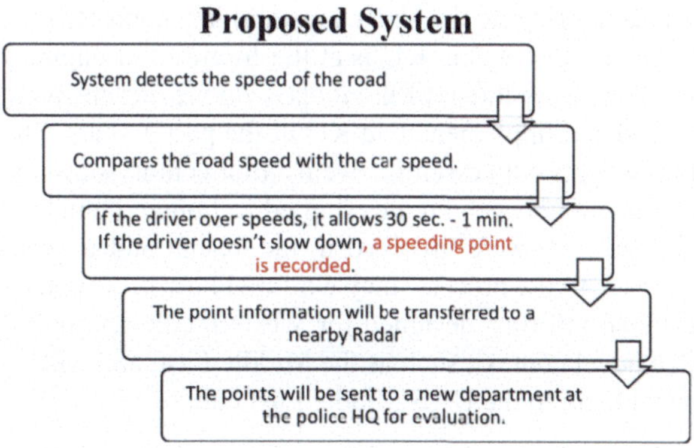

Fig. 4.14 Flowchart of proposed intelligent system

Fig. 4.15 Prototype of the designed intelligent speed control system (ISCS)

annually in dealing with traffic accidents (Every year, the UAE government spends around DHS 3.5 billion on car crashes and injuries in the Emirate of Abu Dhabi). They also worked on a proposed timetable to implement the system in Abu Dhabi through the Police Department by enforcing it as part of the annual registration process of each car.

Mathematical Model
The students also did a mathematical model that predicts the correlation between speed and car accidents and predicts the reduction in accidents as a result of reduction in speed. The model was built based according to a statistics done by the Abu Dhabi Transport Authority and the World Health Organization (WHO).

4.3 Discussion

In the context of the requirements of revamped accreditation criteria and the calls from various engineering education stakeholders regarding the skills needed in modern engineering graduates, it would appear that these demands are not likely to be met by a traditional engineering curriculum and "chalk and talk" pedagogy. Towards bridging the skills gaps and key competencies critical to engineers in the UAE and the region, KU is at the forefront of engineering education reform. This work presents three different classroom experiments in engineering education designed and implemented at KU in the past 3 years. The primary goal was to equip the students with essential twenty-first-century skills and minimize the current misalignment between engineering education and employers in the Gulf. The authors of this work were involved in the design and implementation of these reformed courses. In the process, they explored how these pedagogical innovations and educational reforms, designed for Western cultures, will function when implemented in a novel context, such as the Middle East, and what it would take to successfully adapt them to the UAE/Middle East culture.

In the case of KU, we find that problem-based, project-based, and collaborative learning modalities adopted from Western peer educational models can be implemented with minimal modification to their core features. Our research results show that learning gains or favorable attitudinal gains in the three pilot courses studied are all improved relative to the traditional, lecture-centered version of the same courses and with the same student body. While there is still a gap between these and similar Western counterpart courses, evidence shows that the differences are likely due to pre-instruction factors, particularly second-language issues, and not to the effect of the adopted pedagogy or instructional strategies.

Efficacy

In each of the above case studies, a variety of measures of efficacy were applied to determine the efficacy of the instructional strategy.

For example, the Engineering Design-and-Build course has two significant positive impacts on students: (1) as measured by a newly developed design attitudes survey, course graduates are 20 ± 5 % more likely than other engineering students to express attitudes consistent with professional engineers (Perkins et al. 2004) regarding problem-solving practices in the engineering design process (Fig. 4.16), (2) a measure of teams' adherence to the course's prescribed design cycle is moderate to strongly correlated ($\rho \sim 0.66$, > 90 % confidence) with the quality of the finished design as measured by live, in-class demonstrations (Fig. 4.17).

These two metrics, the Problem Solving Attitudes survey score (Fig. 4.16) and the correlation of design knowledge with design performance, as depicted in

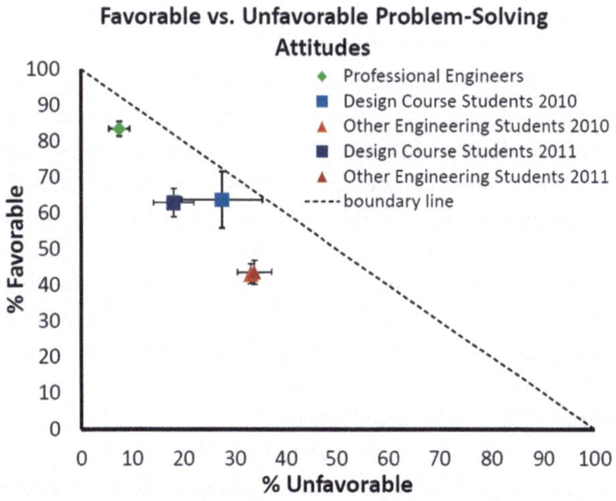

Fig. 4.16 Favorable versus unfavorable plot for the eight-statement problem-solving section of the survey. Favorable score (%) is the percentage of statements where respondents answered as the survey creators would, and vice versa for unfavorable score (%). No score was assigned to neutral or blank responses, which is why the two scores do not sum to 100 % and therefore fall below the diagonal boundary line. Scores are an average for each group surveyed; professional engineers *(blue diamond),* students who have taken the freshman design course *(red squares),* and students who have not taken the course *(green triangles)*

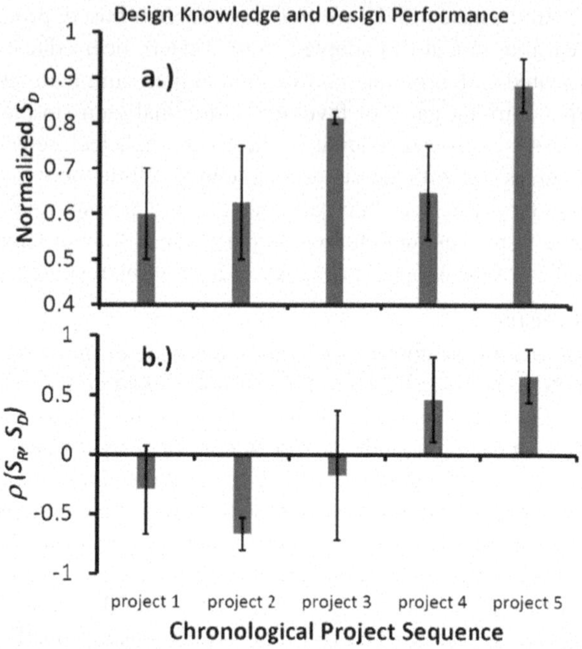

Fig. 4.17 Emerging correlation ρ in the student population, between design demonstration performance S_D and a measure S_R of their adherence to the course's prescribed design cycle as shown on design reports. S_D standard deviation of design scores, S_R standard deviation of report scores

Fig. 4.17, are clear indications that (1) as a result of the course, students perspectives on engineering design are more like expert engineers as compared with their peers who did not take the course. *Engineering Problem-Solving course students are significantly more likely to think like an engineer when faced with a design problem.* And (2), thinking like an engineer and systematically, conceptually following a realistic design cycle, as indicated by report scores, leads to greater demonstration performance in a live demonstration of real, fabricated designs. *For Engineering Problem-Solving course students, design thinking leads to design performance!*

For another example of reform project efficacy in the KU university physics course, the new CWP instructional strategy was specifically designed to (1) improve retention of engineering students, (2) increase conceptual learning, (3) increase traditional problem-solving skill, and (4) be particularly beneficial for "at-risk" (conditionally admitted, CA) students. Table 4.13 below summarizes these metrics for efficacy, the measures used, and a comparison with the same measures of the course prior to the pedagogical reform.

Based on measures of efficacy in Table 4.13, the success of the CWP session reform of KU University Physics I is mixed. Clearly, the new instructional strategy has positively and significantly impacted DFW rates and traditional problem-solving performance, and does so particularly for CA students. However, the improvements in conceptual learning, as measured by FCI normalized gain, are far lower than anticipated. As a reminder, Hake reported that successful alternative pedagogies used in introductory mechanics courses produce FCI normalized gains

Table 4.13 Measures of pedagogical efficacy for the Collaborative Workshop Physics (CWP) session compared to the pre-reform baseline

No.	Measures of efficacy	Metric	Pre-reform value (N=221 students)	Post-reform value (N=78 students)
1	Improve retention by lowering course failure rate	DFW rate (%)	37%	23%
2	Increase conceptual learning	Normalized FCI gain (FCI <g>)	0.08±0.05	0.14±0.04
3	Increase "end-of-chapter"/traditional problem-solving skill	Exam performance, mid-semester (%)	61±3%	77±2%
		Exam performance, final (%)	56±3%	70±2%
4	Differentially beneficial for CA students	CA-DFW rate (%)	50%	24%
		CA-FCI <g>	0.03±0.03	0.13±0.04
		CA-exam performance	47±4%	74±5%

DFW drop/fail/withdraw, *FCI* Force Concept Inventory, *CA* conditionally admitted

of 0.48±0.14 (Hake 1998). The CWP session result of 0.14±0.04 (Hitt et al. 2014) is approximately 2.5 standard deviations below this value.

Sustainability

Ultimately, the CWP session use was ceased because of a failure to adequately plan for the staffing needs in light of attrition and enrollments. After fall semester 2011, it was determined that there were insufficient human resources available to continue its use, and the lab portion of the University Physics I course reverted to its traditional form beginning in the spring semester of 2012. This is in spite of efforts to purposely design the CWP instructional strategy to be a "low effort" methodology and to require no additional equipment purchases or staff hires. Large and unpredictable increases in student enrollments and registrations into the course, however, eventually made new staff hires critically necessary. The easiest way to cope with these changes was to revert to traditional instruction, which allows a larger number of students to fit into the lab room and does not require an instructor to be present for the full duration of the session.

By contrast, both the PDL course in BME and the Freshman Engineering Design-and-Build course were able to adequately deal with the institutional and staffing constraints. In fact, the design-and-build course expanded to include all engineering majors at the university and evolved as the first step towards a vertically integrated design-based curriculum (Fig. 4.2, see Balawi et al. 2013 for further details).

Common Challenges

Gender integration in the Gulf Arab context presents a unique challenge for alternative pedagogies that rely heavily on student teaming and engagement. All three of the course reform case studies presented in this chapter used team-based pedagogies. All three also used instructor-prescribed teams to varying degrees. At KU, 70–80% of students are UAE nationals and as a result of history and culture, have not regularly interacted with members of the opposite gender outside of their family, prior to joining the university. KU policy states that no student may be forced to work in a mixed-gender situation if they are not comfortable with it. As a result, instructor-prescribed teaming can be over-constrained and combinatorically frustrated. For large-enrollment courses, teaming may require software-based solutions to reduce the time required to converge on a set of teams. In the case of the University Physics I course reform, the CWP sessions often registered 25–30 students per section, with 3–4 sections and with approximately 1:1 male–female ratio. Students were reteamed once every 4 weeks, a job that, when done by hand and respecting those students who object to mixed-gender teams, required an entire day of the instructor's time. Eventually, a software-based solution was necessary, bringing the time required for teaming students down to a more tractable 60–90 min.

Language proficiency was a ubiquitous challenge in all of the above three case studies. As shown in Table 4.1, approximately 100 % of KU students are ELLs. However, the language of instruction for the three courses under consideration is English. There is a diversity of ways in which language proficiency was planned for among the reformed course designs. In the case of the CWP session and PDL courses, the teaming recipe explicitly relied on International English Language Testing Service (IELTS) data to place students into teams. Student teams were constructed so as to include at least one member with an IELTS score in the top third of the distribution within the section. The intention behind this was to increase the likelihood that, should a lesson present difficult jargon for a student, someone in their team could assist in giving a clear interpretation into Arabic and that the student, now having a clearer understanding of the task, would be more likely to learn new concepts during instruction.

One would then anticipate that if a measure of conceptual improvement, like FCI normalized gain in the CWP example, were plotted against IELTS score for both pre-reform and post-reform student populations, the two curves would differ by an overall shift, signifying that the new pedagogy and teaming recipe improved learning for all students regardless of language proficiency. Yet, in the final analysis, this approach appears to have had the reverse effect. As shown in Fig. 4.18, only those students with the very-highest-level English language proficiency (IELTS overall score of 7.0) appear to have attained the degree of conceptual learning gains seen in North American case studies (Hake 1998). FCI normalized gain for the CWP case study is 0.08 ± 0.05 and 0.14 ± 0.04, pre-reform versus post-reform respectively, only a moderately significant difference. However, for those students with IELTS overall score 7.0, the pre- versus post-reform performance on FCI is 0.16 ± 0.10 and 0.47 ± 0.08, respectively. One might consider this as evidence that those with high English proficiency were the only students who were engaged effectively in the course, but this overlooks the fact that in the CWP teaming recipe, for example, students with high IELTS scores were always distributed. These students never worked with each other. Given the well-documented importance of rich student–student interactions for producing high conceptual learning gains, it seems unreasonable to imagine that these students were doing all the thinking in their teams alone. This then is reasonable evidence that there is likely a fundamental effect of language ability on measures

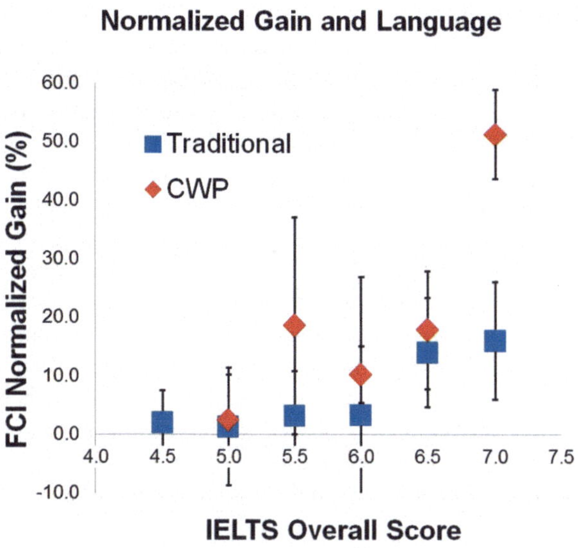

Fig. 4.18 Reproduction of Fig. 4.9 from Hitt et al. 2014. Force Concept Inventory (FCI) normalized gain is plotted for students with a given International English Language Testing Service (IELTS) score. Students taking the university physics I course in the traditional form *(KU-T, blue squares)* and compared to those taking the reformed Collaborative Workshop Physics (CWP) version *(KU-CWP, red diamonds)*

of conceptual learning and that pedagogy that is effective may not register the same "signal" on instruments as FCI does in North America. Presently, it is unclear if there is a primary effect on learning that is connected with language or if IELTS scores are merely a reliable proxy for some other influential factor, such as student motivation or academic aptitude. It is likely that both influences are important to account for.

In implementing the PDL in the BME course, language proficiency was also a particular challenge as the course is more qualitative than quantitative and is the first course where students are inherently exposed to medical jargon and terminology. In the instructional strategy, students were also purposefully teamed to ensure at least one proficient English language speaker per team. The instructor, a native speaker herself, who mainly played the role of coach in the problem-solving instructional studio, often reverted to helping translate the medical terms during the first problem. This translation role was gradually decreased as the students progressed from the first to the third problem moving positively on the skill line and enhancing their independence and self-learning efficacy (Fig. 4.13).

Common Opportunities

Reforming pedagogy in any course presents the opportunity to reconsider what knowledge and skills are central to the discipline of the course. In each of the three cases studies at KU, the considerations of reformers, when in the conception and planning stages of the reform, converged on ways in which the traditional course did not introduce or enforce important knowledge or skills in the associated discipline. In the case of the University Physics I course, in its traditional form, students did not learn how to design physical experiments or apply theoretical knowledge to complex, realistic problems, and yet this is essential for those who practice physics professionally. In the case of the PDL and problem-solving courses, both were created to teach students skills for complex real-world problem-solving and conceptual design, something that engineers must do, but in traditional curricula, many do not learn until senior capstone experiences.

Exportability
The strong dependence of measures of conceptual learning on language proficiency offers some important lessons to pedagogical reformers in the Middle East, particularly to institutions like KU, where instruction is in English. Our data show that it is possible to construct the classroom norms necessary for effective alternative pedagogy. However, measures of conceptual learning may not be the best key performance indicators (KPIs) for this, and reformers are warned to select KPIs for their reform projects carefully as it appears that research-based conceptual assessments are not applicable to ELL students in a straightforward way.

The PDL course in BME may be considered as a conceptual design of an exportable pedagogical system for transfer across cultures (Khalaf et al. 2013c). The system is based on the development of "generic core" problems that are specifically designed to promote, through scaffolded metacognitive apprenticeship, the critical skills uniquely needed for biomedical engineers, but that are then adapted for smooth and effective cross-cultural transfer and relevance by affixing salient surface features or "cultural skins" to these problems.

4.4 Conclusions

This chapter has demonstrated that pedagogical reform in university-level STEM education should and can be done in the Middle East and North Africa (MENA) region. We have examined three course reform projects at KU, UAE as evidence. Certainly, every institution has its own idiosyncrasies, which must be anticipated by educational innovators on the ground, so this result is not a simple "recipe-for-success" in similar institutions in the Gulf or broader MENA region. What is demonstrated is that anticipation of student cultural factors and adaptation to them is possible. In all three cases presented, instruction was carefully redesigned to the particular needs of the student population, and in all three cases, there have been substantial improvements in student attitudes, self-efficacy, conceptual learning, problem-solving performance, and core disciplinary skills. These improvements are consistent over multiple course offerings and multiple year groups of students, with multiple instructors, and on multiple campuses. This is proof of principle that contributes significantly to dispelling the beliefs that alternatives to traditional, lecture-based instruction are impossible to implement or that they could never be fruitful for student learning in the Gulf Arab context. The dramatic differences between US and KU engineering student populations have reliably pointed the way in designing effective pedagogy, but there is little evidence that cultural and gender factors have any further predictive power over improved learning. There is no unreachable part of the student demographic. Indeed, in this regard, second-language acquisition appears far more important than demographics (Table 4.14).

When considering the long-term success and sustainability of course reforms, it is institutional and departmental factors, rather than student cultural factors, that appear most important. There is a clear relationship between the level and quality of

Table 4.14 Indicators of institutional support from Wieman et al. (2010) p. 11 applied to KU case studies

Factors facilitating change	CWP	ED & B	BMED PDL	Factors inhibiting change
Supportive, respected chair/head with authority	×	✓	✓	Unsupportive, inactive, or powerless chair/head
Broad faculty support and involvement	×	✓	✓	Effort limited to a few core faculty
Science education specialists involved and respected	×	×	✓	Science education specialist not integrated into reform
Reward structure for reform-related activities and superior teaching	×	✓	✓	Last-minute teaching assignments
Senior and junior faculty leaders who promote the project	×	✓	✓	Departmental culture that does not respect education research
Newly formed department in need of a curriculum	✓/×	✓/×	✓	Departmental culture that expects individual freedom in teaching

CWP Collaborative Workshop Physics
✓ support factor present, × inhibiting factor present, ✓/× both factors present

institutional support and the health of each of the above three course reforms. There is also strong resonance with the findings of other large STEM pedagogy reform efforts in the USA on these institutional factors, of which Wieman et al. (2010, p. 11) identify six that are critical. In the case of the university physics reform, the CWP project faced all six of the six most negative conditions on these factors: no permanent department chair, only two junior faculty driving the reform, no science education specialist consulting the project, last-minute teaching assignments, high skepticism among other faculty in the department, and an expectation of autonomy in teaching style (Wieman et al. 2010, p. 11). Consequently, CWP did not survive beyond its first year pilot due to staffing concerns. The Engineering Design-and-Build (ED & B) course had many of these negative factors in common with the CWP project and it was expanded to become a college-wide required course, dramatically increasing the enrollment and staff requirements. However, the teaching assignments for the course were planned several months further in advance of the physics course, and the department had a permanent chair to lend weight to the reforms. This course continues at present and is receiving regular attention and reinvestment. At the opposite end of the spectrum, the BMED PDL course is clearly the healthiest of the three course reforms and is running stably at present. In the post-analysis, this course has benefited from a variety of positive conditions on Wieman, Perkins, and Gilbert's six institutional factors: a supportive, respected chair driving the reform vision; broad faculty support within the department; a science education expert advising the reform; and teaching assignments planned months in advance.

The most important conclusion to be drawn about pedagogical reform in the MENA region from the KU case is the importance of departmental and institutional support. Student cultural factors *do* prevent Western pedagogy and instructional strategies from being used in an "off-the-shelf" fashion, *but* with thoughtful analysis of the population and redesign of the instructional strategy, Western-born alternative pedagogies can be reliably adapted for use in the Arab world. If the instructional design is done well, the important determining factor in the long-term success and sustainability of a course reform project is departmental and institutional support. From the KU case, we find that the factors identified by Wieman et al. (2010) are highly valid and predictive of the life span and health of our course reforms. We are certain that these case studies and this result forms a highly valuable report for the course design considerations of persons and institutions considering similar educational reforms in similar contexts.

We recommend that any institution considering similar reforms in the MENA region take several steps to increase the efficacy of their reform process: (1) carefully select a small number of high-impact courses for reform and make sure they are "owned" by chairs or program managers who strongly believe there is a need and a possibility for improvement, (2) gather baseline data on the performance of students in the course prior to the reform so that there is a point of reference for comparison later in the reform project, (3) give faculty *designers* of the new version of the course 6–12 months to create the new course design and engage others (through seminars, colloquia, workshops, etc.) in the department before rolling out the new course or involving the *designing faculty* in that course's teaching rotation, and (4) perhaps most importantly, work as an institution to achieve predictive power over student enrollment and use it to create teaching assignments and teaching rotations for *all faculty in the department or program* that also extend 6–12 months in advance so that there is time for the reform effort to broadly engage them during the course design process.

Acknowledgments This work was partially supported by Khalifa University Internal Research Fund grant number #B4011:210005. The authors wish to gratefully acknowledge Dr. Wendy Newstetter, PhD, director of Learning Sciences Research at the Wallace H. Coulter Department of Biomedical Engineering, Georgia Institute of Technology, for her contributions to the PDL course reform project in KU, Department of Biomedical Engineering. The authors also thank L. E. Hitt for a careful review of the draft version of this manuscript.

References

Adams, W. K., & Wieman, C. E. (2010). Development and validation of instruments to measure learning of expert-like thinking. *International Journal of Science Education, 33*(9), 1289–1312.

Adams, W. K., Perkins, K. K., Podolefsky, N. S., Dubson, M., Finkelstein, N. D., & Wieman, C.E. (2006). New Instrument for Measuring Student Beliefs about Physics and Learning Physics: The Colorado Learning Attitudes about Science Survey. *Physical Review Special Topics—Physics Education Research, 2,* 010101.

Anderson, L. W., Krathwohl, D. R., Airasian, P. W., Cruikshank, K. A., Mayer, R. E., Pintrich, P. R., Raths, J., & Wittrock, M. C. (2000). *A taxonomy for learning, teaching, and assessing: A revision of bloom's taxonomy of educational objectives*. New York: Pearson, Allyn & Bacon.

Atman, C. J., Sheppard, S. D., Turns, J., Adams, R. S., Fleming, L. N., Stevens, R., Streveler, R. A., Smith, K. A., Miller, R. L., Leifer, L. J., Yasuhara, K., & Lund, D. (2010). *Enabling engineering student success: The final report for the center for the advancement of engineering education*. San Rafael: Morgan & Claypool.

Balawi, S., Khalaf, K., Hitt, G. W., Hirsch, B., & Powell, L. (2013). *Cultivating design thinking in freshman: The evolution of the KU freshman design course*. Proceedings of the TALE2013—IEEE International Conference on Teaching, Assessment and Learning for Engineering, 26–29 August 2013, Bali. Indonesia (accepted for publication).

Barrows, H. S. (1998). The essentials of problem-based learning. *Journal of Dental Education, 62*(9), 603–633.

Barrows, H. S., & Tamblyn, R. M. (1980). *Problem-based learning: An approach to medical education*. New York: Springer

Beichner, R. J. (2007). The student-centered activities for large enrollment undergraduate programs (SCALE-UP) project. *Research-Based Reform of University Physics*. http://www.compadre.org/Repository/document/ServeFile.cfm?ID=4517&DocID=183. Accessed 20 June 2013.

Beichner, R. J. (2008). The scale-up project: A student-centered active learning environment for undergraduate programs, a paper presented at the National Research Council Workshop on Linking Evidence to Promising Practices in STEM Undergraduate Education, Washington, DC. http://www7.nationalacademies.org/bose/Beichner_CommissionedPaper.pdf.

Beichner, R. J., Saul, J. M., Allain, R. J., Deardorff, D. L., & Abbott, D. S. (2000). Introduction to SCALE-UP: Student-centered activities for large enrollment university physics. Annual Meeting of the American Society for Engineering Education, St Louis, Mo.

Beichner, R. J., Saul, J. M., Abbott, D. S., Morse, J. J., Deardorff, D. L., Allain, R. J., Bonham, S. W., Dancy, M. H., & Risley, J. S. (2007). The student-centered activities for large enrollment undergraduate programs (SCALE-UP) project. In E. F. Redish & P. J. Cooney (Eds.), *Research-based reform of university physics*. College Park: American Association of Physics Teachers.

Biggs, J., & Tang, C. (2007). *Teaching for quality learning at university* (3rd ed.). New York: Open University Press.

Bloom, B. S. (1956). *Taxonomy of educational objectives, handbook I: The cognitive domain*. New York: David McKay.

Brewe, E., Kramer, L. H., & O'brien, G. E. (2009). Modeling instruction: Positive attitudinal shifts in introductory physics measured with CLASS. *Physical Review Special Topics: Physics Education Research, 5,* 013102.

Brewe, E., Sawtelle, V., Kramer, L. H., O'brien, G. E., Rodriguez, I., & Pamel, P. (2010). Toward equity through participation in Modeling Instruction in introductory university physics. *Physical Review Special Topics: Physics Education Research, 6,* 010106.

Burgoon, J. K. (1978). A communication model of personal space violations: Explication and an initial test. *Human Communication Research, 4,* 129.

Burgoon, J. K. (1995). In R. L. Wiseman (Ed.), *Intercultural communication theory* (pp. 194–214). Thousand Oaks: Sage.

Cheville, A., & Bunting, C. (2011). Engineering students for the 21st century: Student development through the curriculum. *Advances in Engineering Education, 2*(4), 1–37.

Creed, C. J., Suuberg, E. M., & Crawford, G. P. (2002). Engineering entrepreneurship: An example of a paradigm shift in engineering education. *Journal of Engineering Education, 91*(2) 185–195.

Crouch, C. H., Watkins, J., Fagen, A. P., & Mazur, E. (2007). Peer instruction: Engaging students one-on-one, all at once. In E. F. Redish & P. J. Cooney (Eds.), *Research-based reform of university physics*. College Park: American Association of Physics Teachers.

Cummings, K., Marx, J., Thornton, R., & Kuhl, D. (1999). Evaluating innovation in studio physics. *Physics Education Research: American Journal of Physics Supplemental, 67,* S38.

Dancy, M., & Henderson, C. (2010). Pedagogical practices and instructional change of physics faculty. *American Journal of Physics, 78,* 1056.

Domin, D. S. (1999). A review of laboratory instruction styles. *Journal of Chemistry Education, 76,* 543–547.

Dutson, A. J., Todd, R. H., Magleby, S. P., & Sorensen, C. D. (2007). A review of literature on teaching design through project-oriented capstone courses. *Journal of Engineering Education, 76*(1), 17–28.

Dym, C. L. (2004). Design, systems, and engineering education. *International Journal of Engineering Education, 20*(3), 305–312.

Dym, C. L., & Little, L. (2003). *Engineering design: A project based introduction*. New York: Wiley.

Dym, C. L., Wesner, J. W., & Winner, L. (2005a). Social dimensions of engineering design: Observations from mudd design workshop III. *Journal of Engineering Education, 94*(1), 103–120.

Dym, C. L., Agogino, A. A., Eris, O., Frey, D. D., & Leifer, L. J. (2005b). Engineering design thinking, teaching, and learning. *Journal of Engineering Education, 94,* 103.

Elby, A. (1998). http://www2.physics.umd.edu/~elby/EBAPS/idea.htm on 05/17/2010.

Etkina, E., Gibbons, K., Holton, B. L., & Horton, G. K. (1999). Lessons learned: A case study of an integrated way of teaching introductory physics to at-risk students at Rutgers University. *American Journal of Physics, 67,* 810.

European Communities. (2007). Key competences for lifelong learning. European Reference Framework Luxembourg Office for Official Publications of the European Communities.

Evans, D. L., McNeill, B. W., & Beakley, G. C. (1990). Design in engineering education: Past views of future directions. *Journal of Engineering Education, 79*(4), 517–522.

Finkelstein, N. D., & Pollock, S. J. (2005). Replicating and understanding successful innovations: Implementing tutorials in introductory physics. *Physical Review Special Topics: Physics Education Research, 1,* 010101.

Froyd, J. E., & Ohland, M. W. (2005). Integrated engineering curricula. *Journal of Engineering Education, 94*(1), 147–164.

Gaffney, J. D. H., Gaffney, A. L. H., & Beicher, R. J. (2010). Do they see it coming? Using expectancy violation to gauge the success of pedagogical reforms. *Physical Review Special Topics: Physics Education Research, 6,* 010102.

Goodwin, R. A. (1978). Talk and chalk. *The Physics Teacher, 16*(6), 367–373.

HAAD health statistics. (2011). www.haad.ae/statistics.

Hake, R. R. (1992). Socratic pedagogy in the introductory physics laboratory. *The Physics Teacher, 30,* 546.

Hake, R. R. (1998). Interactive-engagement versus traditional methods: A six-thousand student survey of mechanics test data for introductory physics courses. *American Journal of Physics, 66,* 64.

Halloun, I. (1997). *Views about science and physics achievement, The VASS story*. Proceedings of the International Conference on Undergraduate Physics Education, American Institute of Physics Press, College Park, MD.

Heller, P., & Heller, K., (1999). *Cooperative group problem-solving in physics*. Minneapolis: University of Minnesota.

Heller, P., & Heller, K. (2007a). *Cooperative group problem solving in physics*. Minneapolis: University of Minnesota. http://groups.physics.umn.edu/physed/Research/CGPS/. Accessed 10 March 2012.

Heller, P., & Heller, K. (2007b). *Physics 1301 laboratory manual*. Minneapolis: University of Minnesota. http://groups.physics.umn.edu/physed/Research/Lab. Accessed 7 April 2012.

Heller, P., & Hollabaugh, M. (1992). Teaching problem solving through cooperative grouping. Part 2: Designing problems and structuring groups. *American Journal of Physics, 60,* 637.

Heller, P., Keith, R., & Anderson, S. (1992). Teaching problem solving through cooperative grouping. Part 1: Group versus individual problem solving. *American Journal of Physics, 60,* 627.

Henderson, D., & Dancy, M. (2007). Barriers to the use of research-based instructional strategies: The influence of both individual and situational characteristics. *Physical Review Special Topics: Physics Education Research, 3,* 020102.

Hestenes, D. (1987). Toward a modeling theory of physics instruction. *American Journal of Physics, 55,* 440–454.

Hestenes, D., Wells, M., & Swackhamer, G. (1992). Force concept inventory. *The Physics Teacher, 30,* 141.

Hitt, G. W., Isakovic, A. F., Fawwaz, O., Bawa'aneh, M. S., El-Kork, N., Makkiyil, S., & Qattan, I. A. (2014). *Secondary implementation of interactive engagement teaching techniques: Choices and challenges in a Gulf Arab context. Physical Review Special Topics: Physics Education Research, 10,* 020123.

Hsu, L., Brewe, E., Foster, T. M., & Harper, K. A. (2004). Resource letter RPS-1: Research in problem solving. *American Journal of Physics, 72,* 1147–1156.

Kardash, C., & Wallace M. (2001). The perceptions of science classes survey: What undergraduate science reform efforts really need to address. *Journal of Educational Psychology, 93*(6), 199–210.

Khalaf, K., Balawi, S., Hitt, G. W., & Radaideh, A. (2010). *Innovation in teaching freshman engineering design: An integrated approach.* Proceedings of the EduLearn10 Conference, IATED, Barcelona, Spain, 5–7 July.

Khalaf, K., Balawi, S., Hitt, G. W., & Radaideh, A. (2013a). Engineering design education: When, what, how. *Advances in Engineering Education, 3*(5), 1–31.

Khalaf, K., Balawi, S., Hitt, G. W., & Siddiqi, M. M. (2013b). Engineering design education: Effect of mode of delivery. *International Journal of Engineering Pedagogy, 3,* 46.

Khalaf, K., Alsafar, H., & Newstetter, W. (2013c). *Globalization of problem-driven learning: Design of a system for transfer across cultures.* Proceedings of the 4th International Research Symposium on Problem Based Learning (IRSPBL2013), Kuala Lumpur, Malaysia, 2–3 July 2013 (accepted for publication).

Kost, L. E., Pollock, S. J., & Finkelstein, N. D. (2009). Characterizing the gender gap in introductory physics. *Physical Review Special Topics: Physics Education Research, 5,* 010101.

Lasry, N., Mazur, E., & Watkins, J. (2008). Peer instruction: From Harvard to the two-year college. *American Journal of Physics, 76,* 1066–1069.

Laws, P. W. (1991). Calculus-based physics without lectures. *Physics Today, 44,* 24.

Laws, P. W. (2004). *Activity-based tutorials: Introductory physics, the physics suite* (1st ed., p. 1). Upper Saddle River: Wiley.

Litzinger, T. A., Lattuca, L., Hadgraft, R., & Newstetter, W. (2011). Engineering education and the development of expertise. *Journal of Engineering Education, 100*(1), 123–150.

Malone, K. (2008). Correlations among knowledge structures, force concept inventory, and problem-solving behaviors. *Physical Review Special Topics: Physics Education Research, 3,* 020107.

Mazur, E. (1997). *Peer instruction, a user's manual.* Upper Saddle River: Prentice Hall.

McDermott, L., & Reddish, E. (1999). Resource letter: PER-1: Physics education research. *American Journal of Physics, 67,* 755.

McDermott, L. C., & Shaffer, P. S. (2002). *Tutorials in introductory physics* (1st ed.). Upper Saddle River: Prentice Hall.

Mohammed bin Rashid Al Maktoum Foundation. (2012). Arab Knowledge Report 2011–2012. http://www.mbrfoundation.ae/English/Pages/default.aspx. Accessed 23 June 2013.

National Academy of Sciences. (2010). *Exploring the intersection of science education and 21st century skills: A workshop summary.* Washington, DC: National Academies Press.

Newstetter, W. (2006). Fostering integrative problem solving in biomedical engineering: The PBL approach. *Annals of Biomedical Engineering, 34*(2), 217–225.

Newstetter, W., Behravesh, E., & Fasse, B. (2010). Design principles for problem-driven learning laboratories in biomedical engineering education. *Annals of Biomedical Engineering, 38*(10), 3257–3267.

Newstetter, W., Khalaf, K., & Xi, P. (2012). *Problem-driven learning on two continents: Lessons in pedagogic innovation across cultural divides.* Paper presented at the Frontiers in Engineering Education meeting, Seattle, Washington.

Osbeck, L., Nersessian, N. J., Malone, K., & Newstetter, W. (2010). *Science as psychology: Identity and sense-making in science practice.* New York: Cambridge University Press.

Perkins, K. K., Adams, W. K., Pollock, S. J., Finkelstein, N. D., & Wieman, C. E. (2004). *Correlating student beliefs with student learning using the Colorado learning attitudes about science survey.* Proceedings of the Physics Education Research Conference, American Institute of Physics Conference Proceedings, Vol. 790, pp. 61–64, Sacramento, California (USA), 4–5 August.

Prince, M. J. (2004). Does active learning work, a review of the research. *Journal of Engineering Education, 93*(3), 223–231.

Prince, M. J., & Felder, R. M. (2006). Inductive teaching and learning methods: Definitions, comparisons, and research bases. *Journal of Engineering Education, 99*(4), 1–16.

Prince, M., & Felder, R. (2007). The many faces of inductive teaching and learning. *Journal of College Science Teaching, 36*(5), 14–20.

Redish, E., Saul, J. M., & Steinberg, R. N. (1998). Student expectations in introductory physics. *American Journal of Physics, 66,* 212–224.

Reed-Rhodes, T., & Imbrie, P. K. (2008) *Concept inventories in engineering education.* Commissioned Papers of the National Academies, National Research Council, Washington D.C., October 2008, p. 8. (see section "Gaps in Research").

Rosenblatt, R., & Heckler, A. F. (2011). Systematic study of student understanding of the relationships between the directions of force, velocity, and acceleration in one dimension. *Physical Review Special Topics: Physics Education Research, 7,* 020112.

Saul, J. M., & Reddish, E. F. (1997). Final evaluation report for FIPSE grant #P116P50026: Evaluation of the Workshop Physics dissemination project. http://www.physics.ucf.edu/~saul/articles/WP-FIPSE_Rprt.pdf. Accessed 20 June 2013.

Sheppard, S., Macatangay, K., Colby, A., & Sullivan, W. M. (2009). *Educating engineers: Designing for the future of the field.* San Francisco: Jossey-Bass.

The Accreditation Board for Engineering and Technology (ABET). (1997). Engineering criteria 2000 (3rd ed.). http://www.abet.org/EAC/eac2000.htm.

Turpen, C., & Finkelstein, N. D. (2010). The construction of different classroom norms during peer instruction: Students perceive differences. *Physical Review Special Topics: Physics Education Research, 6,* 020123.

UNESCO's EFA Global Monitoring Report. (2012). Learning: Youth and skills. http://www.unesco.org/new/en/education/themes/leading-the-international-agenda.pdf.

UNESCO's Report. (1999). Learning: The treasure within. http://www.unesco.org/delors/delors_e.pdf.

Wieman, C., Perkins, K., & Gilbert, S. (March–April 2010). Transforming science education at large research universities. *Change Magazine.*

Shadi Balawi is currently the associate chair of Aerospace Engineering at the Khalifa University (KU) of Science, Technology, and Research in Abu Dhabi. He earned his PhD degree from the University of Cincinnati in 2007 in Aerospace Engineering, with research in the field of experimental design and testing of solids and structures with a focus on the aluminum honeycomb core materials. Prior to joining the University of Cincinnati, he received his bachelor and master degrees from Jordan University of Science and Technology, where his research was in the areas of experimental heat transfer. Dr. Balawi has taught courses in materials, design, and heat transfer at both the University of Cincinnati and University of South Carolina in Columbia, SC. He has also taught courses in engineering measurements, mechanical systems design, aircraft structures, and engineering problem-solving at the American University of Sharjah during the 2008–2009 academic year and since joining KU of Science, Technology, and Research in August of 2009. He is actively engaged in research of freshman design thinking education. Currently, he is coordinating the freshman Engineering Design course for the whole 300-student freshman class from all engineering departments at KU across two campuses.

Kinda Khalaf is currently the associate chair of Biomedical Engineering at KU. She received her BS (summa cum laude with distinction) and MS (honors) degrees in Mechanical Engineering from the Ohio State University; her PhD, also from Ohio State, is in Biomechanics/Computational Biomechanics. Dr. Khalaf has held faculty appointments in Engineering and Medicine at the University of Miami and American University of Beirut (AUB). She has numerous awards and distinctions including the Lamme Scholarship, the Woman in Engineering Award, the Presidential Fellowship, the prestigious National Merit Scholar, the College of Engineering Teaching Award, and the Faculty Excellence Award. She is also on the International Who's Who List of Professionals. Her research is in the areas of Biomechanics. She has co-established an interdisciplinary education research group at KU focusing on implementing curricular innovative pedagogies and the design of quantitative assessment instruments in engineering education. Her research record is reflected in various national and international research grants and awards, numerous journal publications, and multidisciplinary lectures and conference proceedings. Dr. Khalaf is currently the chair of the UAE branch of the Engineering in Medicine and Biology Society (IEEE, EMBS) and a founding member of the IEEE education chapter in the UAE (IEEE EDUCON) in addition to other professional organizations.

G. Wesley Hitt received his PhD in experimental nuclear physics in 2009 from the Michigan State University. He was a postdoctoral research associate of the National Superconducting Cyclotron Laboratory, East Lansing, MI, USA during that same year. He has authored or coauthored more than 40 peer-reviewed publications in experimental nuclear physics and nuclear astrophysics, which focus on using techniques of charged particle spectroscopy to determine important input parameters for modeling supernova explosions. In August 2009, he joined KU of Science, Technology, and Research where he is currently assistant professor of physics, jointly with the Department of Applied Mathematics and Sciences and the Department of Nuclear Engineering. Since joining KU, he has taken a strong interest in the effective learning of university engineering students taking introductory, calculus-based physics, and project-based engineering design courses. Currently, he and his collaborators are studying the effectiveness of different "interactive-engagement" pedagogies developed in North America and Europe and ways in which they can be modified to meet the needs as per the UAE context in innovative STEM education.

Chapter 5
Model for Transforming Engineering Education Using Technology-Enhanced Learning

Mohammed Samaka and Mohamed Ally

5.1 Introduction

According to the Royal Academy of Engineering in the UK, "No factor is more critical in underpinning the continuing health and vitality of any national economy than a strong supply of graduate engineers equipped with the understanding, attitudes and abilities necessary to apply their skills in business and other environments" (2007, p. 4). This statement also applies to all countries in the Middle East and North Africa (MENA) region. Graduating well-qualified engineers with the proper skills is important for the future development of the countries in the MENA region since most countries in this region have just started to build the infrastructure to become more developed in the twenty-first century. Educational organizations in the MENA region must create a sense of urgency to transform themselves to develop high-quality and flexible engineering programs to produce graduates with the appropriate skills to advance MENA countries in the twenty-first century. Currently, the Middle East recruits engineers from other countries to meet its increasing demands (ASME 2009); however, as the MENA region continues to develop, there will be an increasing need for engineers. This will place tremendous pressure on educational organizations in MENA to transform engineering programs to educate more high-quality engineers to meet the demand. An important question is "How to make the engineering education processes more effective, more quality conscious,

---

Submission Theme: Technology-based Innovations in Engineering Education. This theme has a global scope with reference to Middle East engineering education.

M. Samaka (✉)
Department of Computer Science & Engineering, Qatar University, Al Jameaa Street, 2713 Doha, Qatar
e-mail: samaka.m@qu.edu.qa

M. Ally
Centre for Distance Education, Athabasca University, University Drive, Athabasca, AB T9S 3A3, Canada
e-mail: mohameda@athabascau.ca

M. Abdulwahed et al. (eds.), *Advances in Engineering Education in the Middle East and North Africa*, DOI 10.1007/978-3-319-15323-0_5

more flexible, simpler, and less expensive?" (National Academy of Engineering 2005). Educational organizations will have to be efficient in the curriculum development process to react to the increasing demand for engineers and be flexible in their delivery methods to educate students using a variety of formats. At the same time, engineering programs have to make use of emerging learning technologies to enhance the learning process to reach students in different locations and to prepare engineers to work in a globalized twenty-first-century economy. Some countries in MENA have started to explore the use of emerging technologies in education and training. The state of Qatar has funded a project to research the use of mobile technology in training (Ally and Samaka 2013)[1]. The UAE has implemented the "tablet" initiative where students in colleges and universities are using tablets in their courses to enhance learning (Gitsaki et al. 2013). Allowing students to complete some of their education in their own locations will allow them to learn in their own contexts. Also, the use of innovative learning technologies will make use of the technology to customize the instructional strategies to cater for the different learning styles and the new generation of students, hence improving the completion rate in engineering programs (Denton 1998; Patterson 2011). A recent report by the Organization for Economic Co-operation and Development (Kärkkäinen and Vincent-Lancrin 2013) states that technology-enhanced learning (TEL) could be used for "gaming, virtual laboratories, international collaborative projects, real-time formative assessment, and skills-based assessment" (p. 9).

At a recent forum on engineering education in the Middle East, members concluded that engineering education in the twenty-first century is a key element of the wealth of any country and should be an integral component of national policy and economy (Shanableh and Omar 2007). Hence, engineering education must offer more than science and engineering knowledge but also entrepreneurship and technological advancement. There should be a paradigm shift in engineering education from knowledge-based to learned competencies, skills-based and outcome-based. A problem-based approach in engineering education should be used to allow students to find the relevant information to solve real-life problems. A recent study reported that students who used a problem-based approach to learning performed better than students who did not use a problem-based learning approach (Sahin 2010). The use of innovative learning technologies allows students to conduct research and interact with experts to solve problems. Because of the information explosion, students cannot learn all of the information in an engineering discipline just in case the information is needed later. There must be a shift from emphasizing acquisition of knowledge to developing competencies and skills. Information can be available in time using computers, tablets, and smartphones. Information learned and knowledge created should be based on real engineering applications. Engineering education programs need to provide students with skills to function in the twenty-first century (Silva et al. 2009). Some of these skills can be developed using innovative

[1] This project is funded by Qatar Foundation: Qatar National Research Fund - NPRP Grant # 4-125-5-016

learning technologies. These include information and communication, research, and continuous improvement skills.

A recent report by the Organization for Economic Co-operation and Development (Kärkkäinen and Vincent-Lancrin 2013) raised the following three important questions: "(1) How can technology-supported learning help to move beyond content delivery and truly enhance science, technology, engineering, and mathematics (STEM) education so that students develop a broad mix of skills? (2) Could innovative teaching and learning approaches spark thinking and creativity, enhance student engagement, strengthen communication, and build collaboration? (3) Would they make STEM teaching and learning more effective, more relevant, and more enjoyable?" (p. 9). This chapter will contribute to answering these questions by exploring the challenges facing engineering education and providing suggestions for improvement, the use of TEL in engineering education, and presenting a model to make engineering education more effective and relevant for the twenty-first century.

5.2 Preparing Engineers for the Twenty-First Century

In the twenty-first century, there is globalization, rapid development of emerging technologies in engineering, information explosion, and new generation of students. These trends require engineering education to examine how they train students to function in the twenty-first century. Engineering students will require new skills to be effective in the twenty-first-century workplace. According to Vest (2007), engineering students of the future must have knowledge in the physical, life, and information sciences and know how to *Conceive, Design, Implement, and Operate (CDIO)* complex engineering systems of appropriate complexity. They must also be able to function in a globalized world and contribute to society. This is echoed by Olson (2012) who suggests that engineers should function in a globalized society and there should be a shift from solving technology problems to solving societal problems. Ally (2012) conducted a study to identify the skills that are required by students to function in the twenty-first century. Major categories of skills identified include communication skills, personal skills, project management, continuous improvement, conflict resolution, problem solving, information and communication technology, teamwork, interpersonal skills, emotional/social intelligence, personal well-being, leadership, globalization, research, and critical thinking. Other reports (Silva et al. 2009; Shanableh and Omar 2007; The Royal Academy of Engineering 2007) mentioned similar skills required by engineering graduates in the twenty-first century. These include knowledge of engineering science fundamentals and design and manufacturing processes, knowledge of the context in which engineering is practiced, communication skills, critical thinking, creative thinking, lifelong learning, teamwork, ethical responsibilities, and ability to function in a multidisciplinary context. In addition, Agogino (2010) claimed that skills and characteristics that the 2020 engineers will need to function effectively include analytical and creative skills, communication skills, ability to interact with multiple stakeholders, leadership skills, high ethical standards, high professionalism, lifelong learning, and problem solving. The National Research Council (2012) in the USA identified similar twenty-first-century skills for life and work. These include creativity/innovation, critical thinking, problem solving, decision-making, communication, collaboration, information literacy, research and inquiry, media literacy, digital citizenship, information and communications technology operations and concepts, flexibility and adaptability, initiative and self-direction, productivity, and leadership and responsibility. Engineering education programs in the MENA regions need to examine if their training programs are providing the training for students to develop the twenty-first-century skills required to be competent and productive engineers. If not, then engineering programs must reengineer their programs to produce high-quality graduates.

5.3 Upcoming Generation of Students

The young students who are entering engineering programs and those who will enter in the future have certain characteristics and learning needs that must be catered for when designing engineering programs. Characteristics of new generation of students are described below along with strategies to educate these students so that they can be successful:

- The new generation of students has a high online Internet presence. They conduct everyday business and socialize online from anywhere and at any time. Most of the young generation's peak learning time is in the evenings and they expect education to be available online so that they can learn anywhere and at any time. All engineering courses must have an online presence so that the new generation of students can access course materials from anywhere and at any time.
- The new generation of students is referred to as the "now" generation, which wants information and feedback right away. If young students want to find information, they search for the information on the Internet instantly rather than going to the library or asking the instructor. Also, the new generations want feedback on their performance immediately using technology. This requires that the instructor provide timely feedback using information and communication technologies (ICTs), which require that instructors be proficient in using the ICTs.
- Young students are also referred to as the virtual generation. They are comfortable using social software to interact with each other. Engineering education should include strategies for virtual interaction between students and between instructor and student.
- Young students spend time playing computer games, which are highly interactive and are of high quality. They expect learning materials to be game-like with a high level of interaction. Engineering programs must include education games,

which are appropriate to motivate students to achieve high-level learning outcomes. Turkay and Adinolf (2012) surveyed students after they completed education games to determine how games contributed to learning. Results showed that students learn social skills such as teamwork and negotiation; motor and spatial skills such as getting better at reflexes, navigation, and eye–hand coordination; and cognitive skills such as problem solving, estimation, languages, and strategy development. The skills that are developed using educational games are important for engineering and science students in the twenty-first century. Hence, it is important to include educational games as a learning strategy in engineering education (National Research Council 2011).

- The new generation of students is digital experts. They do not need training on how to use the technology, which is second nature to them, and they adapt to technology very quickly. Instructors in engineering programs must be technology literate to educate the new generations who are ICT experts.

Catering for the new generation of students will motivate these students to complete their programs. Limited research has been conducted on how to attract and retain new generation of students in engineering education. Studies have been conducted on how to attract and retain the new generations in the workplace (Lim 2012) but not in engineering programs.

5.4 Problem-Based Learning for Outcome-Based Engineering Education

An effective strategy to help students make the transition from knowledge to the development of skills is to use problem-based learning, which has been used in engineering and medical education. Engineering programs in MENA must implement problem-based learning for outcome- and skill-based learning. Research and development is being conducted to make it easier and more user-friendly for educators to develop high-quality problem-based learning materials. One example is a research project that is being conducted in Qatar to develop a template-based system to allow educators to develop problem-based learning materials (Samaka et al. 2013)[2]. Research has shown that the use of problem-based learning strategies improved learning (Patterson et al. 2011). Sahin (2010) conducted a study to investigate students' performance in an engineering course that used problem-based learning compared to students who did not use problem-based learning. Results of the study showed that students who used the problem-based learning approach performed better than students in the control group. In a summary report of four engineering education programs in the Middle East, the authors mentioned two important themes that must be addressed to improve engineering education in the twenty-first

[2] This project is funded by Qatar Foundation: Qatar National Research Fund - Grant # NPRP 5–051–1–015

century (Shanableh and Omar 2007). Engineering programs must educate beyond science and engineering knowledge and educate students on entrepreneurship and technological advancement. Also, there should be a shift from knowledge-based to skills-based and outcome-based learning. The most efficient way to encourage skills-based and outcome-based learning is to use problem-based learning strategies.

Problem-based learning and real-life design activities should be introduced early in engineering programs so that students can appreciate the value of engineering and they will be motivated to complete their engineering programs. Rather than having students take theory courses in the first and second year of engineering, there should be some project-based courses in the first and second years to motivate students so that they can see the practical side of engineering and that engineers are important for society (National Academy of Engineering 2005). Allowing students to apply the theory in a practical situation right away will also result in high-level and meaningful learning. This is more efficient rather than having students recall the theory at a later time to apply. Students may not be able to apply the theory at a later time because of cognitive decay and inability to retrieve the theory.

5.5 Formative Feedback on Program Effectiveness

Engineering programs need to obtain ongoing feedback from students to continuously improve the programs. This could include formal feedback by asking students to comment on the course content and delivery or informal feedback where the instructor conducts a debriefing session at the end of the course. Nair et al. (2011) surveyed 1658 students to obtain feedback on their engineering programs. Some of the areas where students said that they need improvements include feedback on progress, more explanation of course content, more feedback on course work, courses that must cater for individual differences, and encouragement of more discussions in the classes. In addition, engineering-program effectiveness can also be determined by interviewing or surveying students after they graduate and start working in the engineering profession. As working professionals, the graduates can provide feedback on whether the engineering program prepared them for the workplace, especially the twenty-first-century skills to function effectively there. Most engineering programs conduct program reviews every 5 years to determine program effectiveness and currency. Some engineering programs may need to conduct program reviews more often because of constant changes in the field. For example, computer engineering is one program that should be reviewed frequently because of the rapid changes in technology. The program reviews usually include businesses and industries that employ engineering graduates, accreditation bodies, graduates of the program, regulatory bodies, etc. It is critical to include businesses and industries in program reviews since they employ the graduates of the program and they can specify the skills that are required for the workplace. According to Chapman and Miric (2009) , one of the drawbacks of engineering education in the MENA region is the lack of cooperation between industry and education. Some engineering programs also go through accreditation review by accreditation bodies to make sure that national and international standards are followed.

5.6 Learning in Context

To encourage outcome-based education and to allow for meaningful learning, it is important to teach engineering in context so that students can apply what they learn right away and can see the benefits of learning engineering (Patterson et al. 2011). Learning in context allows students to experience the real environment, and they can use ICT to communicate with the instructor whose role is that of a tutor or coach. One of the major issues in engineering education is the high dropout rate of students because of lack of motivation. Learning in context will motivate students to finish their programs since they are able to see what they learn is making a difference in the world and they are contributing to society. Students should be given real and contextual experience so that they can be motivated, and they can see how what they are studying is having a positive impact on society. Giving students real-life experience is new to engineering programs and there are some challenges, especially in MENA where there is limited interaction between education and industry. The National Academy of Engineering (2012) identified three major challenges in implementing contextual learning. These include: lack of funding, determining faculty workload, and building appropriate partnerships with industries. However, with careful planning, these challenges can be overcome and the benefits to students will outweigh the investments and effort.

5.7 Technology-Enhanced Learning in Engineering Education

To provide flexibility in learning, facilitate contextual and problem-based learning, and cater for the new generations of students, technology should be used to enhance learning. Connection to the Internet is important to implement learning technologies in engineering education. Internet connectivity is growing faster in the Middle East countries than in other regions of the world (Mirza and Al-Abdulkareem 2011). Educational organizations should take advantage of this growth to develop and deliver online courses in engineering programs. Given below are the benefits of using TEL in engineering education:

- The use of technology to deliver education at a distance removes barriers to learning, especially for students with disability who cannot travel and for female students with family responsibilities.
- Students who live in remote locations can use ICTs to access learning materials and instructor support.

- Students who are working and travel frequently can use ICTs to learn from anywhere and at any time.
- A major benefit for using technology to deliver engineering education is many students already have the technology and they know how to use it. Engineering programs should take advantage of this and deliver learning materials using ICTs.
- Intelligence can be built into the delivery system to detect the learning style of students and deliver the appropriate learning materials based on the student learning style.
- In TEL, learning is more learner-centered since students determine when and where they want to learn, and they can access additional materials to enrich their learning.
- TEL allows for learning at a distance, which is more affordable than campus-based education since learners do not have to travel to a physical location to access learning materials.
- For the instructor, the learning materials are easy to update since they are in electronic format. The instructor can change the learning materials, and students will be able to see the changes right away.
- TEL caters for the new generation of learners since they are technology literate and they are comfortable using the technology.
- Because of information explosion where information changes on a frequent basis, students can use ICTs to access current information.
- There are many initiatives around the world that are developing learning materials as open educational resources. Students can use ICTs to access open educational resources. Also, the trend is for textbooks, journals, and other information to be in electronic format. Students can use ICTs to access the electronic materials.
- Students can use ICTs to learn in context so that they can develop high-level skills.
- Collaborations are facilitated using ICTs. Students can collaborate to solve problems.
- Students can use ICTs to access experts in the field to discuss recent research and develop creative ideas.

Engineering education in the future should use a blended learning approach using TEL to facilitate student-centered education. Blended learning (Fig. 5.1) uses a combination of face-to-face instruction and self-directed, independent learning (Francis and Shannon 2013). Use of technology in learning allows students to be independent learners in a student-centered approach, resulting in higher level learning, which facilitates critical thinking (Kärkkäinen and Vincent-Lancrin 2013; Kashefi et al. 2012; Saadé et al. 2012). For example, a problem-based learning strategy can be used where students are given a problem to solve. Students then use the technology to research solutions on the Internet, interact with experts in the field to obtain current information, collaborate with other students virtually, etc. They can use ICTs to access learning materials so that they can learn at their own pace with

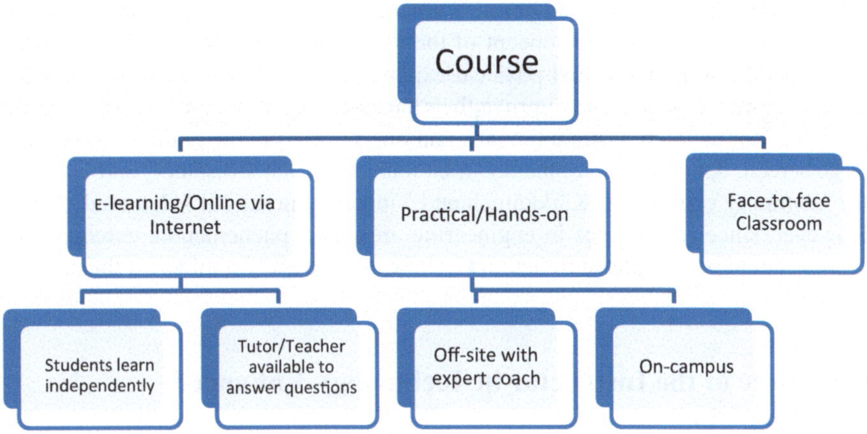

Fig. 5.1 Blended learning approach

access to a tutor for individual coaching when required. They can also use the technology to collaborate with peers and experts in the field, which will allow students to access high-level knowledge for their intellectual development (Norhayati et al. 2012).

TEL allows students to learn in context, which results in deep and meaningful learning. According to Boyle and Ravenscroft (2012), "deep learning design encourages the creative study of a learning problem or opportunity. It applies substantive insights from the learning disciplines to exploit the affordances of the technology in order to develop contexts that empower students to achieve educational goals" (p. 1225).

5.8 Use of Remote/Virtual Labs

The use of remote/virtual labs to teach engineering students is a new phenomenon that has great potential to allow engineering students to access remote equipment from anywhere and at any time using ICT (Domingues et al. 2010; Kennepohl 2010). For example, at Athabasca University in Canada, students studying astronomy can access and manipulate the telescope from anywhere in the world using ICTs. Students can explore the universe from anywhere without having a physical telescope. Another example is students can manipulate a microscope from anywhere and at any time to view microscopic specimens. Use of remote labs is a major benefit to educational organizations since students can practice using the real equipment at a distance to improve their skills. Domingues et al. (2010) conducted a study to determine the effectiveness of virtual labs in a chemical laboratory technology course. The intent of the virtual labs was to provide web support to the traditional lab experiment for data analysis and report preparation. A large majority of students

(95%) said that the virtual lab portal is easy to use, 97% said that the virtual lab helped them understand the concept of the experiment, and 94% said that virtual lab helped in doing the actual physical experiment. Other benefits of remote labs include: disabled people can improve their hands-on skills without having to travel, the instructor can demonstrate the labs and observations to the entire class using a large screen, dangerous equipment can be manipulated at a distance, and less cost for laboratory equipment (Kärkkäinen and Vincent-Lancrin 2013; Ku et al. 2011). However, since remote labs in engineering are a new phenomenon, extensive research should be completed to determine how to effectively implement them.

5.9 Role of the Instructor in Technology-Enhanced Learning

Countries in the MENA region need to use ICT to allow flexibility in learning and student-centered learning. However, for this to occur, instructors must make the shift from instructor-centered pedagogy to student-centered pedagogy (Wiseman and Anderson 2012). The role of the instructor will change from being a presenter of information to a facilitator of learning. Depending on the geographic distribution of students, the instructor can use synchronous or asynchronous communication tools to communicate and interact with students. The support using the technology could be synchronous where the instructor and students interact in real time or asynchronous where the interaction is done at different times. In synchronous learning, support is provided in real time using two-way text, two-way audio, or two-way video. The student and the instructor are able to interact with each other in real time. In the asynchronous mode, there is a delay in the communication between the instructor and students. For example, in computer conferencing, students post their comments and other students and the instructor respond at a later time. Hence, as instructors start using learning technologies, their roles will change drastically to function effectively in the new learning environment. The instructor role will shift from a dominant person in front of the classroom to being a facilitator of learning by managing the learning process, providing one-to-one coaching to students, and supporting and advising students. As a result, instructors will have to be trained in the use of learning technologies in engineering education. Specific skills required by instructors to function in TEL are described below.

Instructors must be trained in how to be good facilitators of learning. The instructor has to facilitate learning by role-modeling behavior and attitudes that promote learning, encourage dialogues, and use appropriate interpersonal skills. The instructor must be trained in how to recognize different learning styles and how to cater for different styles. An effective instructor must recognize that students have different styles when learning and some students prefer certain strategies to others. The instructor should use techniques that will satisfy and develop different learning styles and should include activities for the different styles to allow students to experience all of the learning activities. Also, appropriate learning support should be provided depending on the learning style of the student (Ally and Fahy 2002).

The instructor should be trained in the importance of feedback and how to provide effective and constructive feedback to students. Timely feedback is important in TEL for the instructor to maintain a presence in the learning process. The instructor should make students feel comfortable and should show enthusiasm about the course materials to keep students motivated. As a result, appropriate training on when and how to provide feedback to learning is critical. The instructor has to adapt to the student needs and provide timely feedback to students. Motivating students can be done by letting them know what they are learning is beneficial and then challenging students by suggesting additional learning activities.

The instructor must be a good problem solver to interpret students' problems and provide solutions. This implies that the instructor must have the content expertise to solve content problems. The instructor solves content problems by being up to date in the field, interpreting students' questions, communicating at the level of the student, providing remedial activities, and conducting follow-ups on the help provided. Interaction with students requires good oral and written communication skills. Also, the instructor is required to develop and revise courses on an ongoing basis. As part of the problem-solving process, the instructor needs good listening skills to understand what the student is saying in order to respond to the student. Most important, the instructor must be trained in how to use the learning technologies effectively to promote learning. This is critical since the instructor must model proper use of the technology.

5.10 Barriers to Technology-Enhanced Learning and How to Overcome Them

In the MENA region, TEL strategies such as e-learning, mobile learning, online learning, and blended learning are somewhat new. As a result, there are many barriers that must be overcome to make the implementation of TEL effective in MENA. Below are some critical barriers and suggestions on how to overcome the barriers.

- A major barrier is getting instructors involved in TEL. Instructors are not motivated because they have been involved in campus-based education for many years and they have a difficult time adjusting to TEL. Also, instructors do not think that they have the expertise to implement TEL. Instructors work on other priorities rather than get involved in TEL. To overcome the motivation barrier, instructors must be trained in how to implement TEL, success stories should be shared with them, they should be provided with funding to develop learning materials for TEL, and awards should be given for getting involved in TEL.
- The campus-based education system was developed for face-to-face delivery, which is not appropriate for TEL. As a result, it is difficult to implement TEL

in the campus-based system. The solution is to reengineer the system for both campus-based learning and TEL so that blended learning can be implemented.
- The organization operations do not meet the needs of TEL. There is competition for resources between departments that implement TEL and other departments. This is a result of the budget, which was developed for face-to-face delivery rather than TEL. Also, the TEL instructor workload is based on face-to-face instructor workload. The organization should develop implementation plans for both TEL and face-to-face delivery. Also, enough resources should be provided for TEL.

5.11 Learning Theories to Guide the Development of Engineering Programs

There are different learning theories that should be followed when designing engineering education. This section will discuss learning theories to design effective engineering education programs.

An important predictor of success in engineering education is students' self-efficacy, which is students' belief that they can be successful as an engineer (Patterson et al. 2011). Instructors must build and enhance students' self-efficacy so that they can be successful in their programs. A good model to follow to improve students' motivation and self-efficacy is the motivational Attention, Relevance, Confidence, and Satisfaction (ARCS) model proposed by Keller (1983). Attention: Get students' attention at the start of the course by presenting real-life examples or ask some application questions. Activities should also be used throughout the course so that motivation is maintained while students learn. Relevance: Inform students of the importance of the course and how taking the course could benefit them. Strategies could include describing how students will benefit from taking the course, and how they can use what they learn in real-life situations. This strategy helps to contextualize the learning and make it more meaningful, thereby maintaining students' interest throughout the learning session. Confidence: Use strategies such as designing for success and sequencing from simple to complex or from known to unknown, and use a competency-based approach where students are given the opportunity to use different strategies to complete the course. Inform students of the course outcome and provide ongoing encouragement to complete the course. Satisfaction: Provide feedback on students' performance and allow them to apply what they learn in real-life situations. Students like to know how they are doing, and they like to contextualize what they are learning by applying the information in real life.

Behaviourist theory suggests looking for observable behavior in students and places emphasis on feedback in learning. Some guidelines for designing learning materials based on behaviourist theory include: (1) Inform students of the learning outcomes so that they can set expectations and can judge for themselves whether or not they have achieved the outcome of the course. (2) Evaluate students' performance to determine whether or not they have achieved the learning outcome.

(3) Sequence the training materials appropriately to promote learning. (4) Provide timely feedback to students.

Another psychological theory of learning that one should follow is the cognitivist school of learning. According to the cognitivist, instructors must use learning strategies that allow students to attend to the learning materials so that the information can be transferred from the senses to the sensory store and then to working memory and eventually to long-term memory. The amount of information transferred to working memory depends on the amount of attention that was paid to the incoming information and whether cognitive structures are in place to make sense of the information. Instructors must check to see if the appropriate existing cognitive structure is present to enable the student to process the information. Also, because of the limited capacity of working memory, information must be chunked in pieces of appropriate size to facilitate processing. Information maps that show the major concepts in a topic and the relationships between those concepts should be included in the courses.

Guidelines for designing learning materials based on the cognitivist model of learning include: (1) Use effective strategies that allow students to perceive and attend to the information so that it can be transferred to working memory. (2) Information that is critical for learning should be highlighted to focus students' attention. (3) The difficulty level of the material must match the cognitive level of the student, so that the student can both attend to and relate to the material. This will prevent information overload during the learning process. (4) Strategies that allow students to retrieve existing information from long-term memory to help make sense of the new information must be used. (5) Provide conceptual models that students can use to retrieve existing mental models or to store the structure they will need to use to learn the details of the course. (6) Information should be chunked to prevent overload during processing in working memory. (7) Strategies that promote deep processing should be used to help transfer information to long-term storage. Strategies that require students to apply, analyze, synthesize, and evaluate promote higher level learning, which makes the transfer to long-term memory more effective. (8) Strategies to allow students to apply the information on the job should also be included, to contextualize the learning and to facilitate deep processing. (9) Information should be presented in different modes to facilitate processing by different learning styles and transfer to long-term memory. Whenever possible, textual, verbal, and visual information should be presented to encourage encoding and to cater for different learning styles. (10) Learning strategies that facilitate the transfer of learning should be used to encourage application in practical situations. Simulation of the real situation, using real-life cases, should be part of the course. Also, students should be given the opportunity to complete projects that use real-life applications and information. Transfer to job situations could assist students to develop personal meaning and contextualize the information.

A more recent theory of learning is constructivist, which suggests that students should be active in the learning process since knowledge is not received from the outside or from someone else; rather, it is the individual student's interpretation and processing of what is received through the senses that creates knowledge. The student is the center of the learning, with the instructor playing an advising and facilitating role. A major emphasis of constructivists is situated learning, which sees learning as contextual. Learning must be done in context to allow students to contextualize the materials during the learning process. If the information has to be applied in many contexts, then learning strategies that promote multi-contextual learning should be used to make sure that students can indeed apply the information broadly. According to Tapscott (1998), learning is moving away from one-way instruction to construction and discovery of knowledge. Guidelines for TEL based on the constructivists' school of learning include: (1) Learning strategies should keep the students active and they should apply what they learn in practical situations to facilitate personal interpretation and relevance. (2) Students should be actively constructing their own knowledge rather than accepting what is given by the instructor. (3) For maximum benefit, learning should be made meaningful for students by relating to what they already know or to something personal.

5.12 Model for Developing Engineering Programs

This section presents a model for developing TEL for engineering programs so that students can be trained for the twenty-first-century workforce. The delivery method in engineering programs is primarily lectures, especially in the MENA region. There is the belief that for students to learn, someone must stand up in front of a large class and do one-way lectures. Based on learning theories, this is the least effective way to educate students. The students must be active in the learning process so that high-level learning can be promoted. When developing engineering programs, learning theories must be followed to facilitate high-level learning and to promote transfer of training to the real world. Also, the learning style of students should be catered for and learning materials must be developed for all of the learning styles. According to Kolb (1984), there are four learning style types. (1) Divergers are individuals who have good people skills; when working in groups, they try to cultivate harmony in the group, to assure that everyone works together smoothly. (2) Assimilators like to work with details and are reflective and relatively passive during the learning process. (3) Convergers prefer to experiment with and apply new knowledge and skills, often by trial and error. (4) Accommodators are risk-takers, who want to apply immediately what they learn to real-life problems or situations. For example, if an engineering lesson is about learning how to use engineering software, activities for the four different learning styles can be as follows. For divergers, the lesson should include activities for them to work in teams and inform them why they should learn to use the software. The assimilators would like step-by-step instructions on how to use the software. The convergers like to experiment and learn on their own; so they should be allowed to learn the software by themselves. The accommodators are creative and risk-takers, and they would like to use the software creatively. Hence, accommodators should be allowed to use the software in real-life applications.

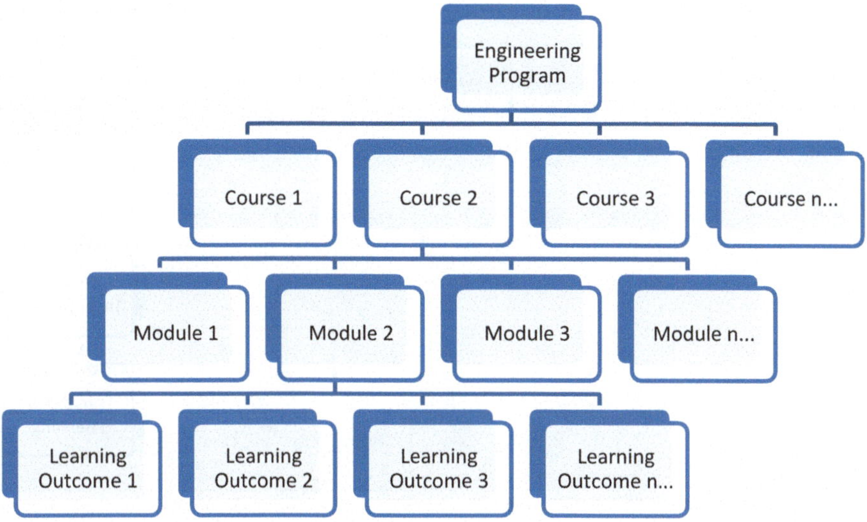

Fig. 5.2 Structure of engineering program

5.12.1 Content Identification

The first step when developing engineering programs is to identify the content, which should be broken down into manageable chunks for effective delivery. One method to break the content into smaller chunks is shown in Fig. 5.2. Engineering programs are broken down into courses, which are broken down into modules. A course should have 5–7 modules. The learning outcomes are then identified for each module. A module should have 5–7 learning outcomes.

5.12.2 Components of the Model

The four major components of the model (Fig. 5.3) are: (1) preparing students for the learning, (2) prescribing learning activities, (3) allow for student support, and (4) design for transfer to the real situation. The components are developed for each module.

1. Prepare Students for Learning

A variety of activities can prepare students for the details of the module, and to connect and motivate them to learn the module. A rationale should be provided to inform students of the importance of taking the module and to show how it will benefit them. For example, in a safety course, students can be shown a video of what will happen if safety procedures are not followed. A concept map is provided to establish the existing cognitive structure, to incorporate the details of the module,

Fig. 5.3 Components of the design model

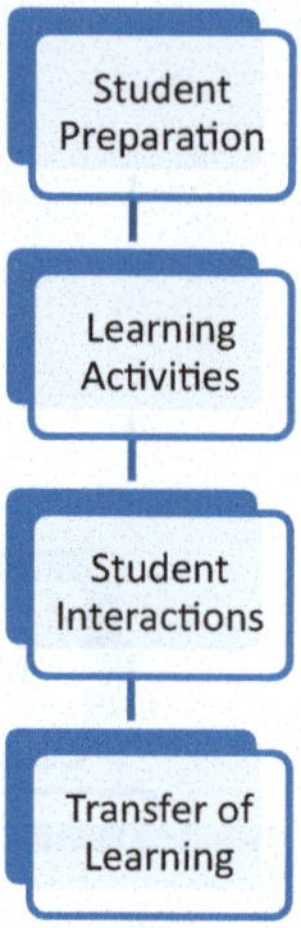

and to activate students' existing structures to help them learn the details in the module. The module concept map also gives students the big picture. Students should be informed of the learning outcomes of the module, so they know what is expected of them and will be able to gauge when they have achieved the module outcomes. An advance organizer should be provided to establish a structure, to organize the details in the module, or to bridge the gap between what students already know and what they need to know. Students must be told the prerequisite requirements so that they can check whether they are ready for the module. Providing the prerequisites to students also activates the required cognitive structure to help them learn the materials. A self-assessment should be provided at the start of the module to allow students to check whether they already have the knowledge and skills taught in the module. If students think they have the knowledge and skills, they should be allowed to take that module's final test. The self-assessment also helps students to organize the module materials and to recognize the important materials in the module. Once students are prepared for the details of the module, they can go on to complete the learning activities and to learn the details of the module (Fig. 5.4).

2. Design Learning Activities

The TEL course should include a variety of learning activities to help students achieve the course learning outcome and to cater for their individual needs (Fig. 5.5). Examples of learning activities include reading textual materials, listening to audio materials, viewing visuals or video materials, games, simulation, virtual experiments, hands-on experiments and activities, etc. Students can conduct research on the Internet or link to online information and libraries to acquire further information. For example, in problem-based learning, students can use the Internet to conduct research to find solutions to solve the problem. Having students prepare a learning journal will allow them to reflect on what they have learned and provide personal meaning to the information. Appropriate application exercises should be embedded throughout the course to establish the relevance of the materials. Practice

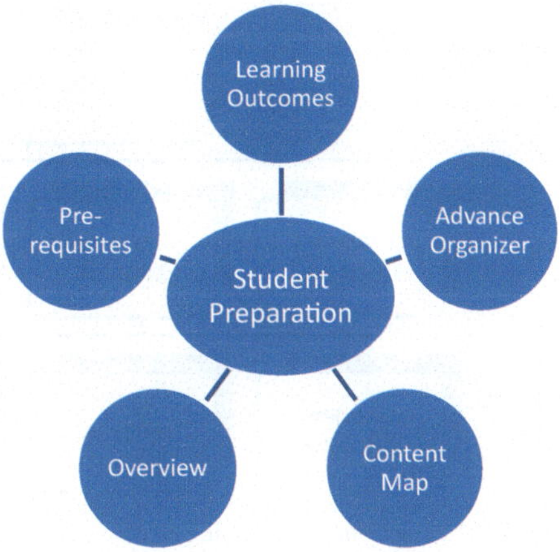

Fig. 5.4 Preparing students for learning

activities, with feedback, should be included to allow students to monitor how they are performing, so that they can adjust their learning method if necessary. To promote higher level processing and to bring closure to the course, a summary should be provided or students should be required to generate a course summary.

3. Types of Interaction in TEL

When engineering programs are designed for delivery using technologies, there are different types of interactions that must be developed (Fig. 5.6). In learner–instructor interaction, the learner must be able to interact with the instructor face-to-face or using communication technologies which will result in high social presence. In

Fig. 5.5 Learning activities to promote learning

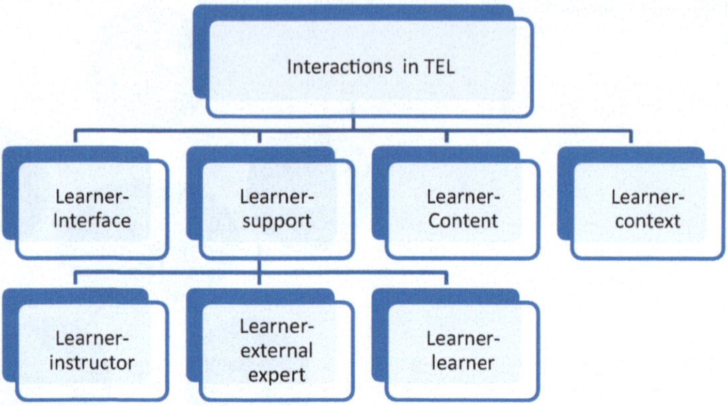

Fig. 5.6 Interactions in technology-enhanced learning *(TEL)*

learner–content interaction, the learner interacts with the content to achieve high-level knowledge and skills. Hence, the content must be interactive for high-level learning. This will increase cognitive presence. In addition, there should be opportunities for learner-to-learner interaction where students can help each other and share ideas. This is very important in problem-based learning and will enhance social presence. The appropriate interface must be developed so that students can interact with the technology to learn. The content should be developed as learning activities where students interact with the content to achieve high-level outcomes. If students are learning in their own context away from the campus, the appropriate interaction with the context should be developed to personalize information and construct their own meaning. Finally, there must be appropriate student support where the instructor or outside experts are available to provide help and feedback to students, there is shared cognition, formation of social networks, and the establishment of social presence. Also, the instructor should arrange for student-to-student interaction where students help each other.

5.13 Design for Transfer of Learning

Opportunities should be provided for students to transfer what they have learned to real-life applications, so that they can be creative and go beyond what was presented in the course (Fig. 5.7).

5.14 Conclusions

With the proliferation of computers and mobile technologies, the globalized society, new generation of students, and information explosion, educational organizations need to reinvent the way they design and deliver engineering education. In a global-

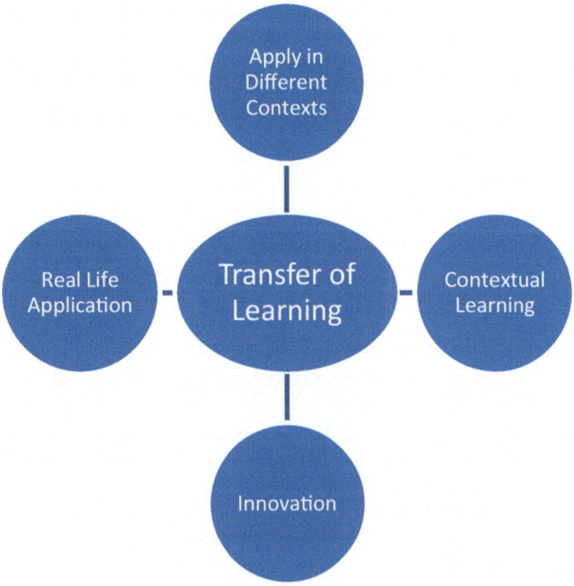

Fig. 5.7 Transfer of learning

ized society, engineering education should train students to work locally but think globally, because how they function locally determines what is happening globally. Engineers have to be aware of advancement in their fields so that they can be effective on the job. Also, to remain competitive and to contribute to their organizations, engineers must be entrepreneurial (Sunthonkanokpong 2011). At the same time, distance education and online learning can be utilized to allow engineering graduates to learn on a continuous basis so that they can be up to date in their fields (National Academy of Engineering 2005).

The output of a program (graduates) will be of high quality if the input (entering students) and processes (quality of the courses and learning experience) are of high quality. Countries in MENA need to examine their K-12 system to make sure that the students graduating from K-12 are ready for engineering education. The education system in K-12 emphasizes rote memorization to pass exams rather than teach students to think critically and to be active in the learning process (Thompson 2008). The K-12 system must be changed to prepare students for engineering programs. Also, the engineering programs should be transformed to prepare high-quality graduates for the twenty-first century.

Research on TEL must continue to better understand how to utilize the electronic multimedia approaches to teaching and learning and how to successfully implement TEL in engineering education (National Academy of Engineering 2005). TEL in some engineering schools is new and strategies for successful implementation must be followed. Steps for successfully implemented TEL are: (1) Create a vision for TEL. (2) Establish a sense of urgency for implementing TEL. (3) Develop the infrastructure for TEL. (4) Implement successful TEL projects. (5) Communicate the successes to the organization.

Acknowledgment This publication was made possible by NPRP Grant # 4–125–5–016 from the Qatar National Research Fund (a member of Qatar Foundation). The statements made herein are solely the responsibility of the authors.

References

Agogino, A. M. (2010). The engineer of 2020: Global visions of engineering practice and education. http://best.berkeley.edu/~aagogino/papers/NSB2005.pdf.

Ally, M. (2012). Competency profile of the 21st century learner. Unpublished competency profile chart.

Ally, M., & Fahy, P. (2002). *Using students' learning styles to provide support in distance education.* Proceedings of the Eighteenth Annual Conference on Distance Teaching and Learning, Madison, Wisconsin.

Ally, M., & Samaka, M. (2013). *Improving communication in the workplace using mobile learning.* Presentation at the UNESCO Second Mobile Learning Week Symposium, Paris.

ASME. (2009). Middle East emerges as an engineering hub. ASME. Strategic Issue. https://www.engineeringforchange.org/.

Boyle, T., & Ravenscroft, A. (2012). Context and deep learning design. *Computers & Education, 59*, 607–618.

Chapman, D. W., & Miric, S. L. (2009). Education quality in the Middle East. *International Review of Education, 55*, 311–344.

Denton, D. D. (1998). Engineering education for the 21st century: Challenges and opportunities. *Journal of Engineering Education, 87*(1), 19–22.

Domingues, D., Rocha, I., Dourado, F., Alves, M., & Ferreira, E. C. (2010). Virtual laboratories in (bio) chemical engineering education. *Education for Chemical Engineers, 5*, 22–27.

Francis, R., & Shannon, S. J. (2013). Engaging with blended learning to improve students' learning outcomes. *European Journal of Engineering Education.* doi:10.1080/03043797.2013.766679.

Gitsaki, C., Robby, M., Priest, T., & Ben-Chabane, Y. (2013). *Researching the UAE iPad initiative.* Plenary presented at the Mobile Learning: Gulf Perspectives Symposium, Abu Dhabi.

Kärkkäinen, K., & Vincent-Lancrin, S. (2013). *Sparking innovation in STEM education with technology and collaboration: A case study of the HP catalyst initiative.* OECD Education Working Papers, No. 91, OECD Publishing. http://dx.doi.org/10.1787/5k480sj9k442-en.

Kashefi, H., Ismail, Z., & Yusof, Y. M. (2012). Supporting engineering students' thinking and creative problem solving through blended learning. *Social and Behavioral Sciences, 56*, 117–125.

Keller, J. M. (1983). Motivational design of instruction. In C. M. Reigeluth (Ed.), *Instructional design theories and instruction: An overview of their current status* (pp. 383–429). Hillsdale: Lawrence Erlbaum.

Kennepohl, D. (2010). Remote control teaching laboratories and practicals. In D. Kennepohl & L. Shaw (Eds.), *Accessible elements: Teaching science online and at a distance* (pp. 167–190). Athabasca: Athabasca University Press.

Kolb, D. A. (1984). *Experiential learning: Experience as the source of learning and development.* Englewood Cliffs: Prentice-Hall.

Ku, H., Ahfock, T., & Yusaf, T. (2011). Remote access laboratories in Australia and Europe. *European Journal of Engineering Education, 36*(3), 253–268.

Lim, H. L. (2012). Attracting and retaining generation Y engineering and business professionals in the Middle-East. *Social and Behavioral Sciences, 62*, 25–29.

Mirza, A. A., & Al-Abdulkareem, M. (2011). Models of e-learning adopted in the Middle East. *Applied Computing and Informatics, 9*, 83–93.

Nair, C. S., Patil, A., & Mertova, P. (2011). Enhancing the quality of engineering education by utilizing student feedback. *European Journal of Engineering Education, 36*(1), 3–12.

National Academy of Engineering. (2005). *Educating the engineer of 2020: Adapting engineering education to the new century*. Washington, D.C: National Academy of Sciences Press.
National Academy of Engineering. (2012). *Infusing real world experiences into engineering education*. Washington, DC: The National Academies Press.
National Research Council. (2011). Learning science through computer games and simulations. Committee on science learning: Computer games, simulations, and education, M. A. Honey & M. L. Hilton (Eds.), *Board on Science Education, Division of Behavioral and Social Sciences and Education*. Washington, DC: The National Academies Press.
National Research Council. (2012). Education for life and work: Developing transferable knowledge and skills in the 21st century. Committee on defining deeper learning and 21st century skills, J. W. Pellegrino & M. L. Hilton (Eds.), *Board on Testing and Assessment and Board on Science Education, Division of Behavioral and Social Sciences and Education*. Washington, DC: The National Academies Press.
Norhayati, A., Yusof, K. M., Kosnin, A. M., & Su, D. M. K. (2012). Comparison between two models of engineering students' intellectual development. *Social and Behavioral Sciences, 56*, 617–626.
Olson, S. (2012). *Educating engineers: Preparing 21st century leaders in the context of new modes of learning*. Washington, DC: National Academy of Sciences.
Patterson, E. A., Campbell, P. B., Busch-Vishniac, I., & Guillaume, D. W. (2011). The effect of context on student engagement in engineering. *European Journal of Engineering Education, 36*(3), 211–224.
Saadé, R. G., Morin, D., & Thomas, J. D. E. (2012). Critical thinking in e-learning environments. *Computers in Human Behavior, 28*, 1608–1617.
Sahin, M. (2010). The impact of problem-based learning on engineering students' beliefs about physics and conceptual understanding of energy and momentum. *European Journal of Engineering Education, 35*(5), 519–537.
Samaka, M., Yongwu, M., Impagliazzo, J., & Ali, Z. (2013). *Implementing problem-based learning using PLATE*. Workshop presented at Qatar University.
Shanableh, A., & Omar, M. (2007). Shaping engineering education in the Middle East—a report on four engineering education conferences held at the University of Sharjah. *World Transactions on Engineering and Technology Education, 6*(2), 313–318.
Silva, A., Henriques, E., & Carvalho, A. (2009). Creativity enhancement in a product development course through entrepreneurship learning and intellectual property awareness. *European Journal of Engineering Education, 34*(1), 63–75.
Sunthonkanokpong, W. (2011). Future global visions of engineering education. *Procedia Engineering, 8*, 160–164.
Tapscott, D. (1998). *Growing up digital: The rise of the net generation*. New York: McGraw-Hill.
The Royal Academy of Engineering. (2007). *Educating engineers for the 21st century*. Report published by the Royal Academy of Engineering, June 2007.
Thompson, R. (2008). Developing engineering education in the Middle East using the North American model—what assumptions are valid? *American Society for Engineering Education*, p. 1–8. https://peer.asee.org/developing-engineering-education-in-the-middle-east-using-the-north-american-model-what-assumptions-are-valid.
Turkay, S., & Adinolf, S. (2012). What do players (think they) learn in games? *Social and Behavioral Sciences, 46*, 3345–3349.
Vest, C. M. (2007). Educating engineers for 2020 and beyond. In E. F. Crawley, J. Malmqvist, S. Östlund, & D. Brodeur (Eds.), *Rethinking engineering education: The CDIO approach*. New York: Springer Science+Business Media, LLC.
Wiseman, A. W., & Anderson, E. (2012). ICT-integrated education and national innovation systems in the Gulf Cooperation Council (GCC) countries. *Computers & Education, 59*, 607–618.

Dr. Mohammed Samaka PhD is an associate professor of computer science in the Department of Computer Science and Engineering (CSE), College of Engineering at Qatar University. He obtained his PhD and master's from Loughborough University in England and a postgraduate diploma in computing from Dundee University in Scotland. He obtained his bachelor of mathematics from Baghdad University. His current areas of research include wireless software architecture and technology, mobile applications and services, networking, e-learning, and computing curricula development.

Dr. Mohamed Ally PhD is a professor at Athabasca University in Canada. He obtained his PhD from the University of Alberta, Canada. His current areas of research include mobile learning, e-learning, distance education, and use of information and communication technology in training and education. Dr. Ally is a former president of the International Federation of Training and Development Organizations (IFTDO) and is one of the founding directors of the International Association of Mobile Learning (IamLearn). He was also on the board of the Canadian Society for Training and Development. Dr. Ally chaired the fifth World Conference on Mobile Learning and co-chaired the first International Conference on Mobile Libraries. He recently edited four books on the use of mobile technology in education, training, and libraries. His book "Mobile Learning: Transforming the Delivery of Education and Training" won the Charles A. Wedemeyer Award in the USA for significant contribution to distance education. He is currently editing three books in the area of emerging technologies in education. Two of his research papers won the best research paper award at national and international conferences. Dr. Ally has published in peer-reviewed journals, chapters in books and encyclopedia, and served on many journal boards and conference committees. He has presented keynote speeches, workshops, papers, and seminars in many countries.

Chapter 6
Using Forums and Simulation Exercises to Enhance Active Learning in Lean Construction Education

Farook Hamzeh, Christina Teokari and Carel Rouhana

6.1 Background and Purpose

6.1.1 Introduction

The construction industry is growing fast in the Middle East as reflected by the amount of planned construction work estimated to have a budget of \$2.4 trillion (GCC 2010). For example, the contribution of construction to the overall gross domestic product (GDP) in Qatar is \$7.7 billion after a growth of 5.1 % (Delloite 2013). In Lebanon, real estate investments and the construction sector constituted 21 % of the GDP in 2012 (Project Lebanon 2013). This emphasizes the importance of improving construction and implementing new systems to enhance the overall performance of construction projects. The improvement of the construction industry can start by improving the engineering education and teaching methods since the graduating students are the future leaders of construction projects in the Middle East (Fig. 6.1).

"Lean" is a production management philosophy and principle developed by Toyota aimed at adding value, reducing waste, and continuous improvement (Liker 2004). The introduction of lean principles to the construction industry requires a

---

Submission theme
Technology-based Innovations in Engineering Education: c- Adaptive and Intelligent Learning Systems and g- Classroom Technologies.

F. Hamzeh (✉) · C. Teokari · C. Rouhana
Department of Civil and Environmental Engineering, American University of Beirut, 406E Bechtel, Riad El Solh, Beirut 1107 2020, Lebanon
e-mail: Farook.Hamzeh@aub.edu.lb

C. Teokari
e-mail: christina.theokaris@gmail.com

C. Rouhana
e-mail: carel.rouhana@gmail.com

M. Abdulwahed et al. (eds.), *Advances in Engineering Education in the Middle East and North Africa,* DOI 10.1007/978-3-319-15323-0_6

Fig. 6.1 One of many new construction projects in the Middle East and North Africa (MENA) region. (IQPC 2011)

cultural change towards collaboration and continuous improvement since lean is more of a philosophy and not just techniques. Building the leaders, teams, and individuals in learning and applying lean construction are necessary to realize a growth in lean applications and benefits (Drysdale 2013). Results of a study on different cultural clusters show that the Middle Eastern cluster had low scores for the team-oriented and participative attributes (Javidan et al. 2006). Thus, the inclusion of lean construction in the civil engineering curriculum is aimed at improving teamwork, sharing, and participation.

Questioning the effectiveness of the traditional teaching methods leads to addressing the elephant in the room, "What is wrong with a 50-minute lecture?" (Eison 2010). The "traditional" teaching method is defined as "an essentially expository form of teaching, dominated by the teacher; it relegates pupils to a passive role, reduces their classroom activity to the memorization of data to be recited to the teacher, and in particular, leads to the acquisition of skills of a lower taxonomic level" (Gauthier and Dembélé 2004). Teachers who often rely solely on their presentations to give information tend not to rely on learners' life skills, formal and informal knowledge, and interests for meaningful information sharing (Heimlich and Norland 2002). Students of the new generation are growing up playing video games that enhance their visual attention and perception; it teaches them collaboration and multitasking. The educational methods needed should take advantage of the strengths of this generation while compensating for their weaknesses (Taekman and Shelley 2010). Education should be molded to reach students in a way similar to their generation's habits and preferable learning methods.

It can be argued that traditional teaching limited to lecturing may prevent the cognitive evolution of students. It reduces them to passive agents receiving the information from the notes of the lecturer and transferring it to their own notes without passing through the minds of either. The nature of the lecturing technique makes it too easy for students to divert their attention away from the lecturer just 10–15 min into the class time (Eison 2010). The traditional teaching method places the responsibility for learning squarely on students. As for teachers, they are considered as sources of knowledge instead of evolving into a network of facilitators (Taekman and Shelley 2010). The success of teachers is not defined by what students know but by what they do (Cooke et al. 2006). Therefore, it can be deduced that all genuine learning is active, not passive: it involves the use of the mind, not just the memory (Adler 1987).

A curious peculiarity of our memory is that things are impressed better by active than by passive repetition (Roediger and Butler 2011). The importance of class discussion and simulation exercises can thus be considered one of the integral methods to achieve an effective learning process. But a question remains unanswered: How effective are discussion forums and simulation exercises in improving student learning and understanding of lean construction concepts? This study aims at studying these two methods and understanding their impact on student learning.

6.1.2 Discussion Forums

6.1.2.1 What are Discussions forums

The discussion method is defined as "a strategy for achieving instructional objectives that involves a group of persons, usually in the roles of moderator and participant, who communicate with each other using speaking, nonverbal and listening processes" (Gall and Gillet 1980). For the discussion to be fruitful, it is important that the teacher provides the students with specific objectives and procedural guidelines for the session. This does not aim to limit students but rather to guide the discussion forward towards a desired direction with a reduced necessity to intervene and interrupt it (Folkestad et al. 2009). It is crucial that the teacher provides both a model for thinking and a structure for the class discussion. Students can then be asked to recall and discuss some lessons learned in class or through homework; they get the chance to think for themselves and internalize this new way of viewing the world (Ozer 2005).

6.1.2.2 Why Use Discussion Forums

People learn by using what they know to construct new understanding based on previous experience. In the process of a class discussion, the teacher can recognize if the students may have the knowledge to a learning situation that is not activated and activate it by building on students' strengths. Moreover, teachers can identify early on if the students have misinterpreted new information because of previous knowledge they used to construct the new understandings (National Academy of Science 2004). Through discussion, teachers can have access to their students' thought processes and can, therefore, guide them to a higher level of thinking. Once the concepts are rooted in the students' minds, teachers can then aim to extend students' trails of thought to stir up controversy and challenge critical intellect (Ozer 2005).

6.1.2.3 Learning Outcomes

In a study conducted in a large, upper-division biology lecture course, discussions were observed to yield significantly higher learning gains and better conceptual understanding (Chudler 2005). The key concept in discussion is interaction that makes it a two-way communicative process (Pendleton 1981). It requires students and the teacher to talk back and forth at a high cognitive and affective level, both with one another and the subject matter being discussed (Larson and Keiper 2002). To establish a continuous discussion, however, the faculty must create a supportive intellectual and emotional environment that encourages students to take risks (Bonwell and Eison 1991).

6.1.2.4 Challenges

Class activities such as discussion can be counterproductive if not implemented in such a way as to give adequate support for student autonomy in learning and project a mastery orientation to learning. Thus, the way the teacher implements such activities for students to interpret is of paramount importance since this method may either reinforce or undermine students' positive motivational beliefs (Hanrahan 1998). Some teachers are frequently frustrated by their futile efforts to engage the whole class in a good and healthy discussion. To mitigate this, Young (2007) suggests creating student-friendly, teacher-friendly activities that result in prepared students, universal participation, and creative responses to the material through small group discussions. Small groups aim to provide student-led discussions where there is opportunity for legitimate disagreement or interpretation (Young 2007). Finally, an inclusive and democratic class discussion can inspire critical thinking, shared learning, and a deep appreciation and understanding of the course subject matter (Kenny 2008).

6.1.3 Simulation Exercises

6.1.3.1 Simulation Exercises: A Pedagogical Tool

Simulation exercises create a virtual world for users and require them to make active decisions and come up with the best result in a certain situation to better understand and prepare for what happens in real life (Zoroja 2010).

Researchers describe the stages that students pass through when learning. In traditional teaching methods, students receive the information, understand it, link it to a specific situation, and then learn how to use it in a general case. Alternatively, through simulation exercises, students start by discovering the particular application, understand the effect of a certain decision in a particular application, link it to its general principle, and then apply what was learned to a different case (Coleman 1978). However, it should be clear that simulation exercises are there to support lectures, not to replace them. A study done covering seven universities teaching lean construction courses showed that six out of the seven instructors are employing simulation exercises in class as a teaching method to support their lectures (Tsao et al. 2013). These exercises benefit students with their practical side where the "doing" part of this teaching method makes it harder for students to forget the lesson (Gilgeous and D'Cruz 1996). "I hear and I forget, I see and I remember, I do and I understand" (Chinese saying).

6.1.3.2 Why Use Simulation Exercises?

The challenging part of education is making the learner see something in the world in a different and new perspective (Kuhn 1970). The use of exercises is not as expensive as training on the actual job; it helps in experimenting with a variety of strategies by learning from mistakes without the fear of actual losses. The active learning that exercises provide promotes more learning since participants can learn from their mistakes throughout the process. Exercises are close to reality since they require users to consider many variables when making decisions; this teaches them how to behave in real-life situations and gives them a sense of quantities (Gilgeous and D'Cruz 1996). Introducing lean construction to an organization willing to implement it is easier through exercises that enable project managers to better understand the impacts of lean strategies on projects (Hajifathalian 2011). In addition, involving industry members to assist in running lean exercises enhances the educational opportunity for students (Hyatt 2014). A successful understanding affects the student and his/her future organization since he/she will carry the lessons learned from the exercise's situation to his/her normal business (Friblick et al. 2007; Rybkowski et al. 2012).

6.1.3.3 Simulation Exercises in Civil Engineering

"Simulation replaces the 'learning about' by the 'taking part'" (Cullingford et al. 1979). While the "taking part" on large-scale construction projects to gain experience is costly, exercises allow students to experience real-life situations and engineering notions in a short space and time and without monetary risks or losses. Exercises require dividing students into small groups, thus creating a strong teaching approach that includes cooperative learning, team-based learning, and peer-led team learning; it helps students in developing collaborative skills and team spirit which are necessary for future engineering professionals (Froyd 2008).

In civil engineering projects, simulations are more complex and difficult, tasks are highly interdependent, and they include human factors with a high number of staff (Long et al. 2009). Business and finance education programs, for example, employ special simulation exercises that focus on contracting, such as Superbid, the Equipment Replacement Game, the Building Industry Game, and Contract & Construct (AbouRizk 1992; Nasser 2002; Martin 2000; Johnson et al. 2003). Other exercises are also known in civil engineering, such as the Parade of Trades

(Tommelein et al. 1999), LEAPCON (Sacks et al. 2007), and CONSTRUCTO (Halpin 1976). Some computer-based exercises are not as recent as the "Muck Game," for example (Scott and Cullingford 1973), which dates back to the 1970s and is also known as the Dam Game. It consists of modeling a dam construction project from a management and planning point of view (Cullingford et al. 1978). The exercises train students on extensive preplanning, learning by doing, teamwork and cooperation, and handling large construction projects (Long et al. 2009).

6.1.3.4 Learning Outcomes from Simulation Exercises

Many benefits are attributed to simulation exercises. They are motivational, enjoyable, and riskless. In addition to accelerating the learning ability of students especially for complicated lessons, they enhance analytical and problem-solving skills as well as the development of intuition in complex problems (Clarke 2009). Other outcomes of this teaching method are improving students' management and leadership skills by improving the quality of their decisions. Active learning involves students by making them explore, analyze, create, reflect, communicate, and use new information, which increases the retention of the knowledge acquired (Clarke 2009). The improvement in decision-making is shaped by the fact that students observe the consequences of their actions, share their ideas with other participants to arrive at helpful conclusions, or demonstrate their ideas to other members (Clarke 2009). Therefore, simulation is not required to provide intelligence or knowledge more than it is required to give the necessary support that enables learners to get the ultimate benefit and competence from these simulations (Thurman 1993).

6.1.3.5 Challenges

Teaching by simulation exercises is more beneficial for students and teachers; however, it requires more preparation by instructors and enthusiasm to communicate with students. This process needs effort and time. Moreover, exercises should be tailored to meet the class size because with the increasing number of students, the process becomes more difficult (Cullingford et al. 1979). Other challenges for implementing learning technologies are being flexible to pedagogical changes, supporting learning continuity and professional development, fulfilling students' expectations, and providing the appropriate space for learning (Clarke 2009).

6.2 Purpose

Understanding the role of discussion forums and simulation exercises in enhancing the classroom environments, increasing student learning, and improving student satisfaction is crucial to convince educators to implement these methods in modern education. Lean construction theory is based on collaborative techniques and team-based management. Accordingly, it makes a lot of sense to use collaborative and team-based methods to teach lean construction to university students. However, research is required to evaluate these methods.

This research aims at understanding the impact of discussion forums and simulation exercises on improving learning in lean construction education. The specific aims identified in this proposal include:

- Understanding the role of discussion forums and simulation exercises as classroom learning techniques
- Understanding the need for using these methods in the classroom
- Assessing the influence of these methods on stimulating students' participation and active engagement
- Assessing the improvements in learning outcomes when employing these methods

The study answers some general questions:

- How can active learning methods improve student learning in civil engineering education?
- What active learning methods should be employed in teaching lean construction to university students?
- Are discussion forums and simulation exercises appropriate methods for lean construction education?

This proposed research addresses some specific questions:

- Which methods help students express themselves better in class?
- Which methods contributed to student learning most?
- How did discussion forums influence in-class student learning?
- What lean construction concepts did students learn in participating in simulation exercises?

Answering the abovementioned questions can help educators understand the different ways that discussion forums and in-class simulation exercises can be employed to improve learning, enhance student engagement, and increase student satisfaction. The results uncovered reasons for students' engagement and suggested ways and methods to increase their motivation. This study uses students' self-assessment data to advise teaching practice. Thus, it is not intended to measure the overall efficacy of the teaching strategy.

Based on the results of the study, educators can base their future course design plans to blend different teaching methods to better engage students and cater for different student learning styles. Educators would have access to real evaluation data of discussion forums and in-class simulation exercises to model their classes after.

6.3 Application

Discussion forums and simulation exercises have been used at the American University of Beirut in the graduate-level course CIVE 686 titled "Lean Construction Methods & Applications." The different ways those were employed are detailed in this section.

6.3.1 Discussion Forums

In class, discussions were employed to explain new ideas and concepts. Weekly readings were assigned and students were asked to post questions on the course website. The class was organized into a discussion panel and the questions were used to initiate discussion. Students expressed their ideas, listened to others, and learned collectively. To do so, each student was required to answer at least one of the questions posed by another student and contribute to the discussion whenever she/he had an idea to share. If the answers were incomplete, other students joined the discussion until the questions were fully answered. The instructor's role was to facilitate and regulate the discussion and intervene only when questions were not fully answered (Hamzeh and Jacobs 2010).

Online forums were also used for cyber discussions. Those were in the form of assignments where students were required to research, analyze, and discuss three main topics: (1) the theory of constraints and its applications in construction, (2) integrated project delivery (IPD) and its barriers for implementation, and (3) the application of time management skills to student life.

6.3.2 Simulation Exercises

The simulation exercises that were used in class are the Airplane exercise, the Parade of Trades exercise, the Red-Green exercise, the Silent Squares exercise, the Bottleneck Simulation exercise and the Helium Stick exercise. Simulation exercises were integrated into the teaching program to help students understand the real-system behavior in real-life conditions, enhancing students' critical thinking and analysis. All these exercises helped students acquire clearer theoretical understanding of lean concepts and their applications in the construction industry. Given below is a description of the exercises used:

- The *Lean Airplane Simulation exercise* illustrates to students the benefits of creating flow in the work process. By playing many rounds, participants learn to reduce waste of time and increase efficiency in production. It teaches them teamwork, pull production, and the impact of supply chain logistics; it trains them

Fig. 6.2 Students engaging in the airplane simulation exercise

on balancing the workflow, reducing batch size, and reducing work-in-progress inventories (Dukovska-Popovska et al. 2008). Figure 6.2 shows students participating in the airplane simulation exercise.

- The *Silent Squares exercise* is used to learn new problem-solving tools while defining each member's role and responsibility. This design management exercise shows students the importance of sharing incomplete information and applying an integrated design process.
- The *Parade of Trades exercise* allows participants to develop a better intuitive understanding of various basic production concepts, such as variability and throughput. A single-line processing system is used to illustrate the impact that variability has on the performance of construction trades and their successors. Using simulation and computer exercises for this system shows the possibility of reducing waste of resources and time by reducing the variability in work flow between trades (Tommelein et al. 1999).
- The *Red-Green Simulation exercise* introduces students to the world of risk and risk sharing in projects. It teaches them how to take decisions that benefit the whole team and, indirectly, the individuals instead of solely the individual's benefit at the expense of the team.
- The *Bottleneck Simulation exercise* teaches students about the theory of constraints (TOC) which can be applied to project scheduling and can help in managing resources, project costs, and risks. It clearly simulates how the throughput of value is determined by the key constraints of every system (Steyn 2002).
- The *Helium Stick exercise* encourages students to work together as a team for a common purpose. It teaches them cooperation and good communication among each other as well as concentration and patience to attain their purpose.

6.4 Design/Method/Approach

To achieve the aforementioned aims, research was performed in three main phases: (1) background research and preparation of surveys, (2) data collection, and (3) data analysis and reporting.

6.4.1 Background Research and Preparation of Surveys

The first phase of the project involved two main tasks: (1) literature review and (2) developing a survey, including a pilot test, to assess discussion forums and simulation exercises.

6.4.1.1 Background Research

The first task involved performing a detailed literature review on discussion forums and simulation exercises. Due to the limited availability of literature found on specifically assessing these two methods in the classroom, the review addressed the available literature on the topics in general.

6.4.1.2 Survey

Task 2 involved preparing two surveys, initial and final, to assess the role of discussion forums and simulation exercises in improving education and enhancing the quality of learning in lean construction education. The survey addressed different types of learning methods employed in the classroom, such as lecture, homework, projects and case studies, and compared them to the discussion forums and simulation exercises. The survey targeted fourth-year civil engineering students as well as graduate students in civil engineering taking a graduate-level course on lean construction methods and applications. A pilot survey was developed and administered among a small group of students prior to the beginning of the course for assessment and improvement. The survey was then adjusted based on the returned pilot surveys to better address the objectives of the study.

The "initial survey" was distributed at the start of the course session to learn about student preferences at the debut of the term and later compare them with the "final survey" results. The "final survey" was prepared and administered at the end of the course to assess the effectiveness of the different teaching strategies employed. Similar to the "initial survey," there were sections that asked students to grade their overall satisfaction of the various teaching methods. This reflected the preference of each student to the different methods. Moreover, similar to the "initial survey," the "final survey" included a question about the number of hours per week each student worked for the course. This was used to measure the level of effort each student devoted to the course. To measure the effectiveness of the simulation exercises done in class, questions asking students to identify the lean concept learned through each exercise were included. Finally, to assess the level of understanding of the different modern construction delivery methods taught and compare it before and after the course, students were asked to grade how comfortably they can describe and discuss the various methods in a professional environment. Because different teaching styles were used to teach the construction methods, the effectiveness of the former can be compared based on the degree of improvement.

6.4.2 Data Collection and Survey Analysis

As mentioned, the study employs two surveys administered at the beginning of the semester and towards the end of the semester to assess the change in students' perception of their learning styles. The surveys also aim at assessing the impact of discussion forums and simulation exercises on students' understanding, evaluating students' learning styles, and weighing the desirability of active learning methods.

The "initial survey" was conducted in the first week of class to assess students' perception of various learning methods, understand their preferred learning styles, and shed a light on their expectations. After employing the discussion forums for many sessions and administering five simulation exercises throughout the semester, the "final survey" addressing the same topics covered in the first survey was conducted.

6.4.3 Data Analysis and Reporting

In this study, the data collected were analyzed to examine the changes in students' assessment of the various teaching methods used in the class, including discussion forums and simulation exercises. The data were also used to assess the level of student learning when participating in simulation exercises. Comparisons between the levels of understanding of various course topics were performed to assess the impact of simulation exercises on increasing student learning. Moreover, the level of understanding for various course topics implied in simulation exercises was also analyzed to assess student learning.

Performing this analysis requires a close comparison between the student surveys at the beginning and at the end of the course, arranging data in various ways in order to facilitate the analysis, and performing various statistical tests to measure performance and statistical significance.

6.5 Results and Discussion

To shed light on the efficacy of active teaching methods, specifically discussion forums and simulation exercises, various comparisons were performed, including analysis of changes in students' assessment of discussion forums and simulation exercises as teaching methods, quantitative assessment of student learning, and qualitative assessment of student learning using these methods.

6.5.1 *Changes in Students' Assessment of Teaching Methods*

To assess the impact of employing simulation exercises and discussion forums on student learning, a comparison analysis of student views at the beginning of the course and at the end of the course was performed. Students were asked at the beginning of class instruction to rate the best methods for learning in class using a five-point Likert scale. Towards the end of the semester, students were asked to rate the same methods again after they have had enough experience participating in five different simulation exercises and engaging in several discussion forums.

When asked to rate group activities such as simulation exercises in the "initial survey," students rated them as favorable methods for class instruction. The preference of simulation exercises averaged to a rating of 4.22 out of 5. However, the students' assessment at the end of the course was even higher as the students expressed their satisfaction with this method, raising its average rating to a 4.5.

To assess the significance of this change, a paired *t*-test was performed for the rating before engaging in the simulation exercises and after for a sample of 18 students as shown in Table 6.1. The null hypothesis is assumed to be such that there is no improvement in the assessment rating. However, the results show that the test statistic of −2.557 has a probability of 0.0204 of occurring by chance alone even if the samples had the same parametric value of this statistic. It can therefore be concluded that the results are statistically significant at the 0.05 level of significance and the null hypothesis of equality of the parametric statistics is rejected. Thus, the improvement is significant at the 0.05 level.

While the results shown in Table 6.1 assume that the data resemble a normal probability distribution, some may argue the credibility of this assumption. Hence, a nonparametric hypothesis testing method (the Wilcoxon signed-rank test) was employed to confirm the significance of the results, assuming the distribution of data is not necessarily normal. The results for the nonparametric test as shown in Table 6.2 confirm that the results are statistically significant to the 0.01 level.

Similarly, a test was performed to evaluate the change in students' assessment of discussion forums. The results before the beginning of class instruction average to a rating of 3.89 and the final results to an average of 4 out of 5. When performing the paired *t*-test, the results were found to be not statistically significant. However, the Wilcoxon signed-rank test showed that the results were statistically significant.

Table 6.1 Parametric hypothesis test using paired *t*-test for the difference of group activity means before and after participating in the simulation exercises

Hypothesis Test for the difference of two means: dependent sample (Paired t-test)					
	Two tailed test	Use this area to get X̄d and Sd			
	Ho: Ud = 0.0	Before data	After Data	difference	X̄d
	Ha: Ud ≠ 0.00	5	5	0	-0.277777778
Alpha level: α	0.05	5	5	0	Sd
Claimed difference: μ		5	5	0	0.460888599
Mean differences: X̄d	-0.2778	4	4	0	N
Standard deviation: S	0.4609	5	5	0	18
Sample size: n	18	4	4	0	Copy Values?
Test Statistic	-2.5570	5	5	0	
Critical Value	± 2.1098	5	5	0	Yes
P-Value	0.0204	4	4	0	
Decision	Reject Ho	3	4	-1	
		4	4	0	
Confidence Interval for paired difference (using data in column B)		3	4	-1	
		4	5	-1	
Confidence Level	0.95	3	4	-1	
Confidence Interval	-0.5070	5	5	0	
		5	5	0	

Table 6.2 Nonparametric hypothesis test using Wilcoxon signed-rank test for the difference of group activity means before and after participating in the simulation exercises

Group discussion before	5	5	5	4	5	4	5	5	4	3	4	3	4	3
Group discussion after	5	5	5	4	5	4	5	5	4	4	4	4	5	4
Differences	0	0	0	0	0	0	0	0	0	−1	0	−1	−1	−1
Ranks of the differences										3		3	3	3
Working with 18 pairs	You have to ignore any pairs with zero difference													
Sum of the positive ranks	W_+	0.00	Add up the ranks for all the differences that are positive											
Sum of the negative ranks	W_-	15.00	Add up the ranks for all the differences that are negative											
W-value	0.0													
Values from the tables for two-tailed test	5 %	40	1 %	28										
Conclusion	The result is significant at the 1 % level													

Figure 6.3 shows the changes in students' assessment of simulation exercises and discussion forums between the "initial survey" at the beginning of the course and the "final survey" towards the end of the course. It is crucial to mention that the initial assessment was performed after the first week and first simulation exercise. That might have raised the initial assessment and reduced the significance of the improvements. However, the results show enough changes to claim significant improvements.

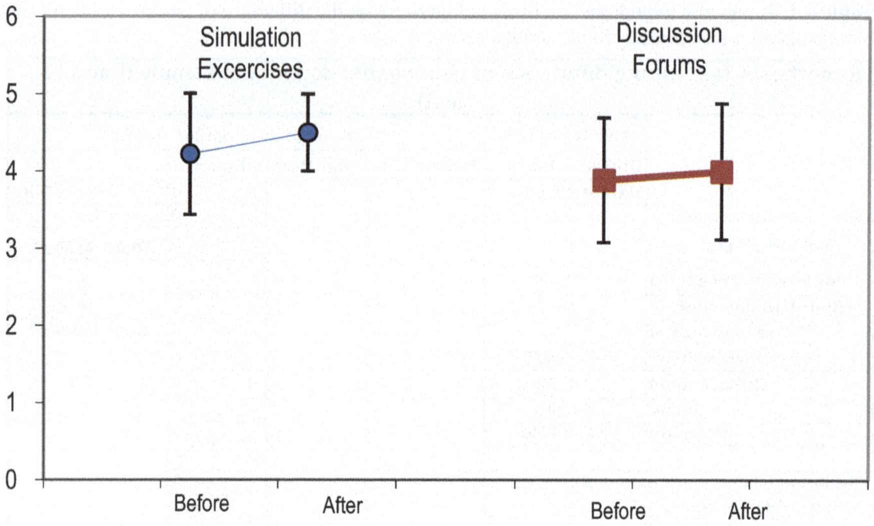

Fig. 6.3 Students' assessment for simulation exercises and discussion forums at the beginning and at the end of the course

In addition to the abovementioned analysis, a test was carried out to assess the impact of simulation exercises on students' understanding. This test involves comparing two course topics: the "Last Planner System" and "Choosing by Advantages." Both methods were clearly explained in class but only the "Last Planner System" was included as an in-class simulation exercise. Students were asked to assess how comfortable they are in describing and discussing each modern construction method within a professional networking event. Figure 6.4 shows a comparison between the students rating for each course topic at the beginning and at the end of class.

To compare the two topics, the difference between before and after is calculated for each survey response. A hypothesis test for the difference between the two means or the unpaired test is performed on the difference results for the "Last Planner System" and "Choosing by Advantages." The null hypothesis assumes no change in the results between the two different topics.

The results show that the test statistic of 2.317 has a probability of 0.027 of occurring by chance alone even if the samples had the same parametric value of this statistic. The results are therefore statistically significant at the 0.05 level of significance, and we reject the null hypothesis of equality of the parametric statistics.

Although this result does not prove any causality between the events, in other words, it does not necessarily prove that students rated the ("Last Planner System") topic higher because of simulation exercises per se, it does show that a simulation-supported course topic was rated higher than a nonsupported class topic.

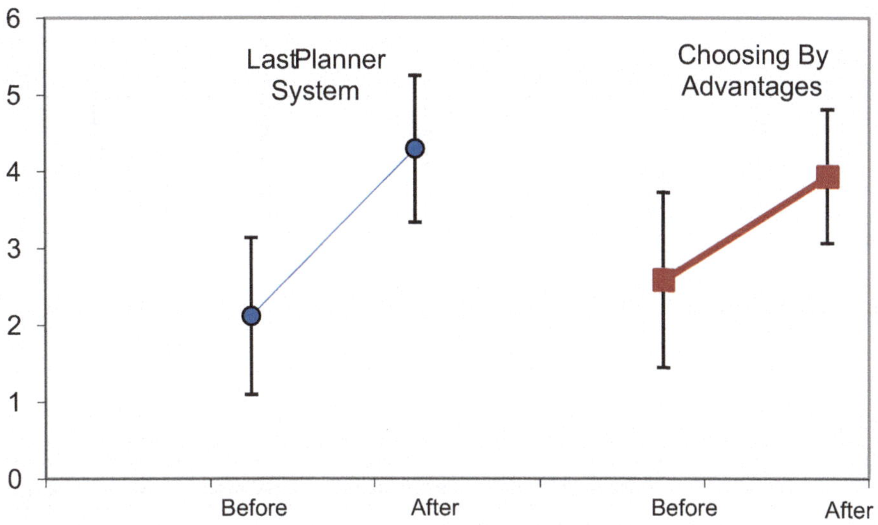

Fig. 6.4 Students' assessment for a simulation-supported class topic (Last Planner System) and a nonsupported class topic (Choosing by Advantages) at the beginning and at the end of the course

6.5.2 Quantitative Assessment of Student Learning When Employing Simulation Exercises

One method to assess learning is to examine students on course topics and find out whether they have grasped certain concepts or not. The survey at the end of class included several questions aimed at evaluating students' understanding of the concepts embodied in the simulation exercises (Airplane, Parade of Trades, Silent Squares, Red-Green, and Helium stick) by asking them to state the concept(s) of each exercise. To analyze the results, the answers were graded on the basis of a grading rubric that gives points for each exercise between 1 and 5. The results for the whole class were then aggregated as shown in Fig. 6.5. The overall means for the previously mentioned exercises were found to be 4.5, 3.55, 4.36, 4.5, and 4.18, respectively.

Although one result (Parade of Trades) leaves much to be desired, the results show a high level of student learning and understanding of course concepts and present yet another piece of evidence to support the efficacy of simulation exercises in classroom teaching.

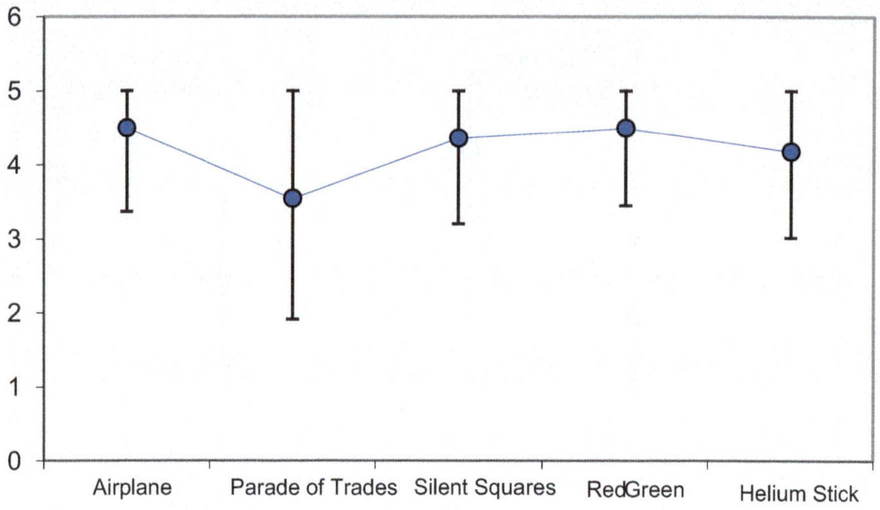

Fig. 6.5 Average and Whisker plot showing the learning assessment for five simulation exercises

6.5.3 Qualitative Assessment of Student Learning

Student reviews of the CIVE 686 "Lean Construction Methods and Applications" that are included in the student evaluations can also be used to qualitatively asses the innovative teaching methods used. The course evaluations and students' comments are good sources of evidence in finding out whether the course was able to achieve the overall teaching objectives, specifically when it comes to developing critical thinking, communication skills, and analytical skills among engineering students.

When students were asked "In what ways did the instructor enhance your learning in the course?", for example, their answers came in support of simulation exercises and discussion forums:

- "In class, exercises enhanced my communication skills and critical thinking which makes learning easier for me."
- "Interaction, taking part in simulation exercises, watching lectures, and having the opportunity to discuss things in class definitely helps enhance the learning experience and helps students better visualizing lean concepts."
- "Exercises are very helpful to illustrate the concepts" (Course Evaluations).

Not only did students acknowledge that simulation exercises and discussion forums improve critical thinking and overall learning skills, they have also voted the class to be ranked first in the Faculty of Engineering and Architecture (FEA; scoring a percentile of 100) on "promoting and encouraging analytical/critical thinking." The overall rating of the course was 4.6 out of 5.

6.6 Conclusions

This study addresses active learning methods and their role in enhancing students' learning in engineering courses. It focuses on two such methods, namely, simulation exercises and discussion forums employed to facilitate class instruction and increase understanding among engineering students in preparing them to enter the workforce equipped with a solid theoretical understanding of lean construction and its real-world applications.

Results from a class-run survey administered at the beginning and at the end of a civil engineering graduate course employing simulation exercises and discussion forums highlight an increased level of student satisfaction. The statistically significant results are rational when factoring in the increase in student enthusiasm and interest when participating in tactile experiences. Similarly, discussion forums stimulate students' imagination and thought process as they prepare, debate, and discuss course topics in a stress-free environment.

The results confirm the results of research performed on active learning methods presented in section 6.1. If the benefits of active learning are realized through more participation, more enthusiasm, and better engagement in learning, then simulation exercises and discussion forums should be seriously considered for addition to the civil engineering curricula in the Middle East and North Africa (MENA) region.

Results also show that simulation exercises may have an impact on real-life projects. For example, dividing the student groups into teams of different subcontractors or trades working together to construct a building helps the students better understand the role of teamwork, collaboration on construction projects, and the impact of variability in production planning and control.

Moreover, the assessment of students' learning for course topics supported by simulation exercises showed a high level of understanding compared to topics that were not supported. Thus, students not only enjoy their participation in simulation exercises but also develop a deeper understating of class topics and their projections into the real world.

Student evaluations administered at the end of the course confirm the survey results and show that simulation exercises and discussion forums improve communication skills, understanding, and critical thinking. These skills are crucial for engineering students in developing their careers and finding future engineering solutions to various world problems.

These methods have profound impact on student satisfaction, engagement, and learning. Accordingly, the authors encourage educators to use multiple active learning methods in classroom education to cater for different student learning styles and reduce the dependence on one-way lectures that tend to limit student interaction.

As educators, we have an obligation to help students develop communication and critical thinking skills that will better prepare them to become the leaders of tomorrow. Each engineering program should evaluate its curriculum to search for possible areas to implement various active learning methods and prepare for the future challenges in engineering education.

References

AbouRizk, S. M. (1992). A stochastic bidding game for construction management. *Proceedings of 2nd Canadian Conference on Computing in Civil Engineering*, Ottawa, Ontario, pp. 576–87.

Adler, M. (1987). *Reforming education*. New York: Collier Macmillan.

Bonwell, C. C., & Eison, J. A. (1991). *Active learning: Creating excitement in the classroom*. ERIC Clearinghouse on Higher Education Washington DC. George Washington Univ., Washington DC.

Chudler, E. (2005). Teaching more by lecturing less.*Journal of Life Science Education, 4,* 298–310.

Clarke, E. (2009). Learning outcomes from business simulation exercises: Challenges for the implementation of learning technologies. *Education and Training, 51*(5), 448–459.

Coleman, J. S. (1978). Information processing and experiential processing: Two modes of learning. In R. Stadsklev (Ed.), *The handbook of simulation gaming in social education, Part 1: Textbook* (pp. 26–32). Tuscaloosa: Institute of Higher Education Research and Services, University of Alabama.

Cooke, M., Irby, D. M., & Sullivan, W., et al. (2006). American medical education 100 years after the Flexner report. *The New England Journal of Medicine, 355,* 1339–1344.

Cullingford, G., Mawdesley, M., & Davies, P. (1978). *Some experience with computer based games in civil engineering teaching*. Symposium on Computer Assisted Learning in Civil and Structural Engineering and Related Subjects, Swansea, p. 10.

Cullingford, G., Mawdesley, M. J., & Davies, P. (1979). Some experiences with computer based games in civil engineering teaching. *Computers and Education, 3,* 159–164.

Delloite. (2013). *GCC powers of construction 2013, construction section overview.* Deloitte & Touche (M.E.), p. 36.

Drysdale, D. (2013). *Introducing lean improvement into the U.K. highways agency supply chain.* Proceedings of the 21st International Group for Lean Construction Conference, conference, IGLC 21: Fortaleza, Brazil, pp. 1067–1074.

Dukovska-Popovska, I., Hove-Madsen, V., & Nielsen, K. B. (2008). *Teaching lean thinking through game: Some challenges.* Proceedings of 36th European Society for Engineering Education (SEFI) on Quality Assessment, Employability & Innovation: Sense Publishers.

Eison, J. (2010). Using active learning instructional strategies to create excitement and enhance learning. Retrieved from http://www.cte.cornell.edu/documents/presentations/Eisen-Handout.pdf.

Folkestad, J. E., Smith, A., & Mortiz, M. (2009). Phenomenology of a multimedia fishbowl: A learning ecosystem that encourages individuals to innovate through collaborative discovery. *i-manager's Journal on School Educational Technology, 5,* 66–80.

Friblick, F., Åkesson, A., & Leigard, A. (July 2007). *Learning lean through lean game—A case from the infrastructure industry.* Proceedings of the 15th International Group for Lean Construction Conference, IGLC 15: Michigan, USA, pp. 475–484.

Froyd, J. (2008). White paper on promising practices in undergraduate STEM education. Paper presented at the National Research Council's Workshop Linking Evidence to Promising Practices in STEM Undergraduate Education, Washington DC. http://www7.nationalacademies.org/bose/Froyd_Promising_Practices_CommissionedPaper.pdf.

Gall, M. D., & Gillet, M. (1980). The discussion method in classroom teaching. *Theory into Practice, 19,* 98–103.

Gauthier, C., & Dembélé, M. (2004). Quality of teaching and quality of education: A review of research findings.

GCC. (2010). *GCC.* http://sites.gcc-sg.org/Statistics/index.php?SID=408. Accessed Sept. 2013.

Gilgeous, V., & D'Cruz, M. (1996). A study of business and management games. *Management Development Review, 9*(1), 32–39.

Hajifathalian, K. (2011). *Effects of working strategy and duration variance on productivity and work in process a simulation-based investigation and oops game: Cost-benefits tradeoff analysis of reliable planning for construction activities.* Master's thesis. Raleigh, North Carolina: North Carolina State University.

Halpin, D. W. (1976). CONSTRUCTO—An interactive gaming environment. *Journal of the Construction Division, 102*(1), 145–156.

Hamzeh, F. R., & Jacobs, F. (2010). *Open forum as an active learning method for teaching lean construction*. Proceedings of the Lean Advancement Initiative, 5th LAI/EdNet Lean Educator Conference, May 19–21, Daytona Beach, Florida.

Hanrahan, M. (1998). The effect of learning environment factors on students' motivation and learning. *International Journal of Science Education, 20,* 737–753.

Heimlich, J. E., and Norland, E. (2002) "Teaching Style: Where Are We Now?" *New Directions for Adult and Continuing Education, 93*, 17–25.

Hyatt, B. (2014). Evaluating the implementation of lean construction into construction education. Construction Research Congress.

IQPC. (2011). International quality and productivity center, 4th Annual Bridges Middle East. http://www.iqpc.com/Event.aspx?id=637760.

Javidan, M., Dorfman, P. W., Sully de Luque, M., & House, R. J. (2006). In the eye of the beholder: Cross-cultural lessons in leadership from Project GLOBE. *Academy of Management Perspectives, 20*, 67–90.

Johnson, H., Borland, J., & Craig, K. (2003). *Building industry game (BIG): A computer simulation for construction management.* Proceedings of the 39th ASC Annual Conference, Clemson University, South Carolina, pp. 79–90.

Kenny, N. (2008). *Encouraging effective discussion from the first day of class*. Centre for Open Learning and Educational Support, University of Guelph, pp. 1–3.

Kuhn, T. S. (1970). *The structure of scientific revolutions*. Chicago: University of Chicago Press.

Larson, B. E., & Keiper, T. A. (2002). Classroom discussion and threaded electronic discussion: Learning in two arenas. *Contemporary Issues in Technology and Teacher Education, 2*(1), 18.

Liker, J. K. (2004). *The Toyota way—14 management principles from the world's greatest manufacturer* (p. 330). New York: McGraw-Hill.

Long, G., Mawdesley, M. J., & Scott, D. (2009). Teaching construction management through games alone: A detailed investigation. *On the Horizon, 17*(4), 330–344.

Martin, A. (2000). A simulation engine for custom project management education. *International Journal of Project Management, 18*(3), 201–213.

Nasser, K. (2002). Simulation gaming in construction: ER, the equipment replacement game. *Journal of Construction Education, 7*(1), 16–30.

National Academy of Science. (2004). Committee on developments in the science of learning. Commission on behavioral and social sciences and education. In J. D. Bransford, A. L. Brown, R. R. Cocking, M. S. Donovan, J. D. Bransford, & J. W. Pellegrino (Eds.), *How people learn: Bridging research and practice* (Expanded edition.). Washington DC: National Academies Press. (With additional material from the Committee on Learning Research and Educational Practice).

Ozer, O. (2005). Using class discussion to meet your teaching goals. *The Center for Teaching and Learning, Stanford University, CTL Newsletter, 15*(1), 1–6.

Pendleton, R. K. (1981). *Program goals and teaching methods of small program industrial arts teachers in Arizona*. PhD Dissertation, Arizona State University, p. 174.

Project Lebanon. (2013). The 18th international construction trade exhibition for Lebanon & the Middle East. Beirut.

Roediger, H. L., & Butler, A. C. (2011). The critical role of retrieval practice in long-term retention. *Trends in Cognitive Sciences, 15*(1), 20–27. http://dx.doi.org/10.1016/j.tics.2010.09.003.

Rybkowski, Z., Zhou, X., Lavy, S., & Fernández-Solís, J. (2012). Investigation into the nature of productivity gains observed during the Airplane Game lean simulation. *Lean Construction Journal, volume 2012, ISSN 1555-1369*, 78–90.

Sacks, R., Esquenazi, A., & Goldin, M. L. (2007). LEAPCON: Simulation of lean construction of high-rise apartment buildings. *Journal of Construction Engineering and Management, 133*(7), 529–539.

Scott, D., & Cullingford, G. (1973). Scheduling games for construction industry training. *Journal of the Construction Division, 99*(6), 81–92.

Steyn, H. (2002). Project management application of the theory of constraint behind the critical chain method. *International Journal of Project management, 20,* 75–80.
Taekman, J., & Shelley, K. (2010). Virtual environments in healthcare: Immersion, disruption, and flow. *International Anesthesiology Clinics, 48*(3), 101–121.
Thurman, R. (1993). Instructional simulation from a cognitive psychology viewpoint. *Educational Technology Research and Development, 41*(4), 75–89.
Tommelein, I., Riley, D. R., & Howell, G. A. (1999). Parade game: Impact of work flow variability on trade performance. *Journal of Construction Engineering and Management, 125*(5), 304–310.
Tsao, C. C. Y., Azambuja, M., Hamzeh, F. R., Menches, C., & Rybkowski, Z. K. (2013). *Teaching lean construction—Perspectives on theory and practice.* Proceedings of the 21st International Group for Lean Construction Conference, IGLC 21: Fortaleza, Brazil, pp. 977–986.
Young, J. (2007). Small group scored discussions: Beyond the fishbowl, or everybody reads, everybody talks, everybody learns. *The History Teacher, 40*(2), 177–181.
Zoroja, J. (2010). Simulation games and their use in education at economics institutions in the Republic of Croatia. *The Business Review, Cambridge, 15*(1), 113–118.

Dr. Farook Hamzeh is an Assistant Professor in Civil and Environmental Engineering at the American University of Beirut (AUB). He earned his Bachelor in Civil Engineering from AUB in 1997 after which he received a Masters in Engineering Management in 2000. Dr. Hamzeh then received a Masters in Engineering Project Management from UC-Berkeley in 2006. He earned his Ph.D. in Civil and Environmental Engineering from the University of California at Berkeley in 2009. He became a faculty member at Colorado State University between 2009 and 2011.

Dr. Hamzeh has worked for more than seven years in the construction industry in Lebanon, Qatar, Dubai, and the United States on several mega projects. These include the $1.7 Billion Cathedral Hill Hospital in San Francisco, the 333 m high Rose Rotana Hotel in Dubai, Losail motorbike racetrack in Qatar, Olympic Tower in Qatar, Al-Amal Oncology Hospital in Qatar, Serail 1374 Building in downtown Beirut, and Sibline Cement factory 2nd production line in Lebanon. Throughout his career, Dr. Hamzeh has developed a passion for improving construction processes to increase productivity, raise customer value, and reduce process waste. This passion has fueled his research in the areas of lean construction, process improvement, production planning, BIM, and supply chain management.

Christina Teokari is a Civil and Environmental Engineering Graduate. She recently earned her Bachelor's Degree with distinction from the American University of Beirut. During her undergraduate years she has adopted an interest in research. She worked as a research intern at the University of Wisconsin Madison in the Center for Freight & Infrastructure for Research & Education on a study addressing environmental justice in transportation. For their final year project, Christina and her colleagues won the best paper award in the 12th FEA Student and Alumni Conference for their paper titled "Revitalizing Beirut Transport Corridors". Moreover, Christina has been working for the past year with Dr. Farook Hamzeh on research about active learning in lean construction education.

Throughout her undergraduate years, Christina has been a main contributor to the Civil Engineering Newsletter. She has been writing articles for each issue since 2010 and became news editor in 2012 editing newsletter articles and giving authors sessions on improving their writing skills.

Carel Rouhana is a Civil Engineering graduate from Notre Dame University. She finished her Graduate studies in Civil Engineering with emphasis in construction engineering and management at the American University of Beirut.

Carel has worked as intern engineer at MIDMAC contracting in Qatar on the new Doha International Airport. She has also worked on design projects for civil engineering concrete and steel structures and on an elevated highway project involving geotechnical, structural, and geometric design.

Her previous work has honed her interest in construction related issues such as resource management, waste management, scheduling, and lean construction. She is pursuing her interests through research in construction management and lean construction. Carel's ambition to improve construction engineering through active learning has motivated her to contribute to the study of lean construction education with Dr. Farook Hamzeh.

Chapter 7
A Framework for Developing Innovative Problem-Solving and Creativity Skills for Engineering Undergraduates

Faris Tarlochan and Abdel Magid Hamouda

7.1 Introduction

Engineering has always been, is, and will always be the backbone of the growth of civilization for a nation. It is a professional activity servicing the societal needs. Engineers are problem solvers who produce benefit solutions while keeping in mind the importance of ethical responsibilities, efficient and ethical usage of resources, and development of cost-effective solutions that are environmentally sound and sustainable while taking into account the entire life cycle of a system. As such, the engineering education of future engineers cannot be taken lightly.

However, it is no more a secret that the engineering education globally has major deficiencies (Felder et al. 2000). With the rapid development of new technologies and development of new economic models such as innovation ecosystems, it is imperative that traditional engineering curriculum undergoes a major overhaul. Besides equipping our undergraduate students with strong fundamentals and proper communication skills, there is a need to foster innovative problem-solving and creativity skills through "real-world" engineering problems. These concerns are not new and have existed for some good 15 years, which led to the development of key graduate attributes such as those in the Engineering Criteria 2000 (Accreditation Board for Engineering and Technology (ABET)) and those listed under the Washington Accord. On comparing the graduate attributes globally, some similarities are found, and two of the most important generic attributes for engineering graduates are: (i) the ability to design a system or component and (ii) the ability to identify, formulate, and solve engineering problems. These, in essence, are what engineers do generally as professionals.

The problem with these two attributes is that the art to design, along with the skill to identify a problem and formulate a solution, is not usually taught to students

F. Tarlochan (✉) · A. M. Hamouda
Mechanical and Industrial Engineering Department, College of Engineering, Qatar University, Doha, Qatar
e-mail: faristarlochan@gmail.com; faris.tarlochan@qu.edu.qa

M. Abdulwahed et al. (eds.), *Advances in Engineering Education in the Middle East and North Africa,* DOI 10.1007/978-3-319-15323-0_7

formally (Felder and Brent 2003). So much so that these skills are nonverbal activities and as such cannot be taught in the traditional sense. Faculty members continue to concentrate on teaching content rather than showing the processes involved in problem-solving and design skills. Design skill is basically linked with creativity or innovation skills. This is because creativity can be defined as an act of combining or synthesizing existing ideas in an innovative way for the purpose of creating economic value. Innovation, on the other hand, can be defined as a change that creates economic value. According to a study by Adams (Adams 2009), deficiencies exist in the engineering curricula, especially with the development and assessment of creativity within engineering. This task can be frustrating, and lack of skills may cause physiological inertia towards innovative solutions. On the other hand, most students are taught how to solve standard textbook problems where the problem has been well defined. Hence, the ability to identify and formulate a problem is still not obvious in most engineering curricula globally.

Another important element that needs to be addressed is why teach problem-solving and creativity skills. In Europe, problem solving and creativity are presented as important competencies in the requirements for European engineer since these two are important drivers in economic prosperity (Lord Sainsbury of Turville 2007). We have to understand that the current market demand is different from what it was 15 years ago. The product life gets shortened every year. High competition and high consumer demands put strong emphasis on fast innovation. A nation that can create more outputs from a given set of inputs will succeed. Besides this, our graduates are not confined to one working environment, but rather exposed to several different working environments, at times individually and at times simultaneously. Looking at these constraints, our graduates need to be versatile and equipped with generic skills such as problem solving and creativity to assimilate different environments (Adams 2009; Pappas and Pappas 2003; Reidsema 2005; Kimmel et al. 2003) .

The next question that arises is what is the best way to instill these skills in students? Literature has shown that traditional teaching, particularly rote learning, is no longer sufficient (and was never efficient) in instilling these skills (Esquivel 1995). An alternative form of instructional delivery is needed. In 2011, a team of engineering academics from Royal Melbourne Institute of Technology (RMIT), Australia (Steiner et al. 2011), conducted a detailed and comprehensive study on problem-solving skills among undergraduates from year 1 to year 4. Their study showed that students found problem-based learning (PBL) as an effective strategy to enhance their problem-solving skills. This finding is supported by the work of Overton reported in 2003 (Overton 2003), where PBL is often used in engineering as a teaching style that enforces deep approaches to learning. Besides this, another study conducted by the Forfás group (Expert Group on Future Skills Needs 2009) showed that increased usage of PBL and inquiry-based learning approaches should be advocated at the tertiary level of education. PBL stresses on a real engineering problem approach towards learning and is an effective pedagogy for improving problem-solving skills (Poikela and Poikela 2005). Many other studies (Hitt 2010; Grolinger 2011; Walsh 2007; Fink 2002; Petruska 2010) have shown the success of PBL in addressing the issue of problem-solving skills among undergraduates. Hence, it can be summarized here that problem-solving skills are important to be "taught" to undergraduate students, and one of the most effective pedagogy tools is PBL.

In the local context of the Middle East and North Africa (MENA) region, especially in the Gulf countries, this issue of lack of problem-solving, innovative, and creativity skills is more apparent. In the Gulf countries, hereditary monarchies have created massive social safety nets such as free education and health care, subsidies per child, no-interest loans for home purchase, wedding grants, and artificially low utility rates, among others (Walters et al. 2010). The consequence of this is that risk-aversion culture rather than risk-taking culture is embraced, leading to safe governmental jobs or silent partnerships with expatriates (Walters et al. 2010). In addition to this, governmental schools, being mostly for citizens, are poorly operated and governed such that the students are not equipped with critical thinking and problem-solving skills to embrace the knowledge-based economy (Walters et al. 2010; Ghabra 2010; Issan and Gomaa 2010) . Education has become more of a training (rote learning) rather than the development of problem-solving, innovative, and creativity skill sets (Loveluc 2012; Sager 2007; Mrabet 2010). In order for the MENA region to succeed in embracing a knowledge-based economy, problem-solving and creativity skill sets are needed.

Many believe that problem solving, creativity, and innovation are inborn traits. In the 1950s, Genrich S. Altshuller and his colleagues in the former Soviet Union proved that the skills of innovation and creativity in problem solving can be taught to anybody and are not inborn traits. Altshuller came up with a theory which he named *Teoriya Resheniya Izobretatelskikh Zadatch* (TRIZ; Russian) or theory of inventive problem solving (TIPS). He found that patented inventions from different scientific and engineering fields usually came about by using one or more of about 40 fundamental principles and those solutions known to one field may be reinvented into another problem. The TIPS methodology claims that problems can be codified, classified, and solved methodically, just like other textbook problems (Stratton et al.).

A study conducted by Iouri Belski of RMIT, Australia (Belski 2007), has shown that TIPS or TRIZ changed positively most of the students' perception of their ability of problem solving. Thirty-six engineers from mechanical and electrical backgrounds were involved in extensive TRIZ training supervised under a study by Iouri Belski. The outcome of the study (Belski 2005) showed that the engineers found that TRIZ helped them develop solutions that had been previously unachievable. They also found that their own thinking had become more systematic. These studies have shown the power and easiness of TRIZ as a problem-solving tool.

The motivation here is how can we instill innovative problem-solving and creativity skills in engineering undergraduates in order to meet the current market demands especially in the MENA region? The objective of this chapter is to develop a theoretical framework that addresses the need to foster innovative problem-solving and creativity skills in engineering undergraduates. Emphasis will be given on three areas—namely, content, delivery, and assessment. For the content, TRIZ will be used as the knowledge-based tool. This tool will be used to educate students about problem-solving skills by using PBL (delivery method) as a pedagogy tool. The complexity of problem solving especially related to design problems arises from the need to integrate fundamental sciences, mathematics, engineering sciences, and complementary and social studies in order to fully derive the solution of a feasible engineering problem. This makes the role of students in preparing the solutions and academics in assessing the outcome fuzzy and complicated (Pop-Iliev and Platanitis 2008). Since problem-solving skills are very subjective due to their open-endedness, assessing them should be on a more qualitative scheme rather than quantitative. Hence, rubrics are typically used to evaluate a broad range of subjects and activities in problem solving; as a result, the outcome of this proposed framework will be measured via rubrics-based assessment (assessment scheme).

7.2 Design of Framework

7.2.1 Defining Problem Solving, Creativity, and Critical Thinking

Terms such as problem solving, creativity, innovation, and critical thinking have been used by many without properly understanding the meaning behind these words. If one would search the literature, various definitions and meanings will be found. To bring light to these definitions, in 2011, the Association of American Colleges & Universities (AAC&U), through efforts by teams of faculty experts representing colleges and universities across the USA, developed sound definitions for problem solving, creative thinking, and critical thinking.

7.2.1.1 Problem-Solving Skills

Problem solving as defined by AAC&U (Rhodes 2010): *is the process of designing, evaluating and implementing a strategy to answer an open-ended question or achieve a desired goal. Problem-solving covers a wide range of activities that may vary significantly across disciplines. Activities that encompass problem-solving by students may involve problems that range from well-defined to ambiguous in a simulated or laboratory context, or in real-world settings. The focus is on the* ***process*** *and not the* ***end-product****.*

The performance indicators for problem solving are shown in Fig. 7.1.

For problem solving, the engineering undergraduates in their final year should be able (Rhodes 2010) to

- construct clear and insightful problem statement with evidence of all relevant contextual factors;
- identify multiple methods for solving the problem;

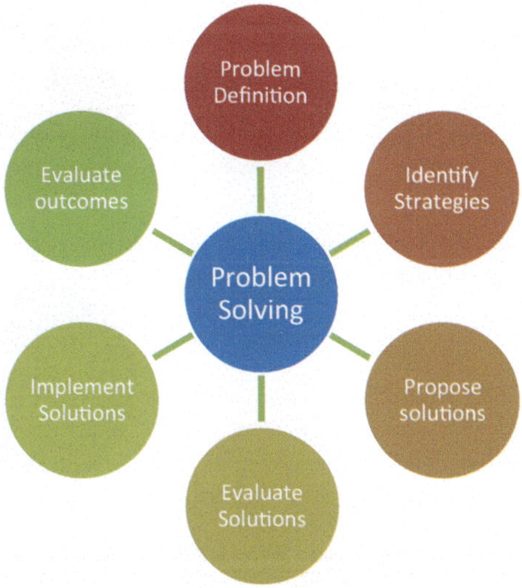

Fig. 7.1 Performance indicators/strategies for problem solving (Rhodes 2010)

- generate one or more solutions/hypotheses through deep comprehension of the problem while keeping in mind the ethical, logical, and cultural dimensions of the problem;
- evaluate thoroughly and weigh impacts of the solution;
- ensure that solution addresses the problem thoroughly; and
- review results/performance of the solution and recommend further work.

7.2.1.2 Creative Thinking Skills

Creative thinking as defined by AAC&U (Rhodes 2010): *is both the capacity to combine or synthesize existing ideas, images, or expertise in original ways and the experience of thinking, reacting, and working in an imaginative way characterized by a high degree of innovation, divergent thinking, and risk taking. The student must have a strong foundation in the strategies and skills of the domain in order to make connections and synthesize recognizing creative risk-taking to achieve a solution.*

The performance indicators for creative thinking are shown in Fig. 7.2.

For creative thinking, the engineering undergraduates in their final year should be able (Rhodes 2010) to

- evaluate creative process and product using discipline-appropriate criteria;
- seek out and follow through potentially risky routes or approaches to the problem;
- develop logical plan to solve the problem;
- integrate alternative, divergent, or contradictory perspectives or ideas fully;
- extend a novel idea or product to create new knowledge that crosses boundaries; and
- transform ideas or solutions into entirely new forms.

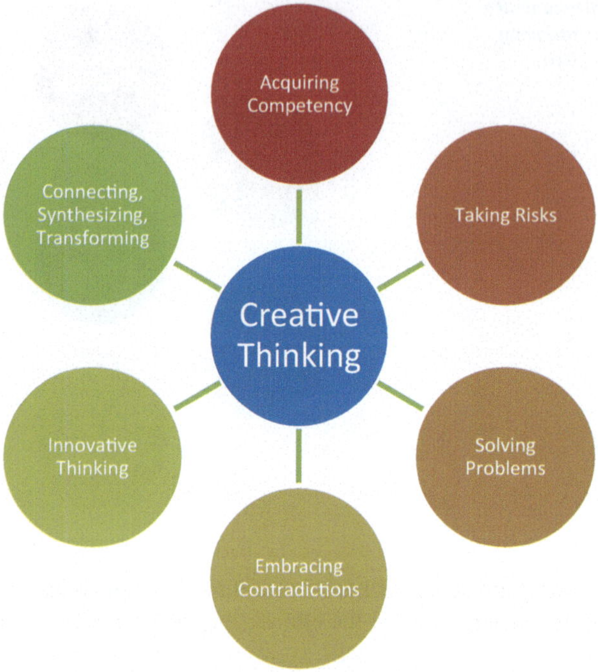

Fig. 7.2 Performance indicators/strategies for creative thinking (Rhodes 2010)

Creativity can be perceived as hierarchal (Liu and Schonwetter 2004). The first level includes *expressive creativity* which is basically the tendency to create without emphasizing on the quality. The next level is *technical creativity* which is the proficiency to create products with consummate skills. The third level is *inventive creativity* which deals with the ability to create new things from existing ideas. The fourth level is *innovative creativity;* here the student is able to "think outside the box" and formulate innovative departures. The final level is called *emergent creativity* such as the ideation of Einstein's work on general relativity. One important element of creativity is the divergent thinking skill which is strongly linked to the cognitive basis of creativity (Liu and Schonwetter 2004). It is the capability of producing new and multiple solutions for a problem.

7.2.1.3 Critical Thinking Skills

Critical thinking as defined by AAC&U (Rhodes 2010): *is a habit of mind characterized by the comprehensive exploration of issues, ideas, artifacts, and events before accepting or formulating an opinion or conclusion. Success in all disciplines requires habits of inquiry and analysis that share common attributes.*

The performance indicators for critical thinking are shown in Fig. 7.3.

Fig. 7.3 Performance indicators/strategies for critical thinking (Rhodes 2010)

For critical thinking, the engineering undergraduates in their final year should be able (Rhodes 2010) to

- consider critically and comprehensively all relevant information necessary for full understanding of the problem;
- develop a comprehensive analysis based on competent and reliable source or information;
- evaluate the relevance of contexts when presenting a stand; and
- conclude logically based on detailed evaluation, and put evidences and perspectives in right order and manner.

Investigating these definitions further, one can come to a conclusion that innovation is actually a subset of creative thinking, which means that creative thinking involves innovation skills. Besides this, critical thinking can also be assumed to be a subset of creative thinking because all the activities or indicators under critical thinking will be carried out directly or indirectly during creative thinking. This leaves us with two important sets of skills that are vital to the engineering curriculum—namely, problem-solving and creative thinking skills. This is further evident by understanding that problem-solving skills are an important outcome in the Washington Accord, and the importance of creativity was further substantiated by a report issued by Royal Academy of Engineering, UK, in 2010 (Engineering Graduates for Industry 2010). Therefore, it has been justified that these skill sets, problem solving and creativity, are important to be taught to engineering undergraduates. The problem-solving skills focus on the process and not the end product; however, the creative thinking skills focuse on the end product. Hence, both these skills work hand in hand, that is, problem solving is a process in which the student selects and uses strategies to find a solution in a creative way. Besides this, as stated at the beginning of the chapter, problem solving and creativity are presented as important competencies in the requirements for European Engineer since these two are important drivers in economic prosperity (Lord Sainsbury of Turville 2007). In addition to this, creativity and problem solving are highly sought-after skills by potential employers (Engineering Graduates for Industry 2010) and are critical in maintaining an engineering edge in the industrial sector (Grand Challenges for Engineering 2012).

7.2.2 Content for Teaching Problem-Solving and Creative Thinking Skills: TRIZ

How would one typically solve a problem, especially a design problem? The answer is the traditional brainstorming, which is a common practice by many. Brainstorming is an activity which involves a group of people from different backgrounds sitting together and churning out ideas. Is brainstorming effective? Sad to say, *traditional brainstorming* is not effective in developing innovative solutions for a problem (Peace 2012). Why so? As we all know, brainstorming involves people; hence, the outcome of the session is only as good as the people that are together in the brainstorming session. If the team comprises experienced people from different fields and expertise, then there might be a possibility that the session may be fruitful provided none of these "experienced people" are experiencing psychological inertia. Psychological inertia is defined as the tendency of an individual to consider only those solutions that fall within his or her familiar solution space as a result of their psychological biases (Samuel and Jablokow 2010).

TRIZ is a "modern brainstorming" method that allows systematic thinking by using generic solutions to guide brainstorming towards feasible/specific solutions, close to ideal. TRIZ facilitates the problem-solving process. A typical generic problem-solving process involves the following key elements:

- Problem identification (identifying the problem and the root cause or governing factors/constraints)
- Solution generation
- Solution implementation
- Evaluation and solution refinement

In general, the problem process as highlighted previously is adopted for generating a specific solution for a given specific problem. What happens is that the solution is found within the same domain. For example, if the problem is mechanical, engineers will try to find solutions within the mechanical engineering domain. This, along with the inherent contradictions and constraints in the problem, will lead to a compromised solution. Here comes the advantage of TRIZ. It takes the problem and converts it into a generalized problem. For this generalized problem, various TRIZ tools are applied to obtain a generalized solution. From this, again TRIZ tools are used to shape the general solutions into specific solutions while resolving any contradictions and conflicts inherent to the specific problem. The beauty is that the solution most of the time will come from a different domain than the problem.

TRIZ tools are relevant to help with the general creative problem-solving stage of any process (Filmore 2008a). TRIZ tools for creativity and problem solving are easy to use for undergraduates, and studies have shown that TRIZ has more potential in problem-solving strategies and enhances the learner's solving abilities (Filmore 2006, 2007). TRIZ tools were appreciated more widely by recognizing their ability to overcome psychological inertia (Filmore 2007). In another study (Filmore 2008b) for producing breakthrough solutions among engineers, TRIZ was shown to have the capability of an extremely effective set of tools/approach to help break mind-sets and psychological inertia, and thus solving problems effectively. This study suggests that TRIZ methodology and its tools have very important contribution to the professional engineering community and that TRIZ should not be taken lightly. This study goes on further to prove that TRIZ has significance in problem solving over tools such as failure mode and effect analysis (FMEA), Taguchi, Six Sigma, quality function deployment (QFD), and total quality management (TQM). Appendix 1 details various TRIZ tools and their capability in problem solving and creativity. In a nutshell, the capability of TRIZ tools (some) can be summarized in Fig. 7.4. The flow of problem solving in TRIZ is shown in Fig. 7.5 (general), and a more detailed flow is given in Fig. 7.6 (San et al. 2011a).

With reference to Fig. 7.6, the problem-solving process starts off with the initial problem at hand. For example, it could be high failure rate of products. This problem is vague and does not really provide enough information to solve the problem. Hence, the problem has to be refined further and TRIZ has tools to assist in this. The first stage is called the *function analysis.* The objective of the function analysis is to map out how components or elements interact with each other in the engineering system as shown in Fig. 7.7. The functions could either be useful, harmful, insufficient, or missing. The second stage is the *cause and effect chain analysis.* The cause and effect chain analysis helps to identify the main root cause or causes of a problem based on the understanding of the function analysis. The problem is eliminated by using another stage in TRIZ called *trimming.* Trimming is used to remove unnecessary functions and components in a system without altering the original function of the device. If trimming cannot be used, then the "heavy artillery" of TRIZ is called in.

These TRIZ stages/tools can be classified as tools for model of problem and tools for model of solution as shown in Fig. 7.6. There are four classes of model of problem—namely, engineering contradiction, physical contradiction, function model, and substance-field model. For the model of solutions, we have the specific inventive principles, specific scientific effect, and specific standard inventive solution. The bridge between the problem and the solution is the TRIZ tools such as contradiction matrix, separation/satisfaction bypass, scientific effects, and system

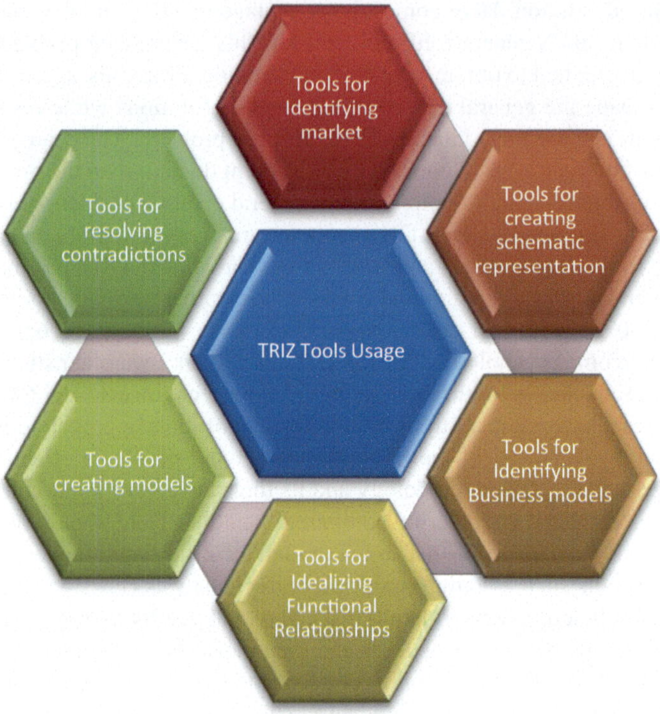

Fig. 7.4 Some of the TRIZ tools characterization

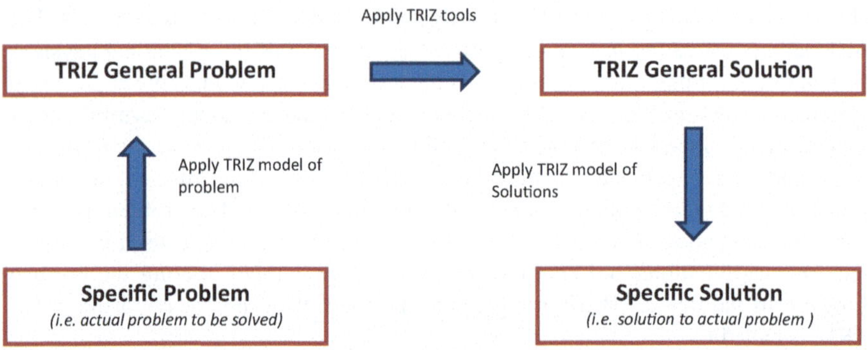

Fig. 7.5 TRIZ problem-solving process (general)

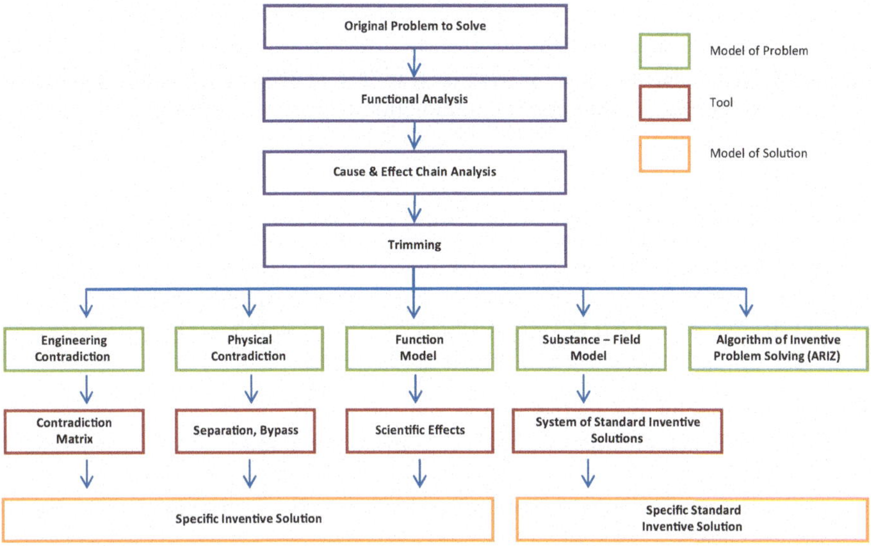

Fig. 7.6 TRIZ problem-solving map (San et al. 2011b)

Fig. 7.7 Functional analysis of a washing machine (Ball et al. 2012)

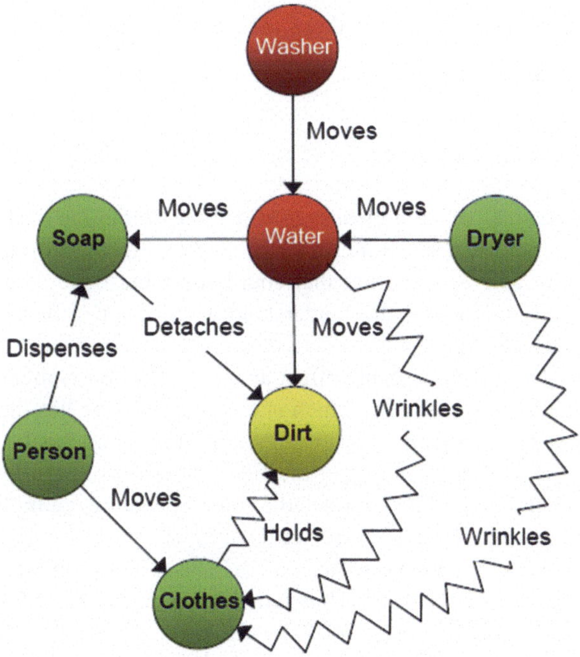

of standard inventive solutions (San et al. 2011a). Only the engineering contradiction route will be discussed further in detail. Model of problem is an important problem-solving step in TRIZ, especially with problems that have contradictions. For the engineering contradiction, contradiction matrix is used to generate potential inventive principles. There are a total of 40 inventive principles derived from a study of 200,000 patents (San et al. 2011b). To illustrate the basic principles of TRIZ in problem solving, an example will be used.

Is problem solving unique? Meaning, each time a problem is solved, a new "product" is created. The answer is *yes* and *no*. *Yes* because the solution is unique to the problem, *no* because the solution may have come from other relevant problems or field; hence, it may not be entirely "new" out from nothing. There are two categories of problem-solving schemes as described below:

- Needs are met by using either new or old systems:
  - New systems: meeting specific needs by creating new technologies and new systems
  - Old systems: bringing a few old systems together in an innovative way to fullfill needs
- Finding new uses for features, technologies, and functions:
  - New technologies: whatever new technology or system has been created, try finding its usage in other fields
  - Old technologies: find new uses for old technologies for different field

TRIZ adopts both of these problem-solving schemes very well as has been described briefly in Figs. 7.4 and 7.6.

7.2.2.1 Example

A typical vacuum cleaner uses a filter to trap the dust. When the vacuum cleaner is used for a long time, it was found that the suction power reduces. The exercise is to derive a solution that could solve the issue of decreasing suction power in the vacuum. To better understand the problem, a function analysis will be done as shown in Fig. 7.8:

From the cause and effect analysis, it is observable that the dust does a harmful function to the filter, which is clogging it. Here lies the contradiction. The filter is needed to perform a useful function of trapping the dust, but the dust does a harmful function by clogging the filter, which results in lower suction power. Hence, the problem will be modeled as an engineering contradiction and the contradiction matrix will be used to device a solution.

What is a contradiction matrix? This is shown in Fig. 7.9 (a portion of the matrix; the complete matrix is shown in the appendix). TRIZ founder Genrich S. Altshuller, while going through 40,000 most inventive patents, found that there are around 1250 technical contradictions which he was able to assemble in a 39×39 matrix called the contradiction matrix. To resolve these contradictions, again through the 40,000 most inventive patents back in 1950, Altshuller invented 40 inventive prin-

Fig. 7.8 Function analysis of a vacuum cleaner

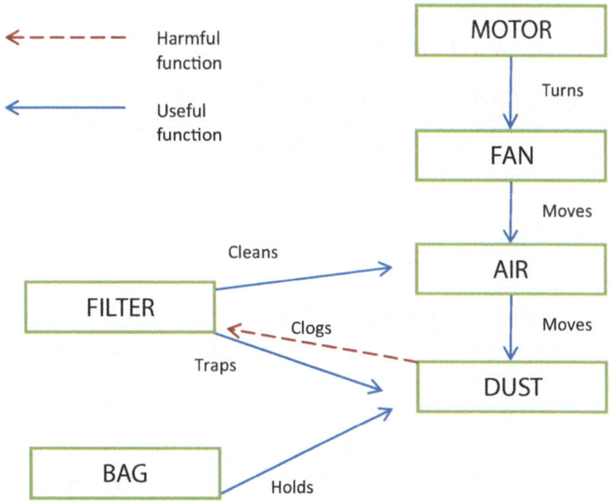

Worsening Feature → / Improving Feature ↓		Length of stationary object	Area of moving object	Area of stationary object	Volume of moving object	Volume of stationary object	Speed
		4	5	6	7	8	9
1	Weight of moving object		29, 17, 38, 34		29, 2, 28, 40		2, 8, 15, 38
2	Weight of stationary object	10, 1, 29, 35		35, 30, 13, 2		5, 35, 14, 2	
3	Length of moving object		15, 17, 4		7, 17, 4, 35		13, 4, 8
4	Length of stationary object			17, 7, 10, 40		35, 8, 2, 14	
5	Area of moving object				7, 14, 17, 4		29, 30, 4, 34

Fig. 7.9 Contradiction matrix to yield the inventive principles

ciples (see appendix for a complete list). These 40 principles are not a direct solution, but rather they guide the designer in the direction that best creates innovative solutions to resolve these contradictions. In our example, the improving parameter in the contradiction matrix is the capability to trap dust, but the worsening parameter is the decrease in suction power.

By using the matrix, the improving parameters that can be used to resemble the function of trapping dust are:

- Parameter 1: weight of moving objects, Parameter 23: loss of substance

The worsening parameters that can be used to resemble the function of decreasing suction power are:

- Parameter 7: volume of moving object, Parameter 9: speed of air flow, Parameter 22: loss of energy

To resolve the contradiction between parameters 1 and 7, the following inventive principles are suggested: 29, 2, 40, and 28. These principles are used to derive solutions to resolve the contradiction, that is, solving the problem of the vacuum cleaner. Similarly, parameters 1 and 9, 1 and 22, 23 and 7, 23 and 9, and 23 and 22 can be used to derive more possibilities of inventive principles. This will create more concepts and potential ideas to solve the problem at hand. This is summarized in Table 7.1.

The example here has highlighted how TRIZ tools can very easily be used to solve problems creatively. The tools have displayed the basic process of problem solving, which is problem definition and solution implementation. These two initial steps are the most critical in any problem-solving process. Compare this with the traditional brainstorming to define the problem and generate solutions. It is evident that indeed TRIZ tools solve problems quick and creatively. Hence, as supported by various studies, it is time to introduce TRIZ to the engineering undergraduates as an effective tool for problem solving and creativity.

7.2.3 Problem-Based Learning As the Delivery Mode

PBL will be adopted as the pedagogy tool to instill problem-solving and creativity skills. The framework consists of a semester-long course where the course will be broken up into four levels. The *first level* is basically learning and applying basic TRIZ tools to solve simple and noncomplex engineering problems. The objective here is to gain the motivation from students. Students need to develop the motivation in order to have confidence that problem solving is easy and fun. Basic TRIZ tools such as *contradiction matrix* and *40 inventive principles* will be used. The instructor will share the information/knowledge by means of case studies and examples. Students will then be asked to solve simple problems individually. Later, the same problem will be solved in groups. The focus here is on the number of ideas and not the quality (divergent thinking). Once a list of solutions has been generated, students will be asked to rank the solutions from the most feasible to the least and to justify their decision.

Table 7.1 Idea generation by using the contradiction matrix tool

Improving feature	Worsening feature	Inventive principles	Ideas/potential solutions
1—Weight of moving object	7—Volume of moving object	29—Pneumatics and hydraulics	Direct stream of air to remove dust from filter/use liquid to remove dust
		2—Taking out	Design the filter such that it is easily removed and cleaned
		40—Composite material	Design the filter material such that it is difficult for dust to stay on it
		28—Mechanic substitution	Replacing the filter with other means of mechanically removing the dust
1—Weight of moving object	9—Speed	15—Dynamic	Filter can rotate. As it gets heavy with dust, it rotates to a new section
		38—Strong oxidant	Ionizing radiation on dust particles
23—Loss of substance	7—Volume of moving object	1—Segmentation	Filter has many segments. Once it gets clogged, it can be rolled to a new segment/section
		30—Flexible shell	Filter made from thin film that can be easily replaced
23—Loss of substance	9—Speed	10—Preliminary action	To wash the filter before use
		13—The other way around	To blow air from the riverside of the filter

Level 2 consists of introduction to *functional analysis, root cause analysis,* and *trimming* along with the tools from level 1. Again similar to level 1, the instructor will share the information/knowledge by means of case studies and real-life engineering problem examples. The objective here is to introduce the students to systematic process for problem solving. The real-life engineering problems that the students have to undertake will not be too challenging, at the same time not trivial. The teaching method can include case studies, role-playing, and team work. For case studies, it will be good to use patents. Students can be taught by using TRIZ trimming tool to circumnavigate patents and to come up with new product without infringing the initial patent. At the end of level 2, students should be motivated and have the understanding of creative problem-solving skills.

Level 3 is similar to level 2, except that here it is important to have problems that include global issues such as sustainability, environmental–economic trade-offs, and safety and regulation issues. TRIZ tools such as *evolution trends* and *substance-field model* will be discussed and explored through real-life open-ended problems. After this level, students will have a clear picture of the problem-solving process, and they will be confident in deriving creative solutions for real engineering problems.

Level 4 is the pinnacle of the process. The objective of this level is to improve students' readiness and effectiveness in solving real-life engineering problems. At this juncture, the instructor will only play the role of a facilitator. Students will have to develop hands-on approach on unsolved real-life problems which are complex in nature (timing, effective cost, effective material usage, optimization, etc.). The problems here have to be developed in consultations with relevant industry players, regulatory bodies, government ministries, and agencies. In this context, universities in the MENA region can develop projects specific to the MENA region and have the students solve them creatively.

The four levels discussed here can be introduced in a course for one semester or combining Levels 1–2 and Levels 3–4 into two semesters long courses where the former is the prerequisite the latter.

7.2.4 Assessment of Problem Solving and Creativity

Academicians involved in teaching creativity and innovation in the context of problem solving have great difficulties in consistently assessing students. One suggested method is the usage of well-defined rubrics (Peirce 2006). These rubrics will aid the faculty members in evaluating students consistently and efficiently, and it also allows students to understand how they will be assessed. Indirectly, it is a positive feedback to students to better understand the elements behind creativity and problem solving. Rubrics can be used to evaluate programs, courses, and individual student assignments and projects. Rubrics in general can be classified as holistic and analytical (Peirce 2006). The holistic and analytical rubrics are shown in Figs. 7.10 and 7.11, respectively.

The analytical rubric has more information than holistic rubric. The holistic rubric combines different kinds of traits into a single category, whereas the analytical rubric separates these traits. Although analytical rubric consumes time to score, they can be useful to departments assessing student's thinking skills in assignments and projects in multi-section courses to determine which areas of student thinking need more attention in the course (Peirce 2006). Hence, the proposed rubric in this framework will be the analytical rubric.

As discussed previously, the two essential skills are creative thinking and problem-solving skills. Hence, the rubrics to be shared here will be covering these two

Fig. 7.10 Holistic rubric

STRONG (4)	Consistently does all or almost all of the following: • Item 1 • Item 2 • Item 3
ACCEPTABLE (3)	Does most of the following: • Item 1 • Item 2 • Item 3
UNACCEPTABLE (2)	Does some of the following: • Item 1 • Item 2 • Item 3
WEAK (1)	Little evidence of achieving the following: • Item 1 • Item 2 • Item 3

PROBLEM IDENTIFICATION	
Emerging	**Mastering**
Does not identify and summarize the problem. Is confused or identifies a different and inappropriate problem Does not identify or is confused by the issue or represents the issue in accurately	Identifies the main problem and subsidiary, embedded or implicit aspects of the problem, and identifies them clearly, addressing their relationships to each other. Identifies not only the basics of the issue, but recognizes nuances of the issue
HYPOTHESIS	
Addresses a single source or view of the argument and fails to clarify the established or presented position relative to one's own. Fails to establish other critical distinctions	Identifies appropriately one's own position on the issue, drawing support from experience and information not available from assigned sources

Fig. 7.11 Analytical rubric

skills only. With reference to Figs. 7.1 and 7.2, the key elements or traits for problem solving and creative thinking will be used to generate the proposed rubrics to assess these skills. The rubrics for creative thinking and problem solving are depicted in Tables 7.2 and 7.3, respectively. These rubrics were developed based on the rubrics developed by AAC&U (Rhodes 2010):

Table 7.2 Assessment rubrics for creative thinking (Adapted from Rhodes 2010)

	Mastering (4)	Emerging (3)	Growing (2)	Infancy (1)
Acquiring competency *(acquiring knowledge on problem solving/ product development processes)*	Reflects, evaluates, and confidently uses the most appropriate problem-solving process	Uses existing methodologies confidently and efficiently	Uses some similar example as a basis for initiating problem-solving methodology	Directly copies an example
Taking risks *(ability to take risk for new strategies of solutions that are not a norm)*	Trying out different (risky) strategies or solutions which an ordinary person will not dare to attempt. Basically, boldly going out of the boundaries of the project or problem	Attempts to incorporate new ideas or strategies which are low in risk, or have low probability of failure	Attempts new directions but within the safety zone where the probability of failure is close to zero	Does not attempt a new solution. Uses existing solutions or strategy with minor tweaking to fit the problem given
Solving problems *(ability to create sound and consistent methodology for the problem at hand, and having the capability to generate and evaluate solutions)*	Creates a good and sound methodology to solve the problem, generates potential solutions, and uses sound evaluation tools to select and defend a final decision on the solution to be adopted	Creates a logical, consistent plan to solve the problem	Somewhat creates a plan to solve the problem	Only a single approach is considered and is used to solve the problem
Embracing contradictions *(Every problem has a contradiction. The best solution is the one which solves the problem and contradiction simultaneously without having to strike a balance via optimization)*	Recognizes the engineering and physical contradictions to a problem, and solves the problem without a compromise (no optimization)	Recognizes the contradictions but obtains a solution that is a compromise between the actual solution and contradictions (optimization)	Recognizes the contradiction but is not able to solve the problem due to the contradiction. If the contradiction was removed, the problem could be solved	Not aware of the existence of any contradictions
Innovative thinking	Creates a new or novel product based on the technologies of solutions from different field or discipline	Creates a new or novel solution from the same field or discipline		Minor modification to existing ideas to solve the problem
Connecting, synthesizing, transforming	Innovatively connecting ideas/solutions to produce a novel stand-alone solution	Creates a new solution by combining ideas	Is able to connect basic ideas/ solutions to form a semi-new solution	Identifies and explains connections between ideas/ solutions

Table 7.3 Assessment rubrics for problem solving (Adapted from Rhodes 2010)

	Mastering (4)	Emerging (3)	Growing (2)	Infancy (1)
Define problem	Ability to identify the problem clearly and design an insightful problem statement	Able to identify the problem and capable of designing a sound problem statement	Able to identify the problem but has difficulty in creating a proper problem statement	Able to identify the problem but cannot create a problem statement
Identify strategies	Ability to identify multiple strategies that can be used to solve the same problem at hand	Able to identify strategies, but not all are capable of solving the problem at hand	Able to identify a strategy, that is, capable of solving the problem at hand	Unable to identify a strategy, that is, not capable of solving the problem at hand
Propose solutions	Proposes one or more solutions/alternatives that indicate a deep comprehension of the problem and sensitive to all ethical, ecological, and cultural dimensions of the problem	Proposes one or more solutions/alternatives that indicate comprehension of the problem and sensitive to *one* of the following: ethical, ecological, and cultural dimensions of the problem	Proposes one solution/alternative that is generic and not specific to the problem	Proposed solution does not reflect the problem statement or the problem at hand
Evaluate solutions	Evaluations are thorough and insightful which take into account feasibility impact of the solution	Evaluations are sufficient which take into account feasibility impact of the solution	Evaluations are brief with minor examination on the feasibility and impact of the solution	Evaluations are superficial with little evidence of examination on the feasibility and impact of the solution
Implement solutions	Solution addresses thoroughly and deeply multiple contextual factors of the problem	Solution addresses superficially multiple contextual factors of the problem		Solution does not address the problem
Evaluate outcomes	Reviews results relative to the problem defined with thorough, specific considerations for the need for further work	Reviews results relative to the problem defined with some/little consideration for the need for further work		Reviews results relative to the problem defined with no plans for further work

For the usage of the suggested rubrics, faculty members are encouraged to assign a zero to any work sample or collection of work that does not meet infancy-level performance.

7.3 Anticipations

It is hoped that the proposed framework will be able to assist teaching faculty members in universities to move out from traditional ways of teaching problem-solving and creativity skills. The proposed framework will be easy to implement and focused on outcome. Besides this, the number of patents generated by undergraduates should increase, creating confidence and removing any sort of physiological inertia towards creativity and problem solving. In particular to the MENA region, the proposed framework is easy to implement and is relevant.

7.4 Conclusions

A framework is proposed to instill the skills of problem solving and creativity in engineering undergraduates in the MENA region. The proposed framework encompasses content, delivery, and assessment means for instilling problem-solving and creativity skills. TRIZ was selected as the content for teaching creativity and problem solving. TRIZ is an efficient problem-solving tool that has been widely used by international companies such as Samsung, Procter & Gamble, and Intel, to name a few. The delivery method will be centered on PBL. This is achieved by dividing the content into four levels. The four levels discussed here can be introduced in a course for one semester or combining Levels 1–2 and Levels 3–4 into two semesters long courses where the former is the prerequisite the latter. The assessment will be achieved via analytical rubrics. The proposed framework will most likely improve the level and quality of solution and reduce the time to generate solutions among undergraduates. This can be used as a basis to translate the findings to secondary and primary schools to start teaching creativity and problem solving at a younger age. The goal of this chapter is not to create Einstein or Thomas Edison, but rather to create students to be more productive and innovative in a meaningful way.

Appendices

Appendix 1: Annotated list of key TRIZ components (Souchkov 2006)

TRIZ tool	Characteristics
Theory of technical systems evolution	Main theoretical foundation of TRIZ. A philosophy behind the theory of technology evolution is that every man-made product which was designed to deliver certain functional value tends to evolve in a systematic way according to generic patterns and trends of evolution
Laws and trends of technology evolution	The TRIZ trends and laws are very powerful knowledge which provides the basis to predict what will happen next with a selected product or a technology from the perspective of internal evolutionary potential of a system
Multiscreen diagram (also known as 9 windows, or 9 screens, or system operator)	The multiscreen diagram of thinking specifies that any specific system (product, technology, organization, etc.) can be viewed at least from three layers: system (the system itself within its boundaries), its subsystems, and supersystem. Although not easy to use, the multiscreen diagram of thinking is a very powerful tool of system analysis and forecast
Ideality	Ideality of a major trend of man-made systems evolution. The degree of ideality is defined as a ratio of the overall performance of a system (everything that creates value) minus harmful effects produced by the system (everything that diminishes its value) to costs necessary to achieve its performance (everything which is needed to create value). Ideality in TRIZ is a qualitative measure which is not directly calculated but serves as a major guideline during processes of problem solving and new idea generation
Ideal final result (IFR)	Enables formulating target solutions in terms of ideality. Formulation helps to correctly set up goals, fight mental inertia, and design cost-effective products and services
Contradiction	A contradiction in TRIZ is a primary problem model which is used to formulate inventive problems. Contradiction is a main feature which distinguishes an ordinary problem from an inventive problem
Resource analysis	During problem solving, resources play a major role in TRIZ. The proper use of available resources helps to obtain more cost-effective and ideal solutions without complicating a system and introducing new expensive components and materials
Function analysis (also known as function-attribute analysis)	Utilizing the same basic approach to modeling existing products in terms of components and functions delivered by the components, FA differs from VEA in a way of how function is defined. In FA, the function is regarded as an effect of a physical interaction between two system components. FA is very useful to conduct a systematic analysis of products and formulate problems in terms required by the other TRIZ problem-solving techniques

TRIZ tool	Characteristics
Root conflict analysis	A technique for casual decomposition of complex problems and invention situations into effects and causing conflicts (contradictions). Helps to map and visualize all system conflicts as well as reveal hidden conflicts. Root cause analysis to identify root problems in inventive situations
40 inventive principles for resolving technical contradictions	Inventive principles for technical contradiction elimination are used to eliminate problems represented in terms of technical contradictions. Inventive principles describe either solution pattern which can be applied to resolve the contradiction or a direction in which a problem has to be solved. There are 40 inventive principles for resolving technical contradictions available in TRIZ
Contradiction matrix	The first technique and still the most popular. TRIZ states that to obtain inventive solution, the contradiction has to be eliminated while no compromise is allowed. The necessity to eliminate contradictions is the driving force of technological progress. The matrix was designed on the basis of 39 generalized parameters. The same lists of parameters are placed along vertical and horizontal axes of the matrix. A point of intersection of two generalized parameters indicates which inventive principle(s) is to be used in each particular situation
Substance-field analysis	Any technical system (product, machinery, technology) or its part can be modeled as a number of substance components interacting with each other via fields. Unlike physics, TRIZ introduces six types of fields: mechanical, acoustic, thermal, chemical, electric, magnetic, and electromagnetic. Abstract physical modeling of the system's part which causes a problem helps to identify and classify a specific interaction which does not meet the required specifications. The unsatisfactory interaction might be of four types: (i) insufficient or poorly controllable to obtain the desired result, (ii) excessive and produces more action than required, (iii) harmful, when the interaction is necessary to obtain a positive effect but results in a negative side effect, and (iv) missing—an interaction is necessary in the system but we do not know how to introduce it. Substance-field modeling and analysis are used for problem modeling and further solving with 76 inventive standards
76 inventive standards	In case a system is modeled in terms of physical components and interactions via substance-field modeling, and a problem is represented as an unsatisfactory interaction, TRIZ recommends to use special rules which contain abstract patterns indicating how the physical model given has to be modified by: (a) replacing the existing components with other components, (b) introducing new components, (c) modifying the existing components, and (d) changing a system structure

TRIZ tool	Characteristics
Algorithm for inventive problem solving (ARIZ)	One of the most powerful and complex analytical TRIZ techniques which helps in solving those problems that cannot be solved with the use of other TRIZ techniques. Since the abovementioned TRIZ techniques operate with direct modeling of a problem and finding a relevant solution pattern or a principle from the TRIZ databases, it is not always possible to formulate the problem directly in the right way. ARIZ helps to extract a core problem through comprehensive analysis of the problem conditions and fighting mental inertia
Trimming (also known as idealization)	A technique which helps to make existing systems and products more ideal by eliminating their components without impairing overall system's performance, functionality, and quality. Usually performed after a system is represented as a function model with the help of function analysis
Alternative system merging (also known as feature transfer, hybridization)	A technique which helps to develop new products on the basis of combining features of two competitive products. Usually competitive products are featured by different sets of advantages and disadvantages. The technique helps to design a new product that inherits advantages of the competitive products while disadvantages are eliminated. However, direct merging of features might be difficult due to a number of contradictions arising when we attempt to develop such product. For this reason, the TRIZ techniques are recommended to use after the contradictions are identified

FA function analysis, *VEA* value engineering analysis

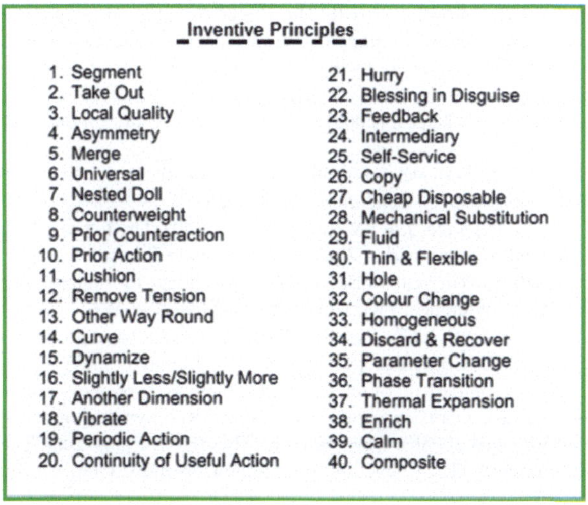

Contradiction Matrix Parameters

1. Weight of Moving Object
2. Weight of Stationary Object
3. Length of a Moving Object
4. Length of a Stationary Object
5. Area of a Moving Object
6. Area of a Stationary Object
7. Volume of Moving Object
8. Volume of Stationary Object
9. Speed
10. Force (Intensity)
11. Stress or Pressure
12. Shape
13. Stability of the ObjectÕs Composition
14. Strength
15. Duration of Action by a Moving Object
16. Duration of Stationary of Moving Object
17. Temperature
18. Illumination Intensity
19. Use of Energy by Moving Object
20. Use of Energy by Stationary Object
21. Power
22. Loss of Energy
23. Loss of Substance
24. Loss of Information
25. Loss of Time
26. Quantity of Substance/Matter
27. Reliability
28. Measurement Accuracy
29. Manufacturing Precision
30. External harm affects the object
31. Object-generated harmful factors
32. Ease of Manufacture
33. Ease of Operation
34. Ease of Repair
35. Adaptability or Versatility
36. Device Complexity
37. Difficulty of Detecting and Measuring
38. Extent of Automation
39. Productivity

References

Adams, J. (2009). Problem solving and creativity in engineering: Turning novices into professionals. *Enhancing the Learner Experience in Higher Education, 1*(1), 4–18.

Ball, L., Troness, D., Ariyur, K., Huang, J., Rossi, D., Krupansky, P., & Hickman, S. (2012). TRIZ power tools: Working with functions, April 2012, Copyright 2012 by Collaborative Authors.

Belski, I. (2005). *Improving the skills of engineers in systematic thinking*. Proceedings of the 2005 ASEE/AaeE 4th Global Colloquium on Engineering Education.

Belski, I. (2007). *Improvement of thinking and problem solving skills of engineering students as a result of a formal course on TRIZ thinking tools*. Melbourne: Royal Melbourne Institute of Technology.

Engineering Graduates for Industry. (2010). The Royal Academy of Engineering. http://www.raeng.org.uk/education/scet/pdf/Engineering_graduates_for_industry_report.pdf.

Esquivel, G. B. (1995). Teacher behaviors that foster creativity. *Educational Psychology Review, 7*(2), 185–202.

Expert Group on Future Skills Needs. (2009). *Skills in creativity, design and innovation*. Ireland: Forfas Group.

Felder, R. M., & Brent, R. (2003). Designing and teaching courses to satisfy the ABET engineering criteria. *Journal of Engineering Education, 92*(1), 7–25.

Felder, R. M., Woods, D. R., Stice, J. E., & Rugarcia, A. (2000). The future of engineering education II: Teaching methods that work. *Chemical Engineering Education, 34*(1), 26–39.

Filmore, P. R. (April 2006). The real world: TRIZ in two hours for undergraduate and master's level students. TRIZ Conference, TRIZCONF2006, Milwaukee, WI.

Filmore, P. R. (23rd–25th April 2007). Teaching TRIZ as a systematic problem solving method: Breaking mindsets. TRIZ Conference, TRIZCON2007, Louisville, KY.

Filmore, P. R. (2008a). A comparison of the problem solving and creativity potential of engineers between using TRIZ and Lean/Six Sigma. The Fourth TRIZ Symposium in Japan 2008.

Filmore, P. R. (13-15th April 2008b). Developing highly effective engineers: The significance of TRIZ. TRIZ Conference, TRIZCON2008, Kent State University, Ohio, USA.

Fink, F. K. (2002). Problem-based learning in engineering education: A catalyst for regional industrial development. *World Transactions on Engineering and Technology Education, 1*(1), 29–32.

Ghabra, S. (2010). Student-centered education and American-style universities in the Arab World. Higher education and the Middle East: Serving the knowledge-based economy, Viewpoint Special Edition, pp 21–26.

Grand Challenges for Engineering. (2012). National Academy of Engineering of the National Academies. The submit series on the grand challenges. http://www.engineeringchallenges.org.

Grolinger, K. (2011). Problem based learning in engineering education: Meeting the needs of industry. *Teaching Innovation Projects, 1*(2), 1–7.

Hitt, J. (2010). Problem-based learning in engineering. Center for teaching excellence. United States Military Academy. West Point, NY. (This paper was completed and submitted in partial fulfillment of the Master Teacher Program, a 2-year faculty professional development program).

Issan, S. A., & Gomaa, N. M. M. (2010). Post basic education reforms in Oman: A case study. *Literacy Information and Computer Education Journal (LICEJ), 1*(1), 19–27.

Kimmel, S. J., Kimmel, H. S., & Deek, F. P. (2003). The common skills of problem solving: From program development to engineering design. *International Journal of Engineering Education, 19*(6), 810–817.

Liu, Z., & Schonwetter, D. (2004). Teaching creativity in engineering. *International Journal of Engineering Education, 20*(5), 801–808.

Lord Sainsbury of Turville. (2007). *The race to the top: A review of government's science and innovation policies*. London: HM Treasury.

Loveluc, L. (March 2012). Background paper education in Egypt: Key challenges. Middle East and North Africa Programme, Chatham House.

Mrabet, J. G. (July 2010). Western education in the Arabian Gulf: The costs and benefits of reform. Higher education and the Middle East: Serving the knowledge-based economy, Viewpoint special edition, pp. 47–51.

Overton, T. (2003). Key aspects of teaching and learning in experimental sciences and engineering. In H. Fry, S. Ketteridge, & S. Marshall (Eds.), *A handbook for teaching and learning in higher education* (2nd ed., pp. 255–277). London: Routledge Falmer.

Pappas, J., & Pappas E. (2003). Creative thinking, creative problem-solving and inventive design in the engineering curriculum: A review. Proceedings of the 2003 American Society for Engineering Education Annual Conference & Exposition.

Peace, N. (9th April 2012). Why most brainstorming sessions are useless. www.forbes.com.

Peirce W. (January 13, 2006). Designing Rubrics for Assessing Higher Order Thinking. Howard Community College Columbia, MD.

Petruska J. (2010). Introduction of the problem based learning to mechanical engineering curricula. 3rd International Symposium for Engineering Education, University College Cork, Ireland.

Poikela, E., & Poikela, S. (Eds.). (2005). *PBL in Context-Bridging Work and Education*. Tampere: Tampere University Press.

Pop-Iliev, R., Platanitis, G. (2008). *A rubrics based methodological approach for evaluating the design competency of engineering students*. Proceedings of the TMCE 2008.

Reidsema, C. (2005). Fostering creative problem solving and collaborative skills through impromptu design in engineering design courses. Proceedings of the 2005 ASEE/AaeE 4th Global Colloquium on Engineering Education.

Rhodes, T. L. (2010). Assessing outcomes and improving achievements: Tips and tools for using rubrics. Association of American Colleges and Universities, Washington DC.

Sager, A. (December 2007). The Private Sector in the Arab World – Road Map towards Reform *, Arab Reform Initiatives, Arab Reform Brief, V 19.

Samuel, P., Jablokow, K. (October 15–16 2010). Psychological inertia and the role of idea generation techniques in the early stages of engineering design. Fall 2010 Mid-Atlantic ASEE Conference, Villanova University.

San, Y. T., Jin, Y. T., & Song, C. L. (2011a). *TRIZ: Systematic innovation in manufacturing* (2nd ed.). Firstfruits Publications, Malaysia.
San, Y. T., Jin, Y. T., & Song, C. L. (2011b). *TRIZ: Theory of inventive problem solving—systematic innovation in manufacturing*. Fruits Publication, Malaysia.
Souchkov, V. (2006). Annotated list of key TRIZ components. ICG Training & Consulting.
Steiner, T., Belski, I., Harlim, J., Baglin, J., Ferguson, R., & Molyneaux, T. (2011). Do we succeed in developing problem-solving skills-the engineering students' perspective? Proceedings of the 2011 AAEE Conference, Fremantle, Western Australia.
Stratton, R., & Mann, D., Otterson, P. (2010). The theory of inventive problem solving (TRIZ) and systematic innovation: A missing link in engineering education? www.system-innovation.com\.
Walsh, P. (2007). *Problem based learning in engineering*. International Symposium for Engineering Education, Dublin City University, Ireland.
Walters, T., Walters, L., Barwind, J. (July 2010) Tertiary education in the Arabian Gulf: "A Colossal Wreck, Remaining Boundless and Bare?" higher education and the Middle East: Serving the knowledge-based economy, Viewpoint Special Edition, pp. 18–20.

Faris Tarlochan is currently holding the position of Professor and Program Coordinator in the Department of Mechanical and Industrial Engineering, Qatar University. He is a Fellow with the Institution of Mechanical Engineering UK and a registered Chartered Engineering with the Engineering Council of UK.

Abdel Magid Hamouda is currently a Professor in the Department of Mechanical and Industrial Engineering, Qatar University. He is the Associate Dean for Research and Postgraduate for the College of Engineering. Hamouda is corresponding member for the American Academy of Mechanics (AAM), member of the American Society of Mechanical Engineering (ASME), and member of the Institute of Electrical and Electronic Engineers (IEEE). He is also a senior member of Institute of Industrial Engineering (IIE), USA, Member of the Institute of Mechanical Engineers (IMechE), UK; and Member of the Institute of Highway Transportation, UK.

Chapter 8
Active Blended Learning to Improve Students' Motivation in Computer Programming Courses: A Case Study

Saleh Alhazbi

8.1 Introduction

Ordinarily, computer programming is a compulsory subject for computing majors. Moreover, it is an integral part of any engineering program to equip students with computer-based problem-solving skills to tackle computational problems using logical and numerical approaches. Recently, there has been increasing interest in using programming as a tool to infuse computational thinking in K-12 curricula as a necessary skill for everyone in the twenty-first century (Grover and Pea 2013; Thomas et al. 2011). However, students usually find computer programming to be a difficult subject. This problem has been a concern for educational institutions since the early 1960s. There has been much research to investigate this problem (Butler and Morgan 2007; Lahtinen et al. 2005; Robins et al. 2003) and to improve student learning through different proposed pedagogical approaches (Esteves et al. 2011; Estefanía et al. 2010; Merrick 2010; Yan 2009; Miliszewska and Tan 2007; Guo 2006; Deraadt et al. 2005), subject curriculum reviews (Summet et al. 2009; Doyle 2005), and assessment methods updates (Simon et al. 2012; Petersen et al. 2011).

8.2 Reasons for Programming Learning Difficulties

There are various reasons behind the difficulties faced by novice students when learning programming; these difficulties can be categorized as follows:

1. *Problems related to students*
   Although students, lately, come from high schools with prior computing experience and maybe some programming knowledge, they are often not well prepared to develop and apply problem-solving techniques to write computer programs

S. Alhazbi (✉)
Department of Computer Science and Engineering, Qatar University, Doha, Qatar
e-mail: salhazbi@qu.edu.qa

M. Abdulwahed et al. (eds.), *Advances in Engineering Education in the Middle East and North Africa,* DOI 10.1007/978-3-319-15323-0_8

(Yan 2009). They lack the required skills to develop correct algorithms and just rush to write the code before thinking carefully (Gomes and Mendes 2007). Additionally, they are not patient enough to continue modifying the program and fixing errors to reach the complete and correct version. Perkins et al. (1989) classify novice students into two types: "stoppers" and "movers." As the names imply, stoppers cannot continue when confronted with a problem, while movers use feedback about errors effectively and are able to progress.

The main problem with students is their unawareness of the nature of this subject. Learning programming not only involves understanding the programming language, for example, syntax and data types, but also includes developing multiple other skills, such as analyzing problems, developing algorithms, writing code, debugging and fixing errors (Jenkings 2002). Developing such skills requires intensive practice of solving problems and writing programs outside the classroom; however, the problem is that many students believe it is enough to read the textbooks and understand language syntax (Gomes and Mendes 2007). Consequently, they are unaware of their own deficiencies (Ala-Mutka 2004). Milne and Rowe (2002) found that students' awareness of their programming limitations is less than what teachers believe it to be; therefore, helping students to recognize their deficiencies results in them applying more effort to overcome their weaknesses and improve their learning.

2. *Problems related to subject nature*

   Novice learners find some programming concepts difficult to grasp, and they struggle to conceptualize programming conventions such as variables, memory addresses, and data types because such constructs do not correspond to things in real life (Miliszewska 2008; Dunican 2002). Another challenge of programming is utilizing abstract design; Qualls (Qualls et al. 2011) defines abstraction as the process of focusing on a problem's major details rather than thinking of the many minor ones. In programming, learners need to think of the problem at different levels of abstraction; otherwise, they will be overwhelmed by the many details and become frustrated.

3. *Problems related to teaching methods*

   Educators bear some responsibility for the difficulties faced by students in programming courses. Many instructors ignore the nature of this subject by focusing more on syntactic details rather than on training students in problem-solving skills. Another problem in this regard is the teaching method: Traditional lecturing seems to be an ineffective method for teaching programming. Because teaching programming mainly focuses on problem solving, students require immediate feedback to help them overcome their problems and learn from that. Consequently, students need more personalization to address different individuals' problems. Programming learners need to develop multiple skills; however, this might be difficult to implement in class because of time constraints and course sizes (Gomes and Mendes 2007). Generally, students need to practice programming on their own outside the classroom, which requires more support beyond the classroom time to keep their progress and not become frustrated when facing problems.

8.3 Motivating Students to Overcome Difficulties

The difficulties in learning programming usually frustrate students and lead many of them to either drop out of the major or continue with fear of programming tasks in subsequent courses, which makes them eventually choose career paths that do not require programming skills (Stamouli et al. 2004). For that reason, motivating students in programming is very essential to improve their retention within computing majors and to enhance their learning of the subject. Motivating students can be defined as the process of improving their willingness to participate and be successful in the learning process (Bomia et al. 1997, p. 8). Generally, motivation is an important cognitive factor for successful learning in any subject; it affects a student's learning efforts, which consequently affect his/her performance. Learning motivation can be classified into two types: intrinsic and extrinsic. Intrinsic motivation refers to the performance of an activity because of its inherent satisfaction rather than external reasons such as rewards, while extrinsic motivation can be defined as doing the activity to attain an outcome (Ryan and Deci 2000).

In programming courses, motivation has a positive impact on students' achievements. It was found that students with high motivation perform better in programming than students with lower motivation levels (Bergin et al. 2005). The importance of motivation in computer programming course is due to the subject nature, which requires students to practice frequently, which cannot be sustained unless they are highly motivated (Jenkings 2011; Law et al. 2010; Jenkings 2002).

To promote students' motivation, different strategies should be utilized to engage students actively in the learning process (Carroll and Leander 2001; Lord 1997; Meyers and Jones 1993; Bonwell and Eison 1991). For example, it was found that improving levels of interaction has been a major factor to increase students' motivation and to create positive attitudes, which consequently enhance their performance (Lopez-Perez et al. 2011; Donnelly 2010; Woltering et al. 2009). Additionally, more support and additional learning materials help students understand the subject and promote their motivation (Lei 2010). Implementing such activities and support might be difficult during the class due to time limits; therefore, a blended learning environment seems the appropriate approach to combine multiple activities and to provide such additional support.

8.4 Blended Learning to Teach Programming

Blended learning is an educational approach that uses technology to combine face-to-face and online teaching strategies to enhance the learning process (Bonk and Graham 2006; Voos 2003). The major advantage of blended learning is that it incorporates the strengths of both synchronous (traditional face-to-face) and asynchronous (online) learning activities (Pahinis et al. 2007; Harding et al. 2005). This model provides the learner with socio-emotional support as a result of social presence and interaction with the teacher and other learners. It also brings more flexibility in learning time and place. Moreover, a well-designed blended learning course can be used to promote personalized learning to fulfill different students' learning styles. Blended learning can be categorized into different models according to the ratio of face-to-face and online roles in the learning process (Horn and Staker 2011). It can be implemented through three different approaches: providing the same class contents through online materials, providing online materials as supplementary resources, and replacing portions of the face-to-face contents with online materials (Alebaikan and Troudi 2010). The common model is where most of the curriculum is delivered by traditional face-to-face interaction, and online learning is used as a supportive tool to improve student learning outside the classroom. This model has been increasingly adopted by higher education institutions (Sharpe et al. 2006).

Many research studies have investigated the effectiveness of using blended learning to tackle learning difficulties with this subject and to enhance students' motivation (Deperlioglu and Kose 2013; Djenic et al. 2011; Mohorovičić and Tijan 2011; Hadjerrouit 2008; Boyle et al. 2003). It has been found that blended learning is the most suitable model for learning programming because it facilitates and advances the learning process better than the traditional face-to-face learning approach. Additionally, students' academic performances and success rates were improved, and their satisfaction was increased (Deperlioglu and Kose 2013; Djenic et al. 2011; Law et al. 2010).

8.5 Teaching Programming in the Middle East

Similarly to others around the world, students in Middle Eastern universities face difficulties when learning computer programming. Zualkernan et al. (2006) found great similarity in the learning styles of computer programming between Middle Eastern and American students. Only a few local research studies have investigated this problem and proposed a variety of approaches to enhance student learning of this subject. Mhashi and Alakeel (2013) conducted a case study to investigate programming difficulties faced by students at Tabuk University, Saudi Arabia. They found that students find programming to be a difficult subject, and most students lack the skills to analyze and solve programming problems. Additionally, the English language represents a challenge for students when learning programming. Because of the nature of this subject, which requires much of an individual's own efforts, the researchers pointed to a cultural-based reason for the difficulties, which was students' level of maturity; this includes students' responsibility to solve homework problems, get feedback from the lecturers, and be involved in interactive class discussions. Hawi (2010a) has investigated causal attribution of success and failure made by novice programming learners in Lebanon, and he found ten causal attributions: "learning strategy," "lack of study," "lack of practice," "subject difficulty," "lack of effort," "appropriate teaching method," "exam anxiety," "cheating," "lack of time," and "unfair treatment." However, "learning strategy" was the most important attribution; many students put much effort into this subject, but they fail because of the inappropriate learning strategy they use. Thus, by the same effort and an appropriate learning strategy, they could have achieved better. Such students need help developing an effective learning strategy.

Different approaches have been proposed to improve students' learning and enable them to overcome the difficulties. Alice, a 3D software programming environment, was used as a teaching tool to help students understand object-oriented programming in Cairo University, Egypt. Although the study shows the effectiveness of this tool in improving student learning, it suggests localizing this software's characters to further improve student learning. In this context, Alice for Middle East, "Alice ME," is a research project funded by the Qatar National Research Fund (QNRF) in Qatar, which aims to enrich this tool with animation objects that are relevant to the local culture. According to his experience of teaching programming in Notre Dame University in Lebanon, Hawi (2010b) argues that a constructivism approach to teaching programming is important to develop a sense of exploration and autonomous thinking. He emphasizes that teaching programming should be dynamically achieved in the classroom through a problem-solving setting and coding in front of students on a computer connected to a projector. Similarly to this, Herath et al. (2013) in the University of Bahrain used a project-based approach as a substitute for the traditional lecture-based classroom to engage students in the process of problem solving. In Qatar, it has been proven that robotics can be an effective tool to help students understand programming concepts; it motivates them and makes learning this subject more enjoyable (Ahmed and Alsaleh 2011). In Zayed University, UAE, Mahmoud et al. (2011) have addressed learning difficulties by redesigning the structure of the programming course using multiple programming tools. Their objective was to motivate students for the subject by using a web context to teach programming concepts; students start learning programming by learning HTML and JavaScript.

Despite the proliferation of e-learning and its suitability for teaching programming, there is not much research in the Middle East on using blended learning to teach programming, only the study by Boyle et al. (2003), which used a blended learning environment to teach programming at the Afyon Kocatepe University in Turkey. This research found that blended learning had improved students' academic performance. Moreover, it increased students' satisfaction because of the support they received through blended learning; thus, student dropout was reduced greatly.

8.6 Challenges of Blended Learning in the Middle East

Although blended learning has increasingly gained interest in the Middle East, the readiness for this type of learning, as part of e-learning, is varied from one state to another. There are still many challenges that should be considered when deciding to utilize this type of learning. Some of these challenges are technical, and some of them are cultural.

The main barrier for adopting blended learning in many countries in the Middle East is the low Internet penetration rate. While many countries enjoy a high rate of

Internet users, such as Israel, UAE, Qatar, and Bahrain, there are many other countries that have very low Internet penetration, such as Syria, Yemen, and Iraq. This is because of the high cost of Internet access, low speed, and quality (Mirza and Al-Abdulkareem 2011). In such countries, usage of blended learning will be limited to the minimum. For example, with low-bandwidth Internet connections, students will not be able to see supportive multimedia or download large materials.

In addition to technical barriers, many cultural factors should be considered when a blended environment is implemented. According to Hofstede (2001), Middle Eastern people maintain a high level of power distance, which is a culture dimension that refers to how people accept and expect that power is distributed unequally. Students in such cultures rely on teachers to lead the learning process, and they view the instructor as the only authority who is responsible for transferring knowledge to students. This should be considered in blended learning, as the instructor usually needs to involve students in active learning by initiating discussion in class and online forums. For group activities, he should pay much attention to how each member in each group is involved in the learning process, and he should attempt different approaches to encourage the participation of each.

Additionally, countries in the Middle East are considered high-context cultures, where communication between the instructor and students tends to be very formal. This might represent a barrier for collaborative learning, especially for online activities where students might need to use the English language; this could result in them finding it difficult to express themselves clearly and formally, so they may prefer to use short sentences to avoid using slang words. It is the responsibility of the instructor to decrease the level of formality with students to make them feel comfortable with discussion and interaction with him. Additionally, the instructor should avoid criticism of a student in front of the group or online discussion to protect their dignity, and critical comments and student assessments should be kept private (Sultan el al. 2012).

Moreover, Hofstede (2001) indicates that people in the Middle East have high uncertainty avoidance, which means that people from these cultures are worried about unknown situations; thus, they need clear rules and instructions. In the learning environment, this requires an instructor to use written guidance for learning activities that explain the objectives and describe possible approaches for conducting these activities. This is very important, especially for activities that students are not familiar with, such as online forums or using a learning journal. In blended learning, it is not enough to post such instructions online; they should also be discussed and clarified during class time.

8.7 Case Study

Currently, most universities use different tools, such as Blackboard, WebCT, or Moodle, to implement blended learning. However, the utilization of such systems is still kept to a minimum and is limited to posting lecture slides and homework submissions. Usually, this is because instructors lack the pedagogical background, especially engineering instructors, to infuse active learning strategies in such blended learning. Our case study was aimed to develop a blended learning environment for teaching computer programming to improve students' motivation and, ultimately, enhance their performance. To achieve this objective, students should be involved actively in the learning process in both aspects of the blended learning: during face-to-face class time and outside the class through online interaction. The blended learning model suits the subject of programming because students need to exert much effort to develop their programming skills, and much of their practice will be performed after the class. Therefore, they need an online supportive environment to help them overcome the difficulties that might face them.

At Qatar University (QU), all courses are managed by Blackboard, and instructors are encouraged to use this tool effectively to promote active learning as one of the strategic objectives of the university. Many instructors limit the use of this tool to posting static materials, such as course material texts or presentation slides, and homework submissions. In this case study, we aim to describe how to exploit some available tools to infuse learning activities online and make it complementary to in-class activities to create an active blended learning environment.

At QU, computer programming is taught using the C++ programming language. The subject is offered as two separate courses: theory and lab; students should enroll in both of them concurrently. This subject is required for students majoring in computer science and computer engineering. Students are expected to have no previous background in computer programming. Improving students' motivation in this course is very important to enhance their retention in the department and to increase their success rate.

8.8 Pedagogical Facets of our Active Blended Learning

Our objective is to design an active blended learning environment to increase students' satisfaction, improve their motivation, and enhance their learning of computer programming. Activities in this environment should address the causes of learning difficulties in this subject. The following pedagogical facets were taken into consideration when developing this environment:

1. Facilitating collaborative learning
   Collaborative learning is defined as a situation where some form of interaction between learners should be facilitated to trigger learning mechanisms (Dillenbourg 1999). In blended learning, technology should be utilized to increase interaction between learners to create a collaborative environment. Learners can discuss the subject problems through either face-to-face or online interaction. This creates positive social relationships that improve their motivation. Discussion in blended learning has the advantage of both face-to-face and online discussion. It frees learners from temporal and geographical constraints, and, at

the same time, they do not feel isolated because discussion also occurs during face-to-face time. Online discussion can occur in a public discussion forum that is open to all students in the class or in a small group where only members of that group can participate (Hiltz and Turoff 2002). Particularly, discussion and collaborative learning address some of the difficulties of learning programming; they engage students in the subject materials, help them learn problem-solving techniques from each other, and increase their feelings of synergy and enjoyment (Teague and Roe 2008). Previous studies (Falkner and Munro 2009; McKinney and Denton 2006) have found that collaborative learning in programming courses leads to higher retention and higher academic performance.

2. Enhancing interaction between the instructor and learners
   A high level of interaction between the instructor and learners is significantly related to student satisfaction in the learning environment (Keeler 2006; Stocks and Freddoline 2000). This type of interaction plays an important role in guiding students to correctly complete their learning activities. It can take place during face-to-face classroom time or online via discussion forum, email, assignment feedback, etc. Because of the nature of the subject of programming, more interaction with the instructor will encourage students to expose their problems and get feedback, which helps them overcome the difficulties they face when learning programming skills.
3. Utilizing timely formative assessment
   Formative assessment is an informal assessment conducted by the teacher during the course to provide feedback to both the teacher and learner about course progress. This type of assessment is different from the summative type, which is part of the course requirements, and its purpose is certifying a student's achievement. Formative assessment style can effectively improve students' learning and increase communication with the instructor; this type of assessment helps students understand expectation and eventually improve their retention (Mantz 2001). In addition to giving students feedback, formative assessments push students to seek help when needed through discussions with the instructor or with peers. This helps create an active learning environment and improves the level of interaction. This type of feedback is essential in computer programming courses to help students recognize their deficiencies.
4. Supporting reflective practicing
   Reflective practicing in learning refers to learning activities that give students opportunities to review their own cognitive process (Lai and Land 2009). It has been used as a pedagogical method to engage students in the learning process in different disciplines. Expressive writing or journaling is a common technique used to train students in reflective practice. In the subject of programming, journaling can be used to help students reflect on their studies to understand their weaknesses and obtain feedback from the instructor to guide them to the appropriate method of studying the subject.

8.9 Environment Implementation

Active learning aims to engage learners in the learning process through different activities, such as discussion, reading, and problem solving, rather than listening to lectures passively. Generally, active learning excites students about the subject matter; therefore, using different strategies of active learning in a programming course can motivate students and help them to overcome the difficulties of the subject. Table 8.1 shows how each pedagogical facet is implemented in our blended learning environment.

The following are the tools used to involve active learning in our blended environment:

1. Group work
   Students were divided to small groups to work together on a small project. Using the group feature on Blackboard, groups were created by the instructor, and students were asked to join these groups. This tool facilitates students' online interactions and supports exchanging materials among group members. The instructor also added himself to each group to make students feel that he is part of the work and ready to help them when needed. This activity was implemented as one of the techniques to infuse collaborative learning and motivate students. Based on Hofstede's (2001) categories of cultural dimension, people in the Middle East are considered collectivists who prefer group interests. Students in such a culture prefer to work within a group, and the purpose of their learning is to maintain social status. Therefore, we aim that this technique will utilize such cultural aspects to enhance students' learning and motivate them to help each other to overcome the difficulties of this subject.
2. Case study
   Students need to know how to analyze problems and design complete solutions, and this is not taught to them when studying individual features of a programming language. To help students overcome this part, a case study that is similar to a project assignment was given to the students. The case study concerns developing a software system to manage a video rental store. Students were given the problem description and requirements. The case study shows them how to analyze the problem and build the solution as a computer program. The main objective of this is to help students utilize all individual features they have learnt separately to build a solution for a specific real problem. From a cultural per-

Table 8.1 Tools to implement activities in blended learning

Pedagogical facet	Tools
Collaborative learning	Group work (in class and online groups), discussion (in class and through an online forum)
Interaction	Online groups, student's journal, class interaction
Formative assessment	Weekly formative assessments
Reflective practicing	Student's journal

spective, and because students in the Middle East need clear guidance for the required task, the case study aims to give students clear direction as to how to build a complete system. Students need to simulate this when developing their programming project as one of the main assessments in this course.

3. Weekly formative assessments
   Over each weekend of the semester, Blackboard's test feature was used to post an exam that covered the materials of the week. These assessments aim to provide students with feedback regarding their understanding of the subject so they can recognize their deficiencies earlier in the semester and work on them. In addition, the instructor uses this tool to emphasize the importance of some of the materials and drive students' attention to them, and students' answers were used as an indicator of the difficulties faced by students so that the instructor can address them again using different strategies. By making students aware of their weaknesses, and with some encouragement, they seek help from the instructor or from their peers, which improves communication and increases learning activities. This ultimately improves student learning in general.
4. Discussion
   Discussion is an active teaching method that motivates students in educational subjects. It makes the class more enjoyable, improves students' critical thinking, and enhances their communication skills. It is the instructor's role to trigger and guide the discussion to achieve a specific outcome. In our environment, usually, discussion is started in the class by questions related to the previous topics to ensure students' understanding of these topics, encourage all students to be part of the discussion, and allow debate between them. Students' answers for the weekly test have been used to raise class discussion to address their mistakes. On the other side, the online forum on Blackboard is used to encourage students to participate in asynchronous discussion. While some of the students have used this tool to seek help, good students have participated by raising new advanced topics and answering others' questions. Like class discussion, online discussion requires the instructor to manage the discussion and direct it toward supporting student learning. Additionally, many of the posts on the discussion forum were related to questions in the weekly tests.
5. Student journal
   Students were required to write a weekly journal entry using the journal feature that was introduced in a recent version of Blackboard. It is a private tool that is only accessible by the instructor and the student. In their weekly journals, students specify the lessons they have studied, their goals before starting to study, difficulties they have faced, how they have overcome them, how they evaluate themselves, and what their reflections were. Journaling was utilized by the instructor to guide students to the correct way of studying programming and to help them adjust their learning method to fit better with the subject nature, which mainly depends on practice rather than on just pure knowledge. Students' journals reflect their attitude and motivation. It was noticed that students start the semester full of enthusiasm and high expectations for their achievement. It is the

responsibility of the instructor to use different strategies to keep this attitude as long as possible by pushing students for greater achievements.

8.10 Evaluation

To evaluate the effectiveness of our active blended learning, we have assessed its impact on students' motivation and performance. The environment above was implemented in the programming concepts course at QU in fall 2012. This subject is compulsory for students in computer science and computer engineering majors, and it is the first major course where students study the C++ programming language. Two types of assessments were conducted: direct assessment to investigate the effectiveness of these activities on students' performance and indirect assessment to evaluate students' motivation and satisfaction after their experience in this course.

Assess Active Blended Learning's Impact on Students' Performance
To assess the impact of our active blended learning environment on students' performance, we use study subjects composed of the 48 students who enrolled in the programming concepts course. All students were female because this study was conducted on a female campus. Students were enrolled in two theory sections and two lab sections. Both lab sessions were taught by the same instructor, while the theory was taught by different instructors. We applied our blended learning environment in one of the lab sections (the treatment group) and used the other group as a control group. Students in the control group had access to Blackboard, but it was used only to post the static materials of the course, so none of the activities were implemented for this section. During lab sessions, both groups were required to solve practical questions with the help of the instructor. At the beginning of each lab, discussion was encouraged to review what students learnt in the theory during the week. Students in the treatment group had access to the online discussion forum, weekly online assessment, online groups, and online student's journal. As a result of using these tools, more discussion was brought to the lab session, and more interaction with the instructor occurred during the lab session.

Because the topics of the lab and theory are the same, we use students' performance in the theory classes as an unbiased indicator of the effectiveness of these strategies. The students enroll in the theory and lab sections freely, that is, there are no restrictions requiring that students who enroll in a specific theory section also enroll in the corresponding lab section. Thus, students who enrolled in the treatment section were enrolled in two different theory sections. At the end of the semester, we compared the performance of students in the treatment group with the others in the control group. We tracked the total grades of those in the theory sections who were taught by different instructors. Figure 8.1 shows that there is a significant difference between students' performance in the treatment group and in the control group in both theory sections. For the first theory section, the average of the treatment group students' grades was 70.6, while the others' average was 63.5. For the

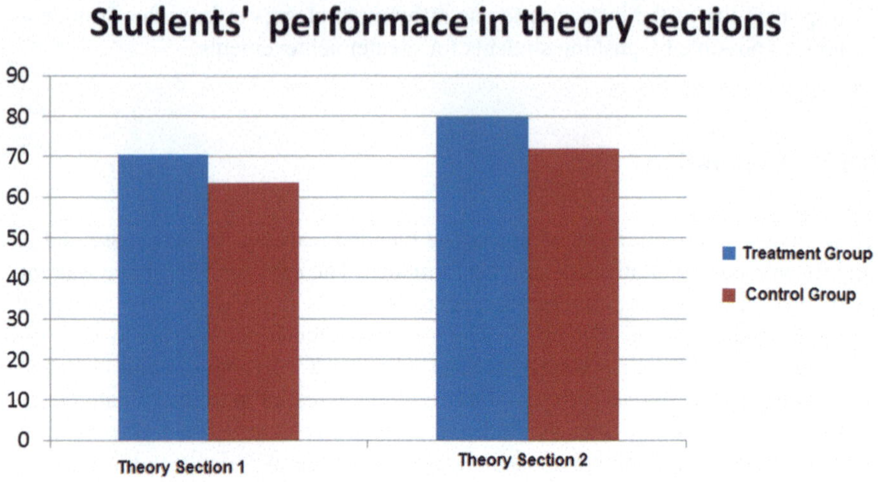

Fig. 8.1 Students' performance

Table 8.2 Independent *t* test, difference between students' performance in my section and others

Group	*n*	Mean	*SD*	*t*	*df*	*p*
Control group	26	68.0	14.6	−2.33	40	.025
Treatment group	22	77.8	12.3			

$*p<0.05$

SD Standard Deviation, *df* degree of freedom

second theory section, the average of the treatment group students' grades was 80, and for the other students, it was 72. This proves the effectiveness of the strategies implemented in the treatment group.

As is shown in Table 8.2, there is a significant difference ($t=-2.33$, $p=0.025$). Students in the treatment group outperformed those in the control group (77.8 vs. 68.0).

Assess students' motivation

At the end of the semester, students in treatment group were asked to fill in a survey to evaluate their attitudes toward their experience during the semester. The survey contains five Likert-scale questions and one open question. Four of the Likert-scale questions are in the form of positive statements. Students respond to each one by selecting one of five options: strongly agree (SA), agree (A), neither agree nor disagree (NAD), disagree (D), or strongly disagree (SD). The following table (Table 8.3) shows the Likert-scale questions:

All students in the treatment group responded to the survey. To interpret students' responses to Likert-scale questions, each response category is assigned a numerical value as follows: SA(5), A(4), NAD(3), D(2), and SD(1), then the average is calculated to represent the overall result for each question. Generally, the respondents have reported very high satisfaction by positive response, with an average value of

Table 8.3 Questions of the survey

Q1	Class discussion during this course helps me to learn better
Q2	I have a feeling of accomplishment from the work I have done in this class
Q3	I am happy that I enrolled in this course
Q4	When I have the opportunity in this class, I choose course assignments that I can learn from even if they do not guarantee a good grade
Q5	The interaction with the instructor in this course was....

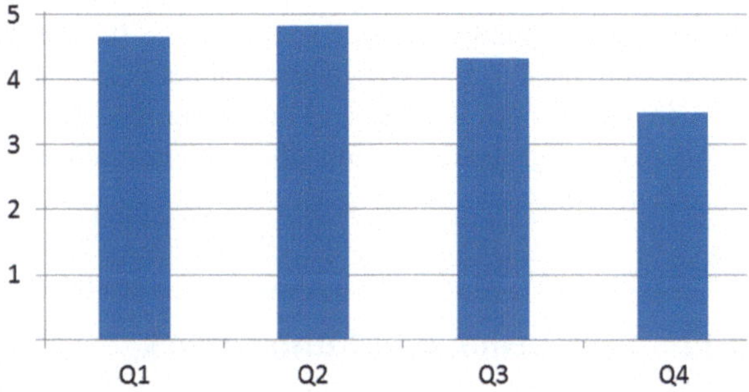

Fig. 8.2 Average of students' responses on the survey

4.3. The following figure depicts the averages of student responses for the questions Q1, Q2, Q3, and Q4 (Fig. 8.2).

The fifth question asked students to rank the interaction between students and the instructor as one of the following: excellent, good, normal, not that good, and no interaction. They all were significantly satisfied by the level of interaction with the instructor. Of the responses, 75 % ranked the interaction as excellent, and 25 % ranked it as good.

8.11 Discussion

The results show that students in the treatment group outperformed others in the control group in both theory sections. The difference between overall grades in both theory sections is due to the different assessments and different theory instructors. However, the superiority of treatment group students in both sections proves the reliability of the results and the effectiveness of this environment on improving students' learning. This is because using different learning activities in blended learning satisfies diversity of students' learning styles; while some students like class discussion, others find themselves more comfortable with online discussion. From their responses to question 1 in the survey, the majority of them agreed that discussion helped them to learn the subject better.

Generally, the activities in this blended learning improve students' satisfaction, which can be deduced from students' responses to questions 2 and 3. This satisfaction at the end of the semester (when the survey was conducted) represents an indicator of the effectiveness of this environment in motivating students during their study, as a result of this motivation, the drop rate was zero in this course. Question 4 aims to measure students' intrinsic motivation which represents students' enjoyment in the task itself and its inherent satisfaction regardless of any external impact. Although students' response to this question has the lowest level of agreement comparing to other questions, the response of this question shows that most students had this type of motivation. Additionally, students were very satisfied by the level of interaction with the instructor, where the majority ranked that as "excellent." This high level of interaction with the instructor encourages students to expose their learning difficulties, so he could address them early in the semester. Different tactics were used to improve interaction between students and the instructor, these include: discussions in class and online forums, through student's journaling, and through online group discussion where the instructor was himself added as member of each group.

8.12 Conclusion and Future Work

The statistical results show that using different strategies to engage students actively in blended learning has a positive effect on their attitude and satisfaction, which motivates students in the subject of programming and improves their performance. Careful design of the blended learning environment is necessary to utilize the advantages of both of its components: face-to-face and online learning. Both components should complement each other to create a supportive learning environment that is collaborative, provides a high level of interaction with the instructor, provides students with good feedback about their standing, and guides students to the proper way of studying the subject. Such a supportive environment is appropriate to the nature of the computer programming subject to help learners overcome its difficulties. From our case study, blended learning improves students' satisfaction and increases their motivation, which ultimately has a positive effect on their performance.

However, involving students in learning activities is not an easy task for the instructor; it requires frequent encouragement and guidance from the instructor. This might be more challenging in the Middle East because of the uncertainty dimension of the people and culture there, where people tend to rely on formalized policies and show a strong resistance to change. The students are in their first year of college, and they come from a traditional learning environment in schools where the teacher was the sole source of learning and knowledge; they are unfamiliar with these learning activities and are used to thinking of the learning process as just studying textbooks and solving assignments. Another challenge was interpersonal communication skills. Many students lack communication skills and find it difficult to express themselves verbally through discussions or in writing through online forums and journaling. It is not only because of the use of the English language; even with Arabic (their native language), they experience difficulties.

Because there is such limited research on the cultural impact of students' learning programming, future research is needed to address these aspects. Additionally, there is a need for research on how to help students adjust quickly to the college environment and various difficulties during their first year on campus.

References

Ahmed, B., & Alsaleh, K. (2011). *Robotics: Its effectiveness as a tool to teach engineering design and computer programming*. Amman, Jordan: IEEE Global Engineering Education Conference. pp. 1018–1021.

Ala-Mutka, K. (2004). Problems in learning and teaching programming—a literature study for developing visualizations in the codewitz-minerva project codewitz needs analysis. Bennedsen, J., & Caspersen, M. E. Failure rates in introductory programming. *SIGCSE Bulletin, 39*(2), 32–36.

Alebaikan, R., & Troudi, S. (2010). Blended learning in Saudi Universities: Challenges and perspectives. *ALT-J. Research in Learning Technology, 18*(1), 49–59.

Bergin, S., Reilly, R., Traynor, D. (2005). *Examining the role of Self-Regulated Learning on Introductory Programming Performance*. First International Computing Education Research Workshop, ICER 2005 Seattle, WA, US.

Bomia, L., Beluzo, L., Demeester, D., Elander, K., Johnson, M., & Sheldon, B. (1997). *The impact of teaching strategies on intrinsic motivation*. Champaign, IL: ERIC Clearinghouse on Elementary and Early Childhood Education. (ERIC Document Reproduction Service No. ED 418925.

Bonk, C. J., & Graham, C. (2006). *Handbook of blended learning: Global perspectives, local designs*. San Francisco: Pfeiffer Publishing.

Bonwell, C., & Eison, J. (1991). *Active learning: Creating excitement in the classroom*, ASHE-ERIC Higher Education Report No. 1.

Boyle, T., Bradley, C., Chalk, P., Jones, R., & Pickard, P. (2003). Using blended learning to improve student success rates in learning to program. *Journal of Educational Media, 28*(2–3), 165–178.

Butler, M., & Morgan, M. (2007). *Learning challenges faced by novice programming students studying high level and low feedback*. Proceedings of the Conference of the Australasian Society for Computers in Learning in Tertiary Education (ASCiLiTE). Singapore, December 2–5. http://www.ascilite.org.au/conferences/singapore07/procs/procs/butler.pdf. Accessed September 2008.

Carroll, L., & Leander, S. (2001). Improving student motivation through the use of active learning strategies. Master of Arts Action Research Project, Saint Xavier University.

Deperlioglu, O., & Kose, I. (2013). The effectiveness and experiences of blended learning approaches to computer programming education. *Computer Applications in Engineering Education, 21*(2), 328–342.

Deraadt, M., Hamilton, M., Lister, R., Tutty, J., Baker, B., Box, I., Cutts, C., Fincher, S., Hamer, J., Haden, P., Petre, M., Robins, A., Simon, & Sutton, K. (2005). *Approaches to learning in computer programming students and their effect on success. University of Otago*. New Zealand: ACM. SIGCSE Special Projects Grant.

Dillenbourg, P. (1999). What do you mean by collaborative learning? In P. Dillenbourg (Ed.), *Collaborative learning: Cognitive and computational approaches* (pp. 1–19). Oxford: Elsevier.

Djenic, S., Krneta, R., & Mitic, J. (2011). Blended learning of programming in the internet age. *IEEE Transactions on Education, 54*(2), 247–254.

Donnelly, R. (2010). Harmonizing technology with interaction in blended problem-based learning. *Computers & Education, 54*, 350–359.

Doyle, J. K. (2005). *Does a 'pre-introduction' course improve subsequent performance in 'introduction to computer programming'?*, 2005 College Teaching & Learning Conference, 2005.
Dunican, E. (2002). *Making the analogy: Alternative delivery techniques for first year programming courses*. In J. Kuljis, L. Baldwin & R. Scoble (Eds), Proceedings from the 14th Workshop of the Psychology of Programming Interest Group, Brunel University.
Estefanía, M., Lázaro, C., Losada, H., & Isidoro. (2010). *Active learning in telecommunication engineering: A case study*. Proceedings of IEEE Engineering Education 2010—The future of global learning in engineering education, 14–16 April 2010. Madrid, Spain.
Esteves, M., Fonseca, B., Morgado, L., & Martins, P. (2011). Improving teaching and learning of computer programming through the use of the second life virtual world. *British Journal of Educational Technology, 42*(4), 624–637.
Falkner, K., & Munro, D. (2009). *Easing the transition: A collaborative learning approach*. In Proceedings of the Eleventh Australasian Conference on Computing Education—Volume 95 (ACE '09), M. Hamilton, T. Clear (Eds.), Vol. 95. Australian Computer Society, Inc., Darlinghurst, Australia, 65–74.
Gomes, A., & Mendes, A. J. (2007). *Learning to program—difficulties and solutions*. In International Conference on Engineering Education—ICEE 2007, pp. 283–287.
Grover, S., & Pea, R. (2013). Computational thinking in K-12: A review of the state of the field. *Educational Researcher, 42*(1), 38–43.
Guo, H. (2006). *Visual material and learning aids in teaching object oriented programming*. In Proceedings of the 7th Annual ICS HE Academy Conference, Dublin, Ireland.
Hadjerrouit, S. (2008). Towards a blended learning model for teaching and learning computer programming: A case study. *Informatics in Education, 7*(2), 181–210.
Harding, A., Kaczynski, D., & Wood, L. N. (2005). *Evaluation of blended learning: Analysis of qualitative data*. In: Proceedings of the Symposium of Blended Learning in Science Teaching & Learning, The University of Sydney, Australia. September 28–30(2005), 56–62.
Hawi, N. (2010a). Causal attributions of success and failure made by undergraduate students in an introductory-level computer programming course. *Computers & Education, 54*(4), 1127–1136.
Hawi, N. (2010b). The exploration of student-centered approaches for the improvement of learning programming in higher education. *US-China Education Review, 7*(9), 47–57.
Herath, S., Herath, A., Siddqui, M., & El-Jallad, K. (2013). A project based CS/IS-1 course with an active learning environment. *International Journal of Advanced Computer Science and Applications(IJACSA), 4*, 31–37.
Hiltz, S., & Turoff, M. (2002). What makes learning networks effective? *Communications of the ACM, 45*(4), 56–59.
Hofstede, G. (2001). *Culture's consequences: Comparing values, behaviors, institutions, and organizations across nations* (2nd ed). Thoudand Oaks: Sage.
Horn, M. B., & Staker, H. (2011). *The rise of K12 blended learning*. Mountain View: Innosight Institute.
Jenkings, T. (2002). *On the difficulty of learning to program*. Proceedings of the centre for information and computer sciences: Information and computing science reprint, University of Leeds, Leeds, UK. pp. 53–58.
Jenkings, T. (2011). *The motivation of students of programming*. MSc Thesis, University of Kent at Canterbury.
Keeler, L. C. (2006). *Student satisfaction and types of interaction in distance education courses*. Ph.D. dissertation. United States—Colorado: Colorado State University. Retrieved September 2, 2008, from Dissertations & Theses: Full Text database. (Publication No. AAT 3233345).
Lahtinen, E., Ala-Mutka, K., & Järvinen, H. (2005). *A study of the difficulties of novice programmers. In ITiCSE '05: Proceedings of the 10th annual SIGCSE conference on Innovation and technology in computer science education Caparica* (pp. 14–18). Portugal: ACM Press.
Lai, T., & Land, S. M. 2009. *Supporting reflection in online learning environments*. Educational Media and Technology Yearbook (pp. 141–154). Springer US.
Law, K., Lee, V., & Yu, Y. (2010). Learning motivation in e-learning facilitated computer programming courses. *Computers & Education, 55*(1), 218–228.

Lei, J. (2010). Quantity versus quality: A new approach to examine the relationship between technology use and student outcomes. *British Journal of Educational Technology, 41*(3), 455–472.

Lopez-Perez, M. V., Perez-Lopez, M. C., & Rodriguez-Ariza, L. (2011). Blended learning in higher education: Students' perceptions and their relation to outcomes. *Computers & Education, 56*(3), 818–826.

Lord, T. (1997). A comparison between traditional and constructivist teaching in college biology. *Innov Higher Educ, 21*, 197–216.

Mahmoud, Q., Abu Talib, M., Maamar, Z., Bataineh, E., & Jololian, L. (2011). *Revitalizing, the introductory programming course*. In Proceedings of IEEE GCC conference and exhibition, Dubai, UAE.

Mantz, Y. (2001). Formative assessment and its relevance to retention. *Higher Education Research & Development, 2*(2001), 115–126.

McKinney, D., & Denton, L. F. (2006). *Developing collaborative skills early in the CS curriculum in a laboratory environment*. SIGCSE 2006 Technical Symposium on Computer Science Education. Houston, Texas, USA.

Merrick, K. (2010). An empirical evaluation of puzzle-based learning as an interest approach for teaching introductory computer science, IEEE Transactions on Education, issue 99.

Meyers, C., & Jones, T. B. (1993). *Promoting active learning*. San Francisco: Jossey Bass.

Mhashi, M., & Alakeel, A. (2013). *Difficulties facing students in learning computer programming skills at Tabuk University*. In Proceedings of the 12th International Conference on Education and Educational Technology (EDU '13), Iwate, Japan, pp 15–24.

Miliszewska, I., & Tan, G. (2007). Befriending computer programming: A proposed approach to teaching introductory programming. *Journal of Issues in Informing Science & Information Technology, 4*, 277–289.

Miliszewska, I., Venables, A., & Tan, G. (2008). Improving progression and satisfaction rates of novice computer programming students through ACME—analogy, collaboration, mentoring, and electronic support-. *Issues in Informing Science and Information Technology, 5*, 311–323.

Milne, I., & Rowe, G. (2002). Difficulties in learning and teaching programming—views of students and tutors. *Education and Information Technologies, 7*(1), 55–66.

Mirza, A., & Al-Abdulkareem, M. (2011). Models of e-learning adopted in the Middle East. *Applied Computing and Informatics, 9*(2), 83–93.

Mohorovičić, S., & Tijan, D. (2011). Blended learning model of teaching programming in higher education. *International Journal of Knowledge and Learning, 7*(1), 86–99.

Pahinis, K., Stokes, C. W., Walsh, T. F., & Cannavia, G. (2007). Evaluating a blended-learning course taught to different groups of learners in a dental school. *Journal of Dent Education, 71*(2), 269–278.

Perkins, D. N., Hancock, C., Hobbs, R., Martin, F., & Simmons, R. (1989). Conditions of learning in novice programmers. *Soloway and Spohrer, 1989*, 261–279.

Petersen, A., Craig, M., & Zingaro, D. (2011). *Reviewing CS1 exam question content*. In Proceedings of the 42nd ACM technical symposium on Computer science education (SIGCSE2011), pages.631–636.

Qualls, J. A., Grant, M. M., & Sherrell, L. B. (2011). CS1 Students' understanding of computational thinking concepts. *Journal of Computing Sciences in Colleges, 26*(5), 62–71.

Robins, A., Rountree, J., & Rountree, N. (2003). Learning and teaching programming: A review and discussion. *Computer Science Education, 13*(2), 137–172.

Ryan, R., & Deci, E. (2000). Intrinsic and extrinsic motivations: Classic definitions and new directions. *Contemporary Educational Psychology, 25*, 54–67.

Sharpe, R., Benfield, G., Roberts, G., & Francis, R. (2006). The undergraduate experience of blended learning: A review of UK literature and practice. http://www.heacademy.ac.uk/assets/documents/archive/blended_elearning_full_review.pdf. Accessed 15 June 2013.

Simon, Sheard, J., Carbone, A., Chinn, D., Laakso, M.-J., Clear, T., de Raadt, M., D'Souza, D., Lister, R., Philpott, A., Skene, J., & Warburton, G. (2012). *Introductory programming: Examining the exams*. 14th Australasian Computing Education Conference (ACE 2012), Melbourne, Australia.

Stamouli, I., Doyle, E., & Huggard, M. (2004). *Establishing structured support for programming students*. In Proceedings of the 34th ASEE/IEEE Frontiers in Education Conference, Savannah, GA.

Stocks, J. T., & Freddoline, P. (2000). Enhancing computer-mediated teaching through interactivity: The second iteration of a world wide web-based graduate social work course. *Research on Social Work Practice, 10*(4), 505–518.

Sultan, N., Bunt-Kokhuis, S. v. d., Davidson, C., Sentini, A., & Weir, D. (2012). E-Learning in the Arab Gulf: Responding to the changing world of education. In M. Ramady (Ed.), *The GCC economies: Stepping up to future challenges* (pp. 33–48). New York: Springer.

Summet, J., Kumar, D., Hara, K., Walker, D., Ni, L., Blank, D., & Balch, T. (2009). *Personalizing cs1 with robots*. In Proc. SIGCSE '09, pages 433–437.

Teague, D., & Roe, P. (2008). *Collaborative learning towards a solution for novice programmers*. In proceedings of the tenth Australasian Computing Education Conference (ACE 2008).

Thomas, M. K., Ge, X., & Greene, B. A. (2011). Fostering 21st century skill development by engaging students in authentic game design projects in a high school computer programming class. *Journal of Educational Computing Research, 44*(4), 391–408.

Voos, R. (2003). Blended learning: What is it and where might it take us? *Sloan-C View, 2*(1), 2–5.

Woltering, V., Herrler, A., Spitzer, K., & Spreckelsen, C. (2009). Blended learning positively affects students' satisfaction and the role of the tutor in the problem-based learning process: Results of a mixed-method evaluation. Advances in Health. *Science Education, 14*, 725–727.

Yan, L. (2009). *Teaching object-oriented programming with games.* In Proceedings of the Sixth International Conference on Information Technology: New Generations, Las Vegas, NV, 969–974.

Zualkernan, I. A., Allert, J., & Qadah, G. Z. (2006). Learning styles of computer programming students: A Middle Eastern and American comparison. *IEEE Transactions on Education, 49*(4), 443–450.

Saleh Alhazbi Saleh Alhazbi received his PhD in Computer Science from university of Science, Malaysia (USM) in 2009, he holds M.Sc. in Computer Science and M.A. in Education from New Mexico State University(NMSU), USA. Dr. Alhazbi is currently a lecturer at Dept. of Computer Science and Engineering at Qatar University(QU). His research interests include software engineering, HCI, computer science education.

Part III
Curriculum Development

Chapter 9
Women's Specific Needs and Urban Planning Practices in the Middle East: The Case of Palestine

Manal A. Al-Bishawi

9.1 Background

Design and planning practices are mainly governed by engineering education, as engineers make their designs and plans according to the knowledge taught to them at engineering faculties. Therefore, engineering education is considered to be the key issue for enhancing the engineers' awareness and knowledge about different issues concerning people's use of their built environment.

In any society, women have specific needs[1], including the need for privacy, security, and comfort. These may be reflected in the urban design, for example, in the physical layout of spaces, furnishings, pavements, lighting, and in the spaces' functions. As an expression of internationalization, modern knowledge in the field of architecture and urban planning is generally based on identifying standard human needs, which are then reflected in the grouping of users according to their functions. For instance, users are grouped into categories such as pedestrians, vehicular traffic and disabled individuals. According to this framework, the role of the designers and planners is to consider and fulfill these standard requirements in their designs and plans.

[1] For this study, specific needs are cultural needs which originate from the local context. These needs are different from basic needs. Maslow (1943) suggests that human needs are classified into two types:
(a) Basic needs which are common to all humans and fundamental in motivating human behavior. These needs are biological and universal; they imply physiological, sociological, and psychological needs (e.g. thirst, shelter, security, belonging, and love needs).
(b) Specific needs which are deep desires and differ from one person to another, from one culture to another, and from one place to another. These needs are learned and influenced by traditions and religion. Therefore, specific needs are cultural needs and always coexist with basic needs; they should not be understood as single determiners of certain kinds of behavior

M. A. Al-Bishawi (✉)
Faculty of Engineering and Technology, Architectural Engineering Department,
Birzeit University-Palestine, Birzeit, Palestine
e-mail: mbishawi@birzeit.edu

M. Abdulwahed et al. (eds.), *Advances in Engineering Education in the Middle East and North Africa,* DOI 10.1007/978-3-319-15323-0_9

Women can be regarded as a user group of public sphere with specific needs, requirements, and values, issues that should be intentionally and specifically addressed in urban design and planning practices. Since women's specific needs related to public sphere are dependent on culture, such needs should be understood and designed according to the local culture. Women's privacy in the presence of outsider (stranger) males is a specific need and a value that has influenced and continues to influence Middle Eastern societies and their built environments.

The present engineering education and accordingly planning practices, however, lack this focus on fulfilling women's need for privacy in design works and planning. This caused women's access to public sphere and many of the employment fields to be controlled. In addition, the esthetic quality of urban environment is influenced negatively due to its adaptation to women's privacy by local inhabitants.

Therefore, this study aims at investigating how women's privacy is considered in design and planning practices in the Middle East, in general, and in Palestine, in particular. The outcome of this study will contribute to enhancing the knowledge in the field of architecture and planning about interaction between engineering and culture, in general, and between engineering, democracy, and equity among citizens in using their urban environment, in particular. Thus, women's privacy in this regard does not mean women's isolation from public sphere; on the contrary, it is an essential requirement for women to achieve their right in access to public sphere. This will help engineers to design and develop sustainable urban environments which are accessible by both males and females. Observations and interviews are the main methods used in data collection.

The traditional fabrics of Middle Eastern cities (as in the rest of the world) were shaped according to the local culture, particularly Arabic Islamic culture, in addition to geographical aspects. Considering women's specific needs is historically part of these cities' urban design process. The outdoor spaces, the buildings, and the elements all reflect different female needs, such as providing security, comfort, and privacy. This way of shaping cities in the Middle Eastern region continued until the nineteenth century, when the political and economic institutions of Middle Eastern cities underwent major transformations. These changes occurred particularly as a result of colonization after World War I. The emergent dominant powers, particularly Britain and France, demolished the existing structures of the defeated Ottoman ruling system. Systemic shifts involved all aspects of society: law, administration, education, and commerce. This paradigm shift transformed local architecture and town planning (Nooraddin 2004).

As a consequence of these changes, a big shift has been experienced in planning practices in the Middle Eastern world. The decentralized planning process controlled by residents, and which was therefore governed by their cultural needs and values including women's privacy, was transformed into a centralized process controlled by the municipalities based on functional and welfare aspects, especially welfare and vehicle access (Bianca 2000). This caused people to introduce their own changes in the physical form of their built environment to comply with their cultural norms, particularly concerning women's privacy. In addition, community's lifestyle has clearly changed, particularly women, as new educational and professional opportunities in different fields were offered to them (Ahmed 1992). Within this change, gender issues are still important in the Middle East and they continue to influence women's access to public sphere, which is still male dominant. In this regard, large gaps between female and male rates are evident in the Middle East despite several job opportunities are offered to women (Document and Publications Production 2013).

In this study, the focus is on Palestine, which has been selected as a representative case of the Middle Eastern countries.

9.2 The Case of Palestine

Palestine is located on the east coast of Mediterranean Sea, west of Jordan, and south of Lebanon (Fig. 9.1). The territory of Palestine covers almost 10,435 sq. miles, which includes 10,163 sq. miles of land area and the rest consists of water. The geography of Palestine is characterized by its diverse topography and terrain. The climate of Palestine, in general, is pleasant. The winter months from mid-December to mid-March are severe. The cool Mediterranean Sea breeze has a cooling effect on the very hot summers.

Fig. 9.1 Location of Palestine

Like any Arab Middle Eastern city, Palestine has a younger population and is evenly split between males and females (Palestinian Central Bureau of Statistics (PCBS) 2012). Palestinian society is very cohesive, and the institution of the family plays an important role. Hence, the family plays an essential role in the economic and social life of the Palestinian cities. Successive occupations, particularly British and Israeli occupations, made the Palestinians famous for their spirit of independence from the earliest ages.

The Palestinians are tied to their traditions and social customs, which are based on Arabic and Islamic values, particularly women's privacy need which is influenced by the culture of gender separation. Because gender separation is a social convention in Palestine, it influences women's behavior in urban and public spaces as women try to expose themselves to outsider men as little as possible (Fig. 9.2).

In Palestine, which is probably not much different from other countries in the Middle East, many public spaces are defined as either for men's (Fig. 9.3) or for

a b

Fig. 9.2 Examples of women's behavior in Palestinian society and privacy

Men's setting for cutting hair is a semi-public space.

Men's setting for social gathering (coffee house)

Fig. 9.3 Examples of public space created for men's use

Women's setting for eating in the upper floor of the restaurant is a semi public space.

Women's setting for cutting hair in the street is a semi-public space.

Fig. 9.4 Examples of public space created for women's and families' use. The upper floor of the restaurant is specialized for families' use which controls men's access to this floor and allows for women to use it comfortably

women's use (Fig. 9.4), such as governmental schools,[2] shops, and sports facilities. Inside public buildings too, such as restaurants, there are parts designated for women's or families' use, and are usually located far from the entrance or on the upper floors (Fig. 9.4).

The close-knit nature of Palestinian society has kept the towns' traditions alive throughout the history. The social conservatism of the Palestinians contributes to women's privacy. For example, men avoid entering to and gazing at women's settings without permission or a prior announcement. They do this for both religious reasons and because of social tradition. In addition, people's emphasis on social rules that influence women's privacy based on religion and tradition is stated explicitly. Because of the importance attributed to women's privacy in the life of Palestinian society, these values influence Palestinians' attitudes toward their physical environment.

Like most Middle Eastern cities, the political and economic institutions of Palestinian cities underwent major transformations, particularly after the British occupation and the increasing commercial exchange with the exterior. Regarding urban planning practices, new standard and formal rules[3] (British ones), which do not consider cultural norms including those about women's privacy, were established to guide buildings and planning activities in the Palestinian cities instead of the infor-

[2] In Palestine, there are two types of schools: the first type is the governmental (public) schools which present the majority and are specialized for either girls or boys. The second type is the private schools, which are mixed. Only a few of these schools exist and they are mainly used by rich people.

[3] Formal rules are standard and written rules as building codes and planning regulations.

mal rules.[4] In addition to municipalities, there are several international groups, in the name of donors, who are involved in design and planning process. In most cases, the designs and the plans are made according to the desires of these donors and by foreign engineers who are unfamiliar to the cultural norms of the local society. Regarding sociocultural aspects of the society, new and modern lifestyle models were introduced to Palestine. In addition, new requirements and needs developed, leading to social, educational, economic, and legal changes within civil society, particularly in relation to women's daily activities and their role in the society. However, Palestinian community was among the first Middle Eastern communities where women assured their rights and liberation. This is reflected in a number of employment fields where women have equal opportunities. In addition, the increasing number of educated women, especially those holders of university degrees, led to enrolment of women in many professional activities, and led to economic growth (Palestinian Central Bureau of Statistics 2008). Moreover, the Israeli occupation has had a major effect on forcing women into the labor force to meet the rising costs of living. In 1967, substantial changes occurred in many Palestinian cities, particularly after the Israeli occupation destroyed large area of the agricultural lands in these cities. During that period, a drop from 14.2 to 20% of the women labor force in agricultural sector was witnessed (Abdulhadi 1990). Women had to bear many of their families' responsibilities, since several men had left for work in Israel, or been deported or imprisoned by the Israeli authorities, or else forced to emigrate.

An important area where the impact of changes in both planning practices and sociocultural aspects of the society can be identified is the urban form of cities. The compact structure of cities was transformed into a scattered one, where the buildings are separated by setbacks. Due to the growth in population and the expansion of the traditional cities, new areas are created with new types and functions of urban spaces. Also, the segregation of uses has led to isolated and low-density zones for commercial and residential areas, which are randomly located and only linked by transportation facilities. On the one hand, this situation created difficulties and restrictions on women's urban life and deprived many of them from their right in access to public sphere, since its design does not comply with cultural norms about privacy. On the other hand, it caused many of the inhabitants to adapt their built environment to comply with their cultural norms and beliefs.

9.3 Women's Needs in Urban Design and Planning

After 1950 when women were integrated into the economy and participated in different fields, they began to face problems in using their cities whose physical layouts were based on men's rather than women's needs (Little 1994; Roberts 1991; Lewis and Foord 1984). Women's experiences and activities today still separate them from men and cause them to have more in common with other women than

[4] Informal rules are cultural norms and beliefs that govern people's behavior and attitudes toward their built environment.

they do with men (Booth et al. 1996). Studies done by Franck and Paxon (1989), Mozingo (1989), Bunston and Breton (1992) show that women have more specialized needs in cities than men. The needs for safety, security, play areas for children, and comfort are found to be important needs to be considered in designing public open spaces for women's use. Several studies clarify how women, particularly older women, are more likely than men to modify their behavior to avoid the risk of crime and these have repercussions for their use of the public realm, return home earlier, or go out at night only with an escort. Women are more fearful of some crimes and locations than others. They feel less safe on public transport than in the general community (Dorfman 1997; Polk 2003; Seedat et al. 2006). Accordingly, fear of crime and other related factors such as darkness in the street, nature of the population, and areas with untrimmed vegetation blocking the view are major contributors to the way women perceive and use the city (Kallu 2001; Tamminga and Luymes 1995).

Accessibility to different uses in the city is an important need for women. Frank and Paxson (1989) explain how the segregated and randomly located functions make it difficult or impossible for the women to put together all that is necessary within the limits of available time. Greed (2005) argues that changes need to be made from the existing suburban pattern of low-density and segregated residential development to zoning that permits the development of shops in close proximity to residential areas. She further argues that this and other land-use changes will increase the accessibility of services for women and, thus, help them balance the multiple roles related to work, household, and children by reducing their need to travel.

The need for affordable, qualified, and conveniently located childcare is one of the pressing concerns of contemporary family life, and much has changed in the life patterns of today's society largely due to the dramatic increase of working mothers. She shows that childcare use has a far-reaching impact on access to care, quality of care, and cost of care (Ritzdorf 1990; Cohen 1987).

In addition to the above-mentioned needs, which are common (standard) for women all over the world, there are other specific needs for women that originated from the cultural contexts, such as women's privacy need in the Middle East. However, engineering education should enhance engineers' knowledge about both women's standard and specific needs. On the one hand, this enhances democracy and equity among citizens in using their urban environment, and increases the efficiency of the spaces and reduces the costs required for adapting these spaces for women's use. On the other hand, this complies with principles of universal design which aim at removing environmental barriers from urban environment and providing sustainable types of public spaces that meet the needs of all individuals of all ages, sizes, genders, and abilities in different sociocultural contexts and for different purposes, "without a need for adaptation or specialized design" (Welch 1995; Aslaksen et al. 1997, p. 5). During the 1980s, there was an emphasis on involving women in design and management of cities as a means of reflecting their needs and values (Lansley et al. 1989), and several studies were done on this topic. Meeting women's specific needs in the use of public spaces requires their effective participation in decision-making about the design of these spaces (The World Commission on Environment and Development 1987, p. 8).

Considering women's needs contributes to a sustainable urban society and improves urban life quality (Parfect and Power 1997), particularly as women form half of the community. The United Nations Conference on Environment and Development that was held in Rio de Janeiro in 1992 affirmed women's critical contributions to environmental management. Agenda 21, the UNCED action plan for sustainable development, proposed actions to strengthen the role of women in sustainable development by eliminating obstacles to their equal participation, particularly in decision-making (Corral 2002).

9.4 Women's Privacy in Middle Eastern Countries

Privacy is an important aspect in all societies. For Middle Eastern societies, privacy is considered a holy tradition and a way of life that should be respected and reflected in everyday life and, as such, adapted in architectural solutions, building and town planning. Although each particular city in the different Islamic nations has developed its own interpretation of privacy, all of them include privacy as a fundamental, shared concern in architectural patterns. Privacy is related to family life, neighborhood life, and women's life.

Privacy involves a diversity of social relationships as individuals in relation to individuals, individuals in relation to a group, and so on (Altman 1975). This study primarily deals with women's privacy in the presence of outsider males. Outsider males are stranger males who do not have a family relationship with the women and are eligible for marriage. Women's privacy in this context is influenced by the culture of gender separation, which is based on local traditions dominant in Middle Eastern cities and Islamic religion. Gender separation is a culturally dependent phenomenon that influences women's behavior. In urban spaces where women are exposed to outsider males, they search for a personal space and territorial control when engaged in different activities. Gaining such personal space enhances women's comfort and feeling of security (based on Altman 1975; Hall 1966; Newman 1972).

Women's privacy is a need and a value that has influenced and continues to influence Arab societies, as well as other Islamic societies. The rules that govern this need are socially defined and mutable (based on Hakim 1986; Arkoun 1990; Roald 2001). Therefore, they are practiced differently according to the cultural context, the period, the situation, and the women's social status: their age, education, and income level. For example, in Saudi Arabia, the rules which govern women's privacy in public spaces are set out in legislation. In Palestine, on the other hand, these rules are defined socially. Also, younger, higher income, and better educated women are less concerned about privacy rules. This does not mean that these women do not need privacy, but the required privacy for them is less than that for other women.

Even for the same woman, her required privacy is not fixed and depends on the situation. For example, for activities where woman spends long time as in recreational activities, she will care about privacy more than for activities where she spends short time as passing by. However, the influence of these status variable stays small and less than the influence of the physical form of the built environment (Al-Bishawi 2008).

Women's privacy in old traditional Islamic culture implies different terms of privacy: visual, physical, auditory, and olfactory. In other words, it means to what extent women should be exposed visually, physically, or in an olfactory or auditory sense (to what extent her voice is recognizable to outsiders) to outsider males in the various events that occur either in private or in public open spaces during everyday life activities. Many of these terms have been inherited by modern Islamic societies, and reflected in people's behavior and in their built environment, which largely reflects local ways in controlling space. In the contemporary literature, the definition of defensible space shows how communities create their individual spaces that fit their needs and make it possible to defend their physical space (Newman 1972).

Women's privacy need is relevant to their security need. According to Islamic jurisprudence, privacy rules are based on the notion that women arouse men's sexual appetite, which exposes the women to danger. Thus, privacy rules, according to Islam, are precaution mechanisms that decrease women's exposure to men and contribute to their protection. Although the Islamic perspective may not comply with other Muslim or cultural perspectives, particularly as such an issue is dependent on the nature of both women and men, it is relevant to women's security in urban public spaces. Several studies about women's use of cities show that among the major problems which women face in public spaces in the USA and Europe are the fear of harm and sexual attack (Worpole 1992; Karp et al. 1991). In general, results of these studies comply with the concept of women's privacy in Islam.

However, Koranic text and tradition do not mention anything about the physical form of public spaces; they only mention behavioral rules which are related to women's privacy. And because of the importance of these rules for the society, they have influenced the decisions and attitudes of Muslims toward their traditional and modern built environment.

In the traditional urban design of Middle Eastern cities, the Islamic values concerning women's privacy were applied formally, as to regulate the height of adjacent buildings and the placement of windows and doors, in a way allowing women to observe the streets without themselves being observed. Besides, the behavioral values concerning women's privacy have traditionally guaranteed a secure environment for women when using these spaces (Hakim 1986). The absence of public facilities from dead-end streets, and the growth of markets along the main streets, enabled women to use these dead-end streets without being exposed much to the public sphere in the neighborhood structure. Also, the old fabric of Arab-Muslim cities enabled women to use the building's roofs for keeping in contact with each other, enjoying fresh air, and observing the public sphere, all without directly using public open spaces (Ottoman 1988). Additionally, the different hierarchal levels of public spaces, their irregular form, and the separation between streets and squares are all components contributed to women's privacy in public spaces (Bianca 2000). Features such as screened windows and the layout of houses and quarters enabled women to see men, but men could do that just in certain they have relationships with them. Also, the rules governed the element of timing, for instance, with regard to the use of public baths and transitional spaces were used to create divisions between males and females in public spaces (Abue-Lughod 1987).

The modern urban design of Middle Eastern cities is much more influenced by modern values, particularly those related to functionality and general public welfare. These design concepts, largely imported from other cultures, are different from Islamic-influenced ones and often directly contradict the residents' cultural values and norms, particularly those regarding women's privacy. This caused people to introduce their own changes in the physical form of public spaces. People in Saudi Arabia use walls, curtains, and other partitions more than in other societies, in order to create physical boundaries and realize their need for privacy. Abu-Gazzeh (1996) outlines how these boundaries became a concept, which influences the physical form of public spaces in Saudi Arabia. Al-Hemaidi (2001) explains how the residents transformed the walls and openings facing the adjacent streets by constructing plastic and steel structures on the top of the dwellings' fences, or by sealing off the second-floor windows of the dwelling, to create the required privacy for women in residential gardens in Saudi Arabia. He also mentions that most women occupying the new style residences refrain to use the backyard at all, and they keep the street-level bedroom curtains closed all the time. Eben-Saleh (1998) explained how cultural norms about women's privacy influenced the traditional urban form of Saudi Arabia and continues to influence people's behavior in public spaces. Because of the imbalance between local Islamic and Western values (which are adopted by designers and planners), the foreign style of public spaces does not respond to the cultural needs. As a result, inhabitants become isolated.

9.5 Women's Role in Planning Practices and Fulfillment of Their Needs

Involvement of women in design and management of cities is an important tool toward expressing needs and issues that concern women. Darke and Yeandle (1996) explain that involvement of women in design and management of cities is an important tool in tapping into women's experience of the environment, and helps to provide knowledge of the daily patterns of women's lives, which can produce more sensitive polices that take into account socioeconomic change in society and respect the social and cultural diversity of society. Moser (1993, p. 101) points out that involvement of women in development and planning processes is a means toward improving project results. The exclusion of women can negatively affect the outcome of a project. According to Moser, the need to acknowledge women's needs in planning not only creates better solutions but also creates more effective ones. In his paper "Women and the Conserver Society," Peterson (1979) notes the hazards associated with having too few women involved in decision-making with regard to new technologies and in the related area of policy-making. He claims that the lack of female perspective in decision-making may lead to a neglect of basic aspects of women's lives essential for analyzing the impact of new technology.

Greed (2005) has investigated the effectiveness of gender mainstreaming as a means of integrating the needs of women and men into spatial planning in the UK.

She explains that although equalities mainstreaming is widespread within local government, much of the emphasis is upon approaches in which gender is given limited attention. Women's (and men's) needs are not explained or defined, and spatial planning fails to integrate women's actual needs into substantive, spatial policy considerations.

The previous studies maintain the necessity of women's participation in decision-making to express their needs and values and raise the question: Is it true that legal instruments and urban design reflect women's needs when they are decision makers?

Research shows that women continue to face problems in using public spaces.

Worpole (1992) wrote about problems facing women in British towns at night. A study in Woolwich showed that 65 % of women are afraid of going out at night for fear of assault. Thirty-six percent of women are afraid of mugging and robbery during daytime. Another study in Edinburgh showed that women feel dissatisfied with the town center because the streets are dirty and poorly lit, bus services and childcare facilities are inadequate, and the women fear sexual harassment. A study of nightlife in 12 British towns and cities found that women's view of urban life was completely different from that of men. These studies reveal that such problems cannot be solved by better policing and security only, but that solutions require "a genuine choice of activities, entertainment and places where women can meet in towns and cities at night, and provision for children where necessary" (Worpole 1992, p. 65). According to this, women's needs differ according to the local context.

Karp et al. (1991) wrote about women's problems and their use of public spaces in the USA, where the frequency of rape is seven times higher than in Europe.

The results of previous studies maintain that women's needs, in general, are not met properly in public spaces even though they are, to a certain degree, involved in designing and managing of these spaces. This indicates that women in the decision-making process do not themselves reflect on and express their cultural needs.

Based on the previous studies, the following questions are raised: What about women's situation in urban design of cities in the Middle Eastern world? How are their needs and values being met in cities? What role do they have in designing and managing cities? These issues will be investigated in the next section which is concerned with women's needs in designing Middle Eastern cities, with particular attention to women's requirement for privacy.

9.6 Methodological Approach

To investigate how women's need for privacy is considered in design of cities and urban planning practices, the research design is based on two sections:

- The theoretical section includes a theoretical discussion of the main topic (women's needs in urban planning, women's privacy and urban design of Middle Eastern cities, and women's involvement in decision-making process and reflection of their needs).

- The practical section is a case study in Palestine to clarify issues related to the topic of the research as they are discussed in the theoretical section.

For data collection, both observations and interviews are used as two main methods, in addition to statistical analysis concerning women's participation in different activities.

The observations were conducted by the researcher, in different urban and public spaces in different Palestinian cities, to investigate how privacy influences the physical form and women's behavior in urban spaces. Photos and written notes were used to document these observations.

To support the observations, interviews were conducted by the researcher with women and different target groups who are relevant to planning practices as follows:

Interviews with Women These interviews were conducted with female inhabitants, particularly where the observations were conducted. Also, there are interviews which were conducted inside dwellings. The interviews were carried out by the researcher and face to face with women without prior arrangement. The time spent in these interviews was dependent on the situation. For example, interviews with women in the streets only took minutes (5–10 min), while in public buildings interviews took from 15 to 30 min. In these interviews, the researcher introduced herself to the selected women in different situations, and sometimes joined them in their activities, for example, shopping or having coffee together with women in the park or a restaurant. The objective of these interviews was to investigate how women's privacy influences their access to public sphere and different employment fields. The women were asked to explain how they use their urban environments, and problems they face in public sphere. Also, if they participate in different events which take place in the streets or city centers and if they do not participate in these activities, they were asked to explain why. In addition, women were asked if they were employed in any field or not, if not why and if yes, are they comfortable or not, and if not why. The sets of information were recorded on separate sheets. However, after conducting 20 interviews, the process was stopped because most of the information proved repetitive at this point.

Interviews with Inhabitants Interviews with inhabitants of semiprivate spaces (residential courtyards, gardens, etc.). The aim of these interviews was to understand their attitudes toward women's privacy and how this influenced the physical form of the spaces which they inhabit. Inhabitants of ten residential courtyards and ten residential gardens were interviewed.

Interviews with Managers Interviews with managers of different public spaces (restaurants, coffee shops, etc.). The aim of these interviews was to understand their attitudes toward women's privacy and how this influenced the physical form of the spaces which they manage. Four (male) managers, two from restaurants, one from public bath, and one from a coffee house, were interviewed.

Interviews with Planners and Architects Interviews with planners and architects (both male and female) in municipalities of three cities located in the north, middle, and south of Palestine. The aim of these interviews was to understand how women's privacy is considered by planners and architects who are responsible for the design and development of cities. The aim was also to find out whether there is a difference between male and female planners toward women's privacy. These interviews were conducted with six engineers: three males and three females who work in different municipalities.

9.7 Results and Discussion

Data collected from the observations and interviews are concerned with influence of privacy on women's access to public sphere, and attitudes of decision makers, including women, about design and planning process toward women's privacy. These data can be categorized into three themes: The first is concerned with women's access to public sphere and the influence of privacy. The second is concerned with attitudes of decision makers (inhabitants, managers of public spaces, planners, and architects) toward women's privacy. The third is concerned with attitudes of women, as being decision makers, toward their needs, particularly privacy.

9.7.1 First: Women's Participation in Public Life and the Influence of Privacy

Collected data from the observations and interviews with women showed that many women do not use their urban areas comfortably due to privacy reasons. For example, in many situations as in crowded sidewalks, females are obliged to walk in one row behind each other to avoid physical contact with males (Fig. 9.5a). Also, in streets where men's coffee houses exist and men usually sit on the sidewalks in front of these coffee houses, females are obliged to walk in the street and expose their lives to danger (Fig. 9.5b) or avoid passing by in these streets and use other alternatives (Fig. 9.6a)

In other situations, women are completely withdrawn from public sphere when it is dominant by males (Fig. 9.6)

Statistical data about individuals participating in different activities in relation to gender in Palestinian territories, issued by Palestinian Central Bureau of Statistics during the year 2012 (Table 9.1), show that percentage of females participating in employments is less than that of males. Also, the average time spent by females in participating activities outside is less than that of males. However, the table shows that women's participating activities are mainly concerned with household rather than employments.

In Palestine, it seems that privacy need is still important for a large number of women despite change in women's role and influence of modernity. For this reason, women's behavior in public spaces is restricted because the physical form of these spaces does not provide women with their required level of privacy. Observations and interviews with women showed that in spaces which do not comply with

Fig. 9.5 Example of women's use of public sphere uncomfortably. **a** Females walked in one line behind each other to avoid direct contact with males. **b** Men are sitting on the sidewalk of the commercial street and women are coming toward them. Women left the sidewalk and walked in the street to avoid passing by in front of men

Fig. 9.6 Examples of women's withdrawal from public sphere. **a** Palestinians, mainly males, help themselves to a piece from the world's largest knafeh. **b** Males sit in the street in front of coffee house. Women usually avoid passing by in this street

Table 9.1 Percentage of individuals practicing and average time spent in performing activities in hours (H) and minutes (M) by sex and activity for the year 2012. (Source: Palestinian Central Bureau of Statistics (PCBS) 2012)

Activity	Percentage of individuals participating activities (%)			Average time spent in participating activities		
	Males	Females	Both sexes	Males (H:M)	Females (H:M)	Both sexes (H:M)
Employment in establishments	43.9	5.2	24.6	7:39	6:17	7:31
Primary production not in establishments	9.3	5.9	7.6	4:16	2:01	3:24
Services for income and other production of goods not in establishments	17.1	4.4	10.8	4:00	2:18	6:5
Household maintenance, management, and shopping for own household	39.8	90.9	65.2	1:20	4:07	3:16
Care for children, the sick, elderly, and disabled for own household	20.3	50.3	35.4	0:59	2:12	1:51
Community services and help to other household	7.2	10.4	8.8	2:13	1:35	1:51
Learning	23.6	22.7	23.2	6:47	7:08	6:57
Social and cultural activities	89.5	88.8	89.2	3:50	3:14	3:32
Mass media use	84.7	87.3	86	2:28	2:47	2:37
Personal care and self-maintenance	99.9	100.0	100.0	10:50	11:25	11:07
Other activities in the group	9.6	25.2	17.3	0:26	0:26	0:26

women's privacy need, women achieve their required privacy through behavioral components. Although behavioral components help women achieve their privacy, these components are restrictions rather than motivations. For example, walking in the street instead of on the sidewalk, walking in one row behind each other, or looking for alternative streets to pass by comfortably are all components that act as restrictions rather than motivation for making use of public space and participating in public life. This was confirmed during the interviews with women where many of them explained that when the physical layout of the space they use does not comply with their privacy, they are obliged to use additional behavioral components. In other situations, women may also be withdrawn from public life due to privacy reasons. During the interviews with women, many of them explained that they do not participate in several activities and events in streets and city centers because they do not find comfortable spaces, spaces where they can sit or stand up comfortably. These results comply with statistical data reflected in Table 9.1, which indicate that women's access to public life is still limited in comparison with men despite both women and men have equal opportunities in different employment fields (Odeh and Atallah 1999). Although the reason for this could be related to women's personal factors or other reasons, privacy is also an important factor which restricted women's access to several employment fields. This was confirmed during interviews with women, as many of them explained that they were deprived from job opportunities offered to them because of privacy reasons. For example, being male domain or located far away outside their cities caused restriction on women's employment. This deprived women from their right in access to public sphere without restrictions, which contradicts with the principle of universal design about equity among citizens in access to their built environments (Welch 1995). Therefore, involving women's privacy in the design of public spaces enhances women's comfort and right to access these spaces.

9.7.2 Second: Attitudes of Decision Makers Toward Women's Privacy

Observations and interviews with inhabitants of the dwellings showed that most of them care much about women's privacy inside the semiprivate spaces. Inhabitants of the courtyards (Fig. 9.7) explained that they cover and divide their courtyards for privacy reasons, in addition to other reasons such as security and climatic ones, or the need for additional space.

Inhabitants of the building blocks (Fig. 9.8) explained that they divided the entrance to guarantee their women's protection while they stop or clean the entrance, in addition to avoiding conflicts with their neighbors.

To guarantee women's privacy while using the entrances of the houses (Fig. 9.9), some of the inhabitants encroach onto the sidewalks by constructing thresholds

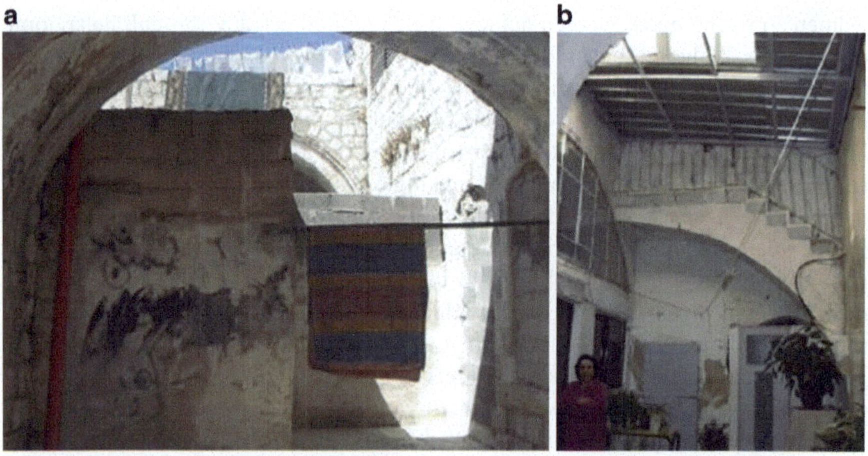

a courtyard which is divided by partitions. a courtyard which is completely covered

Fig. 9.7 Example of women's privacy and inhabitants' attitudes toward residential courtyards. **a** A courtyard divided by partitions. **b** A courtyard completely covered

Fig. 9.8 An example of women's privacy and inhabitants' attitudes toward a shared dwellings' entrance

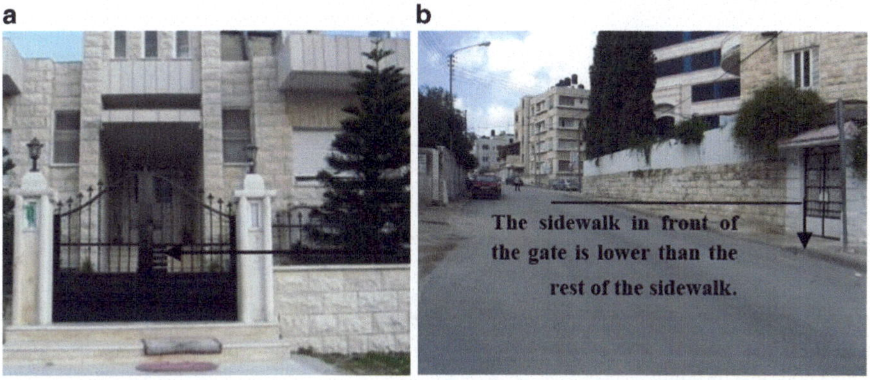

Stairs are constructed on the sidewalk in front of the gate.

An additional part is used above the walls to achieve the desired privacy.

Fig. 9.9 Examples of women's privacy and inhabitants' attitudes toward residential gardens. **a** Stairs are constructed on the sidewalk in front of the gate. **b** An additional part is used above the walls to achieve the desired privacy

(Fig. 9.9a) or changing in the pavement of the sidewalks in front of their entrances (Fig. 9.9b). Also, inhabitants of residential gardens (Fig. 9.9b) explained that they increased the height of walls surrounding their dwellings to guarantee women's protection inside these gardens. Security concerns were an additional reason.

Observations and interviews with managers of public spaces as restaurants and public baths (Fig. 9.10) showed that many of them designate spaces (Fig. 9.10a) or times (Fig. 9.10b) for the use of either women or families in the buildings they manage. They did this to encourage women's use of these spaces and improve their profits.

Observations and interviews with architects and planners showed that women's privacy influenced the decision-making process regarding the physical form of public and urban spaces, despite the fact that building codes and regulations do not mention anything about women's privacy, as shown in the following cases:

A screened wall inside a coffee shop contributes to women's privacy while sitting.

The written sign says that the public bath is reserved for women's use on Tuesdays.

Fig. 9.10 Examples of women's privacy and attitudes of managers of public spaces toward these spaces. **a** A screened wall inside a coffee shop contributes to women's privacy while sitting. **b** The written sign says that the public bath is reserved for women's use on Tuesdays

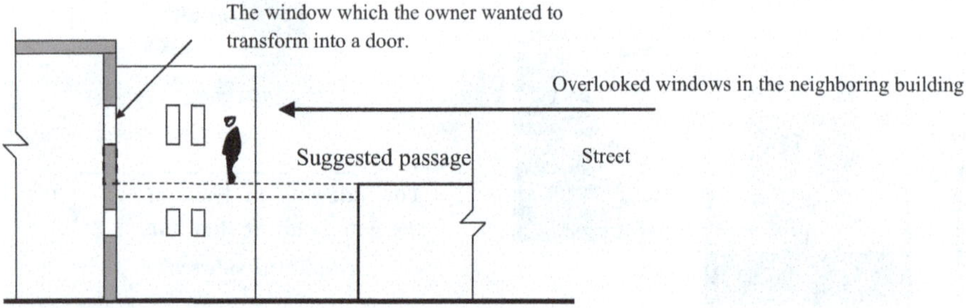

Fig. 9.11 An example of women's privacy and decision made about transformation of window into a door in a dwelling

The First Case (Fig. 9.11**)** An owner of a house wanted to transform a window on the second floor in his house into a door, and connect it with the public street with a passage.[5] After the inspectors of the municipality visited the site, they found that building this passage would cause windows of the neighboring buildings to be overlooked from the passage. Therefore, decision makers in the municipality ordered that the owner should not change the existing situation.

The Second Case (Fig. 9.12**)** Decision makers in the municipality decided to transform a private staircase which was established in the public road in the area 112, plot 2, into a public one. The existing staircase is adjacent to the wall which surrounds the garden of the house. The inhabitants in this building objected as the passersby on the staircase would overlook their women when in the garden. Therefore, decision makers constructed the public staircase adjacent to the existing one, at a 2 m distance from the wall surrounding the garden.

[5] This house is located in an old neighborhood, and according to municipal legislation, any change in buildings of the old neighborhood should not contradict the municipality's regulations which are concerned with protecting the old fabric of the city, and should not cause harm to neighbors. Violating their privacy is considered harmful.

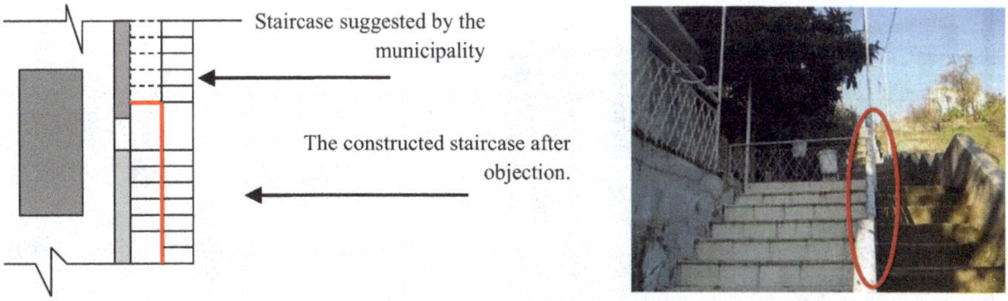

Fig. 9.12 An example of women's privacy and decision made about transformation of a private staircase in a residential area into a public one

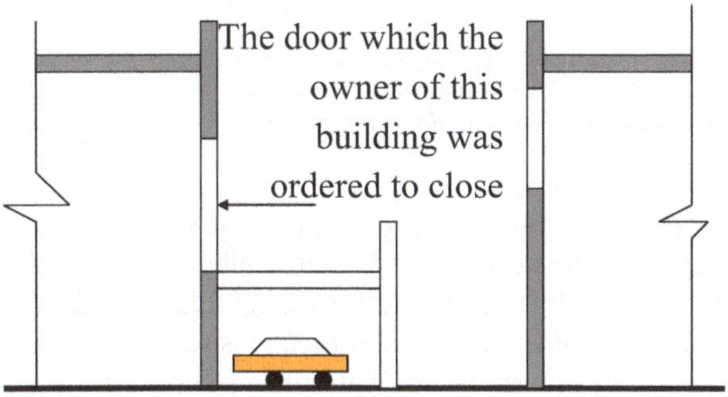

Fig. 9.13 An example of women's privacy and decision made about opening a door in a dwelling that leads to the roof of a neighbor's private garage

The Third Case (Fig. 9.13**)** An owner of a building opened a door that leads to the roof of his private garage.[6] The inhabitants of the adjacent building contacted the municipality to object as they have become overlooked by their neighbors when using the roof of their garage. Decision makers in the municipality ordered the owner of the garage to close the door.

The Fourth Case (Fig. 9.14**)** Inhabitants of one of the neighborhoods were not convinced by the municipal architects' design to improve the square located in their neighborhood. The inhabitants claimed that establishment of a green area and sitting places in this square caused strangers to stay in it, detracting from the privacy of local women. This was a particular problem as women are the main users of the square. Therefore, decision makers in the municipality replaced the green area by a public building which is used for different social occasions.

In addition, the interviews showed that architects and planners consider women's privacy in public spaces designed for children. They explained that when they design any space for children, they think about women's use of this space. Also, when they design mosques, schools, or public toilets, they always separate between males' and females' spaces inside these buildings.

[6] According to municipal legislation, the owner of any house is allowed to construct a garage for his car without a setback area from his neighbors without any access to its roof.

Al-Yasmeeneh yard as the municipality suggested.

Al-Yasmeeneh yard after the objection of inhabitants.

Fig. 9.14 An example of women's privacy and decision made about establishment of sitting places in a residential yard (Al-Yasmeeneh). **a** Al-Yasmeeneh yard as the municipality suggested. **b** Al-Yasmeeneh yard after the objection of inhabitants

Based on the previous results, it can be said that women's privacy is met indirectly (informally) in the design of urban and public spaces, since there are no written rules concerning women's privacy exist in building codes and regulations. The reason for meeting women's privacy indirectly is seen as related to cultural reasons. Planners and architects are mostly Palestinians and are influenced by notions of women's privacy in their way of life. Therefore, they reflect women's privacy indirectly in their design works. But, in the case of projects into which international institutions are involved, the foreign engineers who belong to these institutions will not be influenced by notions of privacy as local engineers. In these cases, informal rules will not contribute to women's privacy and women will be restricted in using these projects, which contradicts with principles of universal design in creating spaces that meet needs of all individuals (Welch 1995). Due to the importance of privacy for the inhabitants, decision makers in the municipalities are obliged to approve changes in the physical form of public spaces concerning women's privacy even if these changes contradict the formal rules. Meanwhile, managers of urban and public spaces, in most situations, specialize spaces for women and families inside public buildings not only to comply with women's privacy need, but also to increase the number of customers. Palestinian society and public habits do not accept mixing males and females, particularly if they are young, in the same building. In the case of public spaces used by both males and females, many families may restrict their daughters' use of these spaces because, according to them, these spaces are "unrespectable." This may decrease the number of women who use these spaces. For this reason, most public buildings include special spaces or times for females' or families' use. Being specialized for families' use means that the space is only accessible to either families or women and inaccessible to males alone.[7] There-

[7] Based on the researcher's knowledge of the local context of Palestine.

fore, specializing the space for families allows for women to use this space either with their family or alone, and in both cases it contributes to women's privacy and encourages women's use of the public space.

Although women's privacy is met indirectly through the informal rules, results of this study showed that privacy need is met partially in different ways in urban and public spaces. This influenced negatively the esthetic quality and the sustainability of urban spaces. The results showed how people make changes in different urban and public spaces to adapt it to women's privacy. For example, in urban spaces that do not comply with women's privacy, people encroach on to these spaces and change its design. This led to esthetic degradation in the physical appearance of these spaces. Observations and interviews showed that the height of many walls surrounding the residential gardens has been increased by using metallic or textile materials for privacy reasons. Also, many residential courtyards are covered or divided for privacy reasons. According to this, including women's privacy in the design of urban and public spaces control encroachments on to these spaces and thus enhances their esthetic quality and sustainability (Al-Hemaidi 2001; Eben-Saleh 1998). Moreover, adaptation of different spaces to women's privacy costs money, which contradicts with the principle of universal design concerning sustainability of the built environment, as explained by Aslaksen et al. (1997).

Therefore, meeting privacy through the informal rules does not assure women's right in access to public sphere, on the one hand, and causes degradation in the physical form of the built environment, on the other hand. This reveals the importance of including women's privacy in the formal rules.

9.7.3 Third: Women's Involvement in Design and Management of Public Space and Fulfillment of Privacy

Interviews with female planners and architects showed that they have the same attitudes as males toward their privacy needs. In other words, women do not reflect their privacy need in professional practice, although they are involved in the design process. Meanwhile, some scholars explained that involving women in design and development of cities helps women to reflect their needs (Darke and Yeandle 1996; Greed 2005; Lansley et al. 1989; Moser 1993), and others explain that women continue to face problems in using public spaces (Karp et al. 1991; Worpole 1992).

The reason for this is seen as related to two factors: first, standard knowledge taught to planners and architects at the universities that do not differentiate between standard and specific needs, including privacy.[8] Second, women's needs, including privacy, are not included in building codes and regulations.

The results also revealed that the number of female architects and planners is less than that of males, and females implement their designs and plans according to the

[8] Professional experience of the researcher as being an architect and a teacher of architectural engineering at Birzeit University.

city's master plan and in accordance with the desire of the heads of engineers in the municipalities, who are mostly males. This indicates that males are still dominant in the field of urban design and planning in both decision-making and numbers. This result complies with relevant literature which explains that design and management of cities is dominated by males (Richter 1982). The background for this situation is seen as related to women's traditional role being considered housekeeping, while men are considered responsible for work outside the house. This probably also gives men the power to make decisions about designing public spaces. Females' situation in the field of architecture and urban planning requires more emphasis on providing more opportunity for female employment similar to males, and empowering of females' roles in decision-making process about the physical environment, particularly as they form half of the society.

Therefore, including women's privacy in the knowledge taught to engineering students, particularly architects and planners, will increase students' awareness of women's privacy and enable them to integrate it into their design works and planning project after they graduate and practice engineering profession. In addition, integration of women's privacy and other cultural needs into engineering education can help in integrating these needs also into formal rules as many of engineering students will be appointed in the municipalities and ministries and become from the decision makers in the future.

9.8 Conclusion: Integration of Women's Needs in Architecture and Urban Planning Curriculum

This study provides us with important findings about how women's specific needs, particularly privacy, are considered in the design of cities and urban planning practices in Palestine in different ways: formally and informally, in traditional and modern urban design. The study revealed that women's privacy is a cultural need that influences the current decisions made about the built environment, informally, since nothing is mentioned about it in building and planning regulations. Therefore, including women's needs, particularly privacy, in the formal rules, which govern the design and development of cities in the Middle East, can help in providing urban environment that meet women's specific needs and enhance women's participation in community and public life.

However, in the Middle Eastern societies where women's privacy is important, architects and planners should be aware of cultural norms concerning this need in order to accommodate women in urban and public life; otherwise, women will be excluded and prevented from their rights in using the public sphere comfortably, including the employment fields. Therefore, it is not correct to apply standard codes and regulations which are imported from Western cultures as they are, and without implementation of our cultural needs, particularly women's privacy, in these codes.

In order to implement women's specific needs in the formal rules, planners and architects should be first aware of these needs and how they are relevant to their design works. This can be achieved only through engineering education and integrating women's needs into architecture and urban planning curriculum which is taught to students at the universities and schools of architecture and planning in general. This knowledge will also help in accommodating women's privacy in formal rules.

If women's privacy is an important need and value for the Middle Eastern societies which influenced their traditional and modern urban designs and women's access to public sphere, the question to be raised here is:

Why women's privacy has not been integrated into engineering education yet?

Engineering education, including architecture and planning, is based on standard knowledge which is transported from the West and is based on functionality and welfare. For example, students in architectural faculties are taught how to design buildings based on standard architects' data available in *Neufert Book*. The required functions and dimensions for any building are considered according to *Neufert*, without linking these data to our cultural and local needs, including women's privacy.

Also, students in planning are taught to design cities and prepare the master plans of these cities according to standard requirements and calculations based on the number of inhabitants and resources without considering the cultural needs of the society, including women's privacy.

Within this situation, the question that will be raised is: "how women's privacy can be integrated into engineering education?" To answer this question, two cases will be explained. The first case is concerned with suggestions about how to reflect women's privacy in design and planning components. The second case is concerned with general approach about how to implement women's specific needs in the curriculum.

Regarding the first case, the following are examples showing how to integrate women's privacy into design and planning concepts:

When dealing with openings' design, students are taught to consider the dimensions and types of windows and doors as mentioned in *Neufert Book* and other architectural books, which are imported mainly from the West and applied internationally. No or little attention is paid to our cultural needs, including women's privacy, for example, how the height and the location of these openings may influence women's privacy.

- Locating windows and balconies of buildings on opposite sides of the road on different sight lines contributes to women's privacy and encourages women to use windows (Fig. 9.15).
- Providing recess areas in the windows and balconies decreases women's exposure to their surroundings (Fig. 9.16).
- Providing deep entrances for dwellings in the modern neighborhoods encourages women to gather in the entrances (Fig. 9.17).
- On urban level, distribution and classification of functions do not consider women's privacy. For example, when dealing with residential areas, students are taught to differentiate between three classifications of housing A, B, and C based on income level of the inhabitants rather than the cultural needs. The setbacks for buildings in zone A are larger than in zones B and C, and the heights of building

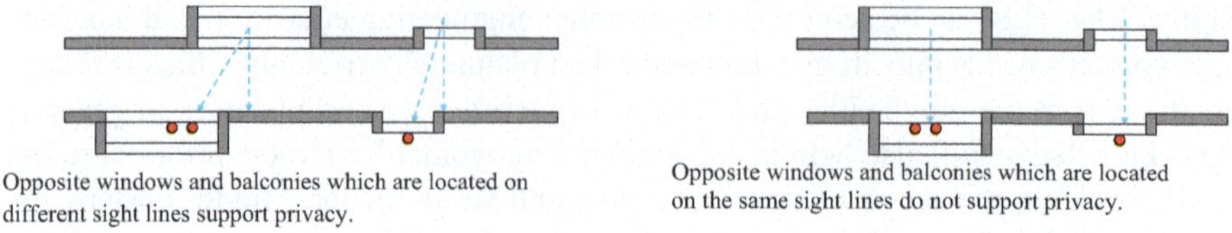

Fig. 9.15 Suggestion for location of windows and balconies to support women's privacy

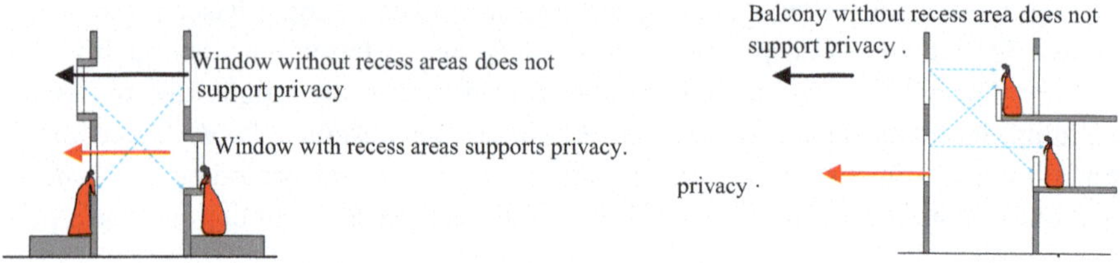

Fig. 9.16 Suggestion for physical layouts of windows and balconies to support privacy

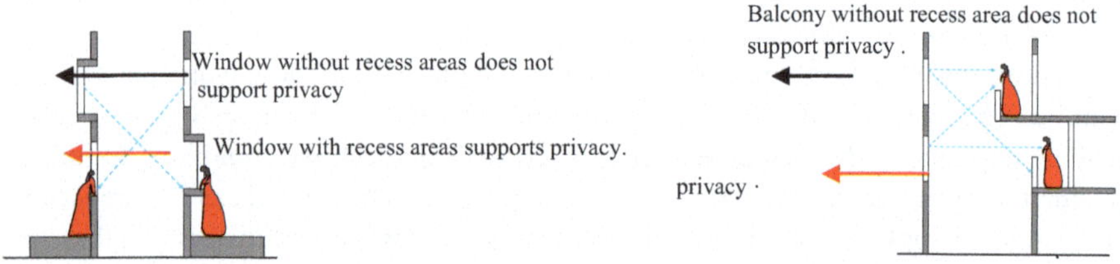

Fig. 9.17 Suggestion for developing entrances of dwellings to support women's privacy

is less than in zones B and C. Also, in zones B and C, it is allowed to include commercial and public facilities; meanwhile, it is not allowed to do in zone A. This is what students are taught to consider in their plans; meanwhile, there is no or very little focus on how the height of buildings and the location of public facilities may influence women's privacy. Also, students are taught to design streets and differentiate between them based on the function and width, and the importance of sidewalks, without considering how the street networks and layout may influence women's privacy. They are also taught how to design sidewalks (height and width) without considering how the width may influence women's privacy, for example:

- More differentiation between streets and less straightness in the layout of these streets provide spaces where women can stop in comfortably without direct contact with passersby.
- Locating main public facilities on the commercial streets instead of in the residential districts controls strangers' access to residential districts and encourages women to use public spaces inside their districts.
- Setbacks (yards) in front of men's coffee houses control men's encroachment onto the sidewalks and streets, allowing women to walk more comfortably (Fig. 9.18).

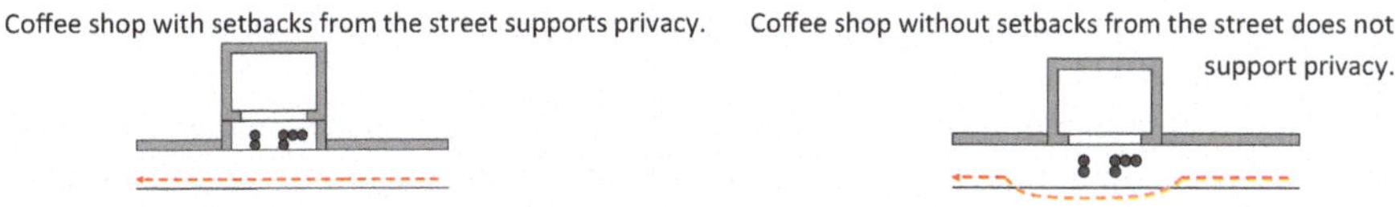

Fig. 9.18 Suggestion for development of men's coffee houses to support women's privacy

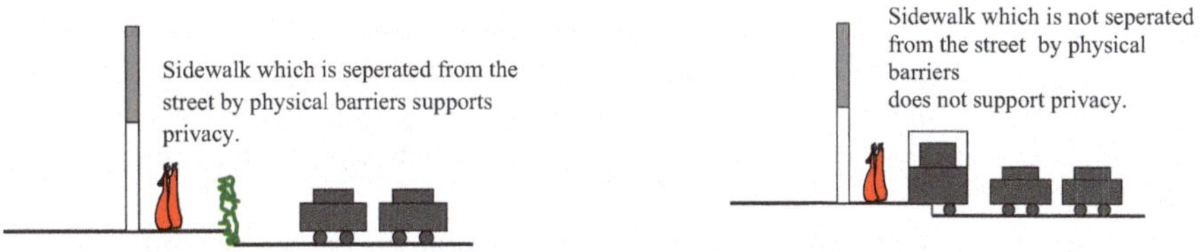

Fig. 9.19 Suggestion for controlling vehicles' encroachment onto sidewalks of streets to support women's privacy

- Physical barriers between sidewalks and streets prevent drivers from parking their cars on the sidewalks and allow for women to use the sidewalks more comfortably for privacy (Fig. 9.19).
- Specializing markets in commodities, particularly those with products for women, controls the access to these markets and encourages women to use these markets.

Regarding the second case, integrating women's privacy into engineering education is concerned with suggesting an approach to teach women's privacy.

For example, the architectural and urban planning curriculum in all contexts in the world should include compulsory courses about cultural needs of the local societies, particularly specific needs of women. This is important for raising the awareness of graduate planners and architects (both male and female) regarding women's specific needs. Accordingly, this will help in implementing these needs in design and planning project, which enhances women's public life and participation in different employment fields.

In the Middle Eastern context, there is a need to include compulsory courses in the architectural and urban planning curriculum about women's specific needs, particularly privacy, which presents an important part of the culture in this context.

Students should be educated about how women use cities and in which ways they differ from men in Middle East and North Africa (MENA) region countries. This will lead to the understanding of women's specific needs. Moreover, reflection of women's need in the physical layouts and functions of spaces on different urban levels, such as land use, neighborhoods, streets, spaces, and elements in spaces, is also an important aspect that should be taught to the students. In other words, students should consider women's needs, particularly privacy, in all their design and planning projects. For this purpose, the theoretical knowledge, which is taught to the students, should include how women's needs can be reflected in different components of the space, such as physical (design), social (use), and cultural (rules) components, and the relationship between these components (Al-Bishawi and Ghadban 2011).

Design components (physical components) include the spaces where women's activities occur, their physical layouts, boundaries, shape, location, and objects. Use components (social components) include the activities in the spaces (type and time), in addition to the users of the spaces and the social relations between them. Rules components (cultural components) include both the formal and informal rules that govern the physical form of urban and public spaces. Formal rules include written rules that are concerned with the design and function of the space, such as written signs that govern the use of the space, building codes, and regulations. Informal rules include rules that are concerned with people's behavior, such as religion and family.

Regarding the relationship between these components, students need to know how design meets the requirements of the rules. The most important ways are to include the rules which are concerned with women's needs, particularly privacy, in planning codes and regulations (formal rules), so that women's needs become an urban design concept. Students also need to know how design makes possible appropriate and convenient use (spaces and design elements appreciated by women). Although the relationship between rules and use components is most appropriate to sociologists and anthropologists, students should also be aware of it. Rules are interpreted in practice, and get reinterpreted over time. And not all women behave the same way at any given time. Architects and planners need to be aware of current changes and research needs to be done regularly to contribute to the development of the urban design discourse.

References

Abdulhadi, R. S. (1990). Land use planning in the occupied Palestinian territories, I.

Abu-Gazzeh, T. (1996). Privacy as the basis of architectural planning in the Islamic culture of Suadi Arabia. *Arch & Comport/Architecture & Behaviour, 11*(3, 4), 269–288.

Abu Lughod, J. (1987). The Islamic city: Historic myth, Islamic essence and contemporary relevance. *International Journal of Middle East Study, 19*, 155–176.

Ahmed, L. (1992). *Women and gender in Islam: Historical roots of a modern debate*. New Haven: Yale University Press.

Al-Bishawi, M. (2008). Women's behavior in public space and the influence of privacy/the case of Nablus/Palestine. PhD Thesis, Department of Spatial Planning and Landscape.

Al-Bishawi, M., & Ghadban, S. (2011). A methodological approach for reading urban open space. *Archnet-IJAR, International Journal of Architectural Research, 5*, 73–85.

Al-Hemaidi, W. K. (2001). The metamorphosis of the urban fabric in an Arab-Muslim city: Riyadh, Suadi Arabia. *Journal of Housing and the Built Environment, 16*, 179–201.

Altman, I. (1975). *Environment and social behavior. Monterey*, CA: Brooks-Cole.

Arkoun, M. (1990). *Expressions of Islam in buildings: "Islamic cultures, developing societies, modern thoughts", the Aga Khan trust for culture on behalf of the Aga Khan award for architecture*. Paper presented at the International Seminar Sponsored, Jakarta.

Aslaksen, F., Bergh, S., Bringa, O. R., & Heggem, E. K. (1997). *Universal design, planning and design for all*. Oslo: The Norwegian State Council on Disability.

Bianca, S. (2000). *Urban form in the Arab world: Past and present*. London: Thames & Hudson.
Booth, C. et al. (1996). *Changing places: Women's lives in the city*. London: Paul Chapman Publishing.
Bunston, T., & Breton, M. (1992). Homes and homeless women. *Journal of Environmental Psychology, 12*, 149–162.
Cohen, A. (1987). *Planning for child care: A compendium for child care advocates seeking the inclusion of child care in the land use plan*. San Francisco: Child Care Law Center.
Corral, T. (2002). Women's sustainable development agenda. *Natural Resources Forum, 26*(3), 249–253.
Darke, J., & Yeandle, S. (1996). *Changing places: Women's lives in the city*. London: Paul Chapman.
Document and Publications Production. (2013). Global employment trends for youth 2013: A generation at risk/international labour office—Geneva: ILO, 2013.
Dorfman, R. (1997). Taking a walk: No longer safe for elders in urban America and Asia. *Journal of Aging and Identity, 2*(2), 139–142.
Eben Saleh, M. A. (1998). Transformation of the traditional settlements of southwest Saudi Arabia. *Planning Perspective, 13*, 195–215.
Franck, K. A., & Paxson, L. (1989). *Women and urban public spaces: Research, design, and policy issues*. In E. Zube & G. Moore (Eds.), *Advances in environment, behavior and design* (Vol. 2, pp. 122–146). New York: Plenum.
Greed, C. (2005). An investigation of the effectiveness of gender mainstreaming as a means of integrating the needs of women and men into spatial planning in the United Kingdom. *Progress in Planning, 64*(4), 243–321.
Hakim, B. S. (1986). *Arabic–Islamic cities: Building and planning principles*. London: Kegan Paul International Limited.
Hall, E. T. (1966). *The hidden dimension*. New York: Anchor Books and Doubleday.
Ghannam, F. (2002). *Remaking the modern: Space, relocation and the politics of identity in a global Cairo*. Unpublished PhD dissertation, University of California Press, Sweden.
Kallu, R. (2001). From abstract to concrete: Subjective reading of urban space. *Urban Design, 6*(2), 129–150.
Karp, D., Stone, G., & Yoels, W. (1991). *Being urban: A sociology of city life*. New York: Praeger.
Lansley, S., Goss, S., & Woolmar, C. (1989). *Council in conflict: The rise and fall of the municipal left*. London: Macmillan.
Lewis, J., & Foord, J. (1984). New towns and gender relations in old industrial regions: Women's employment in peterlee and east kilbride. *Built Environment, 10*(1), 42–52.
Little, J. (1994). Women's initiatives in town planning in England: A critical review. *Town Planning Review, 65*(3), 261–276.
Maslow, A. H. (1943). A theory of human motivation developed as classics in the history of psychology by Christopher D. Green, 2000. An internet resource. Toronto: York University.
Moser, C. (1993). *Gender planning and development: Theory, practice and training*. New York: Routledge.
Mozingo, L. (1989). Women and downtown open spaces (Vol. 6). http://repositories.cdlib.org/ced/places/vol6/iss1/LouiseMozingo.
Newman, O. (1972). *Defensible space: Crime prevention through urban design*. New York: Macmillan.
Nooraddin, H. (2004). Globalization and the search for modern local architecture: Learning from Baghdad. In Yasser Elsheshtawy (Ed.), *Planning middle eastern cities: An urban kaleidoscope in a globalizing world*. London: Routledge.
Ottoman, M. (1988). *The islamic city (in Arabic: Al- Madinah Al- Islamiyyah)*. Al-Kuwit: Al-Majles al-Watani le-Athaqafah wal-Funoon wal-Adab.
Odeh, K., & Atallah, M. e. (1999). *Nablus Bayn Al-Madi w Al-Hader*. Nablus: Documentation Centre at An-Najah National University.
Palestinian Central Bureau of Statistics (PCBS). (2008/2012). Guide to Palestinian Statistics A–Z. http://www.pcbs.gov.ps/site/lang__en/507/default.aspx.

Perfect, M., & Power, G. (1997). *Planning for urban quality: Urban design in towns and cities*. New York: Routledge.

Peterson, R. (1979). Women and the conserver society. *Conserver Society Notes, 2*(1), 4–7.

Polk, M. (2003). Are women potentially more accommodating than men to a sustainable transportation system in Sweden? *Transportation Research, D*(8), 75–95.

Richter, L. (1982). The ephemeral female: Women in urban histories. *International Journal of Women's Studies, 5*, 312–328.

Ritzdorf, M. (1987). Planning and the intergenerational community: Balancing the needs of the young and the old in American communities. *Journal of Urban Affairs Association, 9*(1), 79–89.

Roald, A. S. (2001). *Women in Islam: The western experience*. London: Routledge.

Robert, M. (1991). *Living in a man-made world.* London: Routledge.

Seedat, M., Mackenzie, S., & Mohan, D. (2006). The phenomenology of being a female pedestrian in an African and an Asian city: A qualitative investigation. *Transportation Research, F*(9), 139–153.

Tamminga, K., & Luymes, D. T. (1995). Integrating public safety and use into planning urban greenways. *Landscape and Urban Planning, 33*, 391–400.

United Nations. (1987). Report of the world commission on environment and development. General Assembly Resolution 42/187, 11 December 1987. Retrieved: 2007-04–12.

Welch, P. e. (1995). *Strategies for teaching universal design*. California: MIG Communication.

Worpole, K. (1992). *Town for people: Transforming urban life*. Buckingham: Open University Press.

Dr. Manal Al-Bishawi is an assistant professor in architecture & spatial planning at the Department of Architectural Engineering and Master Program of Urban Planning and Landscape Architecture in Faculty of Engineering and Technology at Birzeit University, Palestine, and member of the Palestine Engineers Association. She worked and participated in several local and international workshops and conferences. Her academic and research activities are oriented towards the spatial planning and landscape architecture, relationship between physical environment and individual's behavior, gender issues in urban planning, particularly in Arab-Muslim cities. In addition, she worked in design and supervision of several architectural projects in Palestine.

Chapter 10
Improving and Expanding Engineering Education in the Middle East and Africa Using Mobile Learning Technology and Innovative Pedagogy

Yacob Astatke, Jumoke O. Ladeji-Osias, Petronella James, Farzad Moazzami, Craig Scott, Kenneth Connor and Abdurrahim Saka

10.1 Introduction

The Middle East and North Africa (MENA) region is in a period of significant educational growth and transformation according to the World Bank and others (The World Bank 2008; Chapman and Miric 2009). Over the past 50 years, these countries have dedicated economic resources to education at rates that are higher than other developing countries. This has resulted in increased enrollment at all levels, with a fivefold increase in higher education between 1970 and 2003. However, there remains much work to be done in bringing the educational achievement of many MENA countries on par with developing countries in East Asia and Latin America. A framework for proposed reforms addresses the need to improve the outcomes by improving inputs to the educational system, incentives, and public accountability.

Y. Astatke (✉) · J. O. Ladeji-Osias · P. James · F. Moazzami · C. Scott
Morgan State University, 1700 East Cold Spring Ln, Schaefer Engineering Building #339, Baltimore, MD, USA
e-mail: yacob.astatke@morgan.edu

J. O. Ladeji-Osias
e-mail: Jumoke.Ladeji-Osias@Morgan.Edu

P. James
e-mail: petronella.james@morgan.edu

F. Moazzami
e-mail: farzad.moazzami@gmail.com

C. Scott
e-mail: craig.scott@morgan.edu

K. Connor
Rensselaer Polytechnic Institute, Troy, NY, USA
e-mail: connor@rpi.edu

A. Saka
Yeni şafak, Istanbul, Turkey
e-mail: abdurrahimsaka@gmail.com

M. Abdulwahed et al. (eds.), *Advances in Engineering Education in the Middle East and North Africa,* DOI 10.1007/978-3-319-15323-0_10

This chapter will highlight the potential of mobile technology and innovative pedagogy to catalyze the reform process, with a focus on engineering education.

The use of technology in the classroom has greatly impacted engineering education during the past 20 years. The expansion of the Internet and the use of computers, tablet PCs, smart boards, and other wired and wireless devices have proliferated throughout education. One thing that has not drastically changed is the traditional method of teaching using lectures. Today's students expect their instructors to create an active and engaged learning environment by delivering course content using more than one method (i.e., classroom lecture), so that they can access it from anywhere, at any time (Strauss and Howe 2007). Universities face an increasing interest in providing students with recorded class material because today's students expect to have 24/7 access to online learning materials. "Millennial" students see technology integration in higher education as a key component of their learning environment because they have grown up using technology in every facet of their lives (Monaco and Martin 2007). The use of new portable laboratory instrumentation and other lecture capture technology has the potential to fundamentally change the way instructors and students interact in the classroom (Dale 2007; Educause Learning Initiative 2014; Harley et al. 2003). It also has the potential to change the classroom environment and the relationship between instructors and students, by shifting the role of the instructor from the "sage on a stage" to one that incorporates more interaction. This is one of the main reasons behind the proliferation of online courses and online programs in higher education. This shift is becoming evident in the MENA region as well as where students are using online learning for informal and formal education. For example, universities in Saudi Arabia are seeing increasing adoption of hybrid learning, which combines classroom and online instruction in a single course. There is also significant government investment in e-learning and recognition of faculty excellence in this area (Yusuf 2013). Other countries such as Jordan, Oman, and Algeria report that e-learning is in the early stages. Adoption of best practices centered on cultural sensitivities, institutional adoption, and faculty training will ensure quality in delivery and implementation (Al-Harthi 2010; Benchicou et al. 2010; Altarawneh 2011).

Internationally, trends in higher education over the past 10 years have shown that enrollments in online courses or online degree programs have been growing substantially faster than overall higher education enrollment. A survey of online learning conducted in 2009 by the Sloan Consortium indicated that enrollment in one or more online courses reached 4.6 million students in 2008 (Allen and Seaman 2010). The 17% growth rate for online enrollments is significantly higher than the 1.2% growth rate of the overall higher education student population during the same time period (The Sloan Consortium 2010). A follow-up report published in 2011 (Allen and Seaman 2011), and other papers (Saba 2005; Joerns and Leinhardt 2006; Bach et al. 2007), seeks to address and provide answers to some of the fundamental questions related to the nature and extent of online education. Some of the questions addressed in the report are as follows: whether retention of students is harder in online courses; if the learning outcomes in online courses are comparable to face-to-face (F2F) courses; whether faculty acceptance of online education has increased; and

Table 10.1 Results from survey regarding online learning. (Allen and Seaman 2011)

Survey question	Survey results
Is online learning strategic?	Sixty-five percent of all reporting institutions said that online education is a critical part of their long-term strategy
Are learning outcomes in online courses comparable to F2F?	Sixty-seven percent of academic leaders rated their learning outcomes in online education to be the same to those in F2F. This number was 57 % in 2003
Has faculty acceptance of online learning increased?	Less than one third of chief academic officers (VPs or provosts) believe that their faculty accepts the value and legitimacy of online education. This number has not changed since 2003
What training does faculty receive for teaching online?	Seventy-two percent of institutions conduct internally run training courses, and 58 % provide informal mentoring
What is the future of online education-enrollment growth?	There is growth in fully online programs by disciplines in public institutions. However, private and for-profit institutions that currently have the largest enrollment are showing a slight decline

F2F face-to-face

the impact of the current economic conditions on online education. The results of the surveys conducted in the study by Allen and Seaman (2011), based on the responses from 2500 colleges and universities, are summarized in Table 10.1. The authors of the survey conclude their report by stating that "online enrollments in U.S. higher education show no signs of slowing." Similar interest in online education (e-learning) is growing in the MENA region. E-Learning provides an avenue to make educational programs more internationally competitive at a lower cost. Countries making significant investments must have the requisite Internet infrastructure and include Israel, Saudi Arabia, United Arab Emirates, and Qatar. Studies conducted on the effectiveness of e-learning in some MENA countries have shown this to be effective in improving learning outcomes and student motivation. However, there is caution about infrastructure, non-English interface, and limited technology literacy in cases where Web-based learning has not been effective (Weber 2010). The rest of the chapter will focus on new educational technologies, the development of an online course with mobile technologies at an American university, and a model for adaptation in developing countries.

10.2 New Educational Technologies and Paradigms

One of the transformations in higher education is possible due to the increase in computational resources and the increasing access to Internet and computer technologies around the globe. Several educational platforms and pedagogies support the delivery of e-learning and the demonstration of learning outcomes on par with on-campus students. Students are now able to learn from leaders in the field, without leaving their home countries. They can also access instrumentation in smaller groups, creating a more personalized learning environment.

10.2.1 Online Courses

Globally, engineering has lagged behind all others in the development and delivery of online education. While there are over 340 engineering schools in the USA that have received accreditation from Accreditation Board for Engineering and Technology (ABET) (2005) for their undergraduate programs, only a handful of these schools offer engineering programs that are completely online at the graduate and/or undergraduate level. Most online engineering programs are at the master of science (MS) level and are targeted at engineers in professional practice, many of whom have received their undergraduate degrees from a campus-based program. The trend has started to change lately, and each year more engineering programs add an online component to their curriculum. The main obstacle impeding adoption is that undergraduate engineering curriculums often require intensive hands-on laboratory components that can be challenging to implement and deliver completely online due to cost and inability of students to manipulate equipment remotely.

Many of the universities that offer an online engineering curriculum indicate to their students that their online program is separate and different from their on-campus program. Some universities, on the other hand, make no differentiation between their online and their on-campus degree programs. For example, the MS degrees in chemical engineering offered at Kansas State and North Carolina State and the BS degree in chemical engineering offered at the University of North Dakota are equivalent to the on-campus degree programs (The Sloan Consortium 2010). One way that universities are ensuring that their online courses and degree programs are on par with their on-campus programs is by implementing rigorous quality assurance standards from the development to the final delivery of the courses (Endean et al. 2010; Oliver 2003). This implies that they have to be able to provide answers to the following questions (Oliver 2003) : "What quality assurance policies and practices does the institution have in place or in the process of development to assure the quality of its teaching and learning performance? How effective and how fully deployed are these? What processes does the institution have to evaluate and monitor the quality of its outcomes? What quality-related indicators does the institution use and why?"

Universities also utilize various types of technologies to increase and enhance the learning experience of their online students (Bonk 2004). In addition to the use of learning management systems (LMSs), some institutions provide online courses using live video from on-campus courses via Web-conferencing and synchronous streaming technology, providing a degree of interaction between local and remote students with limited scheduling flexibility. Others provide asynchronous prerecorded lectures, allowing online students to have a very flexible academic schedule, while providing them with a learning experience that closely mirrors that of on-campus students. Some institutions also take into account the fact that online students might not have a stable and reliable Internet connectivity by providing them with CD-ROMs and videotapes of all the course contents. Massive open online courses (MOOCs) are a new type of online courses that have taken higher education by storm for the past few years. The first MOOC called "Connectivism and Connective Knowledge" (CCK8) was offered in 2008 by educators from the University of Manitoba in Canada. It was attended by about 2000 students. MOOCs gained national and international attention when two Stanford Professors Sebastian Thrun and Peter Norvig offered the course "Introduction to Artificial Intelligence" in 2012 to about 160,000 students, from 190 countries. This course led to the formation of a new company, Udacity (2014), by the two Stanford professors. Within 1 year, two other start-up MOOC providers were formed: EdX (2014) and Coursera (2014). Other start-ups that provide MOOCs have been started in Europe, Asia, and Australia and include Open2Study (Australia), Veduca (Brazil), FutureLearn (Britain), Iversity (Germany), and School (Japan; International MOOC providers 2014). Even current providers of LMS that are used by the majority of higher education institutions, such as Blackboard, are joining the MOOC race. The business model for MOOCs is changing rapidly as it is still an evolving educational paradigm. Although all MOOC providers started by offering courses free of charge, they have started charging fees for proctored exams and certifications. MOOCs provide an opportunity for students to access educational resources that may be unavailable in their home countries. For example, a study conducted by the Wharton Business of the University of Pennsylvania found that 78 % of registered participants in the online business course came from outside the USA. In comparison, an executive master of business administration (MBA) program in 2012 only attracted 14 % of foreign students. Part-time or flexible MBA programs attracted 10–32 % of foreign students, depending on the type of program. Developing countries are turning to MOOCs for accessible world-class training, as there are limited sources for high-quality training and costs are prohibitive (MOOCs Won't Replace Business Schools 2014).

10.2.2 Virtual, Remote, and Mobile Laboratories

The use of laboratory facilities in engineering education has been evolving. Local and remote students are now routinely working with physical instrumentation, virtual interfaces to physical instrumentation, virtual instrumentation, and mobile instrumentation. The use of physical instrumentation is the traditional model where the student has a fixed physical location where they interact with the equipment. As more instrumentation is becoming Web-enabled, students are able to access the instrumentation via a server or virtual interfaces. These remote laboratories are often designed to allow multiple users to access the instrumentation simultaneously (Maiti and Tripathy 2013). While the student may not be able to set up the experiment in this remote laboratory configuration, they are often able to manipulate experimental variables, collect data, and perform data analysis. Virtual laboratories use software tools such as Simulink or LabVIEW to create user interfaces that simulate specific equipment. These laboratory configurations are often utilized because of the expense or safety considerations. Within the past 5 years, the field of electrical engineering (EE) has seen the development of inexpensive, universal serial bus (USB)-powered mobile laboratories that can replace much larger instrumentation such as digital multimeters, oscilloscopes, and power supplies. These pocket-sized laboratories can now allow students to individually build and test circuits in situations where access to equipment occurred in groups. This has allowed the adoption of more active learning pedagogies in the classroom. Within the MENA region, the use of remote laboratories is well documented. While some institutions access remote laboratories in the USA and Europe, other countries such as Jordan and Saudi Arabia have developed remote laboratories for use by their students and others in the region (Salah et al. 2014; Almarshoud 2011).. Use of the mobile laboratories is not yet well documented. Thus, there continues to exist a need to build partnerships between MENA and international institutions around the use of evolving laboratory interfaces.

10.3 Online Course Development

In an effort to provide access to a broader segment of society, the Department of Electrical and Computer Engineering (ECE) at Morgan State University developed an online curriculum of the bachelor's degree for the past two years. Components utilized to support this process were an LMS, mobile laboratory hardware, lecture capture, video conferencing, and a robust outcomes assessment process (Fig. 10.1). The approach taken in the development of this curriculum includes best practices from other authors such as the use of a systematic process that includes analysis, design, development, implementation, and assessment (Sözcü and İpek 2013). The ECE courses offered to online students at Morgan State University utilizes commercially available software and a mobile laboratory. However, we developed

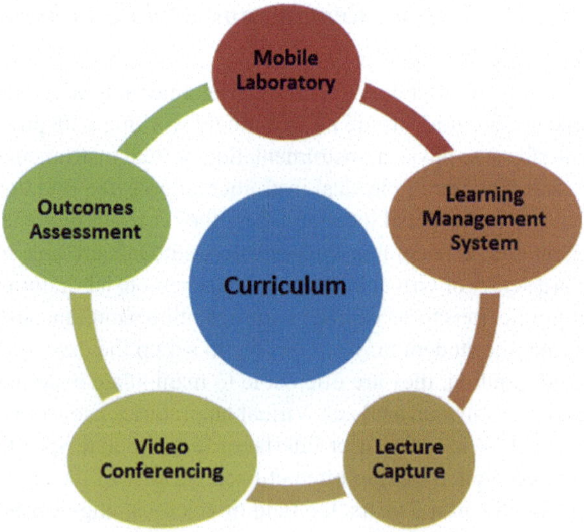

Fig. 10.1 A quality-driven model for online course development

a unique approach by utilizing a new technology—the Mobile Studio IOBoard™—developed at Rensselaer Polytechnic Institute (RPI), to implement the laboratory and design components of our undergraduate courses (The Mobile Studio Project 2014). It should be noted that the Mobile Studio IOBoard™ has now been replaced by the new Analog Discovery™, analog design kit that has been designed and built by Digilent Inc. (Digilent Analog Discovery™ 2013). We also supplement our online courses with captured lectures of our F2F on-campus courses, using Panopto Focus™ software. Finally, we use the Adobe Connect™ software for video conferencing, to allow online students to remotely demonstrate their projects and laboratory assignments to their instructors. The results of our implementation in several ECE courses over a 2-year period are discussed in the rest of the chapter. The use of this new technology and pedagogy has the potential to drastically alter the way engineering courses is taught and how it opens the door for exciting collaborations between higher education institutions in the Middle East and North Africa.

10.3.1 Pedagogy and Implementation of the Mobile Laboratory

The development of online ECE courses discussed in this chapter started about 10 years ago with the addition of Web-based course supplements for the F2F courses (Astatke and Mack 1998a, b). The Web-based course supplements consisted of additional course materials such as PowerPoint slides, animations, short video, and other website links that were there to help students better understand the course material. Students then came to the EE laboratories on campus to use the measurement equipment to design, build, and demonstrate their projects and experiments to the course instructor. This changed in 2008, when our university joined the Mobile Studio project that was funded by a 5-year National Science Foundation (NSF) grant at RPI. The addition of the Mobile Studio IOBoard™ allowed us to redesign our ECE F2F laboratory courses, such that students can complete some of their laboratory work off campus. It should be noted that these portable devices allow students to tinker and experiment with their designs anywhere, and at any time. A similar approach called "lab-in-a-box" has been implemented at two other institutions, Virginia Polytechnic Institute and State University (Virginia Tech) and Georgia Institute of Technology (Hendricks et al. 2005; Williams et al. 2008). Various papers on the use of the Mobile Studio IOBoard™ technology and pedagogy have been published by the other members of the Mobile Studio project (Millard et al. 2007; Conner et al. 2011; Astatke et al. 2011a, b). The faculty at our institution worked with peers at other institutions involved in the Mobile Studio project to redesign various laboratory and design experiments, so that they can be completed by the students using the Mobile Studio IOBoard™. The key issue that we had to address was the fact that the Mobile Studio IOBoard™ was limited to very low output voltage (+/−4 V) because it draws its power from the USB port of the laptop it is connected to. This implied that laboratory experiments that required a "power supply" or "function generator" with more than 4 V had to be redesigned to achieve the same learning outcomes at lower voltages. This constraint was primarily applicable to analog circuits since the use of complementary metal-oxide semiconductor (CMOS) components would satisfy this requirement for digital circuits. The instructors involved in the Mobile Studio project at the various institutions worked closely together, to develop sound pedagogy to deliver the educational content using the Mobile Studio IOBoard™ technology.

In the first phase of our University's implementation, students enrolled in F2F courses were allowed to use the Mobile Studio IOBoard™ to complete laboratory and design projects in their dorms or the library. However, they were still required to see the course instructor in the office or in the ECE laboratory, to conduct a live demonstration of their final projects. The F2F circuits course is now taught with a combination of lecture and hands-on experiments that are conducted using the Mobile Studio IOBoard™ technology and pedagogy. This is different from a typical electric circuits course where students learn the theory in the classroom and perform hands-on experiments in the laboratory course. The addition of the Mobile Studio IOBoard™ technology and pedagogy allowed us to teach both theory and applications of circuits concepts at the same time, in any classroom setting. This approach proved beneficial to all students because the hands-on experiments allowed them to verify and better understand theoretical concepts covered in the course, without the need to wait until they conducted separate lab experiments in the lab course that is held on another day, at a different time. Another benefit of this new technology and pedagogy is that it allows universities in the MENA region that have a shortage in laboratory equipment to offer a very sophisticated and immersing laboratory experience for their students, at a fraction of the cost required to deliver a laboratory course. Figure 10.2 shows the contents of a sample laboratory experiment that was redesigned using the Mobile Studio IOBoard™ technology and pedagogy.

All students have to initially complete a laboratory experiment that teaches them how to use the software and hardware of the Mobile Studio IOBoard™ and how it can be connected to circuits they design on their breadboards. Also shown in Fig. 10.2 is a typical laboratory experiment that has been redesigned, in order to be conducted, using the Mobile Studio IOBoard™. It should be noted that the concepts covered in this lab experiment are the same as those conducted using the laboratory equipment on campus, where a power supply and a multimeter are used. The laboratory experiments in the digital course did not have to be redesigned, because all the experiments could be conducted using the Mobile Studio IOBoard™ as long as the integrated circuits are CMOS, which can be powered by 3.3 V, which is available on the digital side of the IOBoard™. To help students transition to the use of function generators and oscilloscopes found in most ECE laboratories, we utilize manufacturer-developed local area network (LAN) connection modules, to allow students to control every feature of the oscilloscope remotely using a Web browser on their personal computer (PC). The use of a virtual front panel looks and operates the same way as the physical front panel of the oscilloscopes with the same associated keys and knobs. This implies that students who are conducting ECE laboratory experiments online will have access to the same type of equipment that is used by the students enrolled in the F2F laboratory courses. We also developed a

<u>EEGR 202: Dr. Yacob Astatke</u>

Mobile Studio (MS) Lab 4– Thevenin Equivalent Circuit and Max Power Transfer

1. Mobile Studio and Instrumentation Board

MS is a technology-based new learning tool comprising a tablet PC (or any PC) and an instrumentation board, which replaces most of the lab equipment. Therefore, MS allows for small foot -print, mobile laboratory experiments any place any time. The measurement by MS is possible by Windows-based software, Mobile Studio Desktop, which is already installed in the tablet PCs. The icon for the Mobile Studio Desktop is illustrated below.

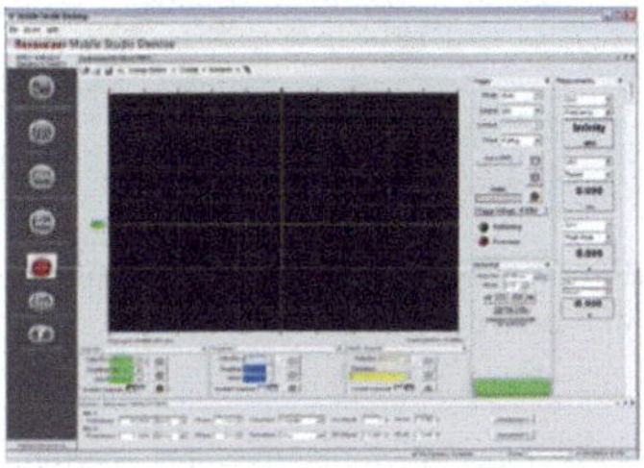

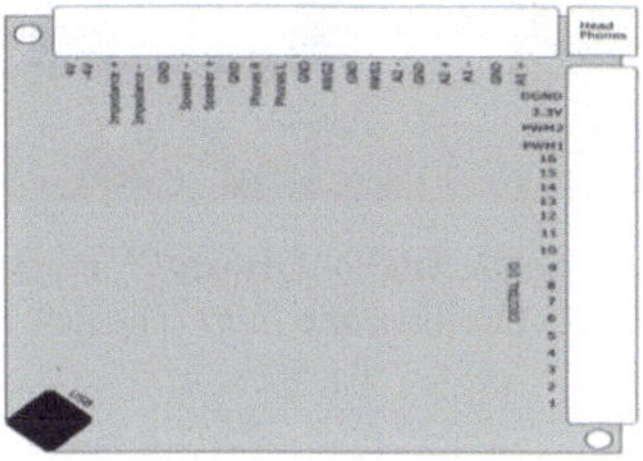

Figure #1: Mobile Studio Desktop™ Software and Hardware Pin Layout

The instrumentation board can function as : (a)Oscilloscope, (b)Digital Multi Meter (DMM), (c)Power Supply, and (d)Function Generator.

There are, however, important limitations in using the Red board

(1) No direct measurement of current-- You get current indirectly (by measuring voltage across a resistor, etc), or by using a DMM.

(2) No direct measurement of resistance-- You get it indirectly or by using hand held DMM.

b

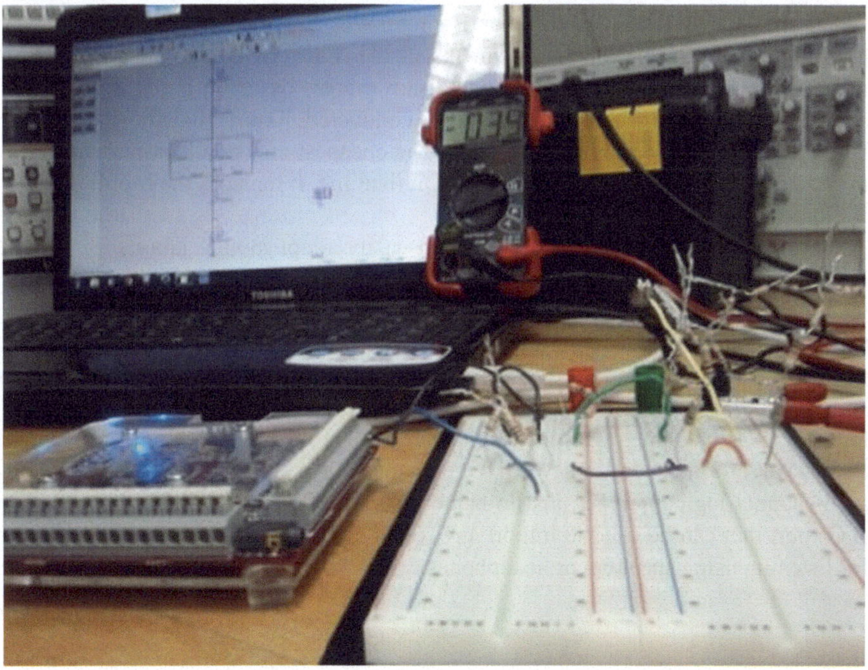

Fig. 10.2 Laboratory experiment showing partial laboratory description and experimental setup. **a** Laboratory experiment—Mobile Studio IOBoard™. **b** Typical setup of Mobile Studio IOBoard™

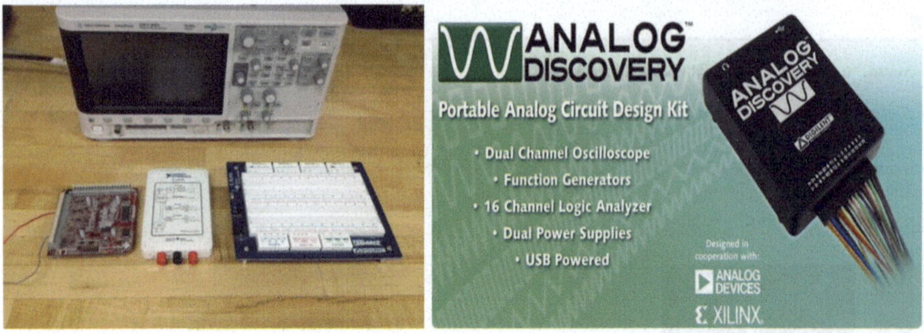

Fig. 10.3 Mobile Studio IOBoard™, myDAQ™ board, Digilent Electronics Explorer™, and Agilent X-Series Oscilloscope *(left)*. The new Analog Discovery™ board by Digilent *(right)*

laboratory experiments that can be conducted completely online, using the Agilent XSeries Oscilloscopes. Several undergraduate students tested the new laboratory experiments by conducting the experiments completely online.

Although we implemented the ECE online courses using the Mobile Studio IOBoard™ technology, there are currently several other similar portable equipment that can be used for the same purpose. Similar results can be obtained by conducting the same experiments using other portable ECE laboratory kits such as the Digilent "Analog Discovery™," "Electronics Explorer™," and National Instrument (NI) myDAQ™ boards (National Instruments 2014). Each ECE laboratory measurement instrument has its advantages and disadvantages based on the method of use and the type of experiments that need to be conducted. Figure 10.3 shows the ECE laboratory equipment next to each other. The cheapest and smallest instrument by far is the Analog Discovery™ offered by Digilent Inc. It replaces the Mobile Studio IOBoard™ developed by RPI.

This implies that ECE departments can easily adopt mobile laboratory equipment and make it part of the lab kit for their students because the cost is less than the typical engineering textbook. One of its main advantages is that it is very small (6 × 8 cm) and does not require an external power supply because it draws all of its power from the USB port of the computer or laptop that it is connected to. The software is free and very easy to run on any type of laptop or tablet PC. The Analog Discovery™ comes with a two-channel 5 MHz differential oscilloscope, two 5 MHz waveform generators, two fixed power supplies + 5 V or − 5 V (50 mA), and a 16-channel logic analyzer, just to name a few. It also has an audio input/output (I/O) port that can be used to import any sound signal for processing and export it for listening using speakers or headphones.

10.3.2 Converting Face-To-Face Courses to Online Courses

Once the laboratory experiments were redesigned to be offered completely online, the next step was to convert our F2F ECE courses into online courses. All of the online courses were offered to a select number of students as pilot courses for at least two semesters, to evaluate their effectiveness. All online courses were designed so that students would be able to access course resources asynchronously; however, the pace at which the course material is to be reviewed is established by the instructor using the course calendar. All assignments and examinations were to be completed by a certain deadline. Students enrolled in the online ECE courses were expected to complete the same amount of course material and assignments as the F2F students. Course materials for all courses are available to students through Blackboard Learn™, a course management and delivery platform. It is used to store the course documents online, such as PowerPoint lecture slides, lecture videos, handouts for homework and laboratory assignments, announcements, etc. It is also used to collect documents submitted by students electronically, to record and share student's grades, and for student–teacher or student–student communication. Features of Blackboard Learn™ such as blogs, discussion boards, and virtual chat rooms are not utilized in the F2F sections. They are only used in the online version of the courses.

Morgan State University (MSU) requires all online course builders to attend an "Online Course Design Workshop" that is offered on campus. This course is delivered online, via Blackboard Learn™. The course topics include the online teaching environment, creating modules, the role of discussion, technology integration, and assessment. All online courses developed at our institution have to undergo a thorough evaluation process to assure that they conform to "the 2008–2010 Quality Matters™ (QM) Rubric" (Quality Matters 2014). This rubric outlines many of the practices that are generally accepted for teaching engineering courses and includes some items that are critical for an online student's success (Endean et al. 2010; Oliver 2003; Ragan and Sax 2005). The rubric assigns points to several aspects of an online course to ensure a student's success and a minimum score is required for approval. This is done to ensure that all online courses meet the minimum course development standard for the success of the students who will be enrolled in it. Samples of the components of the rubric used to evaluate all online courses at MSU are shown in Table 10.2.

10.3.3 Integrating Technology for Online Delivery

The last step in the development of online courses was the implementation of the lecture capture and the video conference systems. We had to carefully evaluate the advantages and disadvantages of synchronous and asynchronous modes of content delivery for our online students. Since our goal was to offer the online ECE courses to students from within the USA or abroad, we decided to use a tool that can offer both synchronous and asynchronous course contents to the online student. The use of lecture capture technology has the potential to fundamentally change the way instructors and students interact in and out of F2F and online courses (Dale 2007; Educause Learning Initiative 2014). The ability to access additional video course material from anywhere at any time plays a key role in all F2F, online courses, or MOOCs. In general, faculty members have the choice of using lecture capture

Table 10.2 Components of a Quality Matters™ (QM) Rubric. (Quality Matters 2014)

1	*Course overview and introduction:* Ensure that all instructions for students are easy to find including establishing expectations for the course and how to use the modules developed for the course
2	*Learning objectives:* Students are provided measurable learning objectives for each module and information on how to meet the objectives
3	*Assessment and measurement:* The course assessment must be aligned with the course objectives and at a level appropriate for the course. Grading criteria must be explicitly stated
4	*Instructional materials:* Course materials must allow students to meet the course and module objectives
5	*Learner interaction and engagement:* Interactions that occur between the student and the teacher must foster interaction between course participants and instructors
6	*Course technology:* The tools and media must support student learning and be accessible to students. Students must have access to all tools, and instructions must be provided on how to use these resources
7	*Learner support:* Students must be aware of technical, academic, and student support services available for the course and at the university
8	*Accessibility:* The course should be accessible by all students and provide alternate means of access

technology to change and enhance their classroom environment in one of three ways (Astatke et al. 2013a). First, they can use it to allow their students to preview the course material before class by posting prerecorded course materials. This will allow instructors to free up more lecture time and spend more of their classroom "face-to-face" (F2F), engaging students with group discussions and hands-on activities. Second, they can use the lecture recordings to allow students to review and understand difficult concepts in the course outside of the classroom as often as they want. Finally, lecture capture software can be used by faculty to provide additional instructions such as demonstrations and step-by-step instructions that are important to understand the primary course material. After implementing and evaluating the new lecture capture technology in our F2F courses, we used them to enhance the content of our online courses.

Choosing an appropriate lecture capture system can present some challenges because there are many stakeholders, such as instructors, students, tech support staff, instructional designers, and others, who need to be heard in the course of making the proper choice. There are currently several tools in the market that can be used to record and stream various types of lecture content that range from free to expensive, such as Tegrity, Echo360, Mediasite, and Panopto. Table 10.3 shows several criteria to consider during selection including the type of information captured, ease of use, system compatibility, and support required. One of the most important factors related to the choice of any lecture capture software package, but is always difficult to quantify, is ease of use. This is a very important issue that has to be taken into consideration by university administrations, because the choice of the cheapest or most cost-effective feature-rich system will not be widely adopted if instructors find it too difficult to use. More information related to the different types of lecture capture software packages can be found in (Astatke et al. 2013a).

Table 10.3 Comparison of video capture and streaming software. (University of North Texas 2012)

Category	Features
Information captured	Video, audio, screen activity, document camera, electronic whiteboard, records full motion video
System type	Web-based (no software or hardware required), server supports simultaneous recording, software-based
OS compatibility	Mac, PC, Linux
Accessibility	Closed captioning, screen reader
Hardware	Camera, computer, microphone, server
Campus and administrative support	Installation, scheduling
System integration	LDAP, Active Directory, CMS, SIS, Blackboard Vista

OS operating system, *PC* personal computer, *LDAP* lightweight directory access protocol, *CMS* content management systems, *SIS* software installation script

10.3.4 Content Delivery—Video Recording

The course instructors at our institution enhanced the PowerPoint-based lesson files by recording lectures for each sub-module using the Panopto Focus™ lecture capture software. Some instructors used the lecture capture software to record "live," the daily course lectures of their regular ECE courses, while others recorded separate lectures for each PowerPoint lesson file outside of the regular classroom. Each week, the instructor evaluates the students' performance on homework assignments and quizzes; records additional video updates, if necessary, to provide feedback; and stress the importance of certain concepts to the class. The regular video updates allow the students to review the instructor's feedback from anywhere at any time, and make the class environment more interactive and dynamic. The lecture recordings were initially available to the students enrolled in the F2F ECE courses as a supplementary material, in order to help them learn and understand the course material better. The students had access to daily course lectures through their Blackboard Learn™ software. Most students downloaded podcast versions of the daily course lectures and watched them as often as they wanted until they understood the topics covered in each lecture.

Online students were expected to complete all of their course requirements without being physically present on our campus. Adobe Connect™ software, a real-time video conferencing suite, was selected to facilitate live laboratory demonstrations because of the available features and existing installation on campus. It provides instructors with a virtual classroom environment for sharing their presentations and desktop applications with remote participants such as online students anytime, anywhere. The main advantage of the software is that it works via a Web browser and does not require users to download any special software to join a meeting. It also allows instructors to go beyond simple PowerPoint and screen sharing by providing with additional options such as interactive chat, quizzes/polls, and breakout rooms for individual interactions. Students were successful in conducting their laboratory and project demonstrations completely online. The main issue we faced during the online demonstrations was the difficulty in getting the software/hardware to work on the students' computers due to the low quality of the webcams.

10.3.5 Results of Online Pilot Courses

The success of the pilot program was evaluated by comparing student's performance in pilot and F2F classes. The "Electric Circuits" and "Introduction to Electrical Lab" online courses were offered for the first time as pilot courses to six ECE students in the summer of 2010. The two courses were also offered during the fall 2010 semester as hybrid courses and the spring 2011 semester as completely online courses to 12 students. More information related to the different types of lecture capture software packages can be found in (Astatke and Mack 1998b). The students enrolled in the pilot online courses were given the same projects, homework assignments, and tests as the students enrolled in the F2F courses. Although the local students enrolled in our two pilot ECE online courses had access to all the facilities on our campuses, the courses were conducted as if the students were completely online. Therefore, they were not required to come to our campus on a daily basis to download and view the lecture notes, to complete their project and lab assignments, or even demonstrate their project results. We took this approach to ensure that the online courses that we developed could indeed be offered completely online to any student, as long as they had access to the Internet.

The data in Fig. 10.4 show the comparison of results for the 12 students enrolled in the online or hybrid courses offered during the fall 2010 and spring 2011 semesters (green bars), versus those enrolled in the F2F courses in the fall 2010 semester (blue bars - N=28), and the spring 2011 semester (red bars - N=33). Our observations were that the success of students enrolled in online or hybrid courses depend

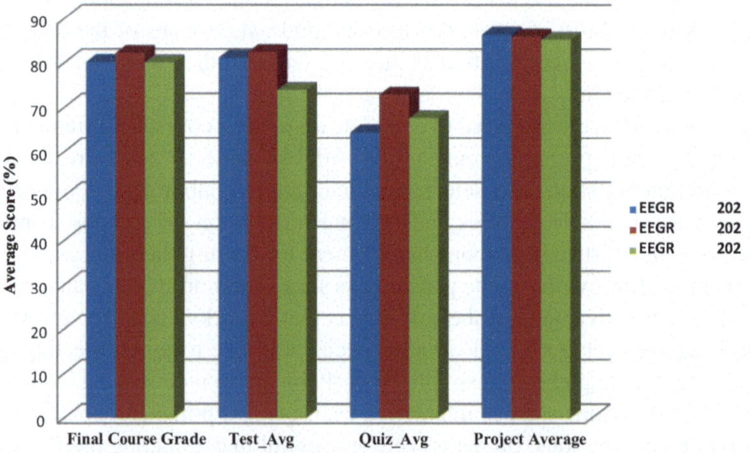

Fig. 10.4 Grade comparison between F2F and online students (fall 2010 and spring 2011)

strongly on the background of the student and their commitment to the course. The students who performed poorly in the online courses did so because they either fell behind in completing the course work or missed a lot of weeks of course work because of various reasons. The students who performed well in the online courses proved that the courses were well designed and delivered because they were able to perform as well or even better than the students enrolled in the F2F courses.

The two pilot courses allowed us to evaluate the various strengths and weaknesses of our online courses. The students enrolled in the online courses had access to lecture recordings from the spring 2010 and fall 2010 semesters. The statistics of the usage of the course lectures recordings suggest that the students enrolled in the two online courses found the Panopto Focus™ lecture recordings very valuable to their success in the courses.

The course design and building phase is extremely integral to developing a successful online program. It is important to create high-quality online courses to protect the university's brand which means faculty need the time and training and support to plan and design courses that will be both rich in content and instructional strategies, with a variety of multimedia tools and technologies to encourage student engagement and learning. One of two strategies that needs to be considered during the course-building stage is to create a cadre of course builders assigned to plan and develop the required courses. Alternatively, a university could consider providing resources to hire teaching fellows or adjunct faculty, while freeing up full-time faculty to build their respective online courses.

10.3.6 Lessons Learned/Best Practices

There are a few lessons learned from the development, implementation, and piloting of these undergraduate courses:

1) Some type of certification process is essential for maintaining a consistent standard of quality.
2) It is optimistic to have faculty build and teach courses, especially when adjunct instructors are involved on a part-time basis. It is a good practice then to pair seasoned faculty with adjuncts or junior faculty, if this approach is used.
3) Resource commitment is important to ensure that the technology and equipment are adequate and available to support the online course delivery.
4) Pre-course training on specialized laboratory equipment and software is important to keep the pace of online learning manageable.
5) The pace of course building must be consistent to maintain the momentum of an online delivery mode.
6) Course and program assessment is important to validate and authenticate online course content.
7) Assessment tools should be considered parallel to the course development and piloting phase, to allow the collection and analysis of course outcomes and objectives, in a seamless manner.

8) A Web-based assessment tool offers greater flexibility and accessibility for faculty and administrators with the added advantage of tracking student learning.
9) Online programs offer a unique opportunity to offer courses during the summer, giving the students an extra window to stay on track and meet their graduation goals.
10) A two-plus-two EE program benefits from the cooperation of two-year colleges by offering at least some of the sophomore courses outside of the four-year institution. This approach also allows four-year institutions to focus resources on the upper-level core and elective courses.

10.4 Assessment

In the process of developing and delivering online courses/programs, there are four main components to be considered: content (course-based or program-based), delivery, technology, and assessment. When applied to the delivery of online courses, and/or programs, the role of assessment becomes crucial especially to validate and authenticate online course content. There is a parallel need for assessment tools that allow the collection and analysis of course outcomes and objectives, in a seamless manner (James et al. 2012). A Web-based assessment system has more potential in terms of access and flexibility for teachers and administrators which, in turn, improve the effectiveness and efficiency of producing learning analytics to guide pedagogical improvements (Baepler and Murdoch 2010). Assessment is about improvement and program assessment, which ultimately helps to focus on improving student learning. The introductory phase of assessment involves building a base of understanding about the benefits and tools of program assessment and what you want your department to gain from assessment. It is during this phase that you identify why you want to assess. Early in the process, the rationale for the importance of engaging in assessment must be clearly articulated, if success is to be achieved.

10.4.1 Performance Assessment Framework

Performance assessment refers to a variety of tasks and exercises in which students are given opportunities to demonstrate their understanding and to thoughtfully apply cognitive skills in a variety of contexts (Marzano et al. 1993). The desired result is to measure the attainment of standards-based outcomes. Performance assessments are designed to help educational institutions effectively utilize data to drive decision-making (Searchlight Performance Assessment Software 2014). A performance assessment framework (PAF) offers the means to perform program assessments through both direct and indirect measures of course outcomes, robust data collection, and analysis and the storage of observations and actions resulting

Table 10.4 Sample program outcomes and associated performance indicators (PI)

a) An ability to apply knowledge of mathematics, science, and engineering
Identifies mathematical and physical concepts needed to solve the problem (M, S)
Explains the role of mathematics as a tool for modeling systems and processes (M)
Expresses problem in terms of mathematical and scientific statements (M, S)
Uses fundamental engineering principles to solve engineering-related problems (E)
b) An ability to design and conduct experiments as well as analyze and interpret results
Develops a hypothesis and a plan (experimental method) to evaluate it using engineering principles and practice (D)
Collects data using software and electronic test and measurement equipment (C)
Analyzes results and components of the design using engineering models (A)
Explains experimental results as they relate to theoretical results (I)

from the analysis. Any performance assessment software should have four main functions—allows data importation, creates performance assessment metrics, implements performance assessments, and generates performance reviews and reports. The software should provide educational institutions with an assessment tool that will allow them to enter, generate, and analyze course or program-based performance rubric. Formative and summative assessment should be used to help faculty evaluate how students are meeting the learning goals in the program, coupled with both direct and indirect means. Both direct and indirect methods can be mapped to program outcomes to ensure program objectives are being met.

10.4.2 Outcome Assessment—Program Assessments

An essential part of performance-based assessment is the use of standards upon which all learning outcomes are based. Accredited engineering programs typically follow ABET standards and guidelines. For engineering programs, ABET has defined outcomes ***a–k***, as standards for accreditation (ABET Website 2014). A sample of the program learning outcomes and the associated performance indicators (PI), used to assess student learning for the ECE department at MSU, is shown in Table 10.4.

A comparison, by outcomes, of the online courses versus the F2F courses for the 2010–2012 period is shown in Fig. 10.5. These charts portray information pertaining to the equivalency of achieving learning outcomes for both laboratory and lecture online sections (James et al. 2013). The results indicate that, in most cases, the performance of students in the online section is slightly better than that of the F2F cohorts, for the sections evaluated. It is thought that the policy of screening for grade point averages (GPAs) of 3.0 and above gives online class participants a slight advantage over F2F cohorts. The students most likely have greater self-efficacy and independent work habits.

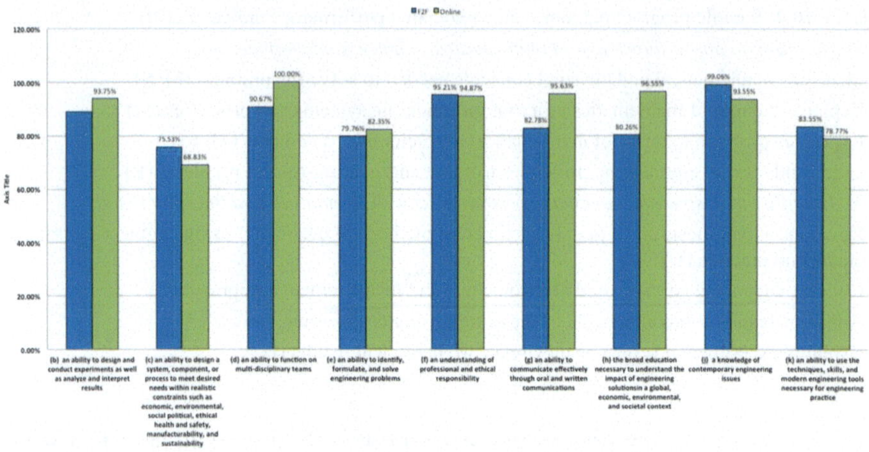

Fig. 10.5 All courses—comparison of F2F & Online 2010–2012. (James et al. 2013)

10.5 Mobile Laboratory for Ethiopian Engineering

The lack of dependable laboratory equipment, especially in engineering education, has led higher education institutions in developing nations to focus more on the theoretical aspects of science, technology, engineering, and mathematics (STEM) education as compared to the practical applications. A possible solution to the problem is to develop collaborations with higher education institutions in other countries to provide targeted training on new pedagogy and state-of-the-art mobile laboratory technology that will allow universities to teach STEM subjects with hands-on activities at a fraction of the cost of the bench equipment. The new inexpensive and portable laboratory instruments can easily be implemented in engineering programs throughout the MENA region. Their small size and low price makes them ideal for widespread adoption and implementation. Universities in the MENA region can collaborate with ECE departments in the USA in order to implement the new technology and pedagogy in their curriculum. This new approach has the potential to significantly change and improve EE education throughout the MENA region.

The ECE department at MSU has been collaborating with five ECE departments in Ethiopia to expand the use of mobile laboratory instruments in their curriculum (Astatke et al. 2013b). The project's major goal was to enable hands-on exploration of ECE principles, devices, and systems that have historically been restricted to expensive laboratory facilities that are not readily available in most engineering schools in developing countries. The results of the collaborations have so far been very promising. The institutions have been able to adopt some of the mobile laboratory exercises from MSU, while others have developed new laboratory exercises. Some ECE departments used the new equipment to allow their graduating senior students (5th year) to work on their capstone design projects in groups of three to five students. The students indicated that the availability of the new technology allowed them to expand the application of their projects and tackle more advanced topics. One example is the "Clamp Activated Switch" capstone project completed by students from Addis Ababa Institute of Technology (AAiT). The students used the spectrum analyzer function of the Mobile Studio Board™ to analyze the clap waveforms picked up by an electret microphone (Fig. 10.6). It allowed them to observe the sampled audio input from the microphone and analyze its general properties in the frequency domain. They were able to correctly determine the frequency range where the clap waveforms were concentrated, and were able to design a bandpass filter that minimized false triggering of the circuit.

The ECE department of another University in Ethiopia used the Mobile Studio Board™ to expand the laboratory equipment used in sophomore and junior level courses. Prior to the availability of the MSIOBoards, students were forced to work in groups of 10–20 students to conduct basic ECE laboratory experiments (Fig. 10.7, *left side*). The addition of ten new Mobile Studio Board™ allowed the ECE department to triple its laboratory workstation benches and accommodate fewer students (4–5) per station to conduct their laboratory experiments (Fig. 10.7, *right side*).

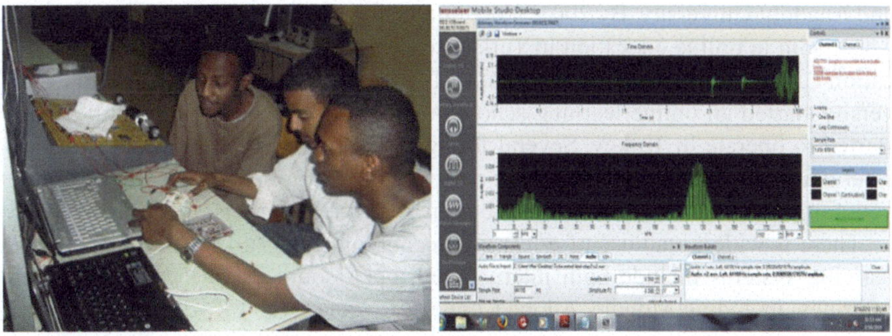

Fig. 10.6 AAiT students working on their Clamp Activated Switch Project (*left side*), Spectrum Analyzer Function of MS-IOBoard showing sound signal in the frequency domain (*right side*)

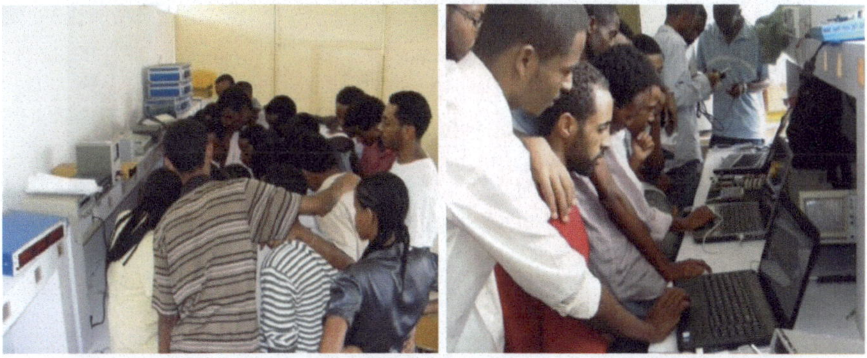

Fig. 10.7 iOTech-HU students working on their laboratory experiments before using the Mobile Studio IOBoard (*left side*), and after using the Mobile Studio IOBoard (*right side*)

10.6 Developing a MENA–US Collaboration Model: Online Education in Turkey

The number of universities offering online programs in Turkey has increased significantly in the past three years. In fact, the notion of online education has resulted in opening of so many new online universities and the shifting of traditional universities toward offering online courses and programs. Many of the universities in Turkey offer distance education. The new and highly competitive market has led to many innovative choices that the online universities are taking towards LMSs and content management Systems (CMS) choices. Some universities are adopting well-known LMSs such as Moodle and Blackboard, while others are developing their own LMSs because of the cost of purchasing commercial software. Three successful locally developed platforms are introduced in this section. Major universities that offer online classes utilize their own in-house LMSs, but there is not a lot of published information.

Since 1996, several universities in Turkey have offered distance education opportunities, such as Istanbul Teknik University, Middle Eastern Technic University, and Anadolu University. By 2001, more universities started offering online courses. This boom in online education in Turkey resulted in the creation of an online education commission by the Department of Education. Later, in 2010, Istanbul University started an online learning school, and by 2013 over 50 online universities started to operate in Turkey. Several foreign partnerships have been developed to offer courses, certificates, and degree programs from European and American universities. Figure 10.8 shows the distribution of online programs in Turkey by 2009.

There is widespread use of e-learning technologies in Turkey. Commonly used learning systems include both international, open source, and locally developed offerings, including Moodle, Adobe Connect, ENocta, Web-CT, BlackBoard, Olat,

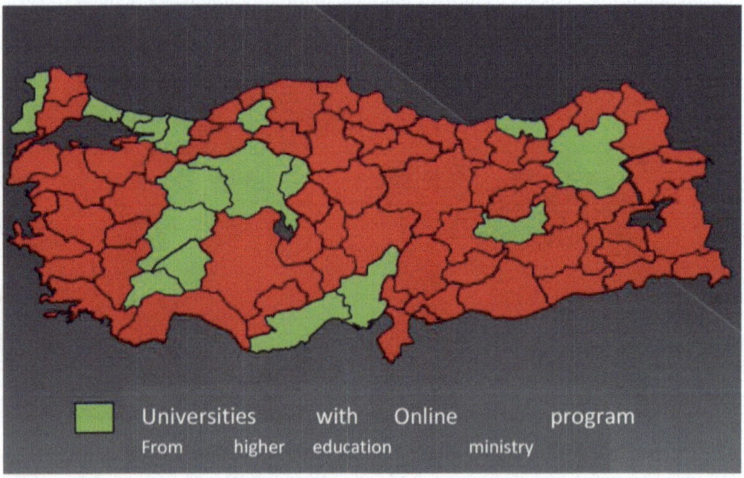

Fig. 10.8 Location of universities with online programs in Turkey

Dokeos, TinyLMS, Preculus, Advancity LMS: ALMS and Vedubox (Learning Management Systems 2014). Several local universities, such as Ankara University and Middle East Technical University, play an important role in online education in Turkey. These universities provide online access to certificate programs—namely associates, bachelors, and masters programs. However, there are limited opportunities for online engineering degrees. This provides opportunities for collaboration with universities and education brokers to provide online engineering education opportunities to local students.

Partnerships between MENA universities around online engineering education are not well documented. One model that is used involves an education broker to assist students with registration and sometimes provides support during enrollment. Kure Educational Institute is an institution that provides students with the opportunity to earn degrees at universities in Turkey and abroad (Academic Partnerships 2014). Students apply for acceptance to the international program via the Kure website. At this stage, prospective students will only provide preliminary documentation to validate their eligibility for application. Once the number of applicants reaches the threshold that is set by the partner university, Kure will request additional and detailed documentation from the students in order to process their applications. Once the application for each student is complete, the documents will be forwarded to the partner university for final evaluation. If the partner university accepts the students, Kure Educational Institute will be notified and the registration process will start. Registration fees and tuitions will be determined and also be collected by the partner university.

Once the students register for the classes, Kure Institute assigns a coordinator for a group of students. This coordinator is tasked with being in close contact with his/her students during their study. This could be accomplished through one of the several available student retention solution platforms, such as Starfish or via a personalized channel such as e-mails or phone calls. The coordinator will also conduct frequent surveys from the students about the quality of the program and instructors and will relay their feedback to the program directors for program adjustments. Assessment methods will be determined with the partner university. Kure has arrangements with test centers throughout Turkey and, if requested, testing could be done within these centers. There will be two program coordinators: one from Kure and the other from the partner university. These individuals will work out the details and will kick start the program using F2F and online meetings. The program coordinators are also responsible for the quality of the program and student and faculty satisfaction.

Although the course notes and textbooks can be in English, the teaching language can be in English or Turkish. This seems to be the new international collaborative approach used between the international and Turkish universities. This model provides a modular application structure that can be easily deployed in any of the MENA countries. As an example, in 2013–2014, the International Struga University started offering courses in Turkish, and 200 Turkish students applied within the first semester. They have an ongoing partnership with three international institutions: the American Heritage University and the Schiller International University from the USA, and the London City College.

10.7 Summary and Conclusion

The growth in higher education enrollment and online learning in MENA provides opportunities to integrate new technologies with a focus on assessing the impact on educational outcomes. Our ECE department has successfully launched its online undergraduate program using new technology and pedagogy. We have learned from the experiences of the pilot courses and have taken additional steps to improve current and future online ECE courses offered at our institution. The availability of inexpensive and portable laboratory instruments has enabled new pedagogical approaches in the teaching of theoretical concepts and design practices in EE. Faculty members at our university are working very closely with their colleagues at universities in the USA and Africa to take advantage of these new tools and incorporate hands-on experimental activities into existing lecture courses. This new approach has led to the implementation of restructured EE courses that utilize problem-based learning with a focus on student-centered learning rather than instructor-centered lectures.

The work we have been conducting for the past 5 years has allowed us to evaluate whether a student-centered learning environment can stimulate a deeper understanding and increase student engagement. Assessment of student's learning at partner institutions has shown that students' response to the hands-on learning activities, regardless of the model of implementation, has been extremely positive. Results indicate that the greatest impact of hands-on learning is the conceptual understanding by students who had the weakest grasp of the material after lecture-based instruction. Students also express a higher degree of confidence in their ability to be engineers because the hands-on activities have helped develop their abilities to design solutions to meet specifications in the projects. This new approach represents a major paradigm shift in the way higher education institutions can approach delivering engineering education. We hope that this contemporary delivery modality method will open the door to many students who are candidates for joining the STEM workforce from around the world or for students from the MENA region who are simply interested in trying out ECE courses from abroad.

Acknowledgment The authors would like to thank the Maryland Higher Education Commission (MHEC), the Department of Education, the National Science Foundation (NSF), and the Smart Lighting Engineering Research Center (ERC) at RPI for their financial support.

References

ABET Website. (2014). Criteria for accrediting engineering programs, 2013–2014. http://www.abet.org/DisplayTemplates/DocsHandbook.aspx?id=3149. Accessed 23 May 2014.

Academic Partnerships. (2014). Kure collaboration model. http://www.academicpartnerships.com. Accessed 28 May 2014.

Accreditation Board for Engineering and Technology (ABET). (2005). http://main.abet.org/aps/Accreditedprogramsearch.aspx.

Al-Harthi, A. S. (2010). Cultural differences in transactional distance preference by Arab and American distance learner. *The Quarterly Review of Distance Education Journal, 11,* 257–267.

Allen, E., & Seaman, J. (2010). *Learning on demand: Online education in the United States, 2009*. The Sloan Consortium.

Allen, E., & Seaman, J. (2011). *Going the Distance: Online Education in the United States, 2011*, The Sloan Consortium, 23 May 2014.

Almarshoud, A. F. (2011). The advancement in using remote laboratories in electrical engineering education: A review. *European Journal of Engineering Education, 36*(5), 425–433.

Altarawneh, H. (2011). A survey of e-learning implementation best practices in Jordanian government universities. *International Journal of Advanced Corporate Learning, 4*(2), 9–17.

Astatke, Y., & Mack, P. L. (1998a). *Creating a distributed learning environment using WebCT.* ASEE 1998 Annual Conference, Seattle, WA, June 28–July 1.

Astatke, Y., & Mack, P. L. (1998b). *Are our students ready for asynchronous learning networks (ALN).* ASEE middle Atlantic section regional conference, Howard University, Washington, D.C., November 6–7.

Astatke, Y., Scott, C., & Ladeji-Osias, J. (2011a). *Electric circuits online—towards a completely online electrical engineering curriculum.* 2011 American Society of Engineering Education (ASEE) Annual Conference, Vancouver, British Columbia, Canada.

Astatke, Y., Ladeji-Osias, J., Scott, C. J., Abimbola, K., & Conner, K. (2011b). Developing and teaching sophomore level electrical engineering courses completely online. *Journal of Online Engineering Education, 2*(2), Article 4.

Astatke, Y., Moazzami, F., Scott, C., & Ladeji-Osias, J. (2013a). *Using online video lectures to enhance ECE courses*. Brooklyn: New York City College of Technology.

Astatke, Y., Scott, C., & Ladeji-Osias, J. (2013b). *Improving ECE education in Sub-Saharan African countries using the mobile studio technology and pedagogy*. Proceedings of the 2013 American Society of Engineering Education (ASEE) Annual Conference, Atlanta, GA.

Bach, S., Haynes, P., & Smith, J. L. (2007). Online learning and teaching in higher education. *Social Science Computer Review, 19*(4), 221.

Baepler, P., & Murdoch, C. J. (2010). Academic analytics and data mining in higher education. *International Journal for the Scholarship of Teaching and Learning, 4*(2), Article 17.

Benchicou, S., Aichouni, M., & Neharic, D. (2010). E-learning in engineering education: A theoretical and empirical study of the Algerian higher education institution. *European Journal of Engineering Education, 35*(3), 325–343.

Bonk, C. J. (2004). *The perfect E-storm: Emerging technologies, enhanced pedagogy, enormous learner demand, and erased budgets*. London: The Observatory on Borderless Higher Education.

Chapman, D. W., & Miric, S. L. (2009). International review of education. *Education Quality in the Middle East, 23*, 311–344.

Conner, K. et al. (2011). *Multi-institutional development of mobile studio based education and outreach*. 2011 American Society of Engineering Education (ASEE) Annual Conference, Vancouver, British Columbia, Canada.

Coursera. (2014). *Take the world2019's best courses, online, for free*. https://www.coursera.org/. Accessed 23 May 2014.

Dale, C. (2007). Strategies for using podcasting to support student learning. *Journal of Hospitality, Leisure, Sport and Tourism Education, 6,* 49–57.

Digilent Analog Discovery™. (2013). USB Oscilloscopes, Function Generators, & Instruments. http://www.digilentinc.com/Products/Catalog.cfm?NavPath=2,842&Cat=17. Accessed 23 May 2014.

Educause Learning Initiative. (2014). Things you should know about lecture capture. http://www.educause.edu/ELI/7ThingsYouShouldKnowAboutLectu/163555. Accessed 23 May 2014.

EdX. (2014). Take great online courses from the world's best universities. https://www.edx.org/. Accessed 23 May 2014.

Endean, M., Bai, B., & Du, R. (2010). Quality standards in online distance education. *International Journal of Continuing Education Lifelong Learning, 3*(1), 53–71.

Harley, D., Henke, J., Lawrence, S., & McMartin, F. (2003). *Costs, culture, and complexity: An analysis of technology enhancements in a large lecture course at UC Berkeley*. Berkeley: University of California.

Hendricks, R. W., Lai, K. M., & Webb, J. B. (2005). Lab in a box: Experiments in electronic circuits that support introductory courses for electrical and computer engineers. ASEE Annual Conference. http://www.asee.org. Accesses 23 May 2014.

International MOOC providers. (2014). The Chronicle of higher education. http://chronicle.com/blogs/wiredcampus/american-mooc-providers-face-internationalcompetition/44637?cid=gn&utm_source=gn&utm_medium=en. Accessed 23 May 2014.

James, P., Nyarko, K. P., Scott, C., Astatke, Y., & Ladeji-Osias. J. (2012). *Maryland Association for Institutional Research (MdAIR)*. Assessment techniques and assessment of student learning outcomes. Baltimore.

James, P. A., Craig, J. S., Astatke, Y., Ladeji-Osias, J. O., Partlow, L. E., Moazzami, F., & Nyarko, K. (2013). *Performance assessment framework for measuring online student learning outcome*. Proceedings of the 120th Annual Conference, Atlanta, Georgia.

Joerns, J. L., & Leinhardt, G. (2006). Review of educational research. *Going the Distance with Online Education, 76*(4), 567–605.

Learning Management Systems. (2014). www.vedubox.com. Accessed 23 May 2014.

Maiti, A., & Tripathy, B. B. (2013). Remote laboratories: Design of experiments and their web implementation. *Educational Technology & Society, 16*(3), 220–233.

Marzano, R. J., Pickering, D., & McTighe, J. (1993). *Assessing student outcomes: Performance assessment using the dimensions of learning model*. Association for Supervision and Curriculum Development: Alexandria.

Millard, D., Chouikha, M., & Berry, F. (2007). *Improving student intuition via Rensselaer's new mobile studio pedagogy*. ASEE 2007 Annual Conference, Honolulu, HW.

Monaco, M., & Martin, M. (2007). The millennial student: A new generation of learners. *Athletic Training Education Journal, 2*, 42–46.

MOOCs Won't Replace Business Schools. (2014). Harvard business review. http://blogs.hbr.org/2014/06/moocs-wont-replace-business-schools-theyll-diversify-them/. Accesses 23 May 2014.

National Instruments. (2014). myDAQ™. http://www.ni.com/mydaq/. Accessed 23 May 2014.

Oliver, R. (2003). Proceedings of the 16th open distance learning association of Australia Biennial Forum, Canberra. Exploring benchmarks and standards for assuring quality online teaching and learning in higher education, 1–13.

Quality Matters. (2014). Quality matters program a national benchmark for online course design. http://www.qualitymatters.org. Accesses 23 May 2014.

Ragan, L., & Sax, C. (2005). *Defining and implementing quality assurance standards in online courses*. Orlando: EDUCAUSE.

Saba, F. (2005). Critical issues in distance education: A report from the United States. *Distance Education Journal, 26*(2), 255–272.

Salah, R. M., Alves, G. R., & Guerreiro, P. (2014). *Porto: Reshaping higher education systems in the MENA region: The contribution of remote and virtual labs*. 2014 11th international conference on remote engineering and virtual instrumentation (REV). doi:10.1109/REV.2014.6784265.

Searchlight Performance Assessment Software. (2014). Insight at your fingertips. http://www.searchlightpa.com. Accessed 23 May 2014.

Sözcü, Ö. F., & İpek, İ. (2013). Considerations for IEL courseware design and the next generation of elearning. *European Journal of Contemporary Education, 6*(4), 204–215.

Strauss, W., & Howe N. (2007). *Millennials go to college* (2nd ed.) Virginia: Life Course Associates.

The Mobile Studio Project. (2014). Learning without limits. Mobile studio project. https://sites.google.com/a/mobilestudioproject.com/mobile-studio-project/. Accessed 23 May 2014.

The Sloan Consortium. (2010). *Class Differences: Online Education in the United States, 2010*, 23 May 2014.

The World Bank. (2008). MENA development report. The road not traveled. Education reform in the Middle East and Africa. http://siteresources.worldbank.org/INTMENA/Resources/EDU_Flagship_Full_ENG.pdf. Accessed 23 May 2004.

Udacity. (2014). *Thanks to UDACITY, I landed my dream job*. https://www.udacity.com/. Accessed 23 May 2014.

University of North Texas. (2012). http://itc.unt.edu/sites/default/files/Lecture_Capture_Systems_Comparison_Chart_rev2010721.pdf.

Weber, A. S. (2010). Review article: Effectiveness of web-based learning in the Middle East and North Africa (MENA) region. The seventh international conference on eLearning for knowledge-based society, 16–17 December 2010.

Williams, D., Jackson, J., Michaels, J., Auerbach, J., & Ferri, B. (2008). *A program for distributed laboratories in the ECE curriculum*. ASEE Annual Conference. http://www.asee.org. Accesses 23 May 2014.

Yusuf, N. (2013). The future of global education. *International Journal of Business and Economic Development, 1*(3), 75–82.

Dr. Yacob Astatke has been a full time faculty member in the Electrical and Computer Engineering (ECE) department at MSU since August 1994 and currently serves as the Interim Associate Dean for Undergraduate Studies in the School of Engineering. Dr. Astatke is the winner of the "2013 American Society for Engineering Education (ASEE) National Outstanding Teaching Award," and the "2012 ASEE Mid-Atlantic Region Distinguished Teacher" Award. He teaches courses in both analog and digital electronic circuit design and instrumentation, with a focus on wireless communication. He has more than 15 years' experience in the development and delivery of synchronous and asynchronous web-based course supplements for electrical engineering courses. He has played a leading role in the development and implementation of the first completely online undergraduate ECE program in the State of Maryland.

Dr. Jumoke O. Ladeji-Osias earned her BS in electrical engineering from the University of Maryland, College Park, USA, and the PhD in biomedical engineering jointly from Rutgers University and the University of Medicine and Dentistry of New Jersey, USA. She is an associate professor and the associate chair for graduate studies in the Department of Electrical and Computer Engineering at Morgan State University. She was awarded the 2014 NTA Technical Achiever of the Year Award by the National Technical Association Baltimore Chapter. She pioneered the use of mobile learning in digital logic courses at the university.

Dr. Petronella James earned both her doctor of engineering and masters of city & regional planning at Morgan State University (MSU), Baltimore, Maryland. She also completed a BS in management studies at the University of the West Indies (Mona), Jamaica. Dr. James currently serves as faculty of assessments and online programs for the Department of Electrical and Computer Engineering at Morgan State University. She has several published works in engineering education and online learning. She is also an adjunct faculty for the Urban Transportation, and City and Regional Planning programs at MSU, where she teaches at the graduate and undergraduate level, using both face-to-face and blended online learning instructions.

Dr. Farzad Moazzami received his BS degree in electrical engineering in 2003 from the University of Tabriz, Iran, and his MEN and DEng degrees in 2009 and 2011 from Morgan State University in Baltimore, MD. He is assistant professor of the Department of Electrical and Computer Engineering at Morgan State University. Dr. Moazzami is currently directing research activities in WiNetS lab at MSU, focusing on developing signal processing algorithms for wireless communication applications and developing information-assurance controls. In addition, he is conducting studies on international collaborations for online educations for engineering students. His research interests are digital signal processing, game theory, and cryptography algorithms. He is a member of the Institute of Electrical and Electronics Engineers (IEEE) and branch counselor for IEEE student branch at MSU.

Dr. Craig Scott received his PhD and BS in electrical engineering from Howard University and an MS in electrical engineering from Cornell University. Dr. Scott currently serves as professor and chairperson for the Department of Electrical and Computer Engineering at the Clarence M. Mitchell, Jr. School of Engineering at Morgan State University, Baltimore, Maryland. His educational scholarly endeavors include conducting pedagogical studies on learning technologies and remedial math preparation for engineering students. He instructs courses in computer vision, computer graphics, computational electrical engineering, electromagnetics, and characterization of semiconductor materials.

Dr. Kenneth Connor is a professor in the Department of Electrical, Computer, and Systems Engineering where he teaches courses on plasma physics, electromagnetics, electronics and instrumentation, electric power, and general engineering. His research involves plasma physics, electromagnetics, photonics, engineering education, diversity in the engineering workforce, and technology-enhanced learning. Since joining the Rensselaer faculty in 1974, he has been continuously involved in research programs at places such as Oak Ridge National Laboratory and the Universities of Texas and Wisconsin in the USA, Kyoto and Nagoya Universities in Japan, the Ioffe Institute in Russia, and Kharkov Institute of Physics and Technology in Ukraine. He was the head of ECSE Department from 2001 to 2008 and served on the board of the ECE Department Heads Association from 2003 to 2008. He is presently the education director for the SMART LIGHTING NSF ERC.

Abdurrahim Saka received his bachelor's degree in 2005 and joined the "Yeni şafak" newspaper team. Since then, he has been part of the editorial team for this paper. After 5 years of research in Turkish and International education models, he instituted "Küre Internatıonal Education" and "Edulinesity eğitim" organizations in 2012. Under the umbrella of these two organizations, he has been connecting international universities with the Turkish universities and other organizations in Turkey for potential international collaborations for online and on-site education. He has been developing road maps and drafting possible collaboration systems for the interested entities.

Chapter 11
Inclusion of Construction Health and Safety in Engineering Programs in the MENA Region: Assessment and Potential Enhancement

Munjed A. Maraqa, Amr M. I. Sweedan and Essam Zaneldin

11.1 Introduction

The construction industry has the most dismal record of safety among all industrial segments (WorkCover NSW 2001; Teo et al. 2005; Choudhry et al. 2008; Al-Humaidi and Tan 2010), with a risk of fatality that is about five times higher than in any other industry (Davies and Tomasin 1996). The higher rate of accidents and fatalities in the construction industry could be due to the nature of the work. However, several factors have been identified in the literature that could affect construction safety (see, for example, Sawacha et al. 1999; Teo et al. 2005; Choudhry and Fang 2008; Choudhry et al. 2009; Maraqa and Mohamed 2013). Among these factors is the provision of construction safety education to engineering students.

Szymberski (1997) noted, "today we are at a juncture where, if owners and engineers will work actively with construction organizations and safety professionals, further meaningful inroads will be made in creating safe working conditions during construction." It is also believed that there is a greater ability to influence safety on a project earlier in the project's life cycle, and such ability diminishes as the schedule moves from conceptual design toward start-up as illustrated in Fig. 11.1 (Szymberski 1997). Thus, provision of construction safety education to engineering students will have benefits both for the project design and later in the procurement stage.

The importance of formal education with regard to construction safety has also been recognized by Davies and Tomasin (1996) when they stated, "It is recommended that time is devoted at universities to the subject of health and safety in the construction industry. This should cover an outline of safety legislation and the duties of employers and employees. It should describe the hazards that professional staff face in their work and outline the problems facing designers in ensuring that their work is safe to build, operate and maintain."

M. A. Maraqa (✉) · A. M. I. Sweedan · E. Zaneldin
Civil and Environmental Engineering Department, United Arab Emirates University,
P.O. Box 15551, Al-Ain, United Arab Emirates
e-mail: m.maraqa@uaeu.ac.ae

M. Abdulwahed et al. (eds.), *Advances in Engineering Education in the Middle East and North Africa,* DOI 10.1007/978-3-319-15323-0_11

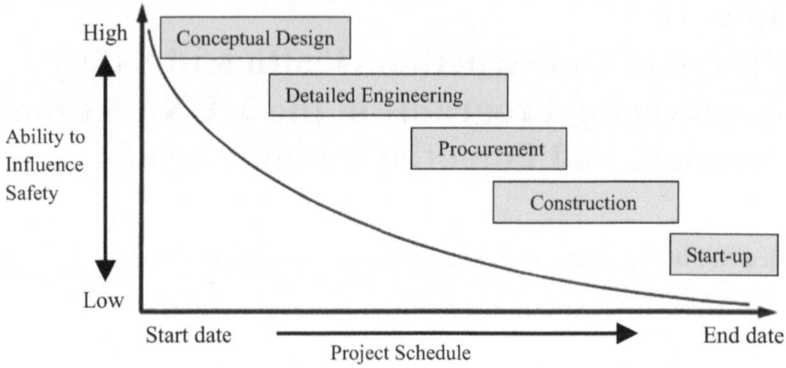

Fig. 11.1 Project schedule versus ability to influence safety. (Szymberski 1997)

While there is a close relationship between education and labor conditions (Haupt 2003), there is generally a lack of specific education in occupational safety for engineering professionals (Kavianian et al. 1993; Bryan 1999; Ferjencik 2007; Petersen et al. 2008; Pellicer and Molenaar 2009). It has been suggested that such deficiency be resolved at the university level. For that, four different approaches have been considered or proposed to enhance safety in engineering curricula including (1) development of a new degree of safety engineering (Alhemood 2004; Vincent 2005), (2) development of a new course in an existing degree (Moccaldi et al. 2005; Ferjencik 2007), (3) development of relevant topics or a cross-field subject in several courses of the current curriculum (Lemkowitz 1992; Al-Mufti 1999; Pellicer et al. 2003; Hill and Nelson 2005; Petersen et al. 2008), or (4) development of relevant topics in one specific course of the current curriculum (Kauffman 1987; Levitzky 1988; Phillis and Wheway 1991).

In general, construction safety education for engineering students involves two main aspects: (1) construction site safety and (2) design for construction safety (DfCS). While there has been some progress made in the developed countries to incorporate safety in the engineering education, it is not known to what extent the engineering students at universities in the Middle East and North Africa (MENA) region receive safety in their construction-related courses. Thus, one of the objectives of this study was to survey construction-related courses in the MENA universities and highlight the extent to which safety has been incorporated in these courses. Another objective was to determine what construction safety knowledge practitioners have received at university and what safety knowledge they would expect in newly hired engineering graduates. Based on the results of the two types of surveys, the authors proposed modification to the current engineering curriculum to enhance the level of construction safety education and outlined approaches that could be undertaken to DfCS. Before starting with the methodology used in this study, an overview of safety in the construction industry is first presented, followed by a review of the literature pertinent to the role of engineering education in construction site safety and DfCS.

11.2 Health and Safety in the Construction Industry

As mentioned earlier, the level of health and safety implied by the construction industry records is significantly lower compared to that in other industries. For example, the incidence of injury in the construction industry throughout Australia is 50 % higher than all other industry rates (Breslin 2004), and the construction-related injuries in the USA reach 21 % of all work-related injuries (Zarges and Giles 2008). Overall, an estimated 350,000 workers die every year in labor accidents. Of these accidents, 60,000 occur in construction (Rubio et al. 2005), representing more than 17 % of all workers. The higher rate of accidents in the construction industry is probably due to the nature of the work, lack of construction safety precautions, or lack of construction safety education. Nonetheless, factors such as poor planning, inadequate safety training, lack of safety incentives, and insufficient incident investigation are likely to amplify the problem. Although the workers' health is invaluable, the estimated cost realities of accidents in the construction industry ranges between 8–15 % of the construction cost (Rowlinson 2003).

Several previous studies have been conducted to identify the important factors that affect health and safety in construction sites (Sawacha et al. 1999; Teo et al. 2005; Choudhry and Fang 2008; Choudhry et al. 2009; Maraqa and Mohamed 2013). While there are some similarities in what controls construction safety in different countries, there are some factors that could be more specific or important than others in certain countries or for certain projects (Maraqa and Mohamed 2013). Nonetheless, the different construction parties could have an influence on the safety of construction workers. Contractors and subcontractors are undoubtedly the pivotal party to control jobsite safety (Levitt and Samelson 1993; Hinze and Figone 1988; Hinze and Talley 1988). Designers can reduce safety hazards in the working environment by considering worker safety issues in their design decisions (Smallwood 1996; Gibb et al. 2004; Behm 2005; Weinstein et al. 2005). The involvement of owners, on the other hand, has been regarded as an essential requirement for a zero-injury objective (Hinze and Gambatese 1996; Liska et al. 1993).

Table 11.1 lists major causes of death in various countries due to construction activities. For the USA, 21.7 % of the major injury cases in the construction industry during 1991–1998 were fall related, and 16.8 % were due to being run over by equipment or a vehicle. The UK Labor Force Survey (LFS) data on self-reported major injury during 2003–2007 cited slipping and tripping as the most frequent cause of major injury (38 %), followed by handling, lifting, or carrying (15 %); falling or moving objects (12 %); and falling from a height (12 %) (UK HSE 2013).

Health and safety in construction is an international concern. In Denmark, the number of registered accidents in the construction sector in 1994 was 27.1 per 1000 construction workers (Arbejdstilsynet 1996). A higher number of 220/1000 workers injured in 1996 were reported in Hong Kong (Wong et al. 1999). As for the fatality rate, Europe had a fatality incidence rate of 10.4 per 100,000 in 2001 (Karjalainen 2007), and for the same year, there were 13.3 deaths per 100,000 construction workers in the USA (US Department of Labor 2001). However, a significant drop in the

Table 11.1 Causes of major injuries (%) in construction. (Maraqa and Mohamed 2013)

Incident Type	1991–1998 USA (ENR 2001)	1999 USA (ENR 2001)	UK LFS (UK HSE 2013)	Kuwait (Kartam and Bouz 1998)	Kuwait (Al-Humaidi and Tan 2010)	UAE[a] (Barss et al. 2009)	UAE (McGrath 2009)
Falls	21.7	23.8	12.0	41.2	33.0	51.0	12.0
Falling objects			12.0	10.2	25.0	15.0	8.3
Run over	16.8	21.1					8.7
Powered machinery	12.5	9.9			18.0	11.0	5.7
Collapse of structure	4.3	5.0		8.5			
Slipping and tripping			38.0				7.5
Lifting or carrying		5.4	15.0				

ENR Engineering News Record, *HSE* Health and Safety Executive, *LFS* Labour Force Survey, *UAE* United Arab Emirates

[a] Includes all occupational safety cases reported at Al-Ain Hospital from March 2003 to April 2005

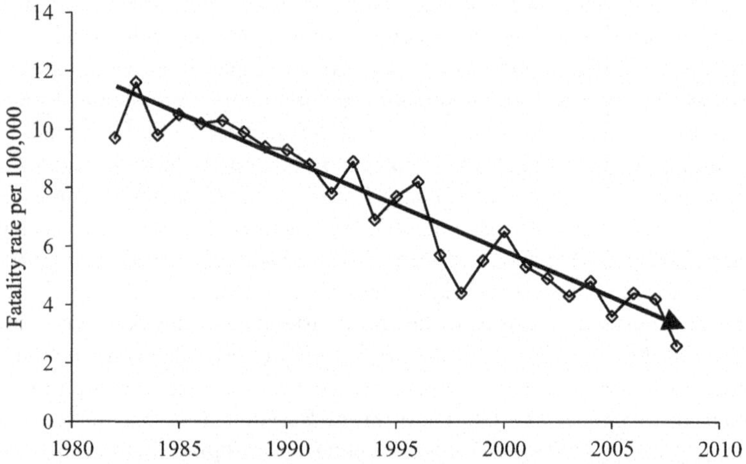

Fig. 11.2 Fatal injury rates in the construction sector in the UK. (Al Palumbo 2010)

fatality rate of construction workers has been observed in the UK as a result of the strict regulations imposed in the last 20 years (see Fig. 11.2).

The construction industry in the MENA region employs a large proportion of the employed population. The percent of employed population in the industry varies among different countries (Fig. 11.3) but averages to 16.5 % of the employed labor force. Records on construction injuries in the MENA region are greatly lacking,

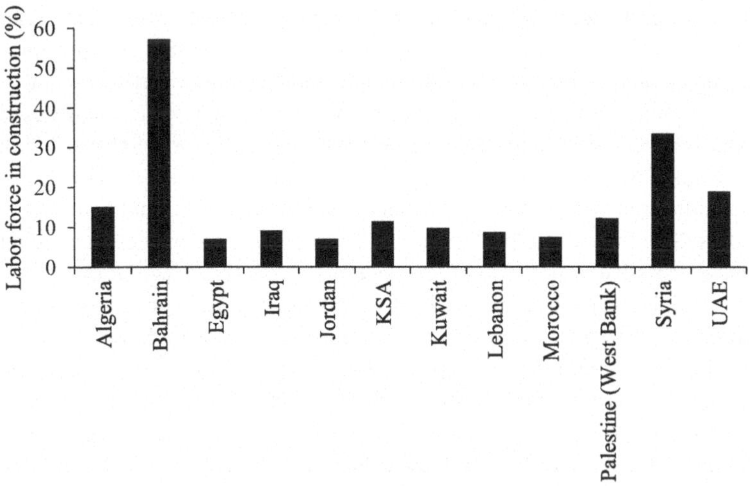

Fig. 11.3 Percentage of construction labor force in some Middle East and North African (MENA) countries. *KSA* Kingdom of Saudi Arabia, *UAE* United Arab Emirates. (Data from Habib 2007)

however. As noted by Habib (2007), not all MENA countries have regular statistics, and some of the countries face a problem of underreporting of work-related accidents because reporting and notification are poorly implemented. Somavia (2005), on the other hand, reported that work-related deaths in the Middle East were estimated at 19,000 in 2005. Meanwhile, Habib (2007) reported that 70 % of all injuries and fatalities that occurred in Lebanon in 2003 were construction related.

Based on an analysis of construction-related accidents in Kuwait city during 1996–2007, Al-Humaidi and Tan (2010) found that falls are the major type of accident (33 %) followed by being crushed or struck by a falling object (25 %), and then by use or misuse of tools (18 %). Furthermore, these authors reported that almost 83 % of the victims of construction accidents in Kuwait sustain permanent disabilities. For United Arab Emirates (UAE), 51 % of the occupational injuries are due to falls, 15 % are due to falling objects, 11 % are due to powered machines, and 6 % are due to burns (Barss et al. 2009).

Although no comprehensive study has been conducted on the status of construction health and safety in the MENA region, the study by Habib (2007) should be useful to have a perspective of the situation. The author studied the status of occupational safety and health (OSH) in 18 Arab countries, comprising Algeria, Bahrain, Egypt, Iraq, Jordan, Saudi Arabia, Kuwait, Lebanon, Libya, Morocco, Oman, Palestine, Qatar, Sudan, Syria, Tunisia, UAE, and Yemen. Results showed that the Arab countries vary in their OSH conditions, with some showing serious deficiencies in OSH mechanisms and performance. The main obstacles for improvements, as reported by the author, include (1) delayed ratification of the OSH conventions set by the International Labor Organization, (2) lack of detailed and comprehensive OSH provisions in the local legislations, (3) absence of national OSH policies and programs, and (4) weak enforcement of OSH regulations. Additional obstacles include insufficient reporting, lack of accurate and comprehensive data related to occupational accidents and diseases, failure to incorporate OSH advisory bodies in the decision-making process on OSH-related matters, and insufficient local OSH expertise in many countries. The author recommended that Arab countries should establish a proper OSH legislative framework necessary for OSH promotion and should evaluate their OSH situation by preparing an exhaustive OSH national profile. The author further recommended that research and educational institutes should undertake OSH studies, and proficient OSH specialists and inspectors should be recruited and provided with the facilities necessary to enforce regulations.

While Habib's report shows lack of a real OSH system in many Arab countries, some initiatives to improve the situation have been undertaken since then. The Abu Dhabi Emirate, for example, issued Decree 42 in 2009 for development of the emirate Environment, Health, and Safety (EHS) Management System across all sectors including the construction sector. All entities in the construction sector in the emirate are expected to implement their developed systems by 2020 as part of their operational activities (Maraqa and Mohamed 2013). According to Maraqa and Mohamed (2013), the Abu Dhabi EHS management system, if successfully implemented, could serve as a model for developing a safety culture in the construction industry in other emirates within the UAE and in other countries in the region. Maraqa and Mohamed (2013) indicated, however, that a successful safety management system should demonstrate leadership and management responsibilities and provide the tools to implement the safety policy. In addition, significant changes in cultural norms and accountability are required among employers and employees where safety considerations in project design and delivery are given a priority, rather than being viewed as a burden.

11.3 Review of Education Role in Construction Health and Safety

11.3.1 Construction Site Health and Safety Education

The need for construction site health and safety education is now a consensus issue among construction educators and the industry for its enormous contribution toward the reduction in number and costs of accidents (Banik 2005). As such, incorporation of construction site safety in university curricula has been the topic of several prior research studies. In 1995, Suckarieh and Diamantes (1995) surveyed construction management programs in the USA that are accredited by the American Council for Construction Education (ACCE). The authors indicated that safety has not influenced construction education to any significant extent. They stated that the time devoted to construction safety for construction engineering students is lacking, but formal training could have a significant impact on their performance as soon as they graduate. The authors found that courses dedicated solely to construction safety existed in only about 50 % of the surveyed construction management programs. The common elements of the lower division safety courses included introduction to the Occupational Safety and Health Act, Occupational Safety and Health Administration (OSHA) Standards, administration in the field, craft education requirements, filing forms and accidents reports, keeping hazardous materials information, and preparing for OSHA inspections. On the other hand, the upper division courses included motivational factors, incentives programs, safety program effectiveness, and new initiatives. The authors concluded that a plan for formal education in construction safety can be either a stand-alone course or, where appropriate, integrated into all elements of the curriculum by covering the material in several courses in a coordinated fashion.

Coble et al. (1999) conducted a survey to obtain information about the extent that safety was integrated into 4-year, university-level, construction programs. The survey covered ACCE-accredited programs in the USA. Data were obtained on how many programs teach construction safety, the materials used to teach safety, the extent and level at which the subject is covered, and the portion of coverage that is devoted to the OSHA Standards for Construction. Based on a total of 55 responses, it was found that some construction programs were very proactive in their safety education while others were not. Of the responding programs, 45 % indicated that a course is offered within their curriculum that is wholly devoted to safety. While not all programs offered a course in safety, many respondents indicated that there was a definite effort to include safety as a mandatory requirement.

The study of Coble et al. also revealed that safety courses are offered primarily at the junior and senior levels, and, if a safety course is offered, it is typically required as part of the curriculum. Meanwhile, all of the safety courses that are offered address the OSHA Standards for Construction, and an OSHA Outreach Training certificate is earned in 44 % of the safety courses. Moreover, 75 % of all programs, including those that offer a separate safety course, address construction safety in other courses within the construction program.

It should be noted that the ACCE did not incorporate specific requirements for safety content and coverage in accredited programs before 2002. However, the current ACCE guidelines (ACCE 2012) require construction safety as part of the curriculum. ACCE requires 4-year programs to devote at least one credit hour of the curriculum to construction safety. The credit can be earned in a single course, or as part of other courses taken in the curriculum. Furthermore, ACCE describes the typical curriculum content on safety to include (1) safe practices; (2) mandatory procedures, training, records, and maintenance; and (3) compliance, inspection, and penalties.

Gambatese (2003) studied the inclusion of construction safety in both construction and civil engineering (CE) academic programs in the USA. All the surveyed programs were either Accreditation Board for Engineering and Technology (ABET) or ACCE accredited. The author found that none of the responding ABET-accredited CE programs offers a course wholly devoted to safety. This was attributed to the fact that ABET accreditation stresses design and does not require coverage of construction site safety. Nonetheless, the author found that construction safety is covered to some extent in other courses in the CE curricula of 64 % of the responding programs. As to the surveyed construction programs, the author found that safety is addressed in the curricula to a greater extent than that of the surveyed CE programs. Again, this was attributed to the accreditation requirements for construction programs as 70 % of these programs were ACCE accredited. Results of the survey of the construction programs also revealed that (1) 90 % of these programs offer a course fully devoted to construction safety, and these programs require such a course to be taken by all their students; (2) safety courses that are offered are typically 3 credit hours and are taught at the junior or senior level about once a year; (3) the OSHA Standards for Construction are utilized in half of the safety courses; (4) OSHA training certification is commonly earned for completing the safety course; (5) the OSHA Standards for construction and safety management topics are covered in all safety courses, with the OSHA Standards being the primary course content; and (6) 80 % of the programs cover safety in other construction program courses. The author concluded that there is specific safety knowledge that the construction industry believes newly hired university graduates should possess, but the means by which this knowledge is communicated, and both the depth and breadth of coverage, differ among institutions.

In the UK, most universities incorporate construction site safety at relevant points throughout the undergraduate CE curriculum as opposed to addressing it in a separate course (Al-Mufti 1999). Coverage of safety in this manner is due in part to the 1994-mandated Construction (Design and Management) Regulations in force throughout the UK. Al-Mufti (1999) indicated that CE programs should provide students with adequate knowledge and an enhanced awareness of safety in construction. The author suggested that the approach to teaching construction safety should comprise safety as an integral part of all other teaching as opposed to an individual course devoted to the topic. The reason for not having a separate course devoted solely to safety, as suggested by the author, is that offering a separate course may lead to problems of timing, fragmentation, and students developing a passive view of the subject as nontechnical, unimportant, and a subject to pass, forget, and never revisit in their career. Al-Mufti further suggested that the integrated approach of teaching construction safety should consider the different components including technical, nontechnical, social, and psychological aspects.

As in the UK, the inclusion of construction site safety into engineering program curricula in Canada is mandated by the Canadian Engineering Accreditation Board (CEAB). CEAB, an organization that accredits Canadian undergraduate engineering programs, requires considerations of worker safety and health in the CE programs (Christian 1999). However, Christian (1999) observed that teaching a class such as safety is sometimes difficult as students regard its regulatory nature as "boring."

In a different approach, Smith and Arnold (1999) conducted a survey to determine what safety knowledge practitioners would expect in newly hired engineering graduates in the USA. The purpose of their study was to establish a baseline for a construction site safety course content to be offered for undergraduate engineering students. Responses were received from 27 construction firms operating predominantly in the northeastern USA. Out of 14 safety responsibility areas identified by the authors, knowledge of how to perform 9 of these areas was identified by the respondents as the most important for newly hired engineers. These areas are (1) conduct a pre-project hazard analysis, (2) prepare accident reports, (3) conduct toolbox meetings, (4) participate in project safety meetings, (5) perform hazard analysis, (6) recognize common hazards, (7) conduct site safety audits, (8) maintain material safety data sheet (MSDS) files, and (9) obtain safety permits.

Banik (2005) noted that safety education on many projects is inadequate and ineffective especially for field managers and workers. As a possible solution to this problem, the author suggested the inclusion of innovations in safety education and more involvement by the industry in the safety education process. As such, the author distributed a questionnaire in the southeastern USA to get input from the construction industry on the content of a safety course for undergraduate construction students. Thirty-five contractors completed the survey form, with a response rate of about 47 %. Based on the survey results, the author concluded that construction safety education cannot be accomplished by teaching only the fundamental requirements of the OSHA 10- h course. The author identified the topic areas that had the greatest amount of agreement (70 % or more) among the contractors surveyed and that they suggested should be covered in the safety course. These areas include (1) general safety and health, (2) personal protective and lifesaving equipment, (3) fire protection and prevention, (4) electrical, (5) scaffolding, (6) fall protection, (7) excavation, (8) stairways and ladders, and (9) toxic and hazardous substances.

Haupt (2003) indicated that graduates from construction-related programs at universities need to be able to recognize, avoid, and prevent unsafe conditions on their construction sites. The author conducted exploratory studies of safety and health education at higher education institutions in the Western Cape province of South Africa. Preliminary results suggest that construction-related programs do not adequately prepare students to be able to ensure the safety and health of workers in construction sites. Courses make scant reference to the provisions of the South Africa Occupational Health and Safety Act and responsibility of employers for worker safety. Later, Smallwood (2010) indicated that South African construction management programs address construction safety to varying degrees. Those that do so to a lesser degree invariably focus on legislation and do not address procurement and other relevant dynamic issues. The author conducted a survey among a group of "health and safety better practice" general contractors to determine the important issues relative to a construction safety curriculum. Findings of the survey indicated that construction management and CE predominate in terms of the importance of the inclusion of construction safety in the tertiary education programs among nine built environment disciplines. Results also revealed that construction safety should be included as a separate subject and as a module in various subjects. Among 24 construction management health and safety curricula subject areas listed in the survey form, 4 areas were predominantly selected by the respondents, namely (1) management of subcontractors, (2) role of management, (3) worker participation, and (4) the Occupational Health and Safety Act.

Pellicer et al. (2003) reported a lack of academic education in construction safety in the CE programs in Spain. To remedy this, the authors suggested adjustment of some courses to incorporate construction safety. The authors also proposed an educational guide in construction safety which introduces the culture of health and safety in the CE curricula. Later, Rubio et al. (2005) analyzed the course requirements of CE programs at eight universities in Spain. The authors found no course specifically devoted to health and safety, but a chapter is dedicated to health and safety taught within a course on procedures of construction and machinery, but not at all universities. In an earlier study conducted by the authors, they sought the opinion of 96 currently employed civil engineers (19 project supervisors, 21 coordinators, and 56 site managers) in Spain concerning the health and safety training they had received. Results showed that on average 61 % of the interviewed personnel did not receive any training. However, more than 95 % of the interviewed personnel thought it would be appropriate to include course material specific to health and safety in the requirements for CE programs. Cortes et al. (2012) conducted a study to define a framework for including occupational risk-prevention education in the new engineering syllabi in Spain. A questionnaire was distributed to a panel of 59 experts. Survey results indicated that education in occupational risk prevention is essential for improving the safety culture within a company or workplace. The experts concurred that this subject should be a separate mandatory course in all engineering degree programs. The participants recommended that an optional course should be considered only if a mandatory course is not approved. It was also deemed desirable to integrate occupational risk prevention as a cross-field subject in other technological courses even if the curriculum already includes some related courses.

11.3.2 Design for Construction Safety

DfCS, also referred to as prevention through design (PtD), is a process that incorporates hazard analysis at the beginning of a design project. The process does not address methods to make construction safer but how to make a project safer to build (i.e., safe constructability). In other words, safe constructability considers worker safety in the design of a facility rather than the traditional design aspects that focus only on the safety of the "end user," such as the building occupant or facility operator. DfCS is a structured process in the planning and design phases of the project that starts with hazard identification. Engineering measures are then applied to eliminate the hazard or reduce the risk. If the hazard cannot be eliminated, then safety devices are incorporated. If some risk still exists, then warnings, instruction, and training should be used as a last resort.

Traditionally, designers have been under no obligation to inform contractors of hazards resulting from the design. However, designers could play an important role in ensuring how safety of construction workers is considered during the project design process. Competent designers can draw from their own experience and published information to comply with the duties of reducing hazards and appreciate the risk in their design. In fact, designing to eliminate or avoid hazards is considered the preferable means for reducing risk in the hierarchy of controls (Manuele 1997). It should be emphasized that reducing injuries and fatalities and improving worker health are not the only benefits associated with DfCS. Addressing safety in the conceptual or early design stages could yield other measurable benefits such as improved productivity, a decrease in operating costs, avoidance of expensive retrofitting to correct design shortcomings, and significant reductions in injuries, illnesses, environmental damage, and attendant costs.

Implementation of DfCS in engineering practice depends on several major factors such as change in designer mindset toward safety, existence of a motivational force to promote designing for safety, incorporation of construction safety knowledge in the design phase, designers' knowledge about design-for-safety concepts, and making design-for-safety tools and guidelines available for use and reference.

There is evidence that careful consideration of safety during the design stage could have eliminated or reduced injuries and fatalities in some construction projects (European Foundation 1991; Smallwood 1996; Christensen and Manuele 1999; Suraji et al. 2001; Gibb et al. 2004; Reese and Eidson 2006). For example, the European Foundation (1991) found that 60 % of the accidents it surveyed could have been eliminated or reduced with careful thought during the design stage. Gibb et al. (2004) reviewed many construction accident cases and found that in 47 % of these cases, changes in the design would have reduced the likelihood of accidents. Similarly, Behm (2005) found that design was linked to accident in about 22 % of 226 injury incidents that occurred in California, Oregon, and Washington during 2000–2002, and in 42 % of 224 fatality incidents occurred in the USA during the period 1990–2003. Moreover, contractors responding to a survey of the construction community in South Africa ranked design as the highest out of all components identified that negatively affect safety (Smallwood 1996).

Weinstein et al. (2005) demonstrated how injury prevention efforts in the construction of a semiconductor manufacturing facility can begin upstream by involving designers, engineers, and trade contractors in preconstruction processes. On the other hand, Driscoll et al. (2008) reported that 37 % of 210 workplace fatalities that occurred in Australia during July 2000–June 2002 definitely or probably had design-related issues involved, and the circumstances in another 14 % were suggestive that design issues were involved.

Recognizing the importance of DfCS, the European Union enacted the "Control of Hazards on Temporary and Mobile Construction Sites" directive that requires member states to adopt national laws to formalize a process to ensure that construction site safety is considered during the design process. This legislation has led to important efforts in design for safety (Gibb 2004). Meanwhile, the UK Construction Regulations of 2007 (UK CDM 2007) place a duty on the designer to identify and eliminate hazards and reduce likely risks from hazards where elimination is not possible. Similarly, the American Society of Civil Engineers (ASCE) Policy Statement Number 350 on construction site safety states that "engineers shall have responsibility for recognizing that safety and constructability are important considerations when preparing construction plans and specifications."

While the merits of DfCS are evident, numerous barriers to its implementation have been cited (Toole 2004). Among these barriers are limited availability of tools, guidelines, and procedures for preventive design, and the limited education which engineers receive on issues of construction worker safety and on how to design for safety. Gambatese et al. (2008) listed several key changes that need to be undertaken which could significantly impact the implementation and the outcomes of design for safety including (1) change designer mindset toward safety, (2) establish a motivational force to promote design for safety, (3) increase designer knowledge of the concept and incorporate construction safety knowledge in the design phase, (4) utilize designers who are knowledgeable about design-for-safety modifications, (5) make the design-for-safety tools and guidelines available, and (6) mitigate designer liability exposure. Despite these challenges, the viability of designing for safety provides an incentive to move forward. Implementation of the concept should be promoted to increase its use in practice. Meanwhile, continued exposure will increase designer knowledge of the concept and initiate the development of best practices for its implementation. Gambatese and Hinze (1999) gathered suggestions for improving construction worker safety while in the design phase. Using these design suggestions, the authors developed a safety design tool to assist designers in identifying project-specific safety hazards and to provide best practices to eliminate the hazards.

The extent to which the engineering academic programs responded to the need for inclusion of DfCS varies but some progress has been noticed over the last 15 years. In 2001, Carpenter et al. (2001) reported that the construction undergraduates in the UK had a poor level of knowledge of and a poor attitude toward health and safety in construction. The authors discovered that the requirements for academic coverage of health and safety risk management in construction-related degrees varied considerably among the accreditation bodies that were responsible for setting the requirements for all construction-related degree programs. These accreditation bodies include the Chartered Institute of Building (CIOB), the Joint Board of Moderators (JBM), the Architects Registration Board (ARB), and the Royal Institute of Chartered Surveyors (RICS). The authors concluded that accreditation requirements of health and safety risk management were neither coherent nor consistent. They suggested that it was unreasonable for academia alone to be expected to respond to the challenge of improving the performance of health and safety in the construction industry without increased support and guidance from the accreditation bodies and the industry.

Later, Carpenter et al. (2004) reported that the requirements of the accreditation bodies had improved significantly in both approach and detail and that academia had developed good links with industry. However, the concept of health and safety risk management had not been sufficiently accepted or integrated into courses because of a lack of support from the heads of construction departments and a lack of cohesion among academic staff. Consequently, the authors concluded that the key issue had shifted to enabling and ensuring that the higher education institutions delivered these requirements. Furthermore, Carpenter et al. (2004) recommended that accreditation bodies should request higher education institutions ensure that the external examiner is sufficiently briefed to assess course content and health and safety risk management.

Progress in DfCS in the USA did not move as fast as it did in the UK. Gambatese (2003) found that DfCS is covered only in less than 7% of the courses of the surveyed construction programs in the USA. Weinstein et al. (2005) indicated that the current political and legal environment in the USA rules out any prospect of a safety-in-design requirement being enacted. These authors believe that the idea is likely to diffuse in the USA when owners, designers, and constructors see tangible evidence that safety-in-design processes lead to reduced risks on construction worksites. Nonetheless, the ASCE has entered into an alliance with OSHA to improve construction worker safety, which would more directly involve design engineers in the process in future. The purpose of the initiative was to raise safety awareness within the construction community and to promote the resources offered by OSHA. A key step, as indicated by Toole (2004), is to develop model safety programs that can be used by engineers and architects in their designs. Gambatese et al. (2008) indicated that the implications of the DfCS for education of design engineers requires (1) a shift in mindset, (2) a holistic view, (3) exposure to DfCS fundamentals, (4) offering of training in system-specific DfCS opportunities, and (5) development of engineering course-specific DfCS modules.

11.4 Construction Health and Safety Requirements by Accreditation Organizations

The ABET is the primary organization for accreditation of university engineering programs in the USA. Several universities outside the USA have been accredited by ABET in recent years. The criteria required of ABET-accredited engineering programs is described in the ABET document titled "2013–2014 Criteria for Accrediting Engineering Programs" (ABET 2012). This document describes the program criteria that should be applied to engineering programs. The general criteria for baccalaureate level programs, Part (c) of Criterion 3, which is related to student outcomes, state that students must demonstrate "an ability to design a system, component, or process to meet desired needs within realistic constraints such as economic, environmental, social, political, ethical, health and safety, manufacturability, and sustainability." However, with the exception of the construction engineering programs, the ABET criteria for curricula for other construction-related programs (such as architectural, civil, electrical, or mechanical engineering) do not explicitly emphasize construction site safety or even suggest it as a curriculum component. Rather, these programs are required to be primarily focused on design and the design professional. The absence of health and safety in the accreditation requirements of these programs is perhaps a reflection of the traditional role that architectural and civil engineers play as designers without responsibility for construction means and methods.

The ACCE (2012), Document 103, defines the standards and criteria for accreditation of construction education programs. According to ACCE (2012), the curriculum of baccalaureate degree programs should address health and safety in the construction courses related to project execution. ACCE (2012) requires that at least one semester credit (1.5 quarter credits) must be devoted to safety. This credit can be covered either in a single course or in multiple courses. Safety content must include safe practices; mandatory procedures, training, records, and maintenance; and compliance, inspection, and penalties.

Contrary to ABET, inclusion of health and safety in engineering programs was mandated by the CEAB that was established in 1965 by the Canadian Council of Professional Engineers to accredit Canadian undergraduate engineering programs that meet or exceed educational standards acceptable for professional engineering registration in Canada. The CEAB (2011) accreditation criteria demand that graduates should be able to produce designs that meet specified needs with appropriate attention to health and safety risks. The graduates are expected to have an understanding of the interactions that engineering has with the economic, social, health, safety, legal, and cultural aspects of society; the uncertainties in the prediction of such interactions; and the concepts of sustainable design and development and environmental stewardship. The CEAB (2011) mandates that the curriculum content and quality should be designed to assure that graduates can produce engineering designs that meet specific needs and constraints that relate to several factors including health and safety. According to CEAB (2011), the curriculum must include studies in different major areas including health and safety.

11.5 Study Methodology

The MENA countries include a total of 22 countries, namely Algeria, Bahrain, Cyprus, Egypt, Iraq, Iran, Israel, Jordan, Saudi Arabia, Kuwait, Lebanon, Libya, Morocco, Oman, Palestine, Qatar, Sudan, Syria, Tunisia, Turkey, UAE, and Yemen. In the current study, two types of surveys were conducted in order to assess the extent to which construction health and safety is incorporated in the engineering curricula in the MENA region and the need for graduating engineers to acquire construction health and safety education during their study. The first type of survey targeted practitioners (mainly engineers) working in the construction sector. Meanwhile, the second survey focused on the extent that construction health and safety is currently integrated in engineering programs in the MENA region. Targeted departments comprise CE, architectural engineering (AE), construction management, and construction engineering as these programs are expected to offer construction health and safety in their curriculum. The practitioners' survey form was distributed either as a hard copy or as a digitized pdf file, while the engineering program survey form was circulated as a digitized pdf file. It should be noted that the survey forms were not distributed to all MENA countries. This issue will be discussed in more detail in Sect. 11.5.2. Additionally, the countries where the survey forms were distributed and collected from will be indicated so that overgeneralization is avoided.

Many of the questions in the survey forms are of a stated-preference nature. In some of these questions, respondents were asked to choose one answer from a given list. In some other questions, respondents were requested to choose all applicable answers from a given list. In the analysis of the results of the former set of questions, the authors will refer to the answers in terms of number of respondents, with the total number of respondents equal to the sample size. However, in the analysis of the second set of questions, the authors will refer to the answers in terms of number of responses, which exceeds the sample size.

11.5.1 Practitioners' Survey

The practitioners' survey included 17 questions and was divided into two main parts. The first part (Questions 1–9) was intended to identify the extent that engineering graduates from the MENA region received construction health and safety education/topics during their study. The second part of the survey (Questions 10–17) was designed to collect opinions on the importance of incorporating construction health and safety education in the engineering curriculum. Some of the respondents did not graduate from the MENA region although they work in the region, or have a nonengineering degree from the MENA region. These respondents (graduates of Australia, France, India, Malaysia, Philippines, South Africa, South Korea, and the UK) were not considered in the analysis of the results of the first part of the survey but were considered in the analysis of the second part. The number of respondents that are considered in the second part of the survey is 401, while the total number of respondents that were included in the analysis of the first part of the survey form is 318.

The first five questions of the practitioners' survey form collected background information about the respondents including nationality, specialization, university, and country from which they graduated and year of graduation. Questions 6–8 collected information about the type of courses, if any, through which the respondents received construction health and safety education during their bachelor degree program, while Question 9 requested practitioners to rank the level of construction health and safety education they received during their undergraduate studies.

Questions 10–12 asked for information regarding the countries the respondents have worked in after graduation, the number of years they have been working in the construction field, and the positions in the field that they have been involved in. Question 13 inquired if the respondent thinks there is a need for more construction health and safety education at the university level in the region, while Question 14 and 15 asked about the form that could be utilized to offer construction health and safety knowledge to university students. Question 16 asked if the respondent thinks graduating students should pass a construction health and safety examination by a certified board in the country before being able to practice engineering. Question 17 requested the respondents to indicate (on a scale from 0 to 10) the need for graduating engineers to acquire some stated safety aspects. Fifteen safety aspects were included in the survey, namely incorporation of safety in design courses

(SA1); hazard identification/risk assessment (SA2); controlling and preventing hazards (SA3); safety regulations (SA4); record keeping (SA5); general safety and health provisions (SA6); electrical safety (SA7); falls-from-height protection (SA8); falling objects protection (SA9); personal protective equipment (SA10); materials handling, storage, use, and disposal (SA11); working on scaffolds (SA12); cranes, hoists, elevators, and conveyors (SA13); excavations/burial under earth falls (SA14); and working/standing on stairways and ladders (SA15). Respondents were also asked to provide comments or suggestions on how to improve construction safety in the engineering curriculum in the MENA region.

11.5.2 Engineering Programs Survey

The engineering program survey form consisted of 15 questions. The first five questions of the form collected information about the name of the department, its location (university and country), average number of students enrolled every year, and the number of faculty members in the department. Question 6 asked if the program is accredited by ACCE, ABET, CEAB, or any other organization. Questions 7–11 requested some details about the part of the curriculum in which construction health and safety is covered (i.e., core course (CC), technical elective, internship, or integral part of offered courses) and the title of the course(s) in which construction health and safety is being fully or partially covered in the curriculum. Question 12 inquired about the extent of overall coverage of construction health and safety in the curriculum. Question 13 asked about the construction health and safety topics (to be selected from a given list) that are covered and the extent of coverage of these topics in the curriculum. These topics are (1) legislations/regulations; (2) safety management; (3) risk identification; (4) safe practices and risk reduction; (5) hazard assessment; (6) hazard control and prevention; (7) mandatory procedures, training, records, and maintenance; (8) compliance, inspection and penalties; and (9) incorporation of construction safety in design courses. Question 14 asked about the materials that are used to teach construction safety in the curriculum (i.e., US OSHA Standards, UK Construction (Design and Management) Regulations, the country's construction safety regulations, materials developed in-house, or others). Question 15 inquired if the construction health and safety courses offered in the department are structured so that students receive certification upon successful completion of these courses.

The survey form was emailed to the heads of the target programs in 15 selected countries among the 22 countries in the MENA region (Table 11.2). It should be noted that seven MENA countries were not contacted due to difficulties and/or unavailability of means of communications. Meanwhile, the contacted programs were selected through an extensive web search to represent the 15 countries listed in Table 11.2. The total number of distributed survey forms was 79. Only 17 responses were received, representing about 22 % of the total programs contacted. Due to this relatively low response rate, additional relevant information was collected by

Table 11.2 Countries and programs that were surveyed (in alphabetical order)

Country	Number of programs contacted	Number of responses received	
		CE program	AE program
Bahrain	4	–	1
Cyprus	7	1	–
Egypt	12	2	–
Iran	12	1	–
Jordan	1	–	–
Kingdom of Saudi Arabia	5	1 + 1[a]	–
Kuwait	1	–	–
Lebanon	8	1	–
Morocco	2	–	–
Oman	4	–	–
Palestine	2	1	–
Qatar	1	1[b]	
Sudan	1	–	–
Turkey	7	1	–
United Arab Emirates	12	4 + 1[c]	1

CE civil engineering, *AE* architectural engineering
[a] A reply has been received indicating inapplicability of the survey to the program
[b] The program combines both CE and AE
[c] The program has been recently launched

searching the web for the courses that include construction safety in the programs from which responses were not received to further substantiate the research findings.

11.6 Results and Discussion

11.6.1 Practitioners' Feedback

11.6.1.1 Health and Safety Education Received

Figure 11.4 shows the countries from which the respondents to the practitioners' survey graduated along with their nationalities. A large portion (about 40 %) of the surveyed practitioners graduated from the UAE. This is attributed to the fact that the authors were able to assign individuals to distribute and collect the survey form in the UAE where they are residing. It should be noted that some of the surveyed engineers are nationals of countries outside the MENA region, but they graduated from universities in the MENA and worked in the region. Figure 11.4 also shows that a large fraction of the respondents (>65 %) graduated in the year 2000 and afterwards. Therefore, the overall opinion of the surveyed engineers about the re-

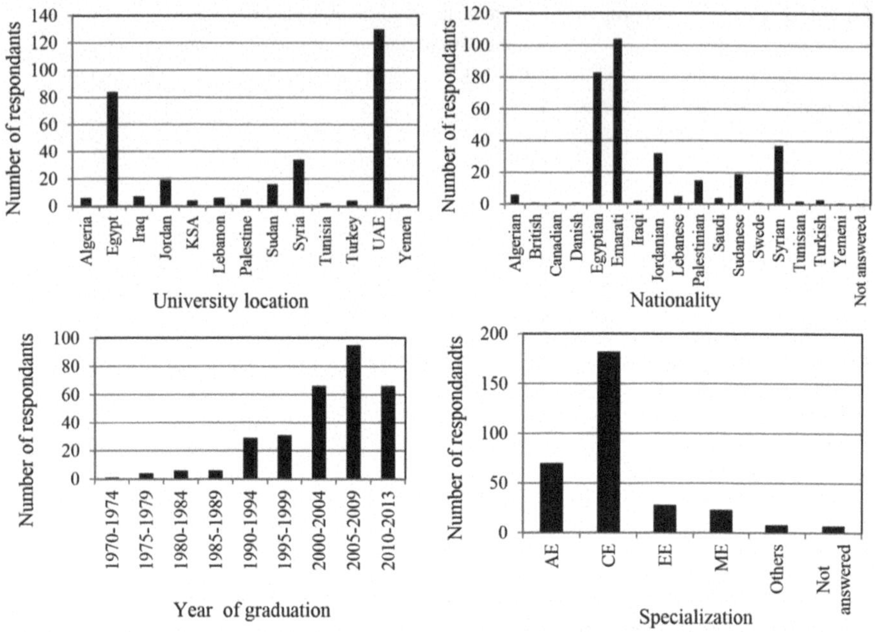

Fig. 11.4 Background information about the surveyed engineers. *AE* architectural engineering, *CE* civil engineering, *EE* electrical engineering, *ME* mechanical engineering, *KSA* Kingdom of Saudi Arabia, *UAE* United Arab Emirates

ceived construction safety education is likely to reflect the current situation in terms of construction health and safety education. With regard to specialization, 57 % of the surveyed engineers have a degree in CE, followed by those with a degree in AE (about 22 %). The remaining respondents (21 %) have a degree in either electrical engineering (EE), mechanical engineering (ME), or other engineering disciplines that were not specified. Moreover, a few respondents did not answer the question related to specialization.

When asked if they have taken any construction health and safety courses or have received construction safety education as part of their courses during their undergraduate education, 69 % of the respondents answered "No," 30 % answered "Yes," and 1 % did not answer this question. This reflects a deficiency of construction health and safety education in the current engineering curriculum in the surveyed MENA countries.

Those who received construction health and safety education during their study (97 out of 318) were requested to mark the type of course(s) in which they received safety education. Almost half of these respondents received such education through internship (training), while less respondents received it through regular university courses (i.e., core, elective, or capstone courses) as depicted by Fig. 11.5.

When requested to rank the level of construction health and safety education received during their undergraduate studies, almost half of the respondents who received safety education ranked such level as being "Average." Meanwhile, about

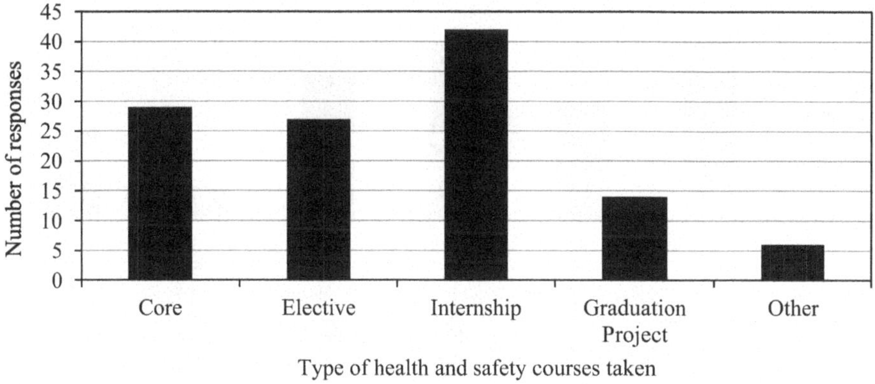

Fig. 11.5 Type of courses through which respondents received health and safety education

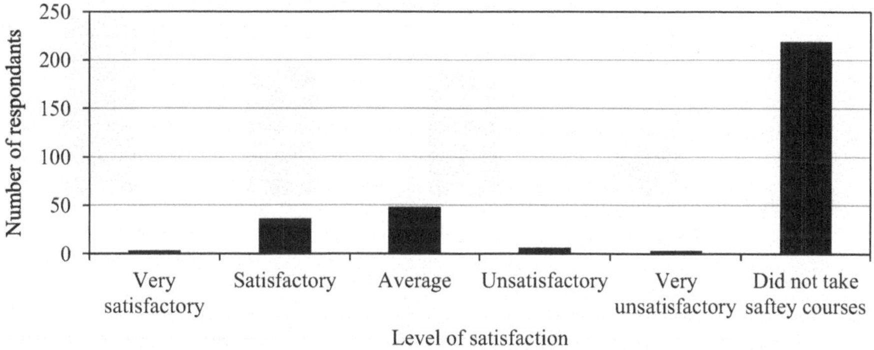

Fig. 11.6 Level of respondents' satisfaction with the construction health and safety education received

40 % of those who received safety education (representing about 12 % of all of the surveyed practitioners) were satisfied with the level received (Fig. 11.6).

11.6.1.2 The Need for Construction Health and Safety Education (Practitioners' Viewpoints)

The 401 surveyed practitioners held different positions, such as being designers/consultants, project managers, site engineers/supervisors, contractors, etc., as illustrated in Fig. 11.7. Other positions held by some of the respondents include safety specialist, project engineer, quality assurance/quality control specialist, inspector, sales engineer, client/owner, project planning/control, urban planner, surveyor, material engineer, supplier, and temporary work designer. As indicated in Fig. 11.7, more than 50 % of the surveyed practitioners have 5 years of experience or more

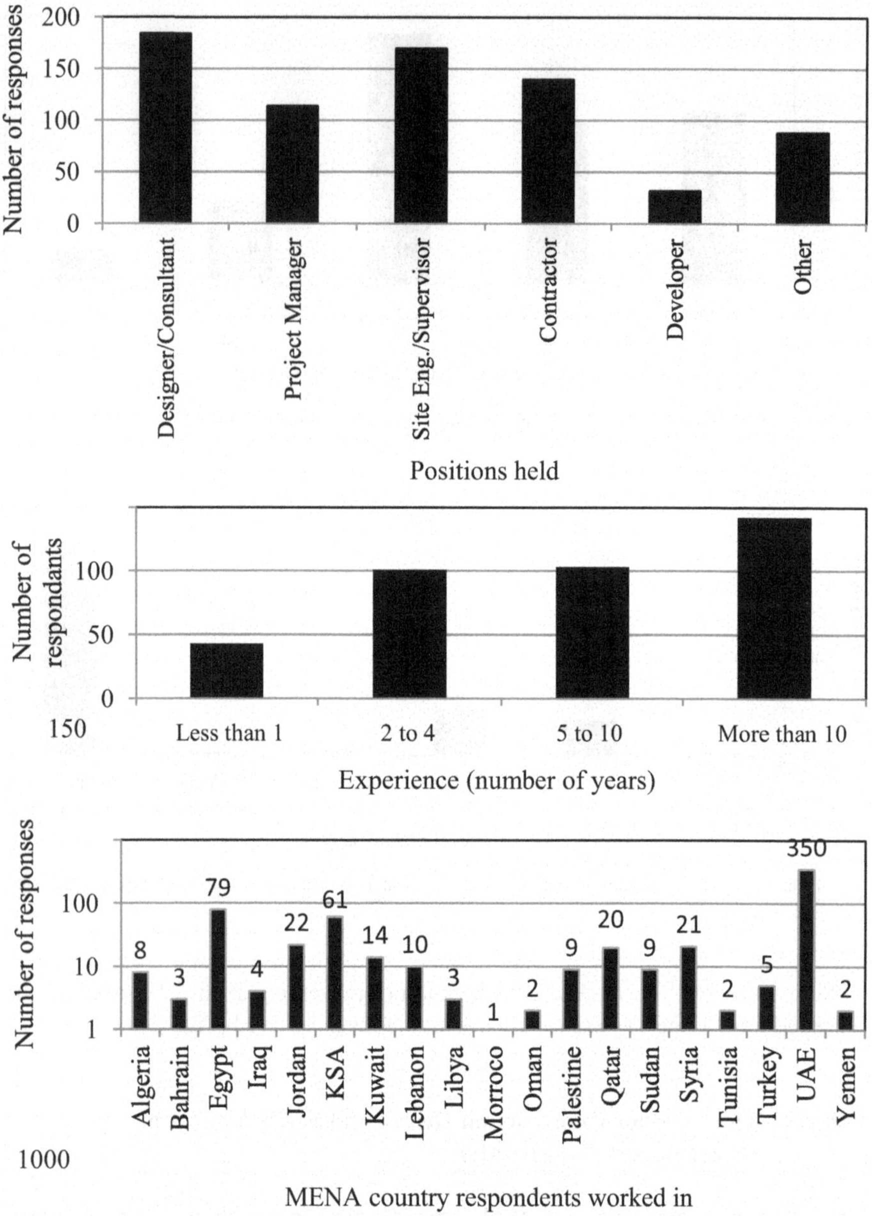

Fig. 11.7 Background information about the surveyed practitioners. *MENA* Middle East and North Africa. *KSA* Kingdom of Saudi Arabia, *UAE* United Arab Emirates

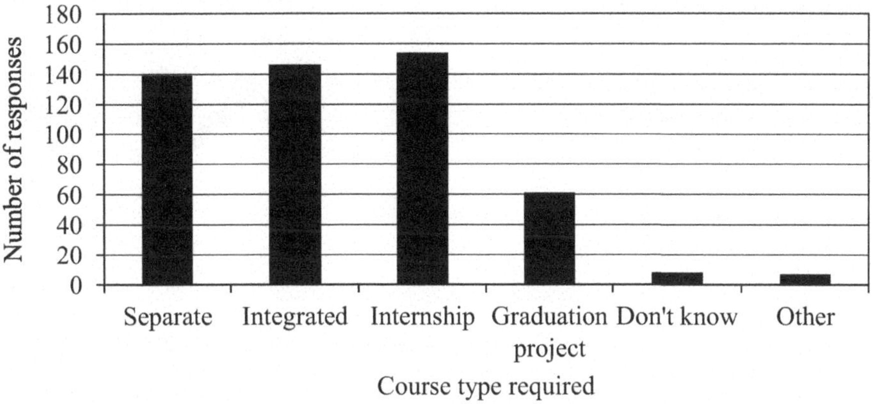

Fig. 11.8 Construction health and safety course type required by students based on practitioners' feedback

in the field. Meanwhile, the surveyed practitioners worked in 18 different countries within the MENA region.

When inquired about the need for more construction health and safety education at the university level, 85 % of the surveyed practitioners reported "Yes," 6 % indicated "No," 8 % answered "Don't know," and 1 % did not answer the question. The large proportion of respondents that agreed that there is a need for more safety education emphasized the importance of incorporating construction safety in the engineering curriculum.

When asked about the method through which construction health and safety should be given to the students, the highest number of respondents indicated that it should be given through internship (Fig. 11.8). A slightly lesser number indicated that construction health and safety should be offered as a separate course or integrated within other existing courses. Only 61 respondents indicated that construction safety should be given as part of the graduation project (GP) course.

The surveyed practitioners were also asked about their opinion of having graduating students pass a construction health and safety examination by a certified body before being able to work. More than half answered "Yes," but a significant proportion (32 %) responded "No," and 11 % replied "Don't know," as presented in Fig. 11.9. The implied support for the idea of having fresh engineers pass a safety exam indicates a necessity for fresh graduates to acquire.

To plan for inclusion of construction health and safety education in the engineering curriculum, one needs to know the safety aspects that are considered important to the engineering program under consideration. As such, surveyed practitioners were asked to rank 15 suggested construction health and safety aspects that should be gained by graduating engineers using a scale of 0–10 (where 0 indicates not needed at all and 10 indicates mostly needed). Fifteen construction health and safety aspects were identified in the survey, comprising incorporation of health and safety

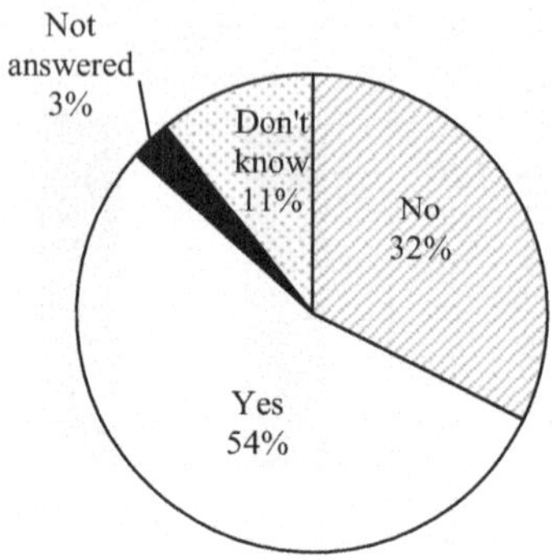

Fig. 11.9 Practitioner's opinions for having engineering students pass a construction health and safety examination before being able to work

in design courses (SA1); hazard identification/risk assessment (SA2); controlling and preventing hazards (SA3); safety regulations (SA4); record keeping (SA5); general safety and health provisions (SA6); electrical safety (SA7); falls-from-height protection (SA8); falling objects protection (SA9); personal protective equipment (SA10); materials handling, storage, use, and disposal (SA11); working on scaffolds (SA12); cranes, hoists, elevators, and conveyors (SA13); excavations/burial under earth falls (SA14); and working/standing on stairways and ladders (SA15). It should be noted that the first three safety aspects (SA1–SA3) are related to DfCS, while the other ones (SA4–SA15) are related to construction site health and safety.

Practitioners' response were averaged and plotted in Fig. 11.10. The upper part of Fig. 11.10 represents the average score on a scale of 0–10, while the lower part represents the number of respondents who answered "Don't know" on a particular

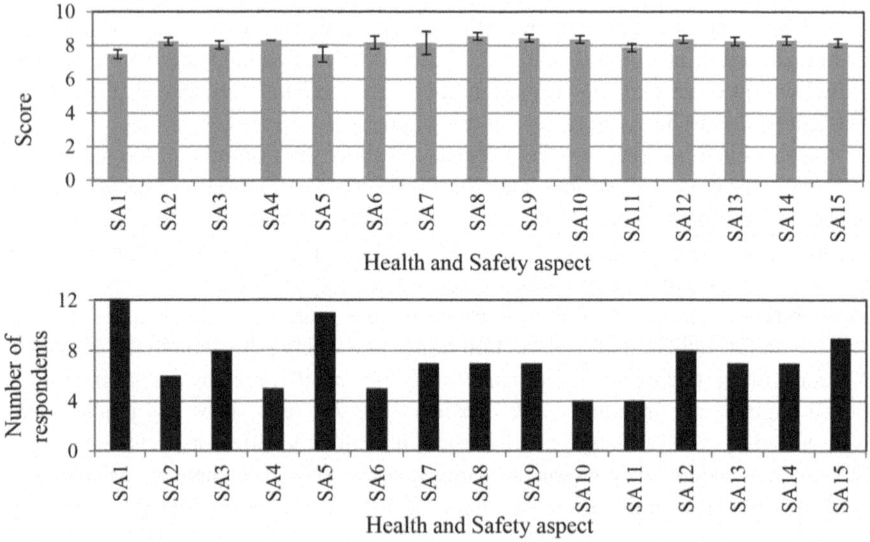

Fig. 11.10 Construction health and safety aspects that should be gained by graduating engineers based on the practitioners' opinion

safety aspect. The average score of any of the stated health and safety aspects fell in the range of 7.5–8.5, with 95 % confidence limits that range from 0.19 to 0.23. The slight variations in the average score and associated confidence limits of the different health and safety aspects suggested that all stated aspects are equally important according to the surveyed practitioners.

Sixty-five respondents provided comments in the survey form. Most of these comments were in support of including construction health and safety in the engineering curriculum. Some practitioners even stated that health and safety should be among the engineer's responsibilities. However, respondents' comments differed in the way that health and safety education should be incorporated (i.e., as a core course, integrated with other courses, or/and as part of the internship). Many of the comments emphasized the need to couple lectures with practice and extensive site visits. Some also suggested that students should attend training workshops and seminars conducted by safety professionals. Few respondents stated that construction health and safety for engineering students should be cultivated as a culture within the curriculum, not just as a topic to pass. This is probably one of the reasons that about 30 % of the respondents were not in favor of the idea of having graduating students pass a safety exam by a certified body.

11.6.2 Engineering Academic Programs Feedback

As was mentioned earlier, a total of 17 programs responded to the survey, among which one response indicated that the questionnaire is not applicable to their program, while another program did not respond as it was recently launched. The 15 engineering programs who filled the survey form indicated that the student enrollment in these programs ranges from 23 to 500. It was noted, however, that the enrollment in the majority of these programs is in the range of 23–100. About 47 % of these programs are accredited by ABET, while 20 % of them are accredited by other accreditation agencies such as the Quality Assurance Agency for Higher Education (QAA) in the UK, the European Credit Transfer System (ECTS), and the UAE Ministry of Higher Education Commission for Academic Accreditation (CAA). The remaining 33 % of the respondent programs are not accredited.

11.6.2.1 Nature and Extent of Construction Health and Safety Coverage in the Curriculum

A major coverage of construction health and safety was shown to be either as an integral part of a core course or through the internship program as reflected by about 67 % of the received respondents presented in Fig. 11.11. A relatively significant coverage is also provided by offering construction health and safety topics as integral parts of technical elective courses with a 40 % response rate (Fig. 11.11). Addressing construction health and safety issues through GPs provides about 13 %

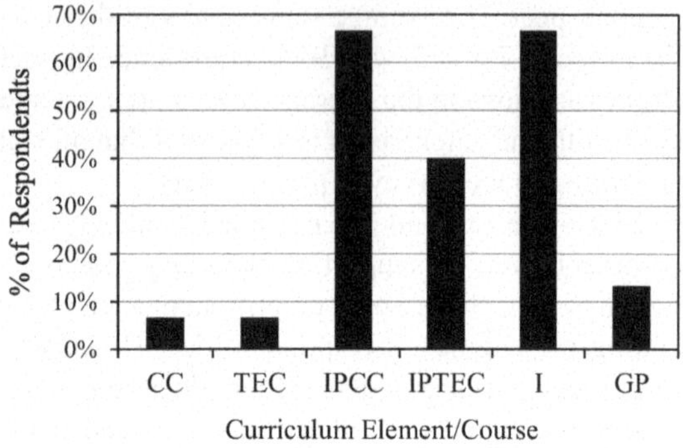

Fig. 11.11 Type of courses covering construction health and safety topics in the engineering curriculum. *CC* core course, *TEC* technical elective course, *IPCC* integral part of a core course, *IPTEC* integral parts of technical elective courses, *I* internship program, *GP* graduation project

coverage as depicted in Fig. 11.11. It is evident from the surveys that assigning an entire course (either as a core course or a technical elective course) for construction health and safety coverage is not common in the MENA region as reflected by the significantly low 7 % response shown in Fig. 11.11 for both core and technical elective courses. In such rare cases, construction health and safety is addressed by 3-credit-hour courses. Whenever the program offers technical elective courses that, fully or partially, cover construction health and safety, it was implied in 70 % of the responses that these courses are offered once a year. By contrast, a very limited number of responses indicated lower course offering frequency of once every 3 or 5 years. The level of enrolment in these courses typically varies between 11–15 students (29 % of the responses) and 21–25 students (57 % of the responses). In rare cases, the enrolment exceeds 25 students (14 % of the responses). Another observation is that construction health and safety is included in less than 15 % of the GP courses offered by the responding programs. This may indicate a high deficiency of inclusion of DfCS in the curriculum.

While 13 % of the responding programs indicated that no coverage of construction health and safety is provided at all, 33 % of the respondents confirmed offering one course to fully or partially cover construction health and safety topics, as shown in Fig. 11.12. Also, 20 % of the responses implied that such coverage is provided by two or three courses. Rarely this coverage is provided by four courses as evident in the 13 % response presented in Fig. 11.12.

Figure 11.13 depicts the extent of overall coverage of construction health and safety in the surveyed curriculum expressed by the total number of lecturing hours. It is worth noting that the total number of lecturing hours for a typical 3-credit-hour course over an entire semester is about 45 h. From Fig. 11.13, it can be seen that in about 60 % of the responses, the coverage extent varies between 1 and 5 credit hours. A more intensive coverage (6–10, 11–15, or more than 30 lecture hours) is provided in only 7 % of the cases. More programs offer 21–25 credit hours coverage with a ratio that reaches 13 % of the responses received.

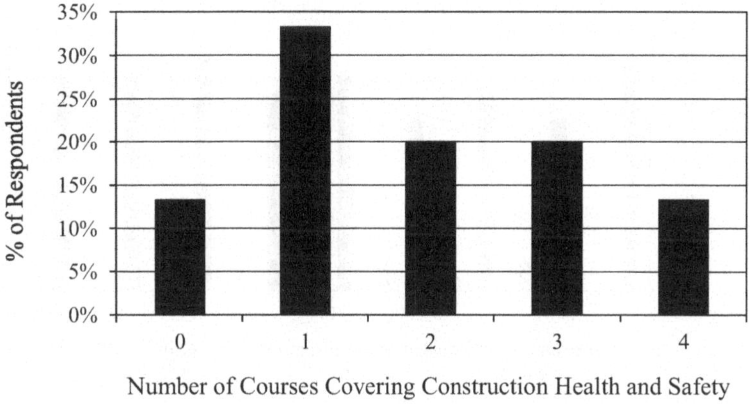

Fig. 11.12 Number of courses addressing construction health and safety topics in engineering curriculum

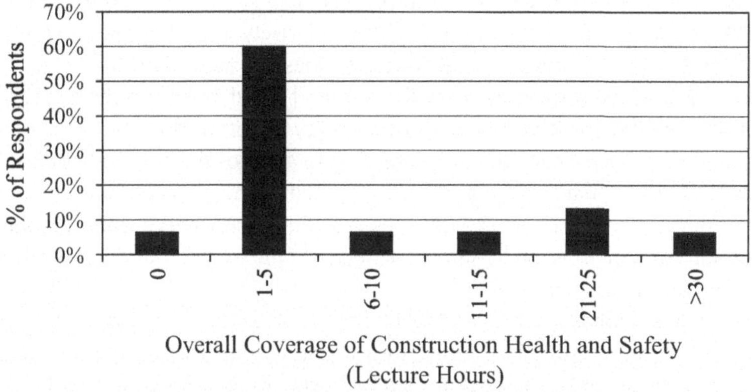

Fig. 11.13 Number of lecture hours for overall coverage of construction health and safety

11.6.2.2 Construction Health and Safety Topics and Standards

Nine basic construction health and safety topics are identified in the survey, namely legislations/regulations (T1); safety management (T2); risk identification (T3); safe practices and risk reduction (T4); hazard assessment (T5); hazard control and prevention (T6); mandatory procedures, training, records, and maintenance (T7); compliance, inspection, and penalties (T8); and incorporation of construction-safety-in-design courses (T9). The extent of coverage of these topics is presented in Fig. 11.14. It can be seen that all respondents indicated that topics T1 and T7 represent no more than 20 % of the covered construction health and safety topics, while about 74 % of the responses indicated that T2 and T4 represent no more than 20 % of the covered topics. However, 13 % of the responses indicated that T2 and

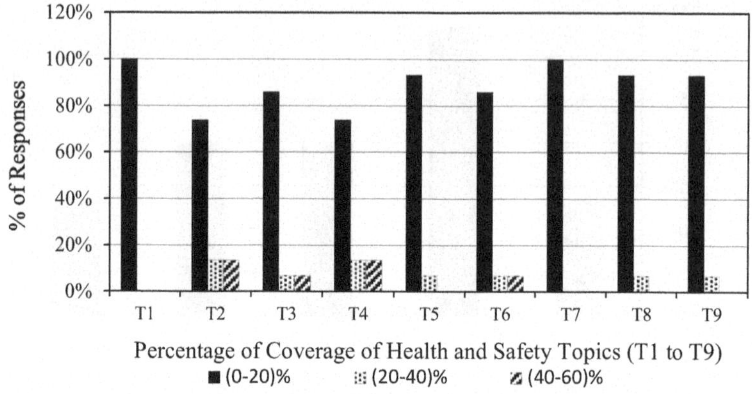

Fig. 11.14 Relative coverage of various construction health and safety topics

T4 may represent up to 40 % or even 60 % of the covered material. Furthermore, the majority of the responses (93 %) imply that T5, T8, and T9 represent no more than 20 % of the covered construction health and safety material, and the remaining responses (about 7 %) indicated that coverage does not exceed 40 %.

About 74 % of the responses show that topics T3 and T6 represent no more than 20 % of the covered construction health and safety material. But 7 % of the responses indicated that these topics may represent up to 40 % of the covered material, and another 7 % believe that these topics may represent as much as 60 % of the covered topics as shown in Fig. 11.14. As obvious, some programs give more emphasis to certain topics of construction health and safety, but the overall coverage of all topics is generally less than adequate.

The survey response shows that OSHA Standards and in-house developed materials are the most common data sources for teaching construction health and safety as evident by the 60 and 53 % response ratios, respectively, shown in Fig. 11.15. The country's local regulations come next as a main source of teaching material with a response ratio of 40 % (Fig. 11.15). The UK construction regulations and others were used the least in teaching the topics, with a response ratio of 7 % for each of them as presented in Fig. 11.15.

11.6.2.3 Certification Received upon Completion of Construction Health and Safety Courses

All programs offering construction health and safety courses indicated that no certification is granted to students upon successful completion of these courses.

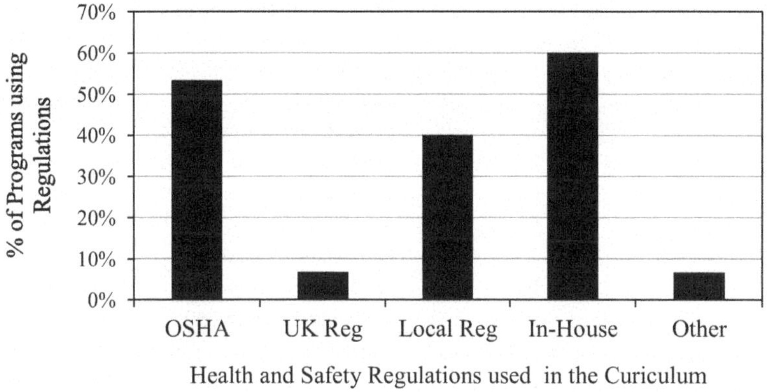

Fig. 11.15 Use of regulations as teaching materials for construction health and safety. *OSHA* Occupational Safety and Health Administration

11.6.3 Construction Health and Safety Courses from Program Websites

A web search was conducted for the CE and AE programs (other than those presented in Table 11.2) in the 15 selected MENA countries to investigate the courses that offer construction health and safety education partially or fully. The number of programs included in the web search is 42. It should be noted that some programs (such as those in Egypt and Cyprus) offer several courses that cover construction health and safety, while the majority of the programs have only one course (Table 11.3). It is also worthwhile mentioning that it was hard to find information related to course description in several MENA countries. This is because either no information was available on the websites or because it was written in a language that the authors were unable to interpret. As shown in Table 11.3, CE programs offer more construction health and safety education than AE programs, with only a few AE programs offering courses that cover the subject in their curriculum. The title and description, when available, of the vast majority of these courses indicate that offered topics focus only on safety of construction site. By contrast, very little attention is given to DfCS. Besides, about 50 % of the construction-health-and-safety-related courses, especially the CCs, are of 100, 200, or 300 level as evident by the posted course number. This implies that these courses are not covering DfCS since students design background, at such levels of courses, does not allow for incorporating relevant design topics in the safety courses.

11.6.4 Proposed Curriculum Modifications

Academic institutions in the MENA region should respond to the needs of their societies and must contribute with their scientific knowledge and experience to the development of better health and safety conditions in the construction sector. Here, the

Table 11.3 Programs that offer construction health and safety education (other than those presented in Table 11.2)

Country	University	Number of courses	
		CE program	AE program
Cyprus	A	2 (1 elective and 1 capstone)	–
	B	1 (core)	–
	C	1 (core)	–
	D	1 (core)	–
Egypt	A	3 (2 core and 1 elective)	–
	B	4 (2 core and 2 electives)	1 (core)
	C	4 (2 core and 2 electives)	–
	D	2 (core)	–
	E	–	2 (core)
Jordan	A	1 (core)	1 (core)
Kingdom of Saudi Arabia	A	2 (core)	3 (2 core and 1 capstone)
	B	1 (core)	–
	C	1 (core)	–
Kuwait	A	2 (core)	–
Lebanon	A	2 (core)	–
	B	1 (core)	–
Qatar	A	1 (core)	2 (electives)
Sudan	A	1 (core)	–
Turkey	A	1 (core)	–
	B	1 (core)	–
United Arab Emirates	A	1 (core)	–

CE civil engineering, *AE* architectural engineering

authors propose several actions that should be implemented to enhance health and safety education for engineering students and to overcome the current deficiency. One action is to incorporate a core course related to construction health and safety. Another action is to integrate DfCS topics in traditional design courses. The main objective of such enhancement is to provide students with the theory and essential elements in the field through which they can identify, select, evaluate, and compare safety provisions in their career.

11.6.4.1 Incorporating a Core Course in Construction Health and Safety

Construction health and safety education for engineering students in the MENA region could be improved by incorporating a core course in the curriculum. The suggested separate core course could be offered to both CE and AE students. A suggested course outline is shown in Table 11.4. The suggested course contains five main topics, which were mainly deduced based on the results of the conducted practitioners' survey. These topics cover safety analysis, legal issues, safety man-

Table 11.4 Suggested outline for a core course in construction safety

Topic	Lecture hours	Topic	Lecture hours
1. Safety analysis	5	4. Hazards on construction sites	
Case histories	2	Ergonomic hazards	2
Causes of accidents	2	Toxic and hazardous material	1.5
Cost of accidents	1	Confined spaces	1.5
		Fire	1
2. Legal Issues	5	Fall	1.5
Laws, regulations, and standards	1.5	Electrical	1
Liability	1.5	Cranes, hoists, elevators, and conveyors	2
Legal implications/ compensation	2	Heat stress, noise, and vibration	1.5
3. Safety management	10	Fieldtrip 2	
Risk assessment/hazard control	2		
Planning	2	5. Temporary work	15
Meetings	1.5	Site procedure, layout, and planning	2
Training	1.5	Excavations and trenches	3
Personal protective equipment	1	Stabilization and piling	2
Record keeping	1	Formwork design	5
Reporting	1	Falsework and scaffolding	2
Fieldtrip 1		Fieldtrip 3	

agement, hazards on construction sites, and temporary work. The course could be offered either as a 3-credit-hour course covering the five suggested topics or as a 2-credit-hour course covering only the first four topics. In terms of course offering within the study plan, it is recommended that students take the course before their internship and before taking design courses.

It should be noted that adding this course to the program may require reviewing the entire curriculum in order to accommodate the extra credit hours of the suggested core course. It may also involve a slow administrative process for adoption at some universities. Another important point to consider is the possibility that some of the current engineering educators may lack the necessary qualifications needed to efficiently deliver this course (Singh 1992). To remedy this, faculty members teaching construction health and safety may seek help from senior engineers and construction safety professionals to improve their knowledge in the field.

Several relevant books have been published that could serve as a textbook for this course including “Construction Safety” by Hinze (2006), “Principles of Construction Safety” by Holt (2001), “Construction Safety Management” by Levitt and Samelson (1993), “Construction Safety Handbook” by Davies and Tomasin (1996), “Construction Safety and Health Management” by Coble et al. (2000),

"Construction Safety Management and Engineering" by Hill (2003), "Construction Safety Manual" by Heberle (1998), among others. A recently published book titled "Construction Equipment and Methods-Planning, Innovation, Safety" by Bernold (2013) provides an overview of how to prevent accidents with used equipment in the construction industry. The book also presents methods for the safe design of trenches, formwork, and temporary structures.

Besides, construction safety videos could be utilized in the course as an integral part of the learning process to enhance students' interest and understanding. Several video clips have been produced by the Construction Safety Council (CSC) including those that demonstrate equipment inspections, operation and maintenance, hazards of working with and around electricity, proper use of personal protective equipment, elimination and control of lead exposures, excavation hazards, fall protection hazards, and oversight of a safe work practice (CSC 2013). Meanwhile, a large number of videos are available on YouTube including "Scaffolding Safety" by Stagehandspace Company, "Emergency Response" by Premier Corporate Training, "Confined Spaces" by WorkSafeBC, "Dangers of Hot Work" by Chemical Safety Board (CSB), "Personal Protective Equipment" by Caterpillar, "Heat Stress in the Construction Environment" by MARCOM, "Hook up to Safety" by Workplace Safety and Health (WSH) Council, among many others.

As noted by Banik (2005), the content and approach of teaching safety courses in different schools vary according to the background of the faculty and the available resources of the department. Meanwhile, Christian (1999) observed that teaching a class like safety is sometimes difficult as the students regard its regulatory nature as "boring." To avoid this, instructors need to conduct a course assessment and collect feedback from students on how to improve the course delivery method and course content. For example, a large number of the surveyed students in the work of Haupt (2003) felt that spending time on construction sites and observing safety practices during the execution of construction activities would be the most beneficial action to take in preparing them to contribute to improving construction worker safety. This is also in line with the suggestions made by the interviewed contractors in the work of Banik (2005). These contractors suggested that a safety class would be made more interesting if students visited the actual site of construction as many times as possible to show the practical nature of work and to identify the common safety hazards. They also suggested demonstration of some safety tools in the class to show how important they are for the specific safety hazards and how effectively they can be used to minimize that hazard. For these reasons, the authors suggest the inclusion of several fieldtrips as an essential component in this course. The number of fieldtrips could be either two or three, depending on the number of credit hours allocated for the course. The first fieldtrip should be related to safety management, the second should focus on hazard identification in a construction site, and the third should be related to safety of temporary work.

To increase students' exposure to the practical nature of the course, students could be divided into groups (each of three to four students) and requested to work on a term project as part of the course activities and requirements. Students should be asked to submit a written report and to do a presentation about their project in class. Topics that could be suitable for a term project include:

- Identify the most common hazards on a specific construction project and outline the corrective measures that can be taken to reduce those hazards.
- Analyze incident scenarios.
- Attend project safety meetings.
- Identify possible safety incentive schemes on a construction site.
- Evaluate the health and safety management system adopted by a construction company.
- Investigate the attitude of workers toward construction site health and safety.

Other possible term projects include:

- Review the impact of accidents on morale and productivity.
- Review the impact of accidents on scheduling and total cost.
- Determine direct and indirect costs of accidents.

In the delivery of the course, it is important that the instructor develops a link with the industry in order to coordinate and facilitate some of the activities of the course including fieldtrips and term projects undertaken by the students in the class. This is emphasized by Banik (2005), who suggested more involvement by the industry in the health and safety education process. Also, it is important to involve the industry in the process of course evaluation and also seek feedback on the level of health and safety education acquired by the trainees or the graduates. Meanwhile, guest lecturers could be chosen among relevant professionals to highlight aspects of construction health and safety that are not sufficiently covered by the program. Guest speakers could include a lawyer specialized in health and safety issues to cover legal aspects of accidents on construction sites, a professional hygienist to cover health issues, or others.

11.6.4.2 Incorporating DfCS in the Curriculum

The authors believe that it is crucial to introduce the DfCS principles to senior undergraduate students. The difference between the traditional design approach that targets end-user safety and DfCS that addresses safety of workers during construction and maintenance stages should be clearly outlined and explained to the students. Nonetheless, DfCS could be incorporated in most of the currently offered traditional design courses. Several suggested topics for possible inclusion in traditional design courses are listed in Table 11.5. The incorporation of the suggested topics requires careful planning and design of the outline of the target courses to ensure that DfCS issues are raised, discussed, and addressed in a smooth manner that does not negatively impact the flow of information that is originally targeting end-user safety. In fact, any DfCS-specific design requirements should be based upon theoretical and design backgrounds of the traditional end-user safety aspects.

Table 11.5 Suggested topics for teaching DfCS. (Gambatese et al. 1997; Behm 2005; Weinstein et al. 2005; Toole et al. 2006)

Suggested topic	Objectives	Course to include the topic
Specify primers, sealers, and other coatings that do not emit noxious fumes	Reduce noxious fumes	Engineering materials
Allow adequate clearance between structure and power lines	Avoid contact of operating cranes with overhead power lines	Construction management
Disconnection, reduction of voltage, or rerouting power lines before work begins	Reduce or eliminate electrical hazards when operating cranes	Construction management
Design of prefabricated units that can be built on the ground and erected in place	Reduce worker exposure to falls and being struck by falling objects	Reinforced concrete design
Ensure that spacing between top reinforcing steel bars does not exceed 0.15–0.20 m	Provide workers with a safe walking surface during construction	Reinforced concrete design
Design roof parapet walls 1.10 m high	Eliminate the need for additional fall protection guardrails	Reinforced concrete design/structural steel design
Design window sills to be at least 1 m above the floor level	Eliminate the need for fall protection during construction and future maintenance	Reinforced concrete design/structural steel design
Design permanent guardrails around skylights	Prevent workers from falling through skylight	Reinforced concrete design/structural steel design
Design floor perimeter beams and beams above floor and roof openings	Prevent workers from falling and hazard of falling objects	Reinforced concrete design/structural steel design
Design permanent anchorage points	Provide tie-off points for attaching fall protection devices during construction and future maintenance	Structural steel design
Design cable type lifeline system for tower structures	Allow workers to hook onto the structure and move up and down during future maintenance	Structural steel design
Shop-welded connections in place of bolted connections	Avoid dangerous positions for welders during construction	Structural steel design
Avoid field welding of connections	Reduce site fire hazard and to avoid dangerous positions for welders during construction	Structural steel design
Carefully locate and detail field connections	Ensure safe access to connections during construction	Structural steel design
Locate column splices at reasonable heights above floor level	Prevent workers from falling and hazard of falling objects	Structural steel design

However, this may require modifications in the already existing courses to accommodate the suggested additional construction safety topics.

Enhancement of DfCS could also be achieved by making it a requirement in capstone courses like GPs, where students demonstrate how the concept is applied during their project design stage. In some universities (like the UAE University), all engineering students undertaking GPs are required to attend supporting lectures related to project management and cost estimation. One or two lectures could be devoted to DfCS where speakers specialized in this area are invited to share their experience with senior students. A major challenge in the incorporation of DfCS in design courses is related to the ability and willingness of the instructor to do so.

Material to support DfCS has been developed by several organizations. For example, the Construction Industry Institute (CII) developed an interactive software by which designers or students can address construction worker safety in their designs (CII 2013). The software contains a database of over 400 design suggestions or "best practices" that can be implemented during the project planning and design phases. Design suggestions not only focus on construction phase safety but can also be applied to the start-up, maintenance, and decommissioning phases of a project. The UK Health and Safety Executive (UK HSE) developed several documents that aid designers in designing for safety (UK HSE 2013). The European Network Education and Training in Occupational Safety and Health (ENETOSH) was established to find and to promote ways of improving the quality of education and training by intensifying and systematizing the exchange of knowledge and experience regarding mainstreaming occupational health and safety into education and training in Europe (ENETOSH 2013). Safety professionals in Australia developed a tool called Construction Hazard Assessment Implication Review (CHAIR) (WorkCover NSW 2001). The goal of this tool is to identify risks in a design at the early stages. The Australian Safety and Compensation Council (ASCC) also developed a document to support elimination of hazards at the design stage as part of an effort devoted to achieve one of the priorities set by the National Occupational Health and Safety Strategy 2002–2012 (ASCC 2006).

11.6.4.3 Emphasizing Health and Safety During Internship

Recall that about 14% of the surveyed practitioners indicated that they received some construction health and safety education during their internships. Construction health and safety education of engineering students could further be enhanced if health and safety become an essential part of the internship. To emphasize this, students (trainees) should be requested to write a section in their final report about all the health and safety aspects they encountered during their training period. These aspects should cover those related to health and safety management, hazard identification, and corrective measures undertaken. Students should also be encouraged to highlight how the design could have been made safer.

11.7 Conclusion

The current study assessed the extent to which engineering students at universities in the MENA region receive construction health and safety education during their undergraduate study. A survey form was designed and distributed to engineering practitioners in the MENA region. The survey was implemented to assess the need for including construction health and safety education in the engineering curriculum. The study showed that only 30 % of the surveyed practitioners received construction health and safety education during their university study. Around 50 % of them indicated that they received this knowledge during internship. Fewer respondents indicated that they received such knowledge through core and/or elective courses. The study results revealed that the majority of the surveyed practitioners believe that it is important to increase the level of construction health and safety education offered at the undergraduate level. The study suggested a total of 15 health and safety aspects to be acquired during undergraduate engineering education. Results also show that 95 % of the surveyed practitioners believe that all suggested health and safety aspects are important to very important. Another survey was conducted to investigate the extent to which construction health and safety is currently integrated in engineering programs in the MENA region. Targeted departments included CE, AE, construction management, and construction engineering. The responses received indicated that about 47 % of these programs are accredited by ABET, 20 % by other accreditation bodies, and the remaining 33 % are not accredited. The study revealed that the majority of the programs do not have a separate course devoted for construction health and safety education. About 40 % of the surveyed programs indicated that construction health and safety topics are included as a part of elective courses. On the other hand, most of the surveyed programs have construction health and safety education either as an integral part of a core course or during the internship. It can be also concluded that construction health and safety is included in less than 15 % of the GP courses offered by the surveyed programs. This indicates a major deficiency in inclusion of DfCS in the curriculum. The study showed also that OSHA Standards and in-house developed materials are the most common sources for teaching construction health and safety, followed by the country's local regulations. Finally, the study suggested several modifications to enhance the current level of construction health and safety education in the MENA region. These enhancements include incorporating a core course in the curriculum that is devoted to construction health and safety. The main topics of this course were also presented. In addition, several topics of DfCS were suggested to be included in traditional design courses in order to address this crucial issue in design courses.

Acknowledgement The authors would like to thank the CE students at the UAE University who participated in the distribution and collection of the practitioners' survey form. Thanks are also due to Eng. M. Shaher Maraqa for the distribution of the practitioners' survey form in Saudi Arabia, Eng. Alaa Tomoum for the distribution of the practitioners' survey form in Kuwait, and Eng. Adella Samaha for the distribution of the practitioners' survey form in Qatar. Thanks are also extended to all engineering departments in the MENA region which provided feedback on the extent of incorporation of safety education in their programs. The authors are grateful to Anas

Issa and Suliman Gargoum for their effort in the data entry. Thanks are also due to Tim Kirk and Elizabeth Whitehouse at the Writing Center at the UAE University for editing an earlier version of the manuscript.

References

ABET. (2012). 2013–2014 criteria for accrediting engineering programs. The accreditation board for engineering and technology. Engineering Accreditation Commission EAC, Baltimore, MD.

ACCE. (2012). Standards and criteria for accreditation of postsecondary construction education degree programs, Document 103. American Council for Construction Education, San Antonio, Texas: http://acce-hq.org/documents/DOCUMENT103REVISIONS0712_000.pdf.

Alhemood, A. (2004). Development of safety engineering curricula in the gulf cooperative countries to meet local manufacturing needs. *Human Factors and Ergonomics in Manufacturing & Service Industries, 14*(3), 321–326.

Al-Humaidi, H. M., & Tan, F. H. (2010). Construction safety in Kuwait. *Journal of Performance of Constructed Facilities, 24,* 70–77.

Al-Mufti, M. A. (1999). Continuous enhancement of health and safety awareness in undergraduate civil engineering courses, in proceedings of the Second International Conference of CIB W99, Implementation of safety and health on construction sites, (Eds., Singh, Hinze and Coble), Balkema, Rotterdam, pp. 277–282.

Al Palumbo, C. C. M. (2010). Safety in design: Enhancing construction safety by implementing safety in the design phase. *CM eJournal.* http://cmaanet.org/files/files/SafetyInDesign060110.pdf.

Arbejdstilsynet. (1996). Exposure limit value for substances and materials. Arbejdstilsynet, Instruction No. 3.1.0.2, Copenhagen, Denmark.

ASCC. (2006). Safe design for engineering students: An educational resources for undergraduate engineering students. The Australian Safety and Compensation Council (ASCC). www.architecture.com.au.

Banik, G. C. (2005). *Content and pedagogy of teaching construction safety at the undergraduate level.* ASC proceedings of the 41st annual conference. In T. Sulbaran, C. Marcum, & M. Brana (Eds.), April 6–9, University of Cincinnati, Cincinnati, Ohio. http://ascpro0.ascweb.org/archives/cd/2005/2005pro/2005/CEUE05_1300_Banik05.htm.

Barss, P., Addley, K., Grivna, M., Stanculescu, C., & Abu-Zidan, F. (2009). Occupational injury in the United Arab Emirates: Epidemiology and prevention. *Occupational Medicine, 59,* 493–498.

Behm, M. (2005). Linking construction fatalities to the design for construction safety concept. *Safety Science, 43*(8), 589–611.

Bernold, L. E. (2013). *Construction equipment and methods: Planning, innovation safety* (p. 594). Hoboken, NJ: Wiley (ISBN 0470169865).

Breslin, P. (2004). Performance versus prescriptive approaches to OHS in the Victorian construction industry. *Journal of Occupational Health and Safety- Australia and New Zealand, 20*(6), 563–571.

Bryan, L. A. (1999). Educating engineers on safety. *Journal of Management Engineering, 15*(2), 30–33.

Carpenter, J., Williams, P., & Smith, N. C. (2001). Identification and management of risk in undergraduate construction courses. Norwich: Health and Safety Executive, Report No. 392/2001.

Carpenter, J., Williams, P., & Smith, N. C. (2004). Identification and management of risk in undergraduate construction courses. Norwich: Health and Safety Executive, Report No. 275/2004.

CEAB. (2011). Accrediting criteria and procedures. Canadian Engineering Accreditation Board, Canadian Council of Professional Engineers (Engineers Canada), Ottawa, Ontario.

Choudhry, R. M., & Fang, D. (2008). Why operatives engage in unsafe work behavior: Investigating factors on construction sites. *Safety Science, 46,* 566–584.

Choudhry, M. R., Fang, D. P., & Ahmed, S. M. (2008). Safety management in construction: Best practices in Hong Kong. *Journal of. Professional. Issues in Engineering. Education and Practice, 134,* 20–32.

Choudhry, R. M., Fang, D., & Lingard, H. (2009). Measuring safety climate of a construction company. *Journal of Construction Engineering and Management, 135,* 890–899.

Christensen, W., & Manuele, F. (1999). *Safety through design*. Itasca: National Safety Council Press.

Christian, J. (1999). Teaching health and safety to construction engineers, in proceedings of the Second International Conference of CIB W99, Implementation of Safety and Health on Construction Sites, (Eds., Singh, Hinze and Coble), Balkema, Rotterdam, pp. 251–256.

CII. (2013). Design for construction safety toolbox, version 2.0. Construction Industry Institute. http://www.construction-institute.org/.

Coble, R. J., Hinze, J. W., McDermott, M. J., & Elliott, B. R. (1999). "Colleges' emphasis on construction safety", in proceedings of the Second International Conference of CIB W99, Implementation of Safety and Health on Construction Sites, (Eds., Singh, Hinze and Coble), Balkema, Rotterdam, pp. 257–263.

Coble, R. J., Hinze, J., Haupt, T. C. (2000). *Construction safety and health management* (p. 234) New Jersey: Prentice Hall (ISBN 0130871737).

Cortes, J. M., Pellicer, E., & Catala, J. (2012). Integration of occupational risk prevention courses in engineering degrees: Delphi study. *Journal of Professional Issues in Engineering Education & Practice, 138*(1), 31–36.

CSC. (2013). Safety videos. Construction safety council. http://store.buildsafe.org/.

Davies, V. J., & Tomasin, K. (1996). *Construction safety handbook* (2nd ed., p. 320), London: Thomas Telford Publishing (ISBN 072772519X).

Driscoll, T. R., Harrison, J. E., Bradley, C., & Newson, R. S. (2008). The role of design issues in work-related fatal injury in Australia. *Journal of Safety Research, 39,* 209–214.

ENETOSH. (2013). The European network education and training in occupational safety and health. http://www.enetosh.net/webcom/show_article.php/_c-29/_lkm-9/i.html.

ENR. (2001). *Construction leads in deaths despite lower fatality rate*. USA: Engineering News Record.

European Foundation. (1991). *From drawing board to building site, European foundation for the improvement of living and working conditions*. London: HMSO Books.

Ferjencik, M. (2007). Best starting point to comprehensive process safety education. *Process Safety Progress, 26*(3), 195–202.

Gambatese, J. A. (2003). Safety emphasis in university engineering and construction programs. *Construction safety education and training—A global perspective*, (Special Issue article), 115.

Gambatese, J., & Hinze, J. (1999). Addressing construction worker safety in the design phase: Designing for construction worker safety. *Automation in Construction, 8*(6), 643–649

Gambatese, J., Hinze, J., & Haas, C. (1997). Tool to design for construction worker safety. *Journal of Architectural Engineering, 3*(1), 32–41.

Gambatese, J., Toole, M., Giles, B. (2008). *The future of design for construction safety*. 18th Annual Construction Safety Conference, February 12, Rosemont, IL.

Gibb, A. G. (2004). Designing for safety and health in construction: A European/UK perspective. Designing for safety and health in construction. In S. Hecker, J. Gambatese, & M. Weinstein (Eds.), UO Press, Eugene, Ore., 44–56.

Gibb, A., Haslam, R., Hide, S., & Gyi, D. (2004). The role of design in accident causality. Designing for safety and health in construction: Proc., research and practice symposium. In S. Hecker, J. Gambatese, & M. Weinstein (Eds.), UO Press, Eugene, Ore.

Habib, R. (2007). *Overview of the occupational safety and health situation in the Arab region*. International Labour Office.

Haupt, T. C. (2003). The state of construction safety and health education at higher education institutions in the Western Cape of South Africa. Special Issue article in: Construction safety education and training—A global perspective, pp. 1–10.

Heberle, D. (1998). *Construction safety manual* (p. 508). New York: McGraw-Hill Inc. (ISBN 0-07-034454).

Hill, D. C. (2003). *Construction safety management and engineering* (p. 717). Des Plaines, IL: American Society of Safety Engineers (ISBN 1885581467).

Hill, R. H., & Nelson, D. A. (2005). Strengthening safety education of chemistry undergraduates. *Chemcal Health and Safety, 12*(6), 19–23.

Hinze, J. W. (2006). *Construction safety* (2nd ed.), Upper Saddle River: Prentice-Hall Inc.

Hinze, J., & Figone, L. (1988). Subcontractor safety as influenced by general contractors on small and medium sized projects. CII Rep. No. 2, Dept. of Civil Engineering, Univ. of Washington, Seattle.

Hinze, J., & Gambatese, J. (1996). Addressing construction worker safety in the project design. CII Research Rep. No. 101–11. Univ. of Texas at Austin, Austin, Tex.

Hinze, J., & Talley, D. (1988). Subcontractor safety as influenced by general contractors on large projects. CII Research Rep. No. 1, Univ. of Texas at Austin, Austin, Tex.

Holt, A. (2001). *Principles of construction safety* (p. 278). Oxford: Blackwell Science Ltd (ISBN 0-632-05682-7).

Karjalainen, A. (2007). A statistical portrait of health and safety in the construction industry. Eurostat, European agency for safety and health at work, http://osha.europa.eu/publications/magazine/7/index_2.htm.

Kartam, N. A., & Bouz, R. G. (1998). Fatalities and injuries in the Kuwaiti construction industry. *Accident Analysis and Prevention, 30,* 805–814.

Kauffman, D. (1987). Health, safety and loss control topics in the senior design courses. *Plant/Operation Progress, 6*(2), 73–76.

Kavianian, H. R., Meshkati, N., Wentz, C. A., & Rao, J. K. (1993). Should engineering schools address occupational and environmental safety and health issues? *Professional Safety, 38*(6), 48–49.

Lemkowitz, S. M. (1992). A unique program for integrating health, safety, environment and social aspects into undergraduate chemical engineering education. *Plant/Operations Progress, 11*(3), 140–150.

Levitt, R. E., & Samelson, N. M. (1993). *Construction safety management* (2nd ed., p. 304). New York: Wiley (ISBN 0471599336).

Levitzky, J. J. (1988). Safety, health, and loss prevention in the undergraduate chemical engineering curriculum. *Plant/Operations Progress, 7*(3), 190–194.

Liska, R. W., Goodloe, D., & Sen, R. (1993). Zero accident techniques. Construction Industry Institute, Source Document No. 86, The Univ. of Texas at Austin, Austin, Tex.

Manuele, F. A. (1997). *On the practice of safety*. New York: Wiley.

Maraqa, M. A., & Mohamed, A. O. (2013). Key drivers for successful safety management system of construction activities in Abu Dhabi Emirate. *International Journal of Advanced Fire, Explosive, Environment Safety and Disaster Management, 1*(1), 1–17.

McGrath, E. (2009). The construction industry's own health, safety, and welfare initiative, 7th Annual GCC Occupational Health and Safety Conference, Dubai, UAE.

Moccaldi, A., Trevisi, R., & Orlando, C. (2005). ISPESL radon programme: A challenge in the academic programme of a safety and security engineering course. *Safety and Security of Engineering, 82,* 143–150.

Pellicer, E., & Molenaar, K. R. (2009). Discussion of 'Developing a model of construction safety culture'. *Journal of Management in Engineering, 25*(1), 44–47.

Pellicer, E., Seron, J., Catala, J., & Jorda, L. (2003). Proposal of a new academic frame for the civil engineering education in construction safety and health. Proc., Int. Conf. on Engineering Education, July 21–25, Valencia, Spain.

Petersen, A. K., Reynolds, J. H., & Ng, L. W. T. (2008). The attitude of civil engineering students towards health and safety risk management: A case study. *European Journal of Engineering Education, 33*(5–6), 499–510.

Phillis, T. W., & Wheway, R. T. (1991). Introducing occupational health and safety issues to undergraduate students: The Wollonong experience. Proc., MECH '91 Int. Mechanical Engineering Congress and Exhibition, The Institution of Engineers, Sidney, Australia.

Reese, C. D., & Eidson, J. V. (2006). *Handbook of OSHA construction safety and health* (2nd ed.), Boca Raton Florida: Taylor & Francis.

Rowlinson, S. (2003). *Hong Kong construction safety management and the law* (2nd ed.), Hong Kong: Sweet and Maxwell Asia Publisher.

Rubio, M. C., Menendez, A., Rubio, J. C., & Martínez, G. (2005).Obligations and responsibilities of civil engineers for the prevention of labour risks: References to European regulations. *Journal of Professional Issues in Engineering Education Practice, 131*(1), 70–75.

Sawacha, E., Naoum, S., & Fong, D. (1999). Factors affecting safety performance on construction sites. *International Journal of Project Management, 17,* 309–315.

Singh, A. (1992). Experienced-based issues in construction education. *Journal of Professional Issues in Engineering Education and Practice, 3*(2), 388–402.

Smallwood, J. J. (1996). The influence of designers on occupational safety and health. Proc., 1st International Conf. of CIB Working Commission W99, Implementation of Safety and Health on Construction Sites, Lisbon, Portugal, 203–213.

Smallwood, J. (2010). Construction management health and safety (H & S) course content: Towards the optimum. Proceedings of the 18th CIB World Building Congress W099: Safety and Health in Construction. In P. Barrett, D. Amaratunga, R. Haigh, K. Keraminiyage, & C. Pathirage (eds.), 1013 May, Salford, UK.

Smith, G. R., & Arnold, T. M. (1999). Safety education expectations for construction engineering and management students, in proceedings of the Second International Conference of CIB W99, Implementation of Safety and Health on Construction Sites, (Eds., Singh, Hinze and Coble), Balkema, Rotterdam, pp. 265–272.

Somavia, J. (2005). Decent work, safe work. Programme on safety and health at work and the Environment (SafeWork), International Labour Organization (ILO).

Suckarieh, G., & Diamantes, J. (1995). *Educating construction management students in safety. Safety and Health in Construction, CIB Publication 209,* Published by CIB Working Commission 99, pp. 154–164.

Suraji, A., Duff, A. R., & Peckitt, S. J. (2001). Development of causal model of construction accident causation. *Journal of Construction Engineering and Management, 127*(4), 337–344.

Szymberski, R. (1997). Construction project safety planning. *Technical Association of the Pulp and Paper Industry (TAPPI) Journal, 80*(11), 69–74.

Teo, E. A. L., Ling, F. Y. Y, & Chong, A. F. W. (2005). Framework for project managers to manage construction safety. *International Journal of Project Management, 23,* 329–341.

Toole, T. M. (2004). Rethinking designers' roles in construction safety. Designing for safety and health in construction: Proc., Research and Practice Symposium, S. Hecker, J. Gambatese, and M. Weinstein (eds.), UO Press, Eugene, Ore.

Toole, M., Hervol, M, & Hallowell, M. (2006). Designing for construction safety. Modern steel construction. http://www.modernsteel.com/uploads/issues/june_2006/30754_safety_web.pdf.

UK CDM. (2007). The construction (design and management) regulations 2007 No. 320. www.legislation.gov.uk/uksi/2007/320/.

UK HSE. (2013). Health and safety in the construction industry. UK health and safety executive. http://www.hse.gov.uk/construction/.

US Department of Labor. (2001). Census of fatal occupational injuries. Bureau of Labor Statistics.

Vincent, J. H. (2005). Graduate education in occupational hygiene: A rational framework. *The Annals of Occupational Hygiene, 49*(8), 649–659.

Weinstein, M., Gambatese, J., & Hecker, S. (2005). Can design improve construction safety?: Assessing the impact of a collaborative safety-in-design process. *Journal of Construction Engineering and Management, 131*(10), 1125–1134.

Wong, K. W., Chan, P. C., & Lo, K. K. (1999). *Factors affecting the safety performance of contractors and construction sites. Implementation of safety and health on construction sites.* Rotterdam: Balkema Publishers.

WorkCover NSW. (2001). CHAIR: Safety in design tool. WorkCover New South Wales. www.workcover.nsw.gov.au.

Zarges, T., & Giles, B. (2008). Prevention through design (PtD). *Journal of Safety Research, 39,* 123126.

Munjed A. Maraqa is an associate professor and Head of the Traffic Safety Division at the Roadway, Transportation and Traffic Safety Research Center at the United Arab Emirates (UAE) University. His research covers a broad range of issues related to modeling environmental systems, water quality, waste management, and traffic safety. Between 2008–2012, Maraqa acted as an advisor to Al Ain Municipality on the development of an Environment, Health, and Safety (EHS) management system for the Building and Construction Sector in Abu Dhabi emirate. Maraqa holds BSc and MSc degrees in civil engineering from the University of Jordan, and a PhD in environmental engineering from Michigan State University.

Amr M. I. Sweedan is an associate professor of structural engineering and advisor to the Provost at the UAE University. He is a registered professional engineer in the province of Ontario, Canada (P.Eng.). Sweedan earned his PhD in structural dynamics from the University of Western Ontario, Canada. He received his MSc and BSc degrees in structural engineering from Ain Shams University, Cairo, Egypt. Before joining the UAE University, Sweedan held the position of a senior engineer at the head office of Rowan Williams Davies & Irwin Inc. (RWDI), Consulting Engineers and Scientists, in Guelph, ON, Canada. Sweedan's areas of research interest include modeling and testing of solid, cellular and castellated steel elements; strengthening of steel structures using fiber-reinforced polymer (FRP) laminates; evaluation of structural response to seismic and wind loading; and design and behavior of shell structures under hydrodynamic and hydrostatic pressures.

Essam Zaneldin earned his PhD in 2000 from the University of Waterloo in the area of construction engineering and management. Zaneldin is a professional engineer currently working as an associate professor of construction engineering and management at the Department of Civil and Environmental Engineering and the head of the Engineering Requirements Unit at the UAE University. In addition to his experience in the academia, Zaneldin has more than 13 years of work experience in areas related to design, construction supervision, and project management of mega size projects in North America and the Middle East. He has authored and coauthored several journal and conference publications in topics related to engineering education, design coordination, change management, site layout planning, constructability, claims and disputes, and simulation of design and construction operations.

Chapter 12
Environmental Engineering Education in the UAE

Mohamed M. Mohamed and Munjed A. Maraqa

12.1 Introduction

Environmental engineering education (E3) is an essential supporting component of environmental protection in any country. The main goal of E3 is to provide an appropriate technical background, but with a vision rather than just skills aiming at producing graduates capable of making positive amendments to the environment for the benefit of mankind. To achieve such a goal, E3 must hold an interdisciplinary flavor to cover various disciplines such as biology, geology, ecology, public health, management, and economics, in addition to the common topics related to water, air, and soil (Tansel 2008). Therefore, assessments of E3 aim achievement must be conducted continuously (e.g., Yaghubi et al. 2006). In general, the E3 programs have the same aim of sustainability (Gutierrez-Martin and Huttenhain 2003). Once viewed as subsets of civil or chemical engineering, they have also mushroomed to include all aspects of human and terrestrial environments. In addition, there are movements to enhance engineering disciplines with an environmental component (Gutierrez-Martin and Huttenhain 2003).

Continuous improvement in the educational systems is a necessity. Many countries are adapting this philosophy to meet the rapid developments in science and technology (Chang and Chiu 2005). Different previous studies focused on comparing E3 in different countries. For example, Ujang et al. (2004) compared E3 among industrialized, fast-industrializing, and slow-industrialized countries. In Japan, for example, students at the University of Tokyo enrolled for the undergraduate programs in sanitary and environmental engineering are expected to spend 1.5 years studying general and fundamental subjects in arts and sciences, before moving to more specialized subjects offered by the Department of Urban Engineering (Mino 2000). In this university, sanitary engineering subjects form a major part of the

M. M. Mohamed (✉) · M. A. Maraqa
Civil and Environmental Engineering Department, United Arab Emirates University, P.O. Box 15551, Al-Ain, United Arab Emirates
e-mail: m.mohamed@uaeu.ac.ae

M. Abdulwahed et al. (eds.), *Advances in Engineering Education in the Middle East and North Africa*, DOI 10.1007/978-3-319-15323-0_12

curriculum, while at Hokkaido University applied chemistry and chemical engineering is the primary focus, yet the main concern within the environmental engineering (EnE) curriculum at Kyoto University is toxicology and radioactive waste management. Mino (2000) concluded that several environmental topics are lacking in all three universities including environmental impact assessment (EIA), environmental economics, and ecology. The basic components of EnE programs at universities in Canada include science (e.g., chemistry, math, biology), applied science (e.g., environmental chemistry, thermodynamics), engineering (e.g., engineering design, hydrology, etc.), EnE (e.g., principles of EnE, wastewater treatment, etc.), public and environmental health (e.g., water quality, environmental protection), and environmental law and policy (e.g., EIA, resource management) (Smith and Biswas 2001; Ghaffari and Talebbeydokhti 2013).

E3 is mostly offered as part of the traditional undergraduate civil engineering program in Europe (Alha et al. 2000). Recently, many educational institutions have started offering full EnE degrees or specialized postgraduate degrees (Alha et al. 2000). In Germany, for example, the system of engineering education is defined to have two types of programs, namely scientific-oriented programs at university level and application-oriented programs at "school of engineers" level (Schmitt and Wilderer 1996).

E3 in North American universities has a similar trend of departmental evolution (Bishop 2000). The criteria of the Accreditation Board for Engineering and Technology (ABET) ensures that graduates of accredited engineering programs are adequately qualified to enter professional work life. EnE is now accredited by ABET, and it is a well-recognized degree program in the USA. ABET goals are to guarantee quality graduates from the engineering programs, to adequately provide graduates with skills, and to motivate improvements in the engineering education. The 2010–2011 ABET EnE program criteria related to curriculum and faculty are listed below (ABET; www.abet.org).

Curriculum The program must demonstrate that graduates have proficiency in mathematics through differential equations, probability, and statistics; calculus-based physics; general chemistry; an earth science, for example, geology, meteorology, soil science, relevant to the program of study; a biological science, for example, microbiology, aquatic biology, toxicology, relevant to the program of study; fluid mechanics relevant to the program of study; introductory-level knowledge of environmental issues associated with air, land, and water systems and associated environmental-health impacts; an ability to conduct laboratory experiments and to critically analyze and interpret data in more than one major EnE focus areas, for example, air, water, land, environmental health; an ability to perform engineering design by means of design experiences integrated throughout the professional component of the curriculum; proficiency in advanced principles and practice relevant to the program objectives; and understanding of concepts of professional practice and the roles and responsibilities of public institutions and private organizations pertaining to EnE.

Faculty The program must demonstrate that a majority of those faculty teaching courses which are primarily design in content are qualified to teach the subject matter by virtue of professional licensure or by education and equivalent design experience.

The main focus of E3 at universities in developing countries, on the other hand, is on the basic infrastructure, water supply, wastewater, and solid-waste management. Ujang et al. (2004) also indicated that there are several universities in Europe and in the USA that have been active in E3 research related to developing countries. Examples of such universities are the Universities of Newcastle, Loughborough, Leeds, and Surrey in England, the International Institute for Infrastructural, Hydraulic and Environmental Engineering (IHE)-Delft in the Netherlands and the Department for Water and Sanitation in Developing Countries (SANDEC) at the Swiss Federal Institute for Environmental Science and Technology in Switzerland (Ghaffari and Talebbeydokhti 2013). On the other hand, several universities in developing countries initiated E3 collaboration programs with universities from Europe, the USA, Australia, and Japan (Ghaffari and Talebbeydokhti 2013). For instance, Universiti Teknologi Malaysia (UTM), Malaysia, has been actively participating with University of Newcastle, England, and the Technical University of Denmark (Ghaffari and Talebbeydokhti 2013). Additionally, in the year 2000, the Malaysian University Consortium on Environment and Development (MUCED) was established, funded by the Danish government, to undertake collaborative effort on E3 at M.Sc. level (Ghaffari and Talebbeydokhti 2013). Such collaboration will indeed contribute to the advances in E3 worldwide.

Gutiérrez-Martín and Dahab (1998) discussed the concepts of sustainability and pollution prevention and their roles in E3. They argued that EnE science and education must be reoriented to focus primarily on pollution prevention technologies as a mechanism for attaining the goal of sustainability (Jassim and Coskuner 2007). Although it is acknowledged that traditional pollution control will remain as an integral part of environmental science and engineering education, the shift toward pollution prevention should be completed in order to achieve the goal of sustainability. Gutiérrez-Martín and Dahab (1998) presented two case studies at the University of Nebraska-Lincoln (USA) and at the Universidad Politécnica de Madrid (Spain) where efforts were made to reorient E3 to promote the concept of sustainability as the primary goal of environmental management.

There seems to be no lack of expertise in either environmental or conventional technology; what is limited is a holistic approach to E3, one that incorporates the environment into the mainstream of technological application (Gutierrez-Martin and Huttenhain 2003). Environmental engineers need skills of integration, communication, and conceptualization in addition to a traditional engineering background (Jassim and Coskuner 2007). Such a complex knowledge and skill mix is best promoted by a specially designed program. The main challenge, therefore, is to integrate these new technologies into E3 (Lemkowitz et al. 1996). An EnE degree program must impart the required knowledge and skill base and promote the intellectual versatility to develop appropriate technical solutions to environmental problems within existing ethical, ecological, and legal frameworks (Jassim and Coskuner 2007). There is also an increased emphasis on minimization and reuse and on some pedagogical methodologies, which are particularly suitable (e.g., project-based learning) (Alha et al. 2000). Gutierrez-Martin and Huttenhain (2003) reviewed efforts to promote the concepts of sustainability in eco-management. They concluded that environmental training is an efficient means to improve graduates' capability of dealing with the multidimensional aspects of the environment.

12.2 Importance of E3 in the UAE

The UAE is located within the arid zone of the southeastern part of the Arabian Peninsula, with a coastline of 1318 km in length (Brook and Dawoud 2005). The annual rate of rainfall in the country averages 100 mm (Ministry of Economics 2007), while the rate of evaporation generally exceeds 2000 mm/year (Jones and Marrei 1982). The country has developed rapidly since its establishment in 1971. This development, which is expected to continue in the future, requires careful consideration of the environment on both the construction and institutional levels. The government of the UAE realizes the importance of the environment to the people of the country and, therefore, shows clear devotion to the environmental awareness and obligation toward the natural world. For example, in March 1999 a law was passed aimed at reducing air pollution. The regulations limit excessive use of harmful gases. Other regulations are also under discussion for controlling the use of leaded fuel which produces harmful emissions. Meanwhile, the UAE enforces strict laws governing the use of chemical insecticides in agriculture to protect public health and reduce the negative impacts on the environment. It has banned the importation of many chemical insecticides, permitting only the importation of those products which are already licensed for agriculture use in the USA, Canada, Japan, and the European Union (EU). Another example is the UAE's participation in the Global Learning and Observation to Benefit the Environment (GLOBE) agreement in June 1999.

Over the years, a number of government organizations have been established in the UAE with the role of studying and protecting the environment. In addition, there are other regional departments which have their own programs for environment protection, wildlife protection, and, most importantly, for increasing public awareness. An indubitable demonstration of this devotion is the words of Late Sheikh Zayed bin Sultan Al Nahyan in his speech on the occasion of the UAE's first Environment Day, 4 February 1998. He said, "With God's will, we shall continue to work to protect our environment and our wildlife, as did our forefathers before us. It is a duty—and, if we fail, our children, rightly, will reproach us for squandering an essential part of their inheritance and of our heritage." Late Sheikh Zayed bin Sultan Al Nahyan also received the Gold Panda award in 1999 from the World Wide Fund (for which he was the first sitting head of state to be so honored) for his services to the environment. The institutional structure, supported by legislation, is now in existence in the UAE, although further legislation is still required. However, in the authors' opinion, the supporting role of environmental education has not yet received the same attention.

12.3 Environmental Challenges in the UAE

The rapid social and economic development in the UAE places a growing pressure on its available resources. According to the Living Planet Report 2006 (WWF 2006), the UAE has the highest ecological footprint of 12 global hectares/person, while the world average is 1.8 global hectares/person. The ecological footprint is an index of global area per head from which a person consumes natural resources.

The UAE, similar to the rest of the Gulf countries, is increasing its oil and gas export capacities to meet the rising global demand. This requires more navigational activities in the Gulf which might increase the incidents of oil spillage and hence puts the marine environment at a greater risk. The estimated oil pollution in the Gulf is about 3 % of the global total or nearly 50 times the average for a marine environment of its size (Golob and Bruss 1984). Pollution of the Arabian Gulf is further complicated due to the physical characteristics and semi-enclosed nature of the Gulf (Smith et al. 2007), which provides ideal conditions for entrapment of pollutants. Aside from oil spillage, several pollutants are induced into the Gulf including those resulting from offshore exploration processes, ballast water discharge, reject brine discharge, dredging activities, and stormwater runoff. Meanwhile, the UAE is developing mega coastal construction projects and coastal industrial facilities. The consequences of such developments on the ecosystem could be severe and long lasting (Van Lavieren et al. 2011). Marine pollution is further aggravated by the fact that most urban developments in the country occurs along the coastal zone. The frequent occurrence of harmful algae blooms on the coastal shores of the UAE in the past few years could be due to enrichment of marine water with nutrients that could have resulted from illegal discharge of untreated sewage or from discharge of ballast water.

Another environmental concern that faces the UAE is related to the substantial increase in water demand as a result of rapid population growth coupled with large foreign population influx. In the past, groundwater was a major source for water supply, but overexploitation of this source caused loss of quantity and deterioration of its quality in some locations (Rizk et al. 1997). Nowadays, desalination is used extensively to cover the domestic-water-supply requirement. The pressure on provision of water supply in the country is further amplified by a high per capita water-consumption rate as compared to that in other countries (Conference Board of Canada 2008). As indicated by Al-Homoud (2003), the three dominant factors that affect water resources in the UAE are water shortages, degradation and depletion of these resources, and public and private sector resource-management performance. Therefore, future plans set to meet the country's rising water demand should incorporate exploration of several alternatives such as adoption of conservation programs, development of groundwater resources, installation of new desalination plants, expansion in wastewater treatment, reuse of treated wastewater, and reuse of greywater. Parallel to that, there should be an expansion in the infrastructure facilities to meet expected water needs. An important issue that also needs to be considered is a change in the water policy from one that is mainly based on increasing supply to a policy that mainly focuses on managing demand.

The third environmental challenge that faces the country lies in the high rate of energy consumption. The country is already facing international pressure to reduce its CO_2 emission rate by adopting energy-conservation programs. This will drive new initiatives to reduce water consumption, improve design for higher energy efficiency in buildings, use of alternative and renewable energy sources, adopt clean technologies in the industrial sector, and shift to a sustainable transportation system. However, successful implementation of these initiatives cannot be achieved without raising public awareness and enhancing public responsibility.

Many of the abovementioned challenges and others have been realized by the UAE government. As reported by the Environment Agency—Abu Dhabi (EAD 2012), the consensus among different stakeholders is that "the seven key environmental system components (biodiversity, marine, land, energy, water, air and climate change, and waste) demonstrate subpar performance, well below global averages, and four are in critical condition." Consequently, environmental priorities have been set in some of the emirates. For example, the EAD has developed the agenda for the Abu Dhabi Environment Vision 2030 policy as an integral part of a holistic policy agenda framework that is intended to steer the emirate toward sustainable development. Five priority areas have been identified that provide the framework for policy development and action, and these form the core of the Abu Dhabi Environment Vision 2030. The priority areas include (1) climate change; (2) clean air and noise pollution; (3) water resources; (4) biodiversity, habitats, and cultural heritage; and (5) waste management.

12.4 E3 in Higher Education Institutes in the UAE: The Review

There are more than 20 public and private universities in the UAE (Table 12.1). Not all of these universities have colleges of engineering; of those few have civil or chemical engineering departments.

The Higher Colleges of Technology (HCT) was established in 1988 and is the largest institution of higher learning in the UAE with over 18,200 students. During the 2009–2010 academic year, there were 11,700 female and 6,500 male students enrolled at 17 campuses and 92 programs throughout the country. More than 30,000 UAE nationals are graduates of the institution. The HCT provides postsecondary education in business, education, engineering technology, information technology,

Table 12.1 Higher education institutes in the UAE

University or college	With environmental subject(s)	With environmental engineering (EnE) program(s)
Abu Dhabi University	√	X
Ajman University of Science and Technology	X	X
Al Hesn University	X	X
Al Ain University for Science and Technology	X	X
Al Ghurair University	X	X
American College of the Emirates	X	X
American University of Asia	X	X
American University in Dubai	√	X
American University of Sharjah	√	X
British University in Dubai	X	X
Birla Institute of Technology and Science, Dubai	X	X
Fujairah College	X	X
Gulf Medical College, now Gulf Medical University	X	X
Hamdan Bin Mohammed e-University	√(NE)	X
Heriot-Watt University Dubai	√(NE)	X
Higher Colleges of Technology	√	X
Ittihad University	X	X
Paris-Sorbonne University Abu Dhabi	X	X
Petroleum Institute in Abu Dhabi	√	X
Rochester Institute of Technology-Dubai	X	X
United Arab Emirates University	√	√(G)
University of Sharjah	√	X
University of Wollongong in Dubai	X	X
Zayed University	√(NE)	X
Canadian University of Dubai	√(NE)	X
NYU Abu Dhabi	X	X
Masdar Institute of Technology (Abu Dhabi)	X	√(G)

NE none engineering courses, *G* Graduate program, *NYU* New York University

communications technology, and health sciences. Examples of courses related to EnE offered by different departments at the HCT are CHEM N2325 Environmental Monitoring and Control; ELME N2130 Occupational Health, Safety, and Environment; SFTY N0215 Health Safety and Environmental Awareness; CIVL N455 Environmental Engineering; MECH N430 Health, Safety, and Environment; EGEN N302 Health, Safety, and Environment; and ECVL N405 Environmental Engineering.

Abu Dhabi University (ADU) was established in 2003. ADU provides both undergraduate and postgraduate study programs and consists of three colleges along with the English Language Institute, the College of Arts and Sciences, the College of Business Administration, and the College of Engineering and Computer Science.

There is a civil engineering program at ADU that offers courses related to water resources and EnE. An example of these courses is the EnE course offered by the Civil and Environmental Engineering Department.

The Ajman University of Science and Technology (AUST) is a technology-oriented university in Ajman. The university was founded in 1988 as a university college. Although AUST contains a college of engineering, this college does not include civil or chemical engineering departments. There is an institute for water environment and energy but offers only M.Sc. degree in groundwater engineering and management. Al Ghurair University is a private higher educational institution that was founded in 1999. Al Ghurair University is licensed by the UAE Ministry of Higher Education and Scientific Research. The university's educational programs include engineering and applied science program with emphases on electrical and computer science only. No EnE courses are offered at this university.

The American University in Dubai (AUD) is a private institution of higher education. AUD was founded in 1995 as a branch campus of the American InterContinental University in Atlanta, Georgia, but it has turned into a private university in 2007. AUD is accredited regionally by the Southern Association of Colleges and Schools. AUD contains engineering schools offering B.Sc. degrees in civil, computer, and electrical engineering. Courses related to EnE are offered through the civil engineering program. An example of these courses is the ECVL 340 Environmental Engineering course offered by the civil and environmental engineering department. This course covers topics related to all aspects of water, air, land pollution, sustainable technologies, risk assessment, and climate change.

The American University of Sharjah (AUS) is an independent higher educational institution in Sharjah. AUS was founded in 1997. The College of Engineering at AUS offers B.Sc. degrees in computer science, chemical engineering, civil engineering, computer engineering, electrical engineering, and mechanical engineering. EnE courses are offered through the civil and chemical engineering departments. Ittihad University (IU) is the first private university located in the emirate of Ras Al Khaimah. IU was founded in 1999. The university is divided into four undergraduate colleges one of which is the School of Engineering. This school offers B.Sc. in computer engineering and computer science only and, therefore, does not have an EnE program—neither does it offer EnE courses. The University of Sharjah is a semigovernmental higher educational institution founded in 1997. More than 1366 students are currently enrolled in the five degree programs at the College of Engineering, namely, architectural engineering, civil and environmental engineering, computer engineering, electrical and electronics engineering, and industrial engineering and management. These five programs are offered through four departments. Both the University of Sharjah and AUS offer an EnE course for the undergraduate students in their civil and environmental engineering departments. This course covers topics in water and air pollution. Another course related to wastewater technologies is offered at the University of Sharjah as an elective course. Meanwhile, a course called physical and chemical processes in environmental engineering is offered as an elective course at the AUS. The Petroleum Institute (PI) was created in 2001. There is a College of Engineering at the PI that offers a B.Sc. degree in chemical engineering but not in civil engineering. A course that covers topics related to water treatment is offered in the Chemical Engineering Department.

The United Arab Emirates University (UAEU) is the oldest of the three government-sponsored institutions of higher learning in the UAE. The other two are the HCT and Zayed University. The UAEU was established in 1976, with probably the largest faculty of engineering in the country. There are five engineering departments including architectural engineering, chemical and petroleum engineering, civil and environmental engineering, electrical engineering, and mechanical engineering. These five departments offer seven programs, namely, architectural engineering, chemical engineering, petroleum engineering, civil engineering, electrical engineering, communication, and mechanical engineering. Both the civil and environmental engineering and the chemical and petroleum engineering departments offer courses related to EnE. Examples of these courses are CIVL 270 Introduction to Environmental Engineering and CIVL 375 Water and Waste Water Technology which are core courses; and CIVL 520 Special Topics in Water Resources and Environmental Engineering, CIVL 522 Advanced Environmental Engineering, CIVL 524 Geo-environmental Engineering, which are elective courses. These courses cover a wide range of topics in EnE such as air, water, solid-waste management, environmental microbiology, environmental chemistry, all aspects of water and wastewater treatment, EIA, air pollution, and solid- and hazardous-waste management. Additionally, there are several master programs at the UAEU, of which the water resources, the environmental sciences, the petroleum sciences, and the civil engineering offer courses related to EnE. In addition, the recently commenced Ph.D. program in civil engineering at the UAEU offers graduate-level courses in EnE.

The University of Wollongong in Dubai (UOWD) is a private university that was established in 1993. This independent institution is affiliated with the University of Wollongong, Australia. UOWD offers a bachelor of engineering and, therefore, does not contain special engineering departments and programs. Zayed University (ZU) is a higher educational institution that was established in 1998 and named in honor of Late Sheikh Zayed bin Sultan Al Nahyan, the country's first president. ZU is organized academically into five colleges: arts and sciences, business sciences, communication and media sciences, education, and information technology. The Canadian University of Dubai (CUD) is an institution of higher education which offers undergraduate and graduate programs. The CUD was established in 2006 with no more than 100 students. Currently, the university serves over 1700 students. The university does not offer degrees in engineering; however, it offers B.Sc. in environmental health management through the School of Environment and Health. The New York University in Abu Dhabi (NYU Abu Dhabi) is a private university that was established in 2007. This independent institution is affiliated with the NYU, USA. The university offers a general degree in engineering but does not have special engineering departments or programs.

The Department of Chemical and Environmental Engineering (CEE) at Masdar Institute is strongly multidisciplinary. The research focus of the department is on advanced desalination technology, high-efficiency water distribution and use, integrated water and energy policy, climate impacts on health, bioinformatics, urban climate, bioprocess engineering and carbon capture and sequestration, and others. The CEE Department does not offer undergraduate degrees but offers two M.Sc. degrees, namely, M.Sc. in water and environmental engineering and M.Sc. in chemical engineering.

It is evident from the previous discussion that none of the higher education institutes in the UAE offers undergraduate program in EnE. Some of the UAE universities integrate E3 into the undergraduate programs through a chemical or civil engineering curriculum. The curriculum of chemical engineering covers non-sanitary components of E3, and the civil engineering syllabus covers subjects related to water supply, sewage distribution systems, and domestic-wastewater-treatment technologies. The Civil and Environmental Engineering Department at the UAEU offers an elective specialization option in EnE and water resources. Other smaller private universities in the UAE focus on information technology and business management with fewer numbers of students.

12.5 UAE Market Needs for E3: The Survey

12.5.1 Survey Form and Respondents

To evaluate the market needs for E3 in the UAE, a survey form (Appendix A) was distributed to target individuals. Recipients of the survey can be categorized into five main groups according to their workplace. These are (1) ministry of environment and water, local environmental agencies in each emirate, city directorates; (2) companies related to water and wastewater management; (3) oil industry; (4) engineering consulting firms; and (5) academic and research institutions. The survey form consists of nine questions. The first four questions were designed to obtain information about the respondent's background and experience. The respondents were then asked about the importance of EnE to their organization/company (Q.5) and how they classify E3 in the UAE (Q.6). The respondents were also asked about the areas of EnE that are highly needed by their organization/company (Q.7). Respondents were then asked about the demand of their organization/company to graduates with different EnE degrees (Q.8). Respondents were asked (Q.9) to scale potential job opportunities in the UAE that require knowledge of EnE at 11 employer groups (see Appendix A).

For most of the questions the respondent selected from a list of choices. However, the respondents were asked (Q.10) to write any comment they may have related to E3 in the UAE. In addition, respondents were allowed to suggest areas of EnE other than those listed in Q.7 that are needed by their organization/company. Respondents were further asked to indicate job opportunities that would exist at other potential recruiters beside those listed in Q.9. The form was distributed to more than 100 individuals, but 70 responses were received from the targeted groups as follows: ministry of environment and water, local environmental agencies, and city directorates (31 %); companies related to water and wastewater management (27 %); oil industry (9 %); engineering consulting firms (21 %); and academic and research institutions (11 %). In the authors' opinion, the response rate is very good, and the collected information should capture the opinion of the wide range of employers with interest in EnE in the UAE. The academic background of the respondents varies, but most of them have a degree in engineering (64 %), followed by management (16 %), science (11 %), and environmental, health, and safety (EHS; 9 %). As to their academic level, 4 % of the respondents have diploma, 47 % have B.Sc. degrees, 31 % have M.Sc. degrees, and 14 % have Ph.D. degrees. Also, the majority of the respondents (91 %) have at least 2-year experience working in the UAE, while 35 % of them have an experience of more than 10 years.

12.6 Results and Discussion

Most of the respondents indicated that EnE is important (32 %) or very important (65 %) to their organization, with only 3 % of the respondents considering EnE not important to their workplace (Fig. 12.1).

Classification of E3 in the UAE varies among the respondents as shown in Fig. 12.2. However, a significant fraction of the respondents (41 %) classified E3 in the country as weak, and some classified it as strong (23 %). Of notice is that 26 % of the respondents do not know how to classify E3 in the UAE.

About 60 % of the respondents considered solid-waste management, water management, wastewater management, and risk assessment as highly needed by their organization/company (Fig. 12.3). Other highly needed EnE areas such as air pollution assessment and control, land pollution, and modeling of environmental systems were selected by a lesser fraction of respondents (about 40 %). The least highly needed EnE area selected was cleaner production (20 %). Overall, these results emphasize the importance of E3 in nearly all its fields to the UAE market. Some respondents indicated other EnE areas that are needed by their organization/

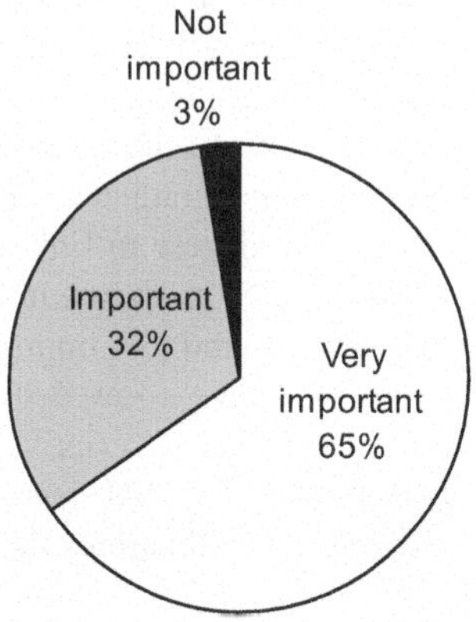

Fig. 12.1 Importance of environmental engineering (EnE) to respondents' organization/company

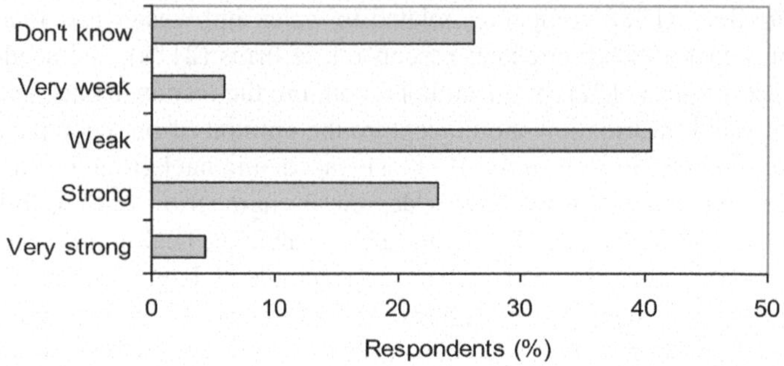

Fig. 12.2 Classification of environmental engineering education (E3) in the United Arab Emirates (UAE)

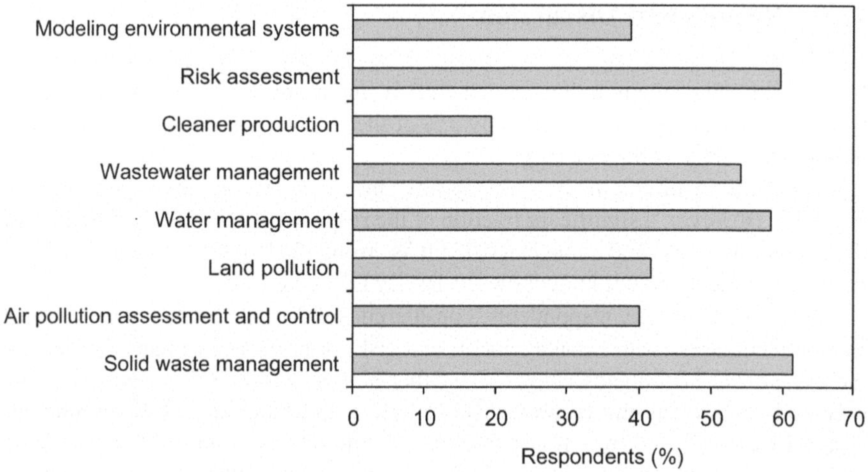

Fig. 12.3 Areas of environmental engineering (EnE) highly needed in the United Arab Emirates (UAE)

company including sustainability design requirements, impact of marine activities on coastal areas, hazardous-waste management, EIA, and environmental management planning.

In response to the question about the anticipated UAE market demand to EnE specializations, graduates with a B.Sc. degree in engineering and a minor in EnE or with a B.Sc. degree in EnE will have a moderate to high demand (i.e., 3.7 and 3.4 on a scale of 5) as shown in Fig. 12.4. Those with M.Sc. degree in EnE will have a moderate demand (3.1 on a scale of 5), and those with a diploma or a Ph.D. degree in EnE will have a low demand (2.3 on a scale of 5). It should be noted that this demand is rather descriptive. Quantitative evaluation of the demand for EnE specializations in the UAE is not known and could be a topic of a future study.

Figure 12.5 illustrates that good job opportunities that require specialization in EnE at the recruiter groups listed in Q9 of the survey form (Appendix A) would

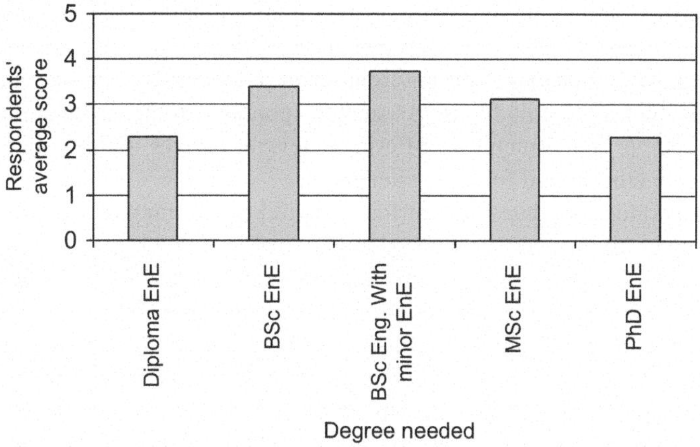

Fig. 12.4 Forthcoming demand to graduates with environmental engineering education (E3; 0 means none, and 5 means very high). *EnE* environmental engineering

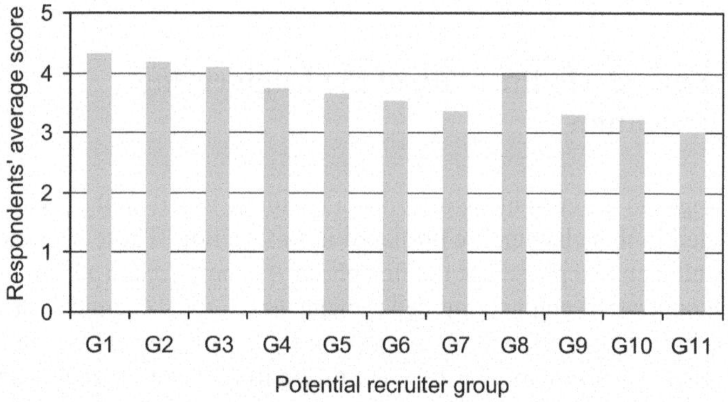

Fig. 12.5 Potential recruiter groups that require knowledge of environmental engineering (EnE) in the United Arab Emirates (UAE; 0 means none, and 5 means very high)

exist. Of these recruiter groups, environmental ministry/agencies, and city directorates (G1), water and wastewater management connected to city municipalities (G2), oil and petrochemical industry (G3), and consulting firms that prepare EIA studies or perform risk assessment (G8) would be the main recruiters. For the other recruiter groups, job opportunities for graduates with knowledge of EnE would be above moderate. Also, respondents added other potential recruiters for graduates specializing in EnE, including energy companies, infrastructure developers, academic sector, and facility management.

Several respondents made some comments about E3 education in the UAE. Some stressed on the need to graduate EnE and EHS professionals to meet current and anticipated demands created by regulatory requirements and several initiatives. Lately, Abu Dhabi emirate recognized EHS as a priority among different sectors in the emirate. Moreover, sustainability is a main pillar of the UAE government policy (i.e., Agenda 2030). Also, top managers in the oil and gas industry call for zero-flaring. All of this makes the environmental field with all its disciplines in high demand for the foreseeable future. As such, respondents requested that universities must respond to this demand by offering relevant courses and graduate capable people to face current and future challenges.

Some respondents suggested that focus should be on innovative environmental education and training systems to develop sustainable technologies and promote the needed cultural changes. Other respondents thought that more EnE courses within the civil engineering curriculum are needed or an EnE division within the civil engineering department should be established. Others thought that all engineering disciplines should receive some E3. Some even thought that the environmental education should expand, by different extents, to all colleges so that graduates are aware of the environmental issues facing the society which then could be transferred to their surroundings. Some respondents suggested that current engineers should seek environment and safety diplomas/certification from recognized institutions as this will add values to their careers.

12.7 Proposed Modification of the Engineering Curriculum

Several changes to the current engineering curriculum, based on the review and survey conducted in this chapter, are proposed in this section. These alterations could be implemented on two phases. In the first phase, it is proposed to add an EnE minor track in the concerned engineering departments (e.g., civil, chemical, or mechanical engineering). This track will focus more on E3 by offering courses to interested students that will help them later to excel as environmental engineers and satisfy the market needs as discussed in the survey section of this chapter. Implementation of this first phase should not require huge alteration to the current curriculum offered by these departments and, in the meantime, shall not require many additional resources. Existing faculty members, laboratories, and other resources could be utilized to suit the requirements of this new track. Therefore, implementation of this track could be a practical immediate response to the market needs of E3 graduates, especially in universities with limited resources. A long-term response, in the authors' opinion, should include establishment of a separate EnE department as a second phase. Implementation of this second phase, however, will require additional resources and, therefore, needs appropriate long-term planning. In the following section, we will propose a plan of a minor in EnE as the first phase of curriculum change in engineering education.

12.8 Minor Track in EnE

This minor in EnE aims at achieving the following objectives:

1. Prepare environmental engineers for industry and government agencies concerned with sustainable engineering technologies coupled with environmental security.
2. Increase the pool of well-prepared students applying for the M.Sc. program in environmental sciences, who have strong background in EnE.
3. Train students in interdisciplinary programs with emphasis on:
   (a) Enhancing the student's abilities in scientific methodologies for collecting and analyzing environmental data
   (b) Preserving environmental resources

To fulfill these specific objectives, it is proposed that the designed program should provide courses that cover the areas of

1. basic fundamentals in science such as engineering computing, chemistry, mathematics, project engineering, fluid mechanics, and geotechnical engineering;
2. data collection and analysis;
3. environmental managements, such as environmental law, policy, impact assessment, and economy; and
4. terrestrial environment such as geo-environmental engineering, groundwater hydrology, and land and water management.

12.8.1 *Additional Elective Courses*

In the final year of the program, more specialized courses should be offered in aquatic processes, coastal and marine waters, environmental management, and terrestrial environment. For example, a student in the EnE track must successfully complete three courses from the specialized elective courses. Examples of these courses could be:

(1) Special topics in EnE
(2) Air pollution Control
(3) Environmental management systems
(4) Environmental law and impact analysis
(5) Solid- and hazardous-waste management
(6) Hydrology
(7) Water structures
(8) Hydraulic systems
(9) Coastal engineering
(10) Water desalination
(11) Groundwater contamination

12.9 Fulfillment of ABET Requirements

To demonstrate the fulfillment of ABET requirements by adopting an EnE track, we considered the cases of the civil engineering program at the UAEU. The ABET requires that civil engineering students acquire proficiency in mathematics through differential equations, probability and statistics, calculus-based physics, and general chemistry. It also requires that students graduate with proficiency in a minimum of four recognized major civil engineering areas. Graduates of the proposed program would have the opportunity to attain proficiency in five major civil engineering areas, namely: structural engineering, surveying and transportation, EnE, water resources, and geotechnical and construction management. Examples of the courses offered in these five areas at the Department of Civil and Environmental Engineering at the UAEU are as follows:

Structural Engineering

1. Statics
2. Concrete technology
3. Computer-aided drawing
4. Structural analysis
5. Structural design
6. Reinforced concrete

Surveying and Transportation Engineering

1. Surveying
2. Highway engineering

Environmental Engineering

1. Basic microbiology
2. Introduction to EnE
3. Geo-environmental engineering
4. Environmental law and impact analysis
5. Water and wastewater treatment
6. Technical elective I
7. Technical elective II
8. Technical elective III

Water Resources

1. Fluid mechanics I
2. Water resources
3. Technical elective IV

Geotechnical and Construction Management

1. Construction management
2. Soil mechanics

12.10 Summary and Conclusions

This chapter presented a review of the current status of E3 in the UAE higher education institutes. The review showed that none of the higher education institutes in the UAE provides specialized undergraduate E3. E3 is rather provided through courses offered in the civil or chemical engineering programs in few of these institutes. The UAE market's need for graduates with E3 is also assessed in this chapter through a survey. It was found that the UAE market is in high need of graduates from this field. Based on the results of the review and the survey, it is recommended to initiate programs of E3 to be offered in major public higher education institutes in the UAE. Such programs could be implemented over the short term in the form of an EnE track within the civil, chemical, or mechanical engineering departments. These tracks could be converted to stand-alone departments on the long term.

Acknowledgments The authors would like to thank all engineers and managers who responded to the market survey and to students who participated in the survey distribution and documentation.

Appendix A

UNITED ARAB EMIRATES UNIVERSITY
Faculty of Engineering
Department of Civil and Environmental Engineering
P.O. Box 17555 • Al Ain • UAE

Dear evaluator: A research team at the Department of Civil and Environmental Engineering at the UAE University is evaluating the market needs for environmental engineering education in the UAE. We would therefore appreciate your valuable input by taking few minutes to fill out this survey.

Opinion Survey*
Needs for Environmental Engineering in the UAE

Organization/Company Name and Location: ____________________
Evaluator Name (*optional*): ____________________
Position: ____________________ Contact e-mail (*optional*): ____________________
Tel (*optional*): ____________________ Fax (*optional*): ____________________

1. How many years of work experience in the UAE do you have? ____________________

2. What is the highest degree you hold?
☐ Diploma ☐ B. Sc. ☐ M. Sc. ☐ Ph. D. ☐ others (please indicate)____

3. What is your area of specialty? ____________________

4. Approximately, how many engineers from all specialties are currently employed in your organization/company? ____________________

5. How important is environmental engineering to your organization/company?
☐ Very important ☐ Important ☐ Not important

6. How do you classify the environmental engineering education in the UAE?
☐ Very Strong ☐ Strong ☐ Weak ☐ Very weak ☐ Don't know

7. Which area of environmental engineering is highly needed by your organization/company (please mark whichever applies)?
☐ Solid waste management ☐ Air pollution assessment and control ☐ Land pollution
☐ Water management ☐ Wastewater management ☐ Cleaner production
☐ Risk assessment ☐ Modeling environmental systems
☐ Others (please indicate)____

8. In your opinion, how will be the demand of your organization/company in the next 5 years to graduates with the following degrees (0 means none and 5 means very high)?

Degree	0	1	2	3	4	5
A diploma in environmental engineering						
A bachelor degree in environmental engineering						
A bachelor degree in engineering (civil, chemical, etc) with a minor in environmental engineering						
A master degree in environmental engineering						
A Ph.D. degree in environmental engineering						

9. In your opinion, what job opportunities that require knowledge of environmental engineering are available in the UAE (0 means none and 5 means very high)

Group	0	1	2	3	4	5
G1. Environmental ministry, environmental agencies, and city directorates						
G2. Water and wastewater management connected to city municipalities and wastewater collection systems						
G3. Oil and petrochemical industry						
G4. Companies which design, construct and operate water and wastewater treatment plants						
G5. Consulting engineers for public and private institutions concerning water management, handling of resources, solid waste, protection and rehabilitation of waters and soil.						
G6. Environmental public laboratories						
G7. Companies which provide necessary materials and equipment for water and wastewater treatment plants						
G8. Consulting firms that prepare environmental impact assessment studies or perform risk assessment						
G9. Industrial enterprises for the management and quality assurance of raw materials, products and waste streams						
G10. Research engineers in the development and application of experiments, models, tools and software to simulate environmental systems						
G11. Institutions (national, regional or international) working on management activities						
Others (please indicate below and rank it needs)						
G12.						
G13.						

10. Please write any comment you may have related to environmental engineering education in the UAE.

Thank you for your corporation

* All information provided in this survey will be kept confidential, and will be used only to evaluate the need for environmental engineering education in the UAE.

References

Alha, K., Holliger, C., Larsen, B. S., Purcell, P., & Rauch, W. (2000). Environmental engineering education—summary report of the 1st European Seminar. *Water Science and Technology, 41*(2), 1–7.

Al-Homoud A. S. (2003). Water resources challenges facing UAE in the 21st century. Paper 20. http://opensiuc.lib.siu.edu/ucowrconfs_2003/20.

Bishop, P.L. (2000). Environmental engineering education in North America. *Water Science and Technology, 41*(2), 9–16.

Brook, M., & Dawoud, M. A. (2005). Coastal water resources management in the United Arab Emirates, Environmental Research and Wildlife Development Agency, Integrated coastal zone management in the United Arab Emirates, 5–8 June 2005, Abu Dhabi, UAE.

Chang, S. N., & Chiu, M. H (2005). The development of authentic assessment to investigate ninth grader's scientific literacy: In the case of scientific cognition concerning the concept of chemistry and physics. *International Journal of Science and Mathematics Education, 3,* 117–140.

Conference Board of Canada. (2008). A report card on Canada, available at http://sso.conferenceboard.ca/HCP/default.aspx.

EAD. (2012). Environment Agency-Abu Dhabi. https://www.ead.ae/en/. Accessed 21 May 2014.

Ghaffari, S., & Talebbeydokhti, N. (2013). Status of environmental engineering education in various countries in comparison with the situation in Iran. *Procedia—Social and Behavioral Sciences, 102,* 591–600.

Golob R., & Bruss, E. (1984). Statistical analysis of oil pollution in the Kuwait Action Plan Region and the implications of selected world-wide oil spills to the region. Combating oil pollution in the Kuwait Action Plan Region. IMO/ROPMEAJNEP. UNEP Regional Seas Reports and Studies No. 44. New York: United Nations Environment Program.

Gutierrez-Martin, F., & Huttenhain, S. H. (2003). Environmental education: new paradigms and engineering syllabus. *Journal of Cleaner Production, 11,* 247–251.

Gutiérrez-Martín, F., & Dahab, M. F. (1998). Issues of sustainability and pollution prevention in environmental engineering education. *Water Science and Technology, 38*(11), 271–278.

Jassim, M., & Coskuner, G. (2007). Environmental engineering education (E3) in the Gulf Cooperation Countries. *European Journal of Engineering Education, 32*(1), 93–103.

Jones, G. P., & Marrei, S. H. (1982). Groundwater resources in the United Arab Emirates. *Middle East Water Science, 6*(1), 41–45.

Lemkowitz, S. M., Bibo, B. H., Lameris, G. H., & Bonnet, B. A. (1996). From small scale, short term to large scale, long term: integrating sustainability into engineering education. *European Journal of Engineering Education, 4,* 353–386.

Ministry of Economics. (2007). Data sheet for the climate of UAE, Metrology Department, Abu Dhabi, United Arab Emirates.

Mino, T. (2000). Environmental engineering education in Japan. *Journal of Water Science and Technology, 41*(2), 17–22.

Rizk, Z. S., Alsharhan, A. S., & Shindo, S. (1997). *Evaluation of groundwater resources of the United Arab Emirates*. The Third Proceedings of the 3rd Gulf Water Conference, March 1997, Muscat, Oman.

Schmitt, T. G., & Wilderer, P. A. (1996). Environmental engineering education in Germany. *Water Science and Technology, 34*(12), 183–190.

Smith, D. W., & Biswas, N. (2001). Environmental engineering education in Canada. *Canadian Journal of Civil Engineering, 28*(1), 1–7.

Smith, R., Purnama, A., & Al-Barwani, H. H. (2007). Sensitivity of hypersaline Arabian Gulf to seawater desalination plants. *Applied Mathematical Modelling, 31,* 2347–2354.

Tansel, B. (2008). Changing the status quo in environmental engineering education in response to emerging markets. *Journal of Professional Issues in Engineering Education and Practice, 134*(2), 197–201.

Ujang, Z., Henze, M., Curtis, T., Schertenleib, R., & Beal, L. L. (2004). Environmental engineering education for developing countries: Framework for the future. *Journal of Water Science and Technology, 49*(8), 1–10.

Van Lavieren, H., J. Burt, D.A. Feary, G. Cavalcante, E. Marquis, L. Benedetti, C. Trick, B. Kjerfve, & Sale, P.F. (2011). Managing the growing impacts of development on fragile coastal and marine ecosystems: Lessons from the Gulf. A policy report, UNU-INWEH, Hamilton, ON, Canada.

WWF. (2006). Living Planet Report 2006. World Wildlife Fund.

Yaghubi, M., Sohrabpour, S., Eslami, M. R., & Ghaffari, M. M. (2006). Scientific and technological development in the field of engineering sciences in Iran in comparison with some countries. *Iranian Journal of Engineering Education, 31*(8), 57–94.

Dr. Mohamed M. Mohamed is an associate professor at the UAE University (UAEU). His research is focused on water resources management, groundwater modeling, contaminant fate and transport in the subsurface, groundwater bioremediation, and environmental impact assessment. Prior to his affiliation with the UAEU, he occupied several positions, including postdoctoral research associate, Sheffield University, UK; assistant/associate professor, Cairo University, Egypt; and research associate, University of Florida, USA. Dr. Mohamed received his B.Sc. and M.Sc. degrees from Cairo University, Egypt, and a Ph.D. degree from the University of Florida, USA. Dr. Mohamed teaches several courses in the fields of water resources and environmental engineering. He is a member of several professional associations.

Dr. Munjed A. Maraqa is an associate professor and Head of Safety Division of the Roadway, Transportation and Traffic Safety Research Center at the UAEU. His research covers a broad range of issues related to modeling environmental systems, water quality, waste management, and traffic safety. Between 2008 and 2012, Maraqa acted as an advisor to Al Ain Municipality on the development of an environmental, health, and safety (EHS) management system for the building and construction sector in Abu Dhabi emirate. Maraqa holds B.Sc. and M.Sc. degrees in civil engineering from the University of Jordan and a Ph.D. in environmental engineering from Michigan State University.

Part IV
Assessment and Accreditation

Chapter 13
Potential and Challenges of Project-Based Learning in Engineering Education at the United Arab Emirates University

Rezaul K. Chowdhury

13.1 Introduction

Several teaching and learning approaches have been trialed, practiced, and modified over the years. These techniques can be grouped into traditional and modern methods. Traditional teaching methods involve conventional lectures followed by tutorial and/or lab sessions, predominately in isolated time segments (Nepal and Jenkins 2011). A direct flow of information from instructors to students is the central aspect of traditional approaches. Students are considered cognitively active, but physically inactive (except for taking notes). According to Cangelosi (2003), most students cannot maintain such attention and behavior for a long period of time. The traditional method lacks sufficient interactions between students and instructors and among students, and considers students as passive learners (Nepal and Jenkins 2011; Steinhorst and Keeler 1995). On the other hand, modern teaching methods are cognitive science based, and they encourage students to construct knowledge rather than piling it on as it is disseminated (Cross 1998, 1999). The modern methods involve contemporary teaching and learning practices using project- and program-based learning, work-integrated learning, and integrating learning approaches. Students are encouraged to use real-world concepts, tools, experiences, and technologies; they work in groups to identify, acquire, and share knowledge to solve real problems. Among modern methods, the project-based learning (PBL) is recognized as a collaborative, progressive, student-centered, interactive, dynamic, and deep-learning approach, especially for engineering education (Nepal and Jenkins 2011). Benefits of PBL approaches in engineering education have been portrayed in several previous studies (Mills et al. 2005; Napal and Jenkins 2011; Perrenet et al. 2000; Gibson 2003; Mills and Treagust 2003; Ribeiro and Mizukami 2005). The PBL approach is thought to promote active and deep learning through

R. K. Chowdhury (✉)
Department of Civil and Environmental Engineering, United Arab Emirates University, Al Ain, PO Box 15551, United Arab Emirates
e-mail: RezaulKabir@uaeu.ac.ae

M. Abdulwahed et al. (eds.), *Advances in Engineering Education in the Middle East and North Africa,* DOI 10.1007/978-3-319-15323-0_13

engaging students in a real-world issue in a collaborative environment. The goal of a PBL approach is to expose students to experiential learning. In engineering education, problem-solving skill is essential for students; therefore, students are required to be exposed to real-world problems (Helle et al. 2006). The driving questions or problems help students to drive activities in creating the final product that addresses the questions or problems (Blumenfeld et al. 1991).

Both approaches of instruction (traditional and modern) possess merits and demerits. Some engineering students dislike the modern PBL approach as they necessitate to conduct a self-directed learning strategy to complete often unclear and open-ended tasks. Their individual learning styles and requirements may be dissimilar to the team learning needs. The challenges required in implementing successful PBL are related to time and efforts, instructor's content knowledge, and lack of experience of instructors and students as well as the demand to develop specialized material for off-campus study (Yam and Rossini 2010). In some instances, students rate the PBL approach lower than the traditional approach in spite of their improved learning and course performance (Nepal and Stewart 2010; Nepal and Panuwatwanich 2011). Because of resource intensiveness, academic institutions often hesitate to embrace a PBL approach (Nepal and Jenkins 2011).

University teaching requires the emergence and evolution of teaching methods consistent with the learning techniques of students (Ewell 1997). Instructors play important roles in the implementation of PBL through motivating students and by creating a classroom environment favorable for students' learning (Yam and Burger 2009). Collaboration among students, instructors, and community is significantly important so that knowledge can be shared and spread among the stakeholders. Knowledge of students' learning styles and instructors' teaching styles is of particular importance for developing successful PBL in engineering education. The savvy about how learning information is treated and presented provides clues on how specific teaching method can be effectively used to maximize students' learning process (Boles et al. 2009). Mismatch between students' learning styles and instructors' teaching approaches provides a barrier for successful learning and PBL implementation.

The United Arab Emirates University (UAEU) is a public university with an enrollment of approximately 16,000 students. The university consists of nine academic colleges (schools) along with the University General Requirements Unit (UGRU) as a foundation-level program within the university. The College of Engineering (COE) was introduced in 1980 and there are five academic departments in COE (civil, mechanical, chemical and petroleum, architectural, and electrical engineering). The full figure of students is about 16,000; the percentage of male and female students is about 24 and 76 %, respectively. The enrollment at COE is about 9.5 % of the total students. The percentage of male and female students at COE is about 53 and 47 %, respectively. About 90 % of the total students are generally considered as first-generation university students, whose parents have no college or university experience (Bielenberg 2005). The total number of faculties at the university (in all colleges, including UGRU) were about 1000 (based on estimation in the fall 2012–2013 semester). The percentage of male and female college faculties is about 84 and 16 %, respectively. About 87 % of these faculties hold a Ph.D.

degree. In terms of age distribution of faculty members, 8 % faculties are below 35 years, 60 % are between 35 and 49 years, and about 32 % are above 50 years of age (CETL 2012a). The UGRU is a developmental program that equips students with the relevant knowledge and skills required to serve as a competent member of the university and society (Gillway and Bielenberg 2006). In the fall 2003–2014 semester (September to January), a set of curriculum competencies was adopted within the UGRU. The selected competencies were learner training, communication, information literacy, thinking skills, and application of knowledge. It is considered that the adoption of PBL in the first-year programs provides a unique opportunity to focus more on the process of learning. In the fall 2004–2005 semester, a team of volunteer teachers from all subject areas (English, mathematics, and Information and Communications Technology (ICT)) was brought together and began work on developing scenarios, materials, and assessment tools for PBL. The initial PBL experience was piloted in the spring 2004–2005 semester (February to June) and the experience was fully implemented in the fall 2005–2006 semester. In the COE, a few engineering courses are instructed based on the PBL approach. Very recently, the university has started implementing smart technologies (iPad, iPhone, iTunes U, etc.) in some engineering courses in order to engage students more interactively. Like many other universities in the Middle Eastern region, the university does not allow coeducation at the undergraduate level.

Achieving quality and excellence in engineering education is high on the agenda of Middle East regional universities, where religion and culture play a crucial role (Shanableh and Omar 2007). According to Al-Maati and Damaj (2010), talking about ethics without linking it in one way or another with religious/cultural affairs is not coherent in the division. Universities in the region must balance tradition and innovation in planning and presenting educational programs. Achieving the balance between invention and tradition must be reckoned as an opportunity that adds unique values to educational programs but not an obstacle that hinders development. Shanableh and Omar (2007) identified four critical issues in the Middle East universities that differ significantly from Western universities. The issues are (a) *pace of change and evolution*—engineering education in the region generally evolves slowly, (b) *technical and generic engineering education*—engineering education at the undergraduate level in this region tends to focus on feeding technical information only, (c) *partnership with stakeholders*—in the region, a serious rift exists between engineering colleges and the community in terms of science and applied technology, and (d) *integration of faculty activities*—the level of integration of teaching, research, and community service in regional universities is reflective of the technology generation activity in the area.

Considering the sociocultural practices of the country and the anticipated shift of teaching approach to a more technology-based interactive way, this chapter provides how a PBL approach in engineering education may be tailored to address the diversity of students' learning and instructors' teaching styles. Learning styles of engineering students, teaching approaches of engineering instructors, role of technologies, and their applications in the implementation of PBL in engineering education are discussed. Overall, the potential and challenges of PBL in engineering instruction at the UAEU are delineated.

13.2 Project-Based Learning (PBL)

The PBL concept promotes students to develop knowledge and skills with their own activities, rather than a transmission of knowledge from instructors to students (Biggs and Tang 2007). PBL organizes learning around projects (Thomas 2000). PBL projects are designed based on challenging questions or problems that involve students in project design, problem-solving, decision-making, and investigative activities, and provide students the opportunity to work independently over an extended period of time (Thomas 2000; Thomas et al. 1999). Thomas (2000) proposed five criteria for a project to be considered in PBL. These are centrality, driving question, constructive investigations, autonomy, and realism.

a. *Centrality*: PBL projects are central to the curriculum. The project serves as the central teaching strategy; it is expected that students learn the central concepts of the curriculum through the project work. According to this criterion, an application project that serves to provide illustrations, examples, and practical applications for the material taught in traditional lecture–tutorial approach is not considered to be the PBL project.
b. *Driving question*: PBL projects focus on questions or problems that drive students to the central concepts and principles of curriculum. Projects are designed in such a way that it makes a connection between student activities and the underlying conceptual knowledge of the curriculum. This can be done with a driving question or a vague problem.
c. *Constructive investigations*: PBL projects engage students in a constructive investigation that involves inquiry, knowledge building, and perseverance. Transformation and construction of knowledge are the central activities of the project. According to this criterion, an exercise project where the students carried out central activities by application of already learned information or skills is not considered a PBL project.
d. *Autonomy*: PBL projects are significantly student-driven projects; they must not be instructor directed, scripted or packaged like laboratory exercises. They do not follow a predetermined path or end up at a predetermined outcome. They incorporate and encourage more student autonomy, unsupervised work time, and responsibility than traditional instruction and traditional projects.
e. *Realistic*: PBL projects need to be realistic, so that it provides a feeling of authenticity to students. The project incorporates real-life challenges and the solutions have the potential to be implemented. The project can include the topic, the tasks, and the roles that students play, the context within which the work of the project is carried out, the collaborators who work with students on the project, the products that are produced, the audience for the project's products, or the criteria by which the products or performances are judged.

Teamwork is the central part of PBL. In teamwork projects, students work together to solve a complex and authentic problem and thereby develop content knowledge and graduate attributes such as problem-solving and communication skills. Students learn from each other while they work together in a team (Bloxham and West 2004).

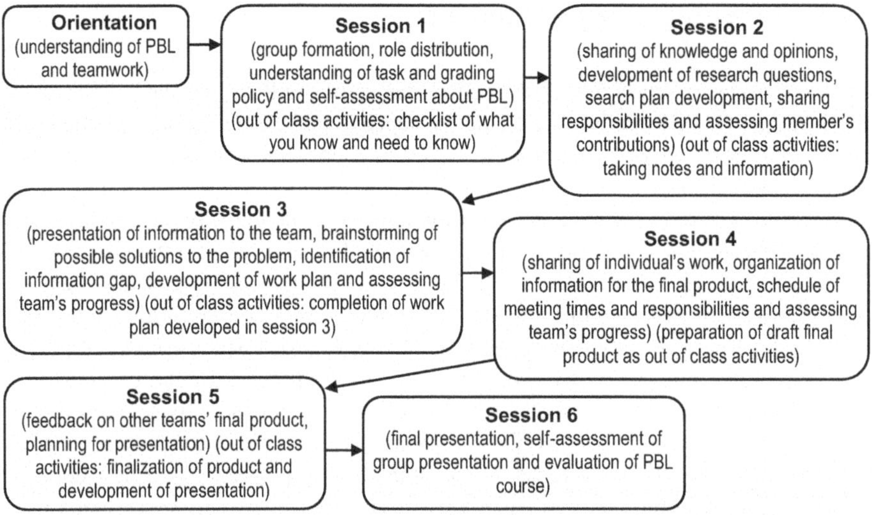

Fig. 13.1 The PBL process adopted in the foundation program of United Arab Emirates University

According to Dyball et al. (2007), benefits of teamwork are opportunities to work on comprehensive assignments; an insight into group dynamics and processes; development of interpersonal and communication skills; exposure to the viewpoints of others; preparation for the real viewpoint; and promoting reflection and discussion. Teamwork has some disadvantages as well. Disadvantages of teamwork are interpersonal conflict among team members leading to the incompletion of the task; unequal workload distribution; possibility of reduction of mental effort through shirking tasks, thus debilitating learning; and students' negative reactions to group-learning experiences (Walker 2001; Spalding et al. 1999).

The PBL experience in a first-year engineering program can move and engage students to a larger level than traditional lecture–tutorial-based courses. Participation in the PBL project enables students to make a variety of connections between different subjects (mathematics, ICT, English, study skills, etc.), connections to the real world, connections among participants, and connections to the knowledge, skills, and dispositions (Gillway and Bielenberg 2006). The drawbacks of teamwork can be minimized by taking up some techniques such as peer assessment and PBL training of students (Willey and Gardner 2009). The PBL process adopted in the foundation program (UGRU) of UAEU is shown in Fig. 13.1. The PBL initiative at UAEU is found to incite and engage students and to offer opportunities to develop curriculum competencies (CETL 2012a). Students were followed to get a number of associations that support their learning throughout their university training and beyond. Both students and instructors identified that the PBL course strongly contributed to the development of the curriculum competencies. Among different curriculum competencies (stated in the introduction section), information literacy (learning how to access information) and learner training (learning to lead in a group, admitting responsibility for one's study, and organizing one's work) were ranked most highly by both teachers and students.

13.3 Learning Styles of Students

Students' engagement is one of the central views of the PBL approach. Therefore, the success of a PBL method significantly depends on students' learning behaviors and techniques. Teaching methods followed in a university need to be consistent with students' learning techniques (Ewell 1997). Students learn in different ways; they focus on different types of information; they operate on the perceived information in different ways; and they achieve understanding at different rates (Boles et al. 2009). Coffield et al. (2004) stated: "A knowledge of learning styles can be used to increase the self-awareness of students and tutors about their strengths and weaknesses of learners. In other words, all the advantages claimed for meta-cognition (i.e. being aware of one's own thoughts and learning processes) can be gained by encouraging all learners to become knowledgeable about their own learning and that of others." Learning styles refer to the ways students prefer to receive, process, and present information and ideas. Some students may prefer to see a new concept by reading a textbook; some others choose a graphical explanation. Students may vary in how they most effectively show their understanding: graphically, verbally, or in writing (Mills et al. 2005). Kolb (1984) identified the four basic types of learning styles and observed that engineering students usually study in a convergent learning style. Kolb's learning styles are:

a. *Convergent style*: good at problem-solving, decision-making, and the practical application of ideas
b. *Divergent style*: good imaginative ability and awareness of meaning and values
c. *Assimilative style: good at inductive reasoning and making theoretical models*
d. *Accommodating style*: efficient in carrying out plans and getting involved in new experiences

Various factors influence students' preferred learning modes. Some of these factors are gender, cultural issues, and age influences on learning styles (Anderson 1991; Beyer 1993; Harding 1996). It is therefore significantly important to understand the gender, age, and cultural differences in a university before implementing a PBL technique in engineering education. This is also important for equity reasons and to support international students having diverse cultural settings. Holt and Solomon (1996) pointed out that since engineering education concentrates on problem-solving and engineering science (convergent and assimilative learning styles), it tends to exclude divergent and accommodative learners from effective learning and limits the opportunities of all learners to develop their skills in design and invention (divergent approach) and business management (accommodative approach).

Felder and Silverman (1988) argued that learning styles of most engineering students and teaching ways of most engineering instructors are mismatched in several dimensions. Engineering students are predominantly visual, sensing, inductive, and active, while most engineering education is auditory, abstract (intuitive), deductive, passive, and sequential. The mismatches lead to poor student performance, professorial frustration, and a loss to society of many potentially excellent engineers.

Felder (1999) has developed a learning style model, particularly for engineering education. The model assesses students on four dimensions of preferences by utilizing a questionnaire study that involves 44 questions. The questionnaire is available at http://www.engr.ncsu.edu/learningstyles/ilsweb.html. The four dimensions of learning preferences are:

a. *Active/reflective*: This dimension is about processing of information. Active learners prefer trying things out and working with others. Reflective learners prefer to think things out and work alone. Active experimentation involves doing something in the external world with the information—discussing, explaining, or testing it in some way, whereas reflective observation involves examining and manipulating the information introspectively. Active learners work well in groups, and reflective learners work better by themselves. Active learners tend to be experimentalists, whereas reflective learners tend to be theoreticians. Engineers are more likely to be active than reflective learners. The reflective observers are the theoreticians, the mathematical modelers, the ones who can define the problems and propose possible solutions. The active experimenters are the decision-makers who evaluate the ideas and concepts, carry out the experiments, and ascertain the solutions that work. An effective technique for reaching active learners is to have students organize themselves in groups and periodically come up with collective answers to questions posed by the instructor.
b. *Sensing/intuitive*: This dimension is about receiving information. Sensors like facts, data, and experimentation, whereas intuitors prefer principles and theories. Sensors like solving problems by standard methods and dislike surprises, whereas intuitors like innovation. Sensors are good at memorizing facts, and intuitors are good at grasping new concepts. Most engineering courses other than laboratories emphasize concepts rather than facts and use primarily lectures and readings (words and symbols) to send information, thereby favor intuitive learners. On the other hand, the majority of engineering students are sensors. Engineering tasks require the awareness of surroundings, attentiveness to detail, experimental thoroughness, and practicality (character of sensors) as well as creativity and theoretical ability (character of intuitors). Therefore, an effective engineering education should reach both types, rather than directing itself primarily to intuitors.
c. *Visual/verbal*: This dimension is about the ways of perceiving information. Visual learners like graphs, images, flowcharts, and diagrams, whereas verbal learners enjoy reading and lectures. Visual learners remember best what they see, and verbal (auditory) learners remember much of what they hear and then say. Verbal learners prefer verbal explanation to visual demonstration, and learn effectively by explaining things to others. Lecture presentations that use both visual and verbal explanations reinforce learning for all students. According to Stice (1987), students retain 10 % of what they read, 26 % of what they hear, 30 % of what they see, 50 % of what they see and hear, 70 % of what they say, and 90 % of what they say as they do something.

d. *Sequential/global*: This dimension is about progress toward understanding. Sequential learners prefer logical steps toward an outcome, whereas global learners grasp the big picture quickly and work out the steps later. Most formal education involves the presentation of material in a logically ordered progression; sequential learners learn by mastering the material more or less as it is presented. Sequential learners follow linear reasoning processes when working problems, and global learners make intuitive leaps and may be unable to explicate how they came up with answers. Sequential learners are strong in convergent thinking and analysis, whereas global learners are better at divergent thinking and synthesis. Global learners are the synthesizers, the multidisciplinary researchers, and the systems thinkers. In order to reach global learners, students should be given the freedom to devise their own methods of solving problems rather than being forced to adopt a specified strategy.

At the UAEU, learning styles of engineering students were assessed during the fall and spring 2012–2013 and spring 2013–2014 semesters for two courses, the CIVL 270 Introduction to Environmental Engineering course (2 credit hours) offered for civil and environmental engineering students and the GENG 315 Engineering Practice and Entrepreneurship, a general engineering course (3 credit hours) offered to all students. The Felder (1999) questionnaire was applied. The total number of students assessed was 118 (males 42 and females 76). The scoring system ranges from 11a to 11b for each of the four dimensions where "a" and "b" indicate first- and second-learning preference, respectively (e.g., "a" is active and "b" is reflective). A learner may be located at any point between these two extremes. Scores from 1 to 3, 5 to 7, and 9 to 11 are considered mild, moderate, and strong preference, respectively. For instance, a score of 9b–11b for the active/reflective dimension indicates a substantial penchant for a reflective learning style. Figure 13.2 depicts the learning styles of surveyed engineering students at the UAEU.

It is evident from Fig. 13.2 that students' learning preferences are diverted toward active (57 %), sensory (71 %), visual (83 %), and sequential (68 %) styles. The median scores for all preferences were found mild except for the visual preference where the median score is moderate. A maximum score of 11 (strong preference) was observed for both intuitive and visual styles, whereas a maximum score of 9 (strong preference) was found for active, sensory, sequential, and global styles; for other styles (reflective and verbal), the maximum score of 7 (mild preference) was found. It is broadly recognized that engineering students experience a variety of learning styles, but in reality, few courses are structured to consider these variables. When teaching is planned in such a way to accommodate all learning styles, student outcomes and satisfaction will increase significantly. In terms of the process of information (active/reflective), balanced learning styles are observed among the respondents. Regarding reception of information, sensory learning style is predominant over the intuitive way. I can state that most students prefer facts, data, experimentation, and problem-solving by standard methods rather than understanding the underpinning principles, theories, and innovation. Approximately 83 % respondents prefer the visual learning style, which is really obvious because most engineering students prefer graphs, images, flowcharts, and diagrams as their means of perceiving information. The majority of respondents were found sequen-

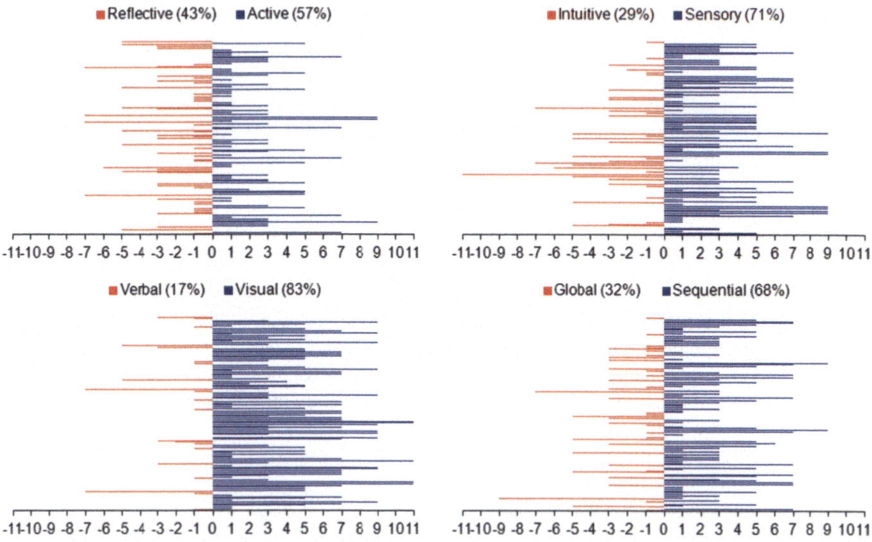

Fig. 13.2 Learning styles of engineering students at UAEU assessed during the fall and spring 2012–2013 and spring 2013–2014 semesters for two courses CIVL 270 and GENG 315; positive and negative scores indicate "a" and "b" preference, respectively (total number of respondents were 118: males 42 and females 76)

tial (68 %), which indicates they prefer logical steps toward an outcome and they are strong in convergent thinking. Engineering projects generally require the awareness of surroundings, attentiveness to detail, experimental thoroughness, and practicality (character of sensors) as well as creativity and theoretical ability (character of intuitors). Therefore, an effective engineering education should reach both sensory and intuitive types, rather than directing to a single preference. Because of students' preferred sequential and sensory learning styles at UAEU, it is very important to emphasize the PBL process so that students can understand how PBL works and how they can set out to solve a problem. The PBL instructions shown in Fig. 13.1 can be considered as an attempt to adopt the PBL in engineering education.

13.4 Teaching Styles of Instructors

Appropriate teaching styles of engineering instructors are needed for motivating students and for creating a classroom environment favorable for students' learning styles. Sufficient interaction between students and instructors, and student engagement are essential for promoting a collaborative learning environment. A mismatch between students' learning styles and instructor's teaching methods provides a barrier for a successful PBL implementation. Grasha (1994) proposed five categories of teaching styles: these are expert, formal auditory, personal model, facilitator, and delegator. All instructors have each of the lineaments of the teaching styles in varying levels. Four clusters of teaching styles were classified in Grasha (1994), and it

Table 13.1 Teaching styles of instructors. (Based on Grasha 1994)

Teaching style	Characteristics	Advantage	Disadvantage
Expert	Instructors possess knowledge and expertise that students need and maintain the status as experts by displaying detailed knowledge. The style focuses on information transmission and ensures that students are well prepared	Information, knowledge, and skills that the instructor possesses	Overuse may be unfavorable to less experienced students, and instructors may not always show the underlying thought processes that produced answers
Formal auditory	The style focuses on offering both positive and negative feedback, establishing learning goals, outlooks, and conventions of behavior for students. Instructors are concerned with the correct, acceptable, and standard ways to do things and with providing students with the structure they need to learn	Clear expectation and acceptable ways of managing things	Can lead to rigid, standardized, and less flexible ways of managing students
Personal model	They believe in teaching by personal example and establish a prototype for how to think and behave. They oversee, guide, and direct students by showing how to do things, encourage students to observe, and then emulate the instructor's approach	Emphasis on direct observation and following a role model is the primary advantage of this style of teaching	Some instructors may believe that their approach is the best way, leading some students to feel inadequate if they cannot live up to such standards
Facilitator	Emphasizes the personal nature of teacher–student interactions. They guide and direct students by asking questions, exploring options, suggesting alternatives and encourage students to develop criteria to make informed choices. The overall goal is to develop students' capacity for independent action, initiative, and responsibility. They work with students on projects in a consultative fashion and try to provide as much information and encouragement as possible	The advantages are the personal flexibility, focus on students' needs and goals, and the willingness to explore options and alternative courses of action	The style is often time-consuming and is sometimes applied when a more direct approach is required. Students become uncomfortable if not used in a confident and affirmative way
Delegator	They are concerned with developing students' capacity to function in an autonomous fashion. Students work on projects independently, and teachers are available at the request of students as resource persons	The style helps students to perceive themselves as independent learners	Disadvantages are misreading of student's readiness for independent work, and some students may become anxious in an autonomous environment

was suggested that instructors prefer one cluster to others or stick with a combination of clusters. Different teaching styles and the clusters are shown in Tables 13.1 and 13.2, respectively.

Table 13.2 Cluster of teaching styles. (Based on Grasha 1994)

Cluster	Primary styles	Secondary styles
1	Expert/formal auditory	Personal model/facilitator/delegator
2	Expert/personal model/formal auditory	Facilitator/delegator
3	Expert/facilitator/personal model	Formal auditory/delegator
4	Expert/facilitator/delegator	Formal auditory/personal model

The Teaching Style Inventory (Grasha 1994) was conducted among a group of 24 engineering faculty members in order to see their teaching styles. Academic staff members were taken randomly from different academic departments within the COE. The inventory involves 40 questions, and the academics were asked to answer each query using a score range of 1–7 (1 and 7 represents strongly disagree and strongly agree, respectively). Teaching styles of surveyed engineering faculty members are depicted in Fig. 13.3. It is noted that all staff members possess high score in expert style. About 71 and 29 % of respondents exhibit high and moderate scores in delegator teaching style, respectively. In terms of facilitator model style, about 42 and 58 % respondents exhibit high and moderate scores, respectively, whereas in the personal model style, about 67 and 33 % of respondents show moderate and high scores, respectively. Regarding the formal auditory style, around 54 % of the respondents exhibit high scores and 46 % exhibit moderate scores.

Figure 13.3 shows that surveyed faculty members display a combination of teaching styles, and they can be categorized as cluster 4 (Table 13.2), where primary styles are expert and delegator. The strong preference for expert teaching style has the benefit of information transmission to well-prepared students. The preference for delegator teaching style is important for developing students' capacity to function in an autonomous fashion. Particularly for the PBL, students work on projects independently, and faculty members are available at the request of students as resource persons. The moderate (58 %) and high scores (42 %) in facilitator teaching style presents the personal nature of faculty and student interactions. Faculty members as facilitators guide and direct students by asking questions, exploring options, suggesting alternatives, and encouraging students to develop criteria to make informed choices. The overall goal of the facilitator-type teaching style is to develop students' capacity for autonomous action, initiative, and responsibility, which are indispensable for successful implementation of PBL in engineering training. Faculty members having formal auditory teaching style (54 % high score) are concerned with the correct, acceptable, and standard ways to do things that may be favorable to sequential students (68 % students), who prefer logical steps and convergent thinking. Moderate preference for the personal model teaching style is observed in about 67 % of respondents. The style is preferable for students who desire to follow a role model. Yet, the function of role model-type teaching style is really limited in the PBL approach.

Felder and Silverman (1988) provided the dimension of learning styles and the corresponding teaching styles as presented in Table 13.3. They recommended a list of teaching techniques in order to address all learning styles of pupils. Further explanations of these techniques are listed below:

Fig. 13.3 Distribution of scores among different teaching styles (based on a group of 24 academic staff members from the College of Engineering)

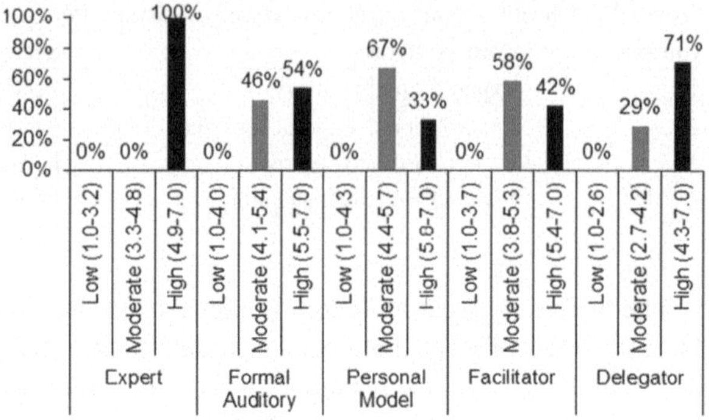

Table 13.3 Dimensions of learning and teaching styles. (Felder and Silverman 1988)

Preferred learning styles		Corresponding teaching styles	
Sensory	Perception	Concrete	Content
Intuitive		Abstract	
Visual	Input	Visual	Presentation
Auditory		Verbal	
Inductive	Organization	Inductive	Organization
Deductive		Deductive	
Active	Processing	Active	Student participation
Reflective		Passive	
Sequential	Understanding	Sequential	Perspective
Global		Global	

a. Motivate learning as much as possible, relate the material being presented to what has come before and what is still to come in the same course, to material in other courses, and particularly with the students' personal experience *(inductive/ global)*.
b. Offer a balance of concrete information (facts, data, real or hypothetical experiments, and their outcomes) (sensing) and abstract concepts (principles, theories, and mathematical models) (intuitive).
c. Balance material that emphasizes practical problem-working methods (sensing/active) with material that emphasizes fundamental understanding (intuitive/ reflective).
d. Provide explicit illustrations of intuitive patterns (logical inference, pattern recognition, and generalization) and sensing patterns (observation of the surroundings, empirical experimentation, and attention to detail), and encourage all students to exercise both patterns *(sensing/intuitive)*. Do not expect either group to be able to do the other group's processes immediately.
e. Follow the scientific method in presenting theoretical material. Offer concrete examples of the phenomena the theory describes or predicts (sensing/ inductive); then develop the theory or formulate the mold (intuitive/inductive/ sequential); establish how the theory or mold can be validated and deduce its consequences (deductive/sequential); and present applications (sensing/deductive/sequential).

f. Use pictures, schematics, graphs, and simple sketches liberally before, during, and after the presentation of verbal material *(sensing/visual)*. Show films *(sensing/visual)*. Provide demonstrations *(sensing/visual)*, hands-on, if possible *(active)*.
g. Use computer-assisted instruction—sensors respond very well to it *(sensing/active)*.
h. Do not fill every minute of class time lecturing and writing on the board. Provide intervals—however brief—for students to think about what they have been told *(reflective)*.
i. Provide opportunities for students to do something active besides transcribing notes. Small-group brainstorming activities that take no more than 5 min are extremely effective for this purpose *(active)*.
j. Assign some drill exercises to provide practice in the basic methods being taught *(sensing/active/sequential)* but do not overdo them *(intuitive/reflective/global)*. Also provide some open-ended problems and practices that call for analysis and synthesis (intuitive/reflective/global).
k. Provide students the choice of collaborating on homework assignments to the greatest possible extent (active). Active learners generally learn best when they interact with others; if they are refused the opportunity to get along, they are being divested of their most efficient learning tool.
l. Applaud creative solutions, even wrong ones (intuitive/global).
m. Talk to students about learning styles, both in advising and in classes. Students are reassured to find their academic difficulties may not entirely be due to personal inadequacies. Explaining to struggling sensors or active or global learners how they learn most efficiently may be an important step in helping them reshape their learning experiences so that they can be successful (all *types*).

During consultation with surveyed faculty members, most of them expressed that they prefer to be a delegator or facilitator in conducting the PBL project. Several of them hinted that an overview of the task at the start of the semester and regular monitoring of the project progress (weekly or fortnightly) are very important for the success of PBL. Some of them preferred that students should use different application software in their project, particularly for engineering design-related courses. Faculty members also expressed that for the success of PBL, students' engagement needs to enhance through collaborative learning and group interaction, by using different communication technologies (blogs, discussion forums, podcasts, etc.).

13.5 Technology-Supported Learning Environment

Very recently, the UAEU has introduced smart technologies (iTune, iPad, etc.) in teaching. Adoption of new technologies in teaching requires support to both students and instructors. The Center for Excellence in Teaching and Learning (CTEL) at the UAEU conducted a survey among faculties in fall 2012–2013 semester in order to understand their skill and knowledge on familiarity with technology in

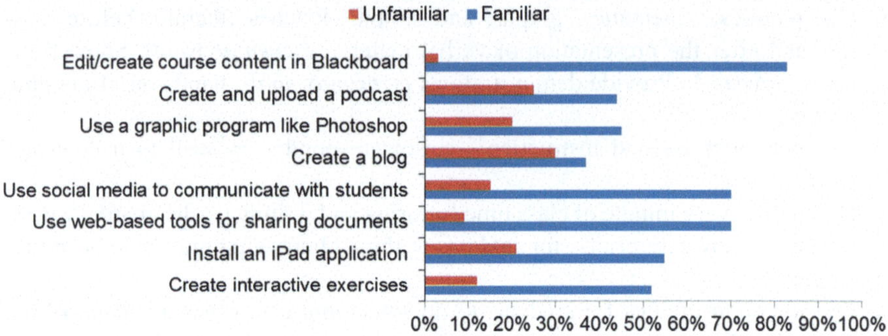

Fig. 13.4 Familiarity of faculty members in using technology in teaching (number of respondents 719, survey conducted by CETL in fall 2012–2013 semester)

pedagogy. A total of 719 faculties were surveyed. Among them, about 70% were college faculties, and the remaining 30% were UGRU instructors. Responses on the familiarity of instructors in using technology in teaching are shown in Fig. 13.4. Their responses on views of using technologies are provided in Table 13.4. It is mentioned that higher familiarity and skill is reported in using smart multimedia Blackboard, web-based sharing of documents, and communication with students using social web tools. Most of these areas are regularly used for teaching purposes or for social purposes. Unfamiliarity is observed in using blogs, podcasts, graphics, programming, and installing iPad applications. Table 13.4 demonstrates that while most faculties enjoy using new technologies, they grow nervous in taking on new technologies. Most faculties (89%) choose to combine traditional and new technologies in their instruction. Nearly 76% of respondents show willingness to apply new technologies after having training courses. In general, it is discovered that faculty members at UAEU showed very positive attitude toward using technology in teaching. This definitely demonstrates a positive advance in developing PBL at UAEU. However, both students and instructors, and particularly students, need to improve their skill in blogs and podcasts, since the PBL approach utilizes these tools for online discussion and seeks feedback from instructors.

Instructors were asked about the aspects of infrastructure underpinning educational technology at the UAEU. Responses from respondents are presented in Table 13.5. Nearly 79% of respondents felt that their approach to technology is sufficient at UAEU. Because of the freely accessible wireless Internet facility, 94% respondents agreed that they had sufficient access to the Internet on campus. About 37 and 21% respondents expressed their satisfaction and dissatisfaction, respectively, regarding the supports they received from the university, and 34% respondents agreed that they do not possess adequate time for discovering new technologies. Nearly half of respondents (53%) agreed that training opportunities are available for the integration of technology into instruction. Regarding the institutionalization of using new technologies in teaching, leadership is critical in motivating and supporting innovation and cultural change within systems. Approximately 87% respondents agreed that the UAEU administration encourages the role of new technologies in instruction.

Table 13.4 Views of faculty members in using technology in teaching (total respondents 719, survey conducted by CETL in fall 2012–2013 semester)

Questions related to technology uses in teaching	Agree (%)	Disagree (%)
Online social networking helps students better their communication skills	65	12
I am nervous using new technology in class	83	7
I find using new technology enjoyable	84	4
Technology can help faculty improve their teaching	55	14
New technology in the class improves instruction	79	5
Online learning is not as effective as classroom learning	55	31
Blackboard is useful for giving class information and assignments	91	4
There is generally too much focus on technology in education	60	18
Mixing new technology and traditional methods increases student–faculty interaction	89	3
I use technology often when I teach	82	7
Experience in online environments increases student motivation for the course	63	26
If there was more support for me to take new technology, I would apply it in my teaching	76	17

Table 13.5 Assessment of technology support to faculties at UAEU. (CETL 2012b)

Questions related to technology support	Agree (%)	Disagree (%)	No view (%)
Sufficient access to the technology that is needed at UAEU	79	9	12
Adequate time to learn technological skills	46	34	20
Sufficient level of technical support at UAEU for iPads and associated technologies	37	21	42
Sufficient access to the Internet on campus	94	6	0
The administration encourages the use of technology in the classroom	87	3	10
Enough training opportunities for the integration of technology into teaching	53	18	29

The success of using technology in teaching also depends on students' attitude and capability to cope with technologies. A total of 1617 students were asked to answer how often they use different technologies in educational purposes. Their answers are presented in Table 13.6. Students were also asked to indicate as many learning resources as they thought were useful to them. The results of their responses are shown in Fig. 13.5. It is observed that students at UAEU have access to and are capable of using technologies in education. In a day, about 80 and 76 % of students use the Internet and smartphone, respectively. About 52 % students use electronic media and technology for their educational resources. From the viewpoint of PBL, these are definitely supportive for the development of PBL at UAEU.

Table 13.6 Frequency of technology uses in educational purposes by UAEU students. (CETL 2012b)

Technologies	Frequency of use (%)				
	Never	Rarely	Sometimes	Often	Daily
Personal computer	11	31	28	17	13
Laptop	6	10	18	29	37
iPad	18	17	18	16	31
iPod	34	24	17	12	13
Internet	1	1	4	14	80
Smartphone	11	3	5	5	76
Blackboard	48	18	16	11	7
Podcast	32	19	20	15	14

It is generally believed that when students use technology as a tool or a support for communicating with others in a PBL project, they are in an active role rather than the passive function of a receiver of information transmitted by a staff member. Students are actively involved in making choices about how to generate, obtain, manipulate, or display information.

13.6 Conclusions

The chapter delineated the potential and constraints of PBL in engineering education at the UAEU by assessing the learning styles of engineering students, teaching styles of engineering instructors, and assessment of technology used in teaching. Discussion on mismatches of learning and teaching styles was included along with

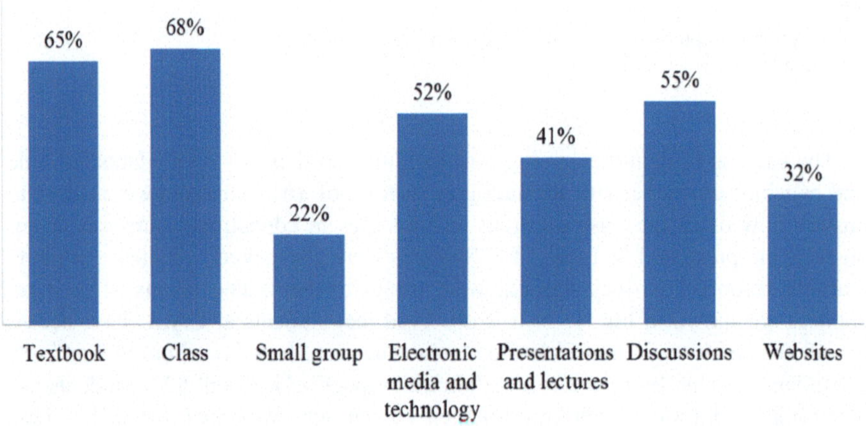

Fig. 13.5 Students' learning resources at UAEU. (CETL 2012b)

a description of how the incorporation of learning and teaching styles in PBL can improve its effectiveness. Research on learning styles suggests that in engineering courses, learning is optimized by the application of different learning styles. Yet, most engineering instructors assume not only that all students adopt (or should adopt) a uniform learning style, they expect the same learning style to be applied to all areas of engineering studies (Felder 1993; Holt and Solomon 1996). There is considerable evidence from previous studies to show that instructors' assumptions about "typical" learning styles of engineering students are inaccurate and that a wide range of learning styles will actually exist within any engineering class. According to Felder and Silverman (1988), how much a given student learns in a class is governed not only in part by that student's native ability and prior preparation, but also by the compatibility of his or her learning style and the instructor's teaching style.

Learning styles of most engineering students and teaching styles of most engineering instructors are incompatible in several dimensions. Many or most engineering students are visual, sensing, inductive, and participating (some exceptional creative students are global), whereas most engineering education is verbal (auditory), abstract (intuitive), deductive, passive, and sequential. These mismatches lead to poor student performance, professorial frustration, and a loss to society of many potentially excellent engineers (Felder and Silverman 1988). The situation of UAEU also reflects this. Most engineering students are observed to be active, visual, sensory, and sequential, whereas a surveyed group of instructors exhibits an expert and delegator style of teaching as their most preferred styles. Teachers also exhibit a firm preference for the facilitator style of instruction. The PBL approach prefers the facilitator style of instruction rather than expert or delegator styles. The sequential and sensory learning styles demand sufficient training of students on the PBL system at the start of the semester. On the other hand, teaching styles of surveyed engineering instructors were found very supportive of the implementation of the PBL approach in engineering education. Both students and staff members were found very effective in applying innovative technologies in teaching. However, there was a lack of the use of online discussion forums, which is really important for collaborative learning through the PBL project.

It is thus important to raise the consciousness of teachers responsible for engineering education about the possible effects of mismatches of students' learning styles and instructors' teaching styles. The primary requirements for improved teaching and learning in this area are improved knowledge and understanding of learning styles, and reflection on practice, followed by any necessary modifications to practice in the light of this knowledge. It is important to include learning styles information in the course booklet along with the course website (Blackboard), giving students and instructors continuing access to information on all aspects of learning styles. Instructors need not apply all the techniques in every class, but rather pick several that look feasible and try them, keep the ones that work, drop the others, and try a few more in the next course. In this way a teaching mode that is both efficient for students and comfortable for the instructor will develop naturally, with a potentially dramatic effect on the caliber of learning that subsequently happens.

Acknowledgements The author is grateful to the Center of Excellence in Teaching and Learning (CETL), UAEU for providing their survey results and for the permission to use their survey results in this book chapter.

References

Al-Maati, S. A., & Damaj, I. (2010). Developing a sustainable engineering education in the Middle East and north Africa region, transforming engineering education: Creating interdisciplinary skills for complex global environments, 2010 IEEE, Dublin, 6–9 April 201, pp. 1–11, doi:10.1109/TEE.2010.5508888.

Anderson, M. (1991). Learning styles of the graduate career change woman engineer: Then and now, Proceedings of GASAT 6th International Conference, University of Melbourne, Victoria, Australia.

Beyer, K. (1993). Gender, science anxiety and learning style. Proceedings of GASAT 7th International Conference, University of Waterloo, Ontario.

Bielenberg, B. T. (2005). Characteristics of the UAE Higher Education Student, Proceedings of the 6th Annual UAEU Research Conference, UAE.

Biggs, J., & Tang, C. (2007). *Teaching for quality learning at university: What the student does* (3rd ed.). Maidenhead: Open University Press.

Bloxham, S., & West, A. (2004). Understanding the rules of the game: Marking peer assessment as a medium for developing students' conceptions of assessment. *Assessment & Evaluation in Higher Education, 29*(6), 721–733.

Blumenfeld, P. C., Soloway, E., Marx, R. W., Krajcik, J. S., Guzdial, M., & Palincsar, A. (1991). Motivating project-based learning: Sustaining the doing, supporting the learning. *Educational Psychologist, 26*(3, 4), 369–398.

Boles, W., Hadgraft, R., & Howard, P. (2009). Exploring synergies between learning and teaching in engineering: A case study approach. *Australasian Journal of Engineering Education, 15*(1), 19–26.

Cangelosi, S. J. (2003). *Teaching mathematics in secondary and middle school.* New Jersey: Prentice-Hall, Inc.

CETL. (2012a). Faculty attitudes towards technology-supported learning environments, center of excellence in teaching and learning, United Arab Emirates University.

CETL. (2012b). Comparative survey of students' attitudes concerning the use of educational technology, center of excellence in teaching and learning, United Arab Emirates University.

Coffield, F., Moseley, D., Hall, E., & Ecclestone, K. (2004). Learning styles and pedagogy in post-16 learning: A systematic and critical review, Learning and Skills Research Centre, www.LSRC.ac.uk.

Cross, K. P. (1998). Opening windows on learning: The cross papers number 2. Mission Viejo, CA: League for Innovation in the Community College and Educational Testing Service.

Cross, K. P. (1999). Learning is about making connections: The cross papers number 3. Mission Viejo, CA: League for Innovation in the Community College and Educational Testing Service.

Dyball, R., Brown, V., & Keen, M. (2007). Towards sustainability: Five strands of social learning. In A. E. J. Wals (Ed.), *Social learning towards a sustainable world; principles, perspectives, and praxis* (pp. 181–195). The Netherlands: Wageningen Academic Publishers.

Ewell, P. T. (1997). Organizing for learning: A new imperative. *AAHE Bulletin, 50*(4), 3–6.

Felder, R. (1993). Reaching the second tier: Learning and teaching styles in college science education. *Journal of College Science Teaching, 23*(5), 286–290.

Felder, R. (1999). Index of learning styles (ILS), http://www.engr.ncsu.edu/learningstyles/ilsweb.html. Assessed Feb 2012.

Felder, R., & Silverman, L. (1988). Learning and teaching styles in engineering education. *Engineering Education, 78*(7), 674–681.

Gibson, I. S. (2003). From solo-run to mainstream thinking: Project-based learning in engineering design. *European Journal of Engineering Education, 28*, 331–337.

Gillway, M., & Bielenberg, B. (2006). Adapting problem-based learning for the first-year experience. *UGRU Journal*, *3*(Fall 2006), 1–10.
Grasha, A. (1994). A matter of style: The teacher as expert, formal authority, personal model, facilitator and delegator. *College Teaching, 42*(4), 142–149.
Harding, J. (1996). Girls' achievement in science and technology: Implications for pedagogy. In: P. F. Murphy & C. J. Gipps (Eds.), *Equity in the classroom*. Falmer Press/UNESCO Publishing.
Helle, L., Tynjala, P., & Olkinuora, E. (2006). Project-based learning in post-secondary education—theory, practice and rubber sling shots. *Higher Education, 51*, 287–314.
Holt, J., & Solomon, F. (1996). Engineering education—The way ahead. *Australasian Journal of Engineering Education, 7*(1), 47–60.
Kolb, D. (1984). *Experiential learning: Experience as the source of learning and development.* Englewood Cliffs: Prentice-Hall, Inc. http://academic.regis.edu/ed205/kolb.pdf.
Mills, J. E., & Treagust, D. F. (2003). Engineering education—is problem-based or project-based learning the answer? *Australasian Journal of Engineering Education, 11*, 2–16.
Mills, J., Ayre, M., Hands, D., & Carden, P. (2005). Learning about learning styles: Can it improve engineering education? *MountainRise, 2*(1), 145–182.
Nepal, K. P., & Jenkins, G. A. (2011). Blending project-based learning and traditional lecture-tutorial based teaching approaches in engineering design courses, Proceedings of the 2011 AAEE Conference, Fremantle, Western Australia.
Nepal, K. P., & Panuwatwanich, K. (2011). Comparative study of project-based learning and traditional lecture tutorial teaching approaches in undergraduate engineering courses, Proceedings of the 2011 AAEE Conference, Fremantle, Western Australia.
Nepal, K. P., & Stewart, R. A. (2010). Relationship between self-directed learning readiness factors and learning outcomes in third year project-based engineering design course, Proceedings of the 2010 AAEE Conference, Sydney.
Perrenet, J. C., Bouhuijs, P. A. J., & Smits, J. G. M. M. (2000). The suitability of problem-based learning for engineering education: Theory and practice. *Teaching in Higher Education, 5*(3), 345–358.
Ribeiro, L. R. C., & Mizukami, M. G. N. (2005). Problem-based learning: A student evaluation of an implementation in postgraduate engineering education. *European Journal of Engineering Education, 30*, 137–149.
Shanableh, A., & Omar, M. (2007). Shaping engineering education in the Middle East—a report on four engineering education conferences held at the University of Sharjah. *World Transactions on Engineering and Technology Education, 6*(2), 313–317.
Spalding, B., Ferguson, S., Garrigan, P., & Stewart, R. (1999). How effective is group work in enhancing work-based learning? An evaluation of an education studies course. *Journal of Further and Higher Education, 23*(1), 109–115.
Steinhorst, K., & Keeler, C. M. (1995). Developing material for introductory statistics courses from a conceptual active learning viewpoint. *Journal of Statistics Education, 3*(3) [online].
Stice, J. E. (1987). Using Kolb's learning cycle to improve student learning. *Engineering Education, 77*(5), 291–296.
Thomas, J. W. (2000). A review of research on project-based learning, The Autodesk Foundation, California. http://www.autodesk.com/foundation.
Thomas, J. W., Mergendoller, J. R., & Michaelson, A. (1999). *Project-based learning: A handbook for middle and high school teachers*. Novato: The Buck Institute for Education.
Walker, A. (2001). British psychology students—perceptions of group-work and peer assessment. *Psychology Learning and Teaching, 1*(1), 28–36.
Willey, K., & Gardner, A. (2009). Developing team skills with self- and peer assessment: Are benefits inversely related to team function? *Campus-Wide Information Systems, 26*(5), 365–378.
Yam, L. H. S., & Burger, P. (2009). Student engagement and teacher's 'self': A case study of inclusive teaching', 15th Annual Conference of the Pacific Rim Real Estate Society, University of Technology Sydney, Sydney.
Yam, L. H. S., & Rossini, P. (2010). Implementing a Project-Based Learning Approach in an Introductory Property Course, 16th Annual Conference of the Pacific Rim Real Estate Society Conference, January 2010, New Zealand.

Dr. Rezaul K. Chowdhury has more than 9 years of university teaching and research experiences. Currently, he is an assistant professor of water resources at the United Arab Emirates University. Previously, Dr. Chowdhury was a research fellow of water engineering at the University of South Australia and was a scientist of integrated water systems planning and management at the Commonwealth Scientific and Industrial Research Organization (CSIRO) in Australia. Dr. Chowdhury started his career as a lecturer of civil and environmental engineering at the Shahjalal University of Science and Technology in Bangladesh. He has taken part in various advanced trainings and workshops on university teaching and curriculum development. He has published more than 45 research articles in leading international journals and conferences. He is an active reviewer of various high impact factor journals. Dr. Chowdhury is involved in supervising postgraduate research scholars.

Chapter 14
Effective Alignment of Disciplinary and Institutional Accreditation and Assessment: A UAE Computing Case Study

Kevin Schoepp, Maurice Danaher and Leon Jololian

14.1 Introduction

As a means of assuring quality in their programs, institutions of higher education in recent times have moved to accreditation and to establishing and assessing learning outcomes. Besides assuring quality and beginning the cycle of continuous improvement, this move has been designed to satisfy government or other legislative bodies, employers, students and prospective students, and other key stakeholders. The major impetus for the shift towards learning outcomes has been the requirement set by accreditation bodies, compelling institutions to adopt that approach (Hernon et al. 2006).

Adopting the learning outcomes approach promulgated by accreditors may involve considerable effort because institutions with multiple accreditations must answer to a number of organizations which have slightly different expectations. If not well managed, these processes can seem overwhelming and, at times, duplicative. In the Middle East and North Africa, the challenge is even greater as, according to the Organisation for Economic Co-operation and Development (OECD 2013), institutions have more quality issues than those in more developed nations. This is partially due to the rapid expansion of education, often at the expense of quality. To set up and maintain an institutional assessment system and also to go through the steps involved to achieve accreditation from a number of accreditors is very demanding on time and resources. Indeed, it is a daunting task to institutions with insufficient resources and can be a deterrent to seeking accreditation.

K. Schoepp (✉) · M. Danaher · L. Jololian
Zayed University, Dubai and Abu Dhabi, United Arab Emirates
e-mail: kevin.schoepp@zu.ac.ae

M. Danaher
e-mail: Maurice.danaher@zu.ac.ae

L. Jololian
e-mail: jololian@hotmail.com

M. Abdulwahed et al. (eds.), *Advances in Engineering Education in the Middle East and North Africa,* DOI 10.1007/978-3-319-15323-0_14

In this chapter, we describe the processes developed at Zayed University (ZU) to assess learning outcomes for institutional quality assurance purposes and for the purposes of accreditation. These processes have provided an effective system for institutional assessment and have enabled the institution to achieve Middle States Commission on Higher Education (MSCHE) accreditation and Accreditation Board for Engineering and Technology (ABET) accreditation for the College of Technological Innovation (CTI).

Based on an analysis of our existing processes, we have designed a new model. The new model, which we are proposing for institutions, accommodates the needs of institutional assessment and various accreditation bodies simultaneously. The goals of the model are efficiency, sustainability, and quality. The model incorporates several key elements:

- An alignment of institutional learning outcomes with the programmatic learning outcomes
- An alignment of the institutional and programmatic learning outcomes with the learning outcomes of the accreditation body
- Standardized internal processes that constitute the assessment program
- A centralized structure that coordinates the various ongoing accreditation and assessment efforts within the institution

The model facilitates accreditation from multiple accreditation bodies and is built around the ethos of continuous improvement. It fosters an institutional culture of assessment as much of it is faculty-driven, a key to faculty buy-in (Gannon-Slater et al. 2014).

14.2 Zayed University

The UAE is a small petroleum-rich nation bordering Saudi Arabia, which has been developing at breakneck speed. Founded in 1971 and one of the last Gulf countries to modernize, in just over one generation it has transitioned from an impoverished nation to a place with immense wealth and privilege (Heard-Bey 2004). The country's growth and development has been mirrored by the rapid development of the higher education sector. The first public university, United Arab Emirates University, was launched in 1976, and ZU was opened in 1998.

ZU is a rapidly expanding public university, with campuses in Dubai and Abu Dhabi, which serves approximately 8500 undergraduate male and female Emirati students in a gender-segregated environment. Though only 15 years old, ZU has been accredited by the MSCHE since 2008 and its CTI accredited by ABET since 2012. Accreditation from the MSCHE was prioritized early on by the institution's leadership because it provided international recognition of quality and because when the university was launched, the country lacked a national accreditation and quality assurance body. As part of the institution's maturation process, the 2009 strategic plan "Destined to Lead" prioritized disciplinary accreditation across the university—again as a way to ensure quality that is international in scope. The CTI was the first of the colleges to obtain disciplinary accreditation, which was accomplished for its bachelor's degrees in *Information Systems and Technology Management* and *Security and Network Technologies*. For both MSCHE and ABET, the assessment and analysis of student learning outcomes, which is part of a cycle of continuous improvement, is the keystone of accreditation.

14.3 Learning Outcomes at Zayed University

The launch of ZU occurred at a time when the learning outcomes and learning outcomes assessment movements were just becoming established. In 1997, ABET released *Engineering Criteria 2000* (2011a) and later Ewell (2001) published the Council for Higher Education Accreditation (CHEA) occasional paper *Accreditation and Student Learning Outcomes: A Proposed Point of Departure*. The shift to a student- and learning-centered ethos was clearly underway, and ZU was influenced by this movement. Though the percentage of programs having student learning outcomes in US undergraduate education is now around 84 % (Kuh et al. 2014), at the time ZU was developing its institutional learning outcomes and associated rubrics, this approach was still in its infancy.

One of the benefits of being a young institution is that there is significant opportunity to achieve milestones because there is a lack of history, sentiment, or bias that can limit progress. Though there is much to do, there is opportunity to have an impact on the organization. The creation of the institution-wide learning outcomes known as the ZU Learning Outcomes (ZULOs) is one such example of where faculty achievements can be substantial at new institutions.

In the 2001–2002 academic year, faculty working groups were formed to develop a set of institution-wide learning outcomes to form the foundation of the university's undergraduate program. Learning outcomes developed in a collaborative manner such as this have a much better chance of being accepted by faculty as meaningful (AAC&U 2010; MSCHE 2007; US Department of Education 2006). The ZULOs represent the best thinking of faculty as to what should be the main outcomes of a quality undergraduate education. Though they have been altered and modified over the years, the major concepts included in them remain very similar today. The ZULOs in their current form are:

- *Language*: ZU graduates will be able to communicate effectively in English and Modern Standard Arabic, using the academic and professional conventions of these languages appropriately.
- *Technological Literacy*: ZU graduates will be able to effectively understand, use, and evaluate technology both ethically and securely in an evolving global society.
- *Critical Thinking and Quantitative Reasoning*: ZU graduates will be able to use both critical and quantitative processes to solve problems and to develop informed opinions.
- *Information Literacy*: ZU graduates will be able to find, evaluate, and use appropriate information from multiple sources to respond to a variety of needs.

- *Global Awareness*: ZU graduates will be able to understand and value their own culture and other cultures, perceiving and reacting to differences from an informed and socially responsible point of view.
- *Leadership*: ZU graduates will be able to undertake leadership roles and responsibilities, interacting effectively with others to accomplish shared goals.

Similar to the model that was later employed in designing the American Association of Colleges and Universities' (AAC&U's) Valid Assessment of Learning in Undergraduate Education (VALUE) rubrics (AAC&U 2010), each ZULO comes with a rubric which has a set of performance indicators and descriptors for the students' level of attainment. The attainment levels are described for first year *(Beginning)*, second and third years *(Developing)*, near or at graduation *(Accomplished)*, and an aspirational level *(Exemplary)*, which is not an expectation for our students.[1] The institutional expectation is that students will graduate at the *Accomplished* level. What this means is that students, faculty, or other stakeholders can read the ZULO rubric at the *Accomplished* level and get an understanding of the abilities of ZU graduates. This closely follows the MSCHE learning outcome guidelines, where they state that well-articulated learning outcomes make it clear to learners, faculty, and the public what graduates will be able to accomplish (MSCHE 2007).

In addition to the ZULOs, each of the colleges or departments has a set of Major Learning Outcomes (MALOs), which are the disciplinary interpretation of the ZULOs. Each MALO is aligned with one or more ZULO and provides the needed disciplinary lens, which enables faculty understanding and acceptance. Within the CTI, the MALOs are:

- *Critical Thinking and Quantitative Reasoning in IT*: College graduates will be able to use critical thinking and quantitative processes to identify, analyze, and solve problems and evaluate solutions in an IT context.
- *Information Technology Application*: College graduates will be able to select existing and cutting-edge IT tools and procedures to develop modules and systems.
- *Information Technology Management*: College graduates will be able to assess and determine information resource requirements to develop solutions suitable for IT and business managers operating in a multinational and multicultural environment.
- *Information Technology Professional Practice*: College graduates will be able to work effectively in individual and group situations, understand how groups interact, be able to assume a leadership role when required, and understand the fundamentals of professional and ethical conduct.
- *Information Technology Systems Theory and Practice*: College graduates will be able to understand and communicate the fundamentals of systems theory in the development of appropriate systems that function in a global environment.
- *Technical Communication (Bilingual)*: College graduates will be able to express themselves effectively and efficiently in both English and Arabic while using the correct IT terms for each language.

[1] Each complete ZULO rubric can be found online at http://www.zu.ac.ae/main/en/_assessment_resource/learning_outcomes.aspx.

Table 14.1 ZULO to MALO mapping

ZULO	Information literacy	Information technology	Critical thinking	Global awareness	Leadership	Language
MALO						
Critical thinking and quantitative reasoning			✓			
IT application		✓		✓		
IT management	✓			✓		
IT professional practice		✓			✓	
IT systems theory and practice	✓		✓			
Technical communication						✓

ZULO Zayed University Learning Outcome, *MALO* Major Learning Outcome

Similar to the ZULOs, the MALOs have rubrics which describe students' expected abilities at three levels—*Beginning, Developing and Accomplished*—again corresponding to entry, mid, and exit points in CTI's bachelor programs. The alignment or mapping between ZULOs and MALOs is shown in Table 14.1, and the rationale for the mapping is provided in Table 14.2.

14.4 International Accreditation and Assessment

International accreditation is considered a seal of approval of quality that many higher education institutions the world over seek to attain in order to demonstrate the effectiveness of their programs. From its most recent data, the CHEA claimed that US accreditors had accredited more than 850 international institutions in 70 countries (University Affairs 2013). Within ABET alone, there are 345 internationally accredited programs (ABET 2013). On one hand, accreditation ensures that an institution is in line with the best practices for the development of their educational offerings, and on the other hand, it demonstrates to the institution's stakeholders the quality of education they are providing to students. By attaining accreditation from one of the many agencies that exist worldwide, an institution is perceived as achieving suitable standards. According to ABET, "accreditation is value. Reaching into our public, private, and professional lives, accreditation is proof that a collegiate program has met certain standards necessary to produce graduates who are ready to enter their professions" (2011b, p. 1).

Table 14.2 Rationale for mapping between ZULOs and MALOs

MALO	Justification of mapping to ZULO
IT critical thinking and quantitative reasoning	ZULO Critical Thinking and Quantitative Reasoning: Both sets of learning outcomes address the development of critical thinking and quantitative reasoning
ITA	ZULO IT: This ZULO strives for the development of basic IT skills in all university graduates. The ITA MALO requires students to develop higher-level IT skills ZULO Global Awareness: Both learning outcomes strive for developing student skills in the use of IT applications that are universal
ITM	ZULO Information Literacy: This ZULO strives for the development of basic information literacy skills for finding relevant information for given tasks. The ITM MALO requires students to develop higher-level IL skills for identifying appropriate resources for IT projects ZULO Global Awareness: Both learning outcomes strive for developing student skills in the use of IT tools and resources based on international standards
ITPP	ZULO Information Technology: This ZULO strives for the development of basic IT skills in all university graduates. The ITPP MALO requires students to engage in the professional practice of IT ZULO Leadership and Teamwork: The ITPP MALO requires students to have developed skills in leadership and teamwork
ITSTP	ZULO Information Literacy: This ZULO strives for the development of basic IT skills in all university graduates. The ITSTP MALO requires students to have knowledge of IT to develop complex systems ZULO Critical Thinking and Quantitative Reasoning: The development and understanding of complex systems required by this MALO demands higher order of critical thinking and quantitative reasoning skills
IT technical communication	ZULO language: Both learning outcomes require the development of the language skills. The MALO focuses on the language skills within an IT context

ZULO Zayed University Learning Outcome, *MALO* Major Learning Outcome, *IT* information technology, *ITA* IT application, *ITM* IT management, *ITPP* IT professional practice, *ITSTP* IT systems theory and practice, *IL* information literacy

As the reach of accreditation has expanded, assessment has expanded at its side, and the two terms have become almost synonymous with each other. According to Ewell (2009, p. 6), "assessment has become a condition of doing business for colleges and universities because of accreditation requirements." Kuh et al. (2014), reporting on a large US study about assessment of student learning, found likewise that the prime driver of assessment is accreditation. Accreditors realize this, so in recognition of the leading role learning assessment plays in accreditation, the CHEA in the USA gives their annual prize in one category only—outstanding institutional practice in student learning outcomes. The award criteria include articulating learning outcomes, assessing the learning outcomes, and using the assessment results to foster improvement (CHEA 2014).

Increasingly, higher education institutions are pursuing multiple accreditations to achieve a much more comprehensive level of alignment with best educational practices that targets different aspects and levels of the institution, from the organizational and operational structure to the individual educational programs. Moreover, institutions will often have to work towards compliance with the various accreditation requirements, including assessment, simultaneously. This can put considerable stress on existing institutional resources that are often already stretched to a great extent. A certain level of coordination must be established between the ongoing accreditation and assessment efforts within an institution not only to ensure operational efficiency by reducing duplication of efforts but also to ensure that the goals and objectives emanating from compliance with various accreditation efforts are aligned and in harmony with one another.

14.4.1 ABET and MSCHE Learning Outcomes

With the release of their EC2000 document in 1997, ABET changed the focus of accreditation criteria from what was taught to what was learned and placed the continuous improvement process at its core (ABET 2011a). This philosophical shift has become the modus operandi of accreditors the world over and is also reflected within the MSCHE. Though there are differences in the language used and individual standards and criteria differ, the underlying principles are the same for both ABET and MSCHE. For example, ABET states through their Student Outcomes Criteria (Criterion 3) that "the program must have documented student outcomes that prepare graduates to attain the program educational objectives" (ABET 2009, p. 9) and through Continuous Improvement (Criterion 4) that "the program must regularly use appropriate, documented processes for evaluating the extent to which both the program educational objectives and the student outcomes are being attained. The results of these evaluations must be utilized as input for the continuous improvement of the program" (ABET 2009, pp. 10–11).

In a very similar manner, MSCHE states that "assessment of student learning demonstrates that, at graduation, or other appropriate points, the institution's students have knowledge, skills, and competencies consistent with institutional and appropriate higher education goals" (MSCHE 2009, p. 63). They describe the four-step assessment process as:

1. Developing clear and measurable learning outcomes
2. Providing learning opportunities where students can achieve the learning outcomes
3. Assessing student achievement of the learning outcomes
4. Using the results of the assessments to improve student learning (MSCHE 2009)

ABET (2009, pp. 2–3) also puts forth a set of learning outcomes that students should be able to achieve by the time of graduation, which in the Computing Accreditation Commission context are:

a. An ability to apply knowledge of computing and mathematics appropriate to the discipline
b. An ability to analyze a problem and identify and define the computing requirements appropriate to its solution
c. An ability to design, implement, and evaluate a computer-based system, process, component, or program to meet desired needs
d. An ability to function effectively in teams to accomplish a common goal
e. An understanding of professional, ethical, legal, security, and social issues and responsibilities
f. An ability to communicate effectively with a range of audiences
g. An ability to analyze the local and global impact of computing on individuals, organizations, and society
h. Recognition of the need for and an ability to engage in continuing professional development
i. An ability to use current techniques, skills, and tools necessary for computing practice
j. An ability to use and apply current technical concepts and practices in the core information technologies
k. An ability to identify and analyze user needs and take them into account in the selection, creation, evaluation, and administration of computer-based systems
l. An ability to effectively integrate IT-based solutions into the user environment
m. An understanding of best practices and standards and their application
n. An ability to assist in the creation of an effective project plan

These outcomes are not truly prescriptive because programs are free to have their own learning outcomes. Programs only need to demonstrate that these ABET learning outcomes are being covered. The assessment program at ZU is able to meet the assessment requirements of both accreditors.

14.4.2 MALOs and ABET Student Outcomes Alignment

The MALOs of the CTI are in line with the 14 student outcomes of ABET given above. The mapping between the MALOs and ABET's student outcomes is shown in Table 14.3, and a justification of the mapping is included. Due to the alignment between the MALOs and ABET's student outcomes, students that achieve a satisfactory standard in the MALOs are automatically meeting the requirements of ABET.

14.5 Assessment Structures and Processes

Of critical importance in having a sustainable, reliable, and valid assessment program is to ensure that proper structures and processes are put in place. The learning outcomes assessment program at ZU is centrally managed through the Office of

Table 14.3 Mapping between MALOs and ABET's student outcomes

MALO	ABET's students outcomes	Mapping rationale
Critical thinking and quantitative reasoning in IT	(b) An ability to analyze a problem and identify and define the computing requirements appropriate to its solution	A clear direct mapping for skills using critical thinking and quantitative reasoning
Information technology application	(a) An ability to apply knowledge of computing and mathematics appropriate to the discipline (i) An ability to use current techniques, skills, and tools necessary for computing practices (l) An ability to effectively integrate IT-based solutions into the user environment	The student outcomes (a), (i) and (l) address the skills involved in the application and integration of technology
Information technology management	(k) An ability to identify and analyze user needs and take them into account in the selection, creation, evaluation, and administration of computer-based systems (m) An understanding of best practices and standards and their application (n) An ability to assist in the creation of an effective project plan	The student outcomes (k), (m) and (n) address the management and deployment of technology solutions
Information technology professional practice	(d) An ability to function effectively on teams to accomplish a common goal (e) An understanding of professional, ethical, legal, security, and social issues and responsibilities (h) Recognition of the need for, and an ability to engage in, continuing professional development	The student outcomes (d), (e), and (h) address the skills required to function effectively in the workplace
IT systems theory and practice	(c) An ability to design, implement, and evaluate a computer-based system, process, component, or program to meet desired needs (g) An ability to analyze the local and global impact of computing on individuals, organizations, and society (j) An ability to use and apply current technical concepts and practices in the core information technologies	The student outcomes (c), (g), and (j) address the skills required to analyze and develop complex systems according to international best practices
Technical communication	(f) An ability to communicate effectively with a range of audiences	A clear direct mapping for communication skills in the workplace

MALO Major Learning Outcome, *ABET* Accreditation Board for Engineering and Technology

Educational Effectiveness (OEE). The responsibilities of the office include accreditation and learning outcomes assessment, and the office is headed by the director of educational effectiveness. The presence of this office is clear evidence that the institution recognizes the importance of supporting assessment and accreditation efforts in line with best practice (Fu et al. 2014). The director chairs the University Learning Assessment Committee (ULAC), which provides oversight and guidance for assessment at the institution. It comprises two faculty representatives from each academic program along with a select few others who participate in assessment, such as student affairs and institutional research professionals. These are the people responsible for assessment within their college or department. This group meets approximately 12 times per year and is the key venue for process and procedural refinement, professional development, and the sharing of best practices. It is these ULAC members who steer the overall assessment program at the university and lead assessment within their academic programs.

Part of what the ULAC has done over the past few years is to establish a set of agreed-upon processes built around an annual assessment calendar. The key events of the assessment calendar are the template-based assessment plan and report. At the start of each academic year, programs submit an annual learning assessment plan that includes information about which outcomes will be assessed, in which courses they will be assessed, and the methods of assessment that will be utilized. To ensure a sustainable assessment regime, we only require the assessment of two learning outcomes per year. According to Blaich and Wise (2011), the Wabash[2] national study found that this is the absolute threshold for the maximum number of outcomes that could be assessed with an expectation of meaningful actions towards continuous improvement being implemented. The majority of the assessments that we utilize are course-embedded direct measures because we believe that the curricular alignment offered by such instruments increases faculty buy-in and ownership (Suskie 2009). However, we also use a number of non-course-embedded direct measures such as the International English Language Testing Service or our College of Business's Assurance of Learning Examination because of the trustworthiness of the instruments (Suskie 2009). Besides these direct measures, the programs also receive the annual results from the Graduating Senior Survey and the Internship Survey for triangulation purposes. These instruments have had their questions mapped to the ZULOs, so they can be a meaningful addition to the assessment of the learning outcomes.

In the CTI, as mentioned earlier, three achievement levels are used with the MALOs—*Beginning, Developing*, and *Accomplished.* A rubric has been defined for each of the MALOs, which provides guidance for instructors to develop assessment tools to measure the learning outcome at each of the three levels.[3] The assessment of a MALO is undertaken by faculty under the guidance of the assessment coordinator (a ULAC member from the college). Faculty agree upon the method of assessment and the task to be used for conducting the measurement. A task-specific rubric is developed, following a standard format within the college, in which the performance indicators are clearly described. An example of a task-specific rubric is shown in Appendix A. This rubric, which was for measurement of Critical Thinking and Quantitative Reasoning at the *Beginning* level, was developed around a Cisco skills exam as part of a networking course.

[2] The Wabash study was a large-scale national project to investigate critical factors that affect outcomes.http://www.liberalarts.wabash.edu/study-overview/.

[3] The MALO rubric can be found online at: http://www.zu.ac.ae/main/files/contents/assessment_resource/malos/IT_MALO_Matrix.pdf.

Every year, near the end of the second semester, programs are required to submit an annual Learning Assessment Report, which is a status update on the assessments conducted the previous academic year. Because a program may choose to utilize a spring semester final exam or project for their learning outcomes assessment, they are given the summer, the fall semester, and some of the spring semester to interpret their data and promulgate actions for improvement. This process usually involves having the data analyzed and presented in a standard format and then having it entered into TracDat, our current institutional assessment software. Once this is done, program faculty are brought together to discuss the results and develop plans for improvement. Similar to the Wabash findings (Blaich and Wise 2011), gaining faculty consensus on how to improve student learning can be an arduous and challenging process. Nonetheless, under the ethos of continuous improvement, actionable solutions are brought forth. As the key events in the assessment calendar, it is the annual assessment plans and reports which provide the major milestones to keep the assessment processes moving forward.

In the CTI, the findings from the MALO assessment are presented to all faculty at a college meeting. The findings are discussed, and then over the next few weeks, faculty give thought to how student performance could be improved in relation to the MALO in question. Faculty need to think deeply about where and how improvements can be made in the program. Faculty are required to draft their proposals and email them to the dean for inclusion in the next college meeting. In their write-up, they should justify their proposal—they must provide sound argument on how the action they are proposing will bring about improvement. It is important that the proposed actions are practical, specific, and achievable. At the college meeting, the proposals are discussed and debated. Following thorough consideration, a number of actions are agreed upon. As an example, two of the actions that were agreed upon in fall 2012 for the Technical Communication MALO were:

1. The International English Language Testing System (IELTS) score will be embedded into a final year course and make up 20 % of the final grade so as to motivate students to improve. The IELTS testing will be moved from semester 8 to semester 7 in order to facilitate this.
2. The Technical Writing course will be modified, and the emphasis will be placed on English communication rather than on only English writing. New topics will be introduced on communication in IT workplaces, new communication technologies, and how to use them effectively. The title will be changed from Technical Writing to Technical Communication.

The actions that are agreed upon, as in the above example, are implemented over the following 12 months. In a subsequent college meeting, these actions are reviewed. Any issues in relation to the implementation of the actions are discussed and follow-up actions, if required, are planned.

14.6 Creating a Culture of Assessment

Establishing an assessment program that is effective, sustainable, and meets the needs of various accreditors has meant that we needed to move beyond a purely bureaucratic process to one which is infused into the culture of the institution. While a bureaucratic process has its merits and certainly assists in making the workload sustainable, to be truly effective, faculty need to engage and understand the assessment process and value its contribution to programmatic improvement. To embed this into the culture of the institution, we turned to Bolman and Deal's (2003) four-frame model, which is composed of the structural, human resource, political, and symbolic frames. Use of most, or all, of the frames is thought to improve decision-making, and knowledge of the frames should offer guidance as to when a particular frame should be dominant as different situations demand different solutions.

A summary of the four frames follows. First, the structural frame is about the organizational chart and organizational hierarchy. Through this frame, issues are examined with the belief that organizational restructuring offers solutions to improve performance. Second, the human resource frame views an organization as a family and tries to match the strengths and desires of the employee to an appropriate role. It can be especially appropriate when morale is low or motivation is an issue. The third frame is political, and although the term often carries with it negative connotations, this perspective recognizes that there are competing factions and agendas within an organization that must be successfully navigated. This is a particularly apt frame when there is institutional conflict or resources are scarce. The final frame is symbolic. This frame views organizations as cultures that are defined by their vision, stories, and ceremonies to a greater degree than their organizational chart or regulations. The symbolic frame is useful when team-building is needed. Integration of these concepts and the understanding that "the use of multiple frames permits leaders to see and understand more…" (Bolman and Deal p. 433) facilitates the development of an assessment program that is effective, sustainable, and meets the needs of various accreditors.

The application of Bolman and Deal's (2003) four-frame model to the development of ZU's assessment program is manifested in many ways and will be demonstrated accordingly. Though this will be presented frame-by-frame, it is clearly representative of multi-frame thinking, which increases the opportunities for success.

Since it remains the dominant lens through which organizations are analyzed, the structural frame is the first to be presented. At the institutional level, the first major structural initiative put forth was the creation of the OEE with responsibilities towards learning outcomes assessment, accreditation, and program review. Once the office was established with a director, institutional effectiveness officer, and administrative assistant, the organization established the ULAC and its steering group to develop and guide the assessment program at the university. At the college level, the CTI has an assessment committee and an assessment coordinator. The assessment committee is concerned with various aspects of the MALOs such as measuring the level of attainment and advising and helping faculty. Two members of the college committee sit on the ULAC. Another way in which the institution has made a commitment structurally is through the launch of the Center for Educational Innovation. The center is the faculty development department of the university, and part of its mandate is to support faculty in their assessment work. Early on in the work of ULAC, it became clear that faculty needed and wanted professional development around effective assessment, so the center was born and is now colocated with the OEE. The final structural decision that was made occurred at the graduate level. Though ZU is primarily an undergraduate institution, there is a rapidly expanding graduate program. The existing structure provided by the graduate program coordinators has been co-opted to also form the graduate assessment committee, which has implemented an assessment program nearly identical to the undergraduate level to aid in sustainability.

Within the human resources frame, many steps have been taken which demonstrate institutional commitment to assessment. Over the previous 2 years, the ULAC has conducted a number of in-house professional development workshops and retreats. These helped educate ULAC members and established the expectations, processes, and procedures that guide the current program. This past year, a major 3-day ABET professional development assessment workshop was hosted by the university that involved upwards of 80 faculty. Though ABET was contracted to lead the workshops, the sessions were in no way specific to engineering and computing. ABET was brought in because of their established leadership role in assessment, which can easily cross disciplinary boundaries. The university's commitment to the financial and temporal costs of these workshops was a demonstration of the priority placed on the development of faculty. Two other ways in which the human resources frame has been acted upon include conference funding and partnering for a publication opportunity. For example, the CTI agreed to fund a faculty member's attendance and presentation at the 2013 ABET Symposium, and the director of the OEE collaborated with another faculty person to write an assessment article submitted for publication.

The political frame is an area where large strides have been made over the past few years, but more needs to be done. The director of the OEE reports directly to the associate provost and has regular access to the provost, so within the political realm, this demonstrates the requisite level of access and authority to be perceived as an institutional priority and have the needed political clout. This is reinforced through regular presentations at university leadership meetings as well as through one-on-one semester-based meetings with the deans to review the progress of their college. New faculty are inculcated with the importance of both assessment and accreditation as part of their orientation process, while veteran faculty have it reinforced through regular presentations at the monthly university faculty and staff meetings. An area that needs to be strengthened is the existing faculty reward structure. Faculty are evaluated according to the traditional metrics of teaching, research, and service, but assessment and accreditation work, even if done in a sustainable manner, can be fairly time-intensive for a dedicated faculty member. Assessment and accreditation work needs to be recognized as a significant service contribution to the institution.

In a busy institution with many competing priorities, the symbolic frame, though often neglected, can take on an especially vital role because it demonstrates institutional priority. At ZU, the symbolic frame has been used to emphasize assessment and accreditation through both awareness and recognition programs. In terms of assessment program recognition, a concerted effort has been put forth to be recognized by external bodies as an international leader and a purveyor of best practices. The university has been recognized by the National Institute of Learning Outcomes Assessment (NILOA) for our outstanding assessment website,[4] by the Commission for Academic Accreditation (CAA) as "Good Practice,"[5] by TracDat as an exemplar case study,[6] and by SunGard Higher Education[7] for our contribution to outstanding assessment of student learning. Within the institution itself, we have had a long-term awareness campaign surrounding the ZULOs through the website, banners, and electronic signboards, and we also produce a biannual assessment newsletter. Suskie's (2009) advice to value and appreciate faculty assessment efforts is done by giving an annual award to deserving faculty. Accreditation has also been given extremely high visibility as it is a standard discussion item at the monthly faculty and staff meetings and always included in the assessment newsletter. It is clear to any visitors to the university that both assessment and accreditation are top institutional priorities.

14.7 Assessment Enhancements

There are a number of steps to be taken to further enhance the assessment program and better weave it into the fabric of the institution. The first major initiative underway is to transition the entire assessment process to a more formal peer review process to increase the rigor and trustworthiness of assessment as well as further develop the internal assessment expertise. Academics are comfortable and accustomed to the concept of peer review, so a small subgroup from the ULAC will be established to review the annual assessment plans, the assessment reports, and the accompanying assessments themselves. Faculty are becoming empowered as assessment leaders, a crucial element of success (Suskie 2009). Second, as part of increasing the rigor of the assessments themselves, we are looking to participate in international benchmark exams such as the recently completed Assessment of Higher Education Learning Outcomes (AHELO)[8] project from the OECD and to implement more internationally recognized externally validated exams. We feel there is an added level of credibility to such exams, and we are interested in the

[4] NILOA http://www.learningoutcomeassessment.org/FeaturedWebsiteZayedU.html.

[5] CAA https://www.caa.ae/caa/Images/gp_8.pdf.

[6] TracDat http://nuventive.com/2013/02/14/zayed-university/.

[7] SunGard is now Ellucian, a higher education software solutions company http://www.ellucian.com/.

[8] The initial phase of AHELO is complete http://www.oecd.org/site/ahelo/, but Zayed University has expressed an interest to participate in a second phase.

international benchmarking process. Benchmarking is one of the key components set forth in the *Guidelines for Assessment and Accountability in Higher Education* (New Leadership Alliance for Student Learning and Accountability 2012). Finally, we plan to continue to build a scholarship of assessment at the institution because we too believe that it "holds great promise for engaging faculty in activities to document and improve teaching effectiveness and student learning quality that are both institutionally and individually valuable" (Angelo 2002, p. 191).

14.8 Proposed New Model

While previous research by Saulnier (2014) has put forth a seven-step process to meet the learning assessment requirements for multiple accreditors, it did not describe the alignment between the different bodies nor what organization might be required to achieve a sustainable and effective program. Our model outlines the preferred approach to manage the diverse institutional and programmatic requirements to achieve the university's goals towards assessment and accreditation. This section describes an approach and a model designed to provide universities with a high-level road map that will help in streamlining all the activities related to accreditation on an ongoing basis. The model advocates a proactive approach that allows universities to respond to the challenges of accreditation in a timely, systematic, and consistent way. The model makes the handling of multiple and simultaneous accreditation efforts manageable and achievable.

At the core of the model is the establishment of an Office of Institutional and Educational Effectiveness (OIEE) to institute, organize, and coordinate all activities within the university that provides context and support for best practices for administration and teaching and learning. As institutions are often dealing with multiple accreditations, oftentimes overlapping, it is important that the simultaneous management of all activities involved in each accreditation is done consistently, efficiently, and in a timely manner. Meeting accreditation standards is often not substantially different from demonstrating that the institution and the academic programs are meeting their mission and goals. Accreditation standards are often defined in such a way so as not to be prescriptive but to allow each institution to demonstrate their effectiveness and adherence to standards by taking the context and sometimes unique circumstances of an institution into consideration as they make their own evaluation. For example, accreditation bodies have a set of core standards that they require almost uniformly of institutions. These standards are aligned and easily mapped into the mission and goals of the university. Similarly, accreditation bodies dealing with academic programs have a set of core requirements and standards to which programs must adhere to. The programmatic accreditation requirements are expectedly aligned and can easily be mapped into the educational objectives of the programs. An example of such programmatic standards is listed in Table 14.4. The table is based on the 14 MSCHE standards. The first seven standards are categorized under the institutional effectiveness column, and the remaining seven are listed under the educational effectiveness column.

Table 14.4 Criteria for institutional and educational effectiveness

Institutional effectiveness	Educational effectiveness
Mission and goals	Admission and retention
Planning and resource allocation	Student support services
Institutional resources	Faculty
Leadership and governance	Educational offering
Administration	General education
Integrity	Related educational objectives
Institutional assessment	Assessment of student learning

14.8.1 Mapping Between Institutional and Educational Effectiveness and Accreditation Requirements

Currently, the need for accreditation drives the process of gathering, evaluating, analyzing, and preparing the required information that will be presented in a self-study report as part of the accreditation requirements. The process that universities follow in most cases is a reactive rather than a proactive one. The process of gathering the information that is required starts at the time that the decision for seeking accreditation is taken. Putting together all the information that is needed becomes a time-consuming process that involves faculty and administrators at all levels. It creates an added burden on everyone involved to meet all that is required in a limited time while they are expected to go on performing their regular duties.

The recommended approach advocated by our model is for institutions to maintain a database of all the information that is relevant to accreditation in general on an ongoing basis. It can be argued that maintaining this database should be considered as an integral part of what institutions should be doing as a way of documenting their own institutional status vis-à-vis their mission and goals. The advantage of having this database created and updated on a regular basis is that it becomes the source of much of the information that is needed for accreditation. The task of writing the self-study report becomes a matter of extracting the required information from the database and formatting it for inclusion in the report.

As an example, if the information listed in Table 14.4 had already been created and maintained in the database, the information required by the ABET accreditation would be readily available by a simple query to the database. Table 14.5 provides a mapping between the information that is to be maintained in the database and the information that is required by MSCHE and ABET.

Maintaining information about institutional and educational effectiveness will allow the institution or the academic unit to respond to their respective accreditation. If and when the institution decides to go after any new accreditation, be it institutional or programmatic, the same database can provide most of the relevant information that will be required. The advantage in this case is that the institution will have the ability to produce this information with minimal effort as it is already stored and maintained in the database. What will be required then is mainly customizing and formatting the information to the task at hand.

Table 14.5 Mapping of institutional and educational effectiveness and accreditation requirements

Institutional data	MSCHE	ABET
Student learning outcomes	✓	Student outcomes (Criterion 3)
Improvement of student learning	✓	Continuous improvement (Criterion 4)
Related educational objectives	✓	
General education	✓	Curriculum (Criterion 5)
Educational offering	✓	Curriculum (Criterion 5)
Faculty	✓	Faculty (Criterion 6)
Student support services	✓	Students (Criterion 1)
Admission and retention	✓	Students (Criterion 1)
Institutional assessment	✓	Program educational objectives (Criterion 2)
Integrity	✓	
Administration	✓	
Leadership and governance	✓	Institutional support (Criterion 8)
Institutional resources	✓	Facilities (Criterion 7)
Planning and resource allocation	✓	Continuous improvement (Criterion 4)
Accreditation planning		
Mission and goals	✓	Program educational objectives (Criterion 2)

MSCHE Middle States Commission on Higher Education, *ABET* Accreditation Board for Engineering and Technology

14.8.2 The Office of Institutional and Educational Effectiveness

Increasingly, universities are recognizing the need to establish an office to facilitate the assessment of institutional and educational effectiveness and to promote and support a culture and processes for continuous analysis and enhancement of institutional effectiveness. Such an office usually goes by the name of Office of Assessment and Educational Effectiveness, or in our model, the OIEE. The mission of the OIEE is to support academic departments and units assessing undergraduate and graduate student learning. It provides technical support and services that promote the implementation of systematic assessment of student learning and institutional improvement. The OIEE includes or collaborates closely with the Office of Institutional Research to document research and findings about institutional and educational effectiveness. Some of the areas of responsibilities under OIEE include support for:

- The analysis of strengths and weaknesses of the curriculum
- Comparing stated learning objectives with evidence of student learning
- Alignment of institutional and programmatic learning outcomes
- Strategic and operational planning and assessment of administrative and academic units

In addition, OIEE helps academic units:

- Collect data throughout the program curriculum for measuring learning outcomes
- Assist faculty in the review and analysis process of the findings from the collected data about student learning

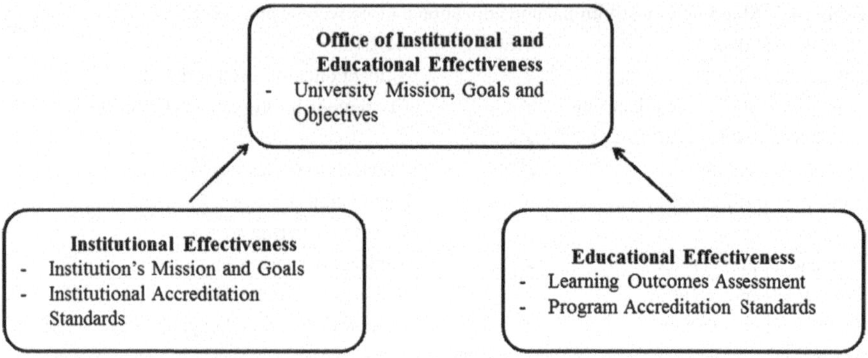

Fig. 14.1 Office of Institutional and Educational Effectiveness

- Help faculty identify the potential areas of improvement to the curriculum
- Develop and maintain a yearly assessment plan that is based on results obtained from the previous year and outlining strategies to improve results for the following year

In order to achieve its objectives, the OIEE must adhere to the following criteria:

- *Alignment*: Whenever possible, do not replicate efforts or create redundancies, especially surrounding accreditation.
- *Sustainability*: Define processes that are simple and an integral part of operational requirements to ensure that these processes do not become a heavy burden or an obstacle for improvement.
- *Effectiveness*: It should be easy to demonstrate that the model provides efficiencies through guidelines which make the process and its outcomes effective.

In relation to the process of accreditation discussed above, the OIEE will act as the coordinator and maintainer of the database of information. Figure 14.1 shows the relationship between OIEE and the university's organizational side on one hand and the academic units on the other. OIEE becomes the focal point for coordinating and keeping track of all data that relate to (1) describing the operation and effectiveness of the university as an institution in supporting academic activities and (2) maintaining a repository for academic units for information that will be of relevance for accreditation.

14.9 Conclusion

The model that has been put forth demonstrates how the assessment and accreditation requirements of diverse accrediting agencies can be made to operate in alignment with one another. Using accreditation as a lever, an assessment program can be implemented that meets the needs of the different disciplines and eliminates duplication of effort. A centrally managed, faculty-driven assessment program can accentuate the similarities across disciplines by being built around sound learning outcomes and emphasizing the cycle of continuous improvement. Indeed "the aim of assessment is not compliance with accreditors—its goal is informed action that enhances student learning" (Walvoord 2010, p. 27).

Appendix A – Assessment Rubric

College of Information Technology, Zayed University
Assessment Rubric

MALO: Critical Thinking and Quantitative Reasoning in IT (Beginning Level)
Course: CIT 255 Networks and Telecommunications

Student Name and Number:

Campus, Semester/year: AUH, Spring 2012 Instructor Name:

MALO Definition: *IT College graduates will be able to use critical thinking and quantitative processes to identify, analyze and solve problems, and evaluate solutions in an IT context.*
Beginning Level*: With assistance recognizes some important basic characteristics of simple problems and produces acceptable solutions.*
Indicator*: Ability to apply install, configure and test computer networks.*
Task*: CCNA1E Skills Exam, including subnetting a class C skills to design, IP address to meet customer specifications, installation, cabling, configuration and testing of the specified network.*
Performance Target: *80% of Students Achieve 3.0 or Greater*

	Unsatisfactory	Developing	Satisfactory	Exemplary
Performance Indicators	**1**	**2**	**3**	**4**
Designs a networking solution	Does not demonstrate any ability to subnet a Class C address.	Demonstrates, but with significant errors, the ability to subnet a Class C address, identifying the subnet mask correctly.	Mostly correctly subnets a Class C address space, identifying the subnet mask and most of the addresses correctly.	Correctly subnets a Class C address space, identifying the first and last host IP address, network address and subnet mask for each subnet
Relates theoretical diagrams to practical solution	Does not select or connect appropriate cabling.	Demonstrates the ability to select mostly appropriate cabling but not the ability to connect the devices according to provided logical diagram.	Demonstrates the ability to select appropriate cabling and mostly correctly connect the devices according to provided logical diagram.	Demonstrates the ability to select appropriate cabling and connect the devices according to provided logical diagram.
Applies design to a practical network	Does not demonstrate the ability to configure host PCs and routers.	Demonstrates, but with significant errors, the ability to configure host PCs and routers.	Demonstrates the ability to configure host PCs and routers, with some minor errors.	Demonstrates the ability to configure host PCs and routers without errors.
Practical problem solving	Does not demonstrate an ability to test for network problems or troubleshoot them.	Demonstrates some ability to test for network problems but only a limited ability to troubleshoot them.	Demonstrates the ability to test for network problems and troubleshoot them with some limitations.	Demonstrates the ability to test for network problems and troubleshoot them.

References

AAC&U. (2010). *VALUE Rubrics*. http://www.aacu.org/value/rubrics/reprint.cfm.

ABET. (2009). *Criteria for accrediting computing programs*. Baltimore. http://www.abet.org/uploadedFiles/Accreditation/Accreditation_Process/Accreditation_Documents/Archive/criteria-cac-2010–2011.pdf.

ABET. (2011a). *Engineering criteria 2000(EC2000)*. http://www.abet.org/History/.

ABET. (2011b). *Why accreditation matters*. http://www.abet.org/why-accreditation-matters/.

ABET. (2013). *Find accredited programs*. http://main.abet.org/aps/Accreditedprogramsearch.aspx.

Angelo, T. A. (2002). Engaging and supporting faculty in the scholarship of assessment. In T. W. Banta (Ed.), *Building a scholarship of assessment* (pp. 185–200). Portland: Jossey-Bass.

Blaich, C., & Wise, K. (2011). *From gathering to using assessment results: Lessons from the Wabash national study*. (NILOA Occasional Paper No. 8). Champaign: National Institute for Learning Outcomes Assessment. http://www.learningoutcomeassessment.org/documents/Wabash_001.pdf.

Bolman, L. G., & Deal, T. E. (2003). *Reframing organizations: Artistry, choice, and leadership* (3rd ed.). San Francisco: Jossey-Bass.

Council for Higher Education Accreditation. (2014). *The CHEA award and learning outcomes*. http://www.chea.org/chea%20award/CHEA_Awards_All.html.

Ewell, P. T. (2001). *Accreditation and student learning outcomes: A proposed point of departure*. http://www.chea.org/award/StudentLearningOutcomes2001.pdf.

Ewell, P. T. (2009). *Assessment, accountability, and improvement: Revisiting the tension*. http://www.learningoutcomeassessment.org/documents/PeterEwell_005.pdf.

Fu, J., Gourley, M., Park, M., Qian, G., Sung, H., & Turner, T. (2014). Obtaining and maintaining ABET accreditation: An experience-based review of the ABET criteria for computer science programs. *Journal of Computing Sciences in Colleges, 29*(4), 13–19.

Gannon-Slater, N., Ikenberry, S., Jankowski, N., & Kuh, G. (2014). *Institutional assessment practices across accreditation regions*. http://www.learningoutcomeassessment.org/documents/Accreditation%20report.pdf.

Heard-Bey, F. (2004). *From Trucial States to United Arab Emirates*. Dubai: Motivate Publishing.

Hernon, P., Dugan, R. E., & Schwartz, C. (Eds.). (2006). *Revisiting outcomes assessment in higher education*. Westport: Libraries Unlimited.

Kuh, G. D., Jankowski, N., Ikenberry, S. O., & Kinzie, J. (2014). NILOA. http://www.learningoutcomeassessment.org/knowingwhatstudentsknowandcando.html.

Middle States Commission on Higher Education. (2007). *Student learning assessment options and resources*. http://www.nova.edu/ie/ice/forms/sl_assessment.pdf.

Middle States Commission on Higher Education. (2009). *Characteristics of excellence in higher education: Requirements of affiliation and standards for accreditation* (12th ed.). Philadelphia.

New Leadership Alliance for Student Learning and Accountability. (2012). *Guidelines for assessment and accountability in higher education*. http://www.chea.org/alliance_publications/committing%20to%20quality-3rd%20edition.pdf.

OECD. (2013). Quality assurance in higher education in 20 MENA economies. *Higher Education Management and Policy, 24*(2), 73–84.

Saulnier, B. (2014). A paradigm for student learning outcome assessment in information systems education: Continuous improvement or chasing rainbows? *Information Systems Education Journal, 12*(1), 4–14.

Suskie, L. (2009). *Assessing student learning: A common sense guide* (2nd ed.). San Francisco: Jossey-Bass.

University Affairs. (2013). *More Canadian universities seek U.S. accreditation*. http://www.universityaffairs.ca/more-canadian-universities-seek-us-accreditation.aspx.

US Department of Education. (2006). *A test of leadership: Charting the future of U.S. higher education*. Washington, D.C. http://www.ed.gov/about/bdscomm/list/hiedfuture/reports/final-report.pdf.

Walvoord, B. E. (2010). *Assessment clear and simple: A practical guide for institutions, departments, and general education* (2nd ed). San Francisco: Jossey-Bass.

Kevin Schoepp is the Director of Educational Effectiveness at Zayed University. His role in educational effectiveness includes learning outcomes assessment, accreditation, and program review. He has a doctorate in higher education leadership from the University of Calgary, and he has presented papers at numerous conferences including the Assessment Institute, the ABET Symposium, and the Arab Network for Quality Assurance in Higher Education. He has published in a variety of journals such as Assessment Update, International Education, and the Near and Middle Eastern Journal of Research in Education. His current research interests are in the development of effective and sustainable assessment and accreditation processes, creating a culture of assessment to foster continuous improvement, and in assessing ABET's professional skills.

Maurice Danaher is an Associate Professor in the College of Information Technology at Zayed University, Abu Dhabi. He received his PhD in Computing and Information Systems from Swinburne University of Technology, Melbourne, Australia, in 2003. His research has been in the areas of Information Technology and Education. In IT he has published in artificial intelligence, computer graphics and multimedia, and graph layout. His PhD research was in the area of application of artificial intelligence to support design decisions. In recent years his focus in Education research has moved towards issues in enhancing the quality of education, with particular emphasis on means of measuring or assessing the quality of education. His research in assessing the quality of education is linked to his work in assessing the quality of the programs in his College at Zayed University. He has taken a lead role in the College in learner assessment and program evaluation and was responsible for outcomes measurement during the ABET accreditation process. Dr Danaher has extensive experience in industry and academia internationally. He has worked with industry in the USA, Canada, England and Ireland. He has held academic positions in Australia and Singapore.

Dr. Leon Jololian is a director in the Center for Systems Integration, at the University of Alabama, in Birmingham. Prior to his current position, he was the Dean of the College of Technological Innovation (2007-2014), at Zayed University, in the UAE. He held faculty positions at William Paterson University, University of Alabama at Birmingham, and New Jersey City University. He has also held the positions of Director of Computing at New Jersey Institute of Technology and Vice President for Research and Development at Thor Technologies in New York. Dr. Jololian's research interests include software component architecture, object-oriented technology, and mobile app technology. Dr. Jololian earned his Ph.D. (2000) in Computer Science from New Jersey Institute of Technology, a Master's degree in Computer Science (1991) from Polytechnic University, a master's degree in Electrical Engineering (1982) from Georgia Institute of Technology, and a bachelor's degree in Electrical Engineering (1980), from Manhattan College.

Part V
Challenges and Sustainability Perspectives

Chapter 15
Engineering Education in Northwest MENA Countries: Challenges and Opportunities for the Twenty-First Century

Mohamed Essaaidi

15.1 Introduction

Higher education, and more specifically engineering education, in northwest Middle East and North Africa (MENA) countries, namely, Algeria, Morocco, and Tunisia, has been considerably influenced by the French higher education system for strong historical reasons (Teferra and Altbachl 2004). These three countries were colonized by France for several decades, and Morocco and Tunisia regained their independence during the mid-1950s and Algeria during the early 1960s, for which the French colonization lasted more than a century. The main engineering education system established in these countries is based on the French "Grandes Écoles" system, which is quite different from other engineering education systems worldwide and more specifically from the one that prevails in the USA and many other countries. There are several dissimilarities between these two systems at different levels. Actually, this system is based on a 5-year program after the high school; in some engineering schools, the first 2 years are spent in special schools' "preparatory classes" during which the students are exposed to extremely intensive math and physics courses besides other courses such as computer science and French. At the end of this 2-year very intensive curriculum, the students have to go through a nationwide competition whose main objective is to rank all the candidates according to their records in this contest, which will make them eligible to enter an engineering program in a prestigious and very elitist engineering school "Grande École." This is a really very selective and elitist process considering the total number of candidates and the number of available seats in the different engineering programs offered in these engineering schools for which they are competing. Since the duration of these programs is 3 years, the overall number of years required to obtain the engineer's degree is 5 years if we consider the other 2 years that are spent in

M. Essaaidi (✉)
National Higher School of Information Technology (ENSIAS), Mohammed V University, Rabat Morocco
e-mail: essaaidi@ieee.org

M. Abdulwahed et al. (eds.), *Advances in Engineering Education in the Middle East and North Africa,* DOI 10.1007/978-3-319-15323-0_15

the "preparatory classes." Therefore, in terms of number of years, this engineering program is equivalent to the master of engineering degree in many other countries. Indeed, it is the case since holders of this degree automatically obtain such recognition in the three countries as is the case in France, from which the system was borrowed, and by doing so, they can directly apply for a PhD program offered by the universities in these countries and in France. However, according to the students and the program outcomes defined by the Accreditation Board for Engineering and Technology (ABET), it is not clear if these programs meet the criteria set even for ABET-accredited bachelor of engineering programs since there are no real quality assurance (QA) systems involved in their accreditation or implementation.

Though this engineering system is very intensive and competitive in math, physics, and other sciences, it is facing many challenges today, such as:

- The national policies of these countries aiming at increasing the number of engineers graduating from these schools to meet the growing demand of domestic and international offshore companies and industries for engineers despite, in many cases, the limited number or even the lack of the required qualified graduates.
- The need to engage in a radical reform of the pedagogical paradigms and approaches of engineering education to embrace novel and more efficient approaches such as project-based and problem-based learning (Felder and Silverman 1988; Mills and Treagust 2004).
- The need to turn the engineering education into a successful technology innovation and entrepreneurial machine (Creed et al. 2002), which will help these countries develop knowledge-based economies that will contribute effectively to harness the competitiveness of their small and medium-sized enterprises (SMEs), industries, and economies and contribute to wealth and jobs generation. This will also help reduce unemployment rates among their young populations, which is one of the big challenges they are facing today.
- The use of French as the main and exclusive learning and communication language in engineering education in these countries is a strong "isolation" factor that drastically limits cooperation and development opportunities for their engineering education system in the mainly English-speaking technology and engineering globalized world. Therefore, the integration of the engineering education system within the global education engineering ecosystem requires a reinforcement of English communication skills among their schools' faculty and graduates.
- The implementation of real QA standards, such as the ones provided through international accreditation from international accreditation bodies such as ABET (ABET accreditation 2013), is a very important challenge that would bring lots of benefits to the engineering programs provided in these schools. In the current situation, no real QA management systems are implemented, and these engineering programs go through a very light accreditation process which lacks any QA management systems and is performed within departments and services of the ministries of higher education of these countries.

- The weakness, and even the lack, of entrepreneurship and innovation-based engineering education activities in this system represents a very limiting factor towards the establishment of entrepreneurial engineering schools in these countries that actively contribute to their economic and social development and welfare, and thereby to their integration in the knowledge-based economy in the framework of globalization.

This chapter will shed some light on the strengths and weaknesses of engineering education in northwest MENA countries. It will also give some recommendations about possible improvement paths for this system to enable it to face the challenges of globalization and the knowledge-based economy raised by the twenty-first century. The most important recommendations in this regard can be stated as follows:

1. The need to establish real engineering education QA and quality management systems, such as the one offered through ABET accreditation
2. The need to implement technology innovation and entrepreneurship education paradigms, such as project-based learning
3. The need to reinforce English communication skills among both faculty and students

15.2 Engineering Education System in Northwest MENA Countries

Engineering education systems in northwest MENA countries (i.e., Algeria, Morocco, and Tunisia; Berrah 2004; Gharbi and Regragui 2004; Elleuch and Bouzid 2004) are based on the French engineering education system. In France, although the *Commission des Titres d'Ingénieurs* (i.e., Engineering Degrees Commission; Commission des Titres d'Ingénieurs 2013) was established in 1934, it was not until 2007 that this organization developed policies and procedures to carry out program-based accreditation of engineering education. Therefore, even in France, real accreditation processes such as QA and quality management systems are relatively new, and they are not as rigorous as the ones implemented in the USA (e.g., ABET) or UK.

The duration of the engineering programs offered through this system is 5 years. Actually, there are two kinds of engineering schools offering these programs, namely:

1. Engineering schools with 3-year programs, which are the main and the most prestigious among the engineering schools in these countries and even in France, such as National Polytechnic School in Algeria, Mohammadia School of Engineering in Morocco, and National Engineering School of Tunisia in Tunisia. These schools, referred to as "Grandes Écoles," offer the last 3 years of the 5-year engineering program in specialized engineering programs (e.g., electrical engineering, computer engineering, mechanical engineering, civil engineering). The first 2 years are provided in/outsourced to high schools in the framework of an undergraduate program called preparatory classes. Access to these classes is

very selective and is open to students holding the baccalaureate degree, which is equivalent to K12 in the US education system. The main objective of these classes is to prepare the students for a nationwide, very selective competition, allowing access to the "Grandes Écoles" through a very intensive, heavy, and strong curriculum whose main components are math, physics, chemistry, engineering science (i.e., electrical and mechanical engineering), and French.

2. Engineering schools with 5-year programs such as the National Schools of Applied Sciences (ENSAs) in Morocco, which follow the French National Institutes of Applied Sciences (INSA) schools' model. These schools offer the whole 5-year engineering program. Access to the first year of this program is open to students holding the high school baccalaureate degree.

15.2.1 Accreditation and Quality Assurance

Contrary to many engineering education systems where accreditation and QA and quality management systems are very rigorous and well developed, in northwest MENA countries, it is relatively a very light process. There are no independent accreditation agencies, and the accreditation of higher education programs in general is carried out and performed in departments or administrations within the ministries of higher education. However, since these programs recruit the best students in math and physics who have in general very good communication skills in French, which is the exclusive language used in teaching science and engineering, many programs' outcomes are satisfactorily achieved. However, the students are still deficient in many skills and competencies.

Accreditation of engineering programs (Berrah 2004; Gharbi and Regragui 2004; Elleuch and Bouzid 2004) is granted for a period of 4–5 years based on relatively light portfolios presenting, mainly, descriptions of the programs' objectives, outcomes, curriculum, faculty members, and admission and evaluation conditions and criteria. Only one relatively very light self-assessment study of the program is required at the end of the accreditation term. However, these countries are aware of the importance of independent accreditation agencies in setting a real framework for QA and quality management of higher education and more specifically for engineering education. All these countries are currently in the process of establishing such agencies.

15.3 Twenty-First-Century Engineering Education Challenges and Opportunities

In the framework of globalization and knowledge-based economy posed by the twenty-first century, the challenges and opportunities of engineering education in these countries are quite similar to what is happening in other countries of the world. A most recent highly quoted report focusing on engineering problems published by the US National Academy of Engineering lists 14 engineering grand challenges (Grand Challenges for Engineering 2008), most, if not all, of them related to engineering, technology, and science, namely:

1. Make solar energy economical
2. Provide energy from fusion
3. Develop carbon sequestration methods
4. Manage the nitrogen cycle
5. Provide access to clean water
6. Restore and improve urban infrastructure
7. Advance health informatics
8. Engineer better medicines
9. Reverse engineer the brain
10. Prevent nuclear terror
11. Secure cyberspace
12. Enhance virtual reality
13. Advance personalized learning
14. Engineer the tools of scientific discovery

In addition to the grand challenges listed above, the world of today and tomorrow confronts other challenges of an ever-changing technology landscape and information, knowledge production, and the result of faster, efficient, and new modes of communication and interconnectedness.

High-quality and pertinent engineering education are imperatives for creating a knowledge-based economy. Engineering education must respond to local challenges as well as global opportunities. During the past 10–15 years, a significant amount of resources have been spent worldwide innovating engineering curricula (Morell 2010). In the USA, for example, some of the most prominent initiatives occurred during the mid-1990s when National Science Foundation (NSF) funded some engineering coalitions such as the "Synthesis Engineering Education Coalition" (www.synthesis.org). These coalitions focused primarily on engineering curricula innovation, integrating outcomes assessment, using complementary technology, and implementing new learning strategies. Although these efforts were successful, they have not permeated the engineering education ecosystem the way the sponsors and participants thought they would. The engineering education culture—the model for training the next generation of engineers—is still the same. Notwithstanding the current global economic crises, the fact remains that the economic progress and achievements of many countries are rooted in their science and engineering talent. The workplace demands new engineers to be technically qualified, flexible, and dynamic thinkers, but their classrooms are not necessarily and systemically supplying them with these tools.

Perhaps the lack of attention to the educators themselves is the key oversight in this system. About 50 % of industry and academia respondents in an Engineering 2020 survey (Palmer et al. 2011) dissent from the assertion that the current undergraduate engineering education is sufficiently flexible to adequately meet the needs of twenty-first-century engineers. The new professional not only needs to be knowledgeable in his/her own discipline but also needs a new set of soft, professional skills and competencies. There are many engineering professional societies' surveys and accreditation criteria which have listed the engineering professional skills and competencies of the engineering graduate. Nevertheless, what the new world order and local/regional challenges require are quite novel. In a world of volatility, uncertainty, complexity, and ambiguity, there will be danger as well as opportunity; thus, leaders must learn new skills to address these and make a new future. Most of the skills envisioned for the twenty-first-century technology professionals are described in Table 15.1 (Palmer et al. 2011).

There are five main actions required to reduce the gap between professionals graduating from universities and the needs of society and private/public enterprises, namely:

1. Innovate: reform the engineering curriculum and the learning experience.
2. Focus on learning, not on teaching.
3. Foster creativity and innovation across the ecosystem.
4. Implement continuous assessment and accreditation to drive excellence.
5. Educate the engineering professor of the future.

According to a survey (Morell 2010), the profile and general characteristics of the ideal engineering professor should be:

- Superior communicator
- Effective teacher and mentor
- Technical expert
- Engineering practitioner
- Committed to global citizenship

Table 15.1 Technical leaders' twenty-first-century skills

Skills	Description
Mobility	Ability to work in large groups; talent for organizing and collaborating with many people simultaneously
Influency	Ability to be persuasive in multiple social contexts and media spaces; understanding that each context and space requires a different persuasive strategy and technique
Ping quotient	Responsiveness to other people's requests for engagement; propensity and ability to reach out to others in a network
Multi-capitalism	Fluency in working with different kinds of capital: natural, intellectual, social, financial, etc.
Protovation	Fearless innovation in rapid, iterative circles
Open authorship	Creating content for public consumption and modification
Emergensight	Ability to prepare for and handle surprising results and complexity
Longbroading	Thinking in terms of higher-level systems and cycles
Signal/noise	Filtering meaningful information, patterns, and commonalities from massively multiple streams of data
Cooperation radar	Ability to sense, almost intuitively, who would make the best collaborators on a particular task

Therefore, the engineering professor of today and tomorrow needs to be a blend of the two professions, engineer and educator. He/she must be an individual who

- is competent in his/her own discipline, engineering fundamentals, and problem solving;
- is current in his/her research, publishes, networks, communicates effectively, and keeps up with trends in his/her discipline; and does all of the above with an entrepreneurial spirit;
- is an effective teacher, knows about learning and outcomes assessment, facilitates learning using learner-centered strategies, keeps up with developments in engineering education, studies and uses them effectively, cares about the students and their learning, and enjoys being a mentor;
- understands the role that the profession has in society, both locally and globally, practices it as a part of his/her career development as well as leads, serves, and participates in forums to promote policy-making and excellence in engineering education and research/innovation; and
- aims at developing the skills and competencies engineers should possess through practice and experience in order to better serve society and be a role model for students.

15.4 Improvement Paths for Engineering Education in Northwest MENA

There are several improvement paths for the engineering education system in northwest MENA countries (i.e., Algeria, Morocco, and Tunisia) that are also valid for many similar systems. These recommendations stem from the comparative study with other successful systems such as the ones based on ABET accreditation which offers a very good framework for QA and quality management system for engineering education. It also provides a reference of skills that should be acquired by the students and the different assessment metrics of the program and students outcomes. Therefore, a key recommendation for these countries is the need to establish independent national accreditation bodies for engineering education which implement real accreditation criteria and processes including QA and management systems.

The other important recommendation concerns the engineering education approach that should be implemented throughout the region to foster innovation and creativity in order to meet the different challenges raised by the twenty-first-century globalized knowledge-based economy. This would also help create jobs and wealth that would contribute to its technological, industrial, economic, and social development. When it comes to promoting innovation and creativity within engineering schools, project-based learning seems to be one of the best engineering education approaches together with the setting up of incubators and technology parks in the engineering schools and/or linking them to existing ones such as Casablanca Technopark and Rabat Technopolis in Morocco and El Gazala Technopark in Tunisia.

Other important measures that may contribute significantly to the improvement of engineering education are based on different solutions and applications enabled by information technology and the Internet such as social networks, e-learning platforms, massive open online courses (MOOC; Massive Open On-line Courses 2013), and web-based experiments (Ko et al. 2001).

Engineering education in Algeria, Morocco, and Tunisia, as is the case with the French engineering education, is based on the very old classroom-intensive approach which requires students, on average, to attend around 35 class hours per week. This is a very excessive class load which leaves very little time for them to digest the material they get in these classes and do more important things, developing innovation and leadership skills such as projects. This might be one of the reasons why universities and engineering schools adapting this system are lagging behind when it comes to innovation and other performance metrics in different international university rankings such as Shanghai Academic Ranking of World Universities (Academic Ranking of World Universities (Shanghai) 2013) and Times Higher Education (Times Higher Education 2013).

15.5 Conclusions

Engineering education in northwest MENA countries (i.e., Algeria, Morocco, and Tunisia) suffers from several deficiencies and shortcomings which are mainly due to the lack of real accreditation processes implementing QA and managements systems. The lack of good English communication skills also contributes to the problems they are facing and limits drastically their ability to benefit fully from the numerous opportunities offered by the quasi-exclusively English-speaking globalized science, technology, industry, and economy of the twenty-first century. In order to overcome most of these problems and to improve engineering education performances in these countries, several recommendations are proposed, namely:

- Implementation of a real accreditation process with a QA management system of engineering education through independent national accreditation bodies that should be established in these countries
- The improvement of technology entrepreneurship and innovation performance of this system by reinforcing these skills in the engineering programs and by establishing incubators and technology parks in engineering schools and/or connecting them with already existing ones
- The reform of the engineering education paradigms through novel education approaches such as project- and problem-based learning and e-learning, MOOCs, and blended learning
- The reinforcement of English communication skills within engineering schools' faculty and students to help engineering education and its graduates in these countries benefit fully from the numerous opportunities offered by the mainly English-speaking globalized technological, industrial, and economic world of the twenty-first century

These recommendations are also valid for any engineering education system wishing to provide world-class engineers for the globalized world of the twenty-first century.

References

ABET accreditation. (July, 2013). www.abet.org.

Academic Ranking of World Universities (Shanghai). (July, 2013). http://www.shanghairanking.com/.

Berrah, M. K. (2004). Engineers training in Algeria, quality evaluation. MEDA Region, Tempus Project Report, 30092-2002.

Commission des Titres d'Ingénieurs. (July, 2013). http://www.cti-commission.fr/.

Creed, C. J., Suuberg, E. M., & Crawford, G. P. (2002). Engineering entrepreneurship: An example of a paradigm shift in engineering education. *Journal of Engineering Education, 91*(4), 185–195.

Elleuch, B., & Bouzid, J. (2004). Engineers training in Tunisia, quality evaluation. MEDA Region, Tempus Project Report, 30092-2002.

Felder, R. M., & Silverman, L. K. (1988). Learning and teaching styles in engineering education. *Journal of Engineering Education, 78*(7), 674–681.

Gharbi, M., & Regragui, F. (2004). Engineers training in Morocco, quality evaluation. MEDA Region, Tempus Project Report, 30092-2002.

Grand Challenges for Engineering. (2008). National Academy of Engineering.

Ko, C. C., Chen, B. M., Chen, J., Zhuang, Y., & Tan, K. C. (2001). Development of a web-based laboratory for control experiments on a coupled tank apparatus. *IEEE Transactions on Education, 44*(1), 76–86.

Massive Open On-line Courses (MOOC). (July, 2013). www.mooc-list.com.

Mills, J. E., & Treagust, D. F. (2004). Engineering education—is problem-based or project based learning the answer? *Australasian Journal of Engineering Education, 3,* 2–15.

Morell, L. (2010). Engineering education in the 21st century: Roles, opportunities and challenges. 4th NEA ICETE Conference Proceedings, Taichung, Taiwan, October 2010.

Palmer, B., Terenzini, P. T., McKenna, A. F., Harper, B. J., & Merson, D. (2011). Design in context: Where do the engineers of 2020 learn this skill? Proceedings of American Society for Engineering Education (ASEE) Annual Conference and Exposition.

Teferra, D., & Altbachl, P. G. (2004). African higher education: Challenges for the 21st century. *Springer, 47*(1), 21–50.

Times Higher Education (THE) World University Ranking. (July, 2013). http://www.timeshighereducation.co.uk.

Mohamed Essaaidi Prof. Dr. and IEEE Senior Member, is the Dean of National College of IT (ENSIAS) of Mohammed V University, Rabat, Morocco since December 2011 and he was a Professor of Electrical & Computer Engineering at Abdelmalek Essaadi University, Morocco from 1993 till 2011. He is the founder and Chairman of the IEEE Morocco Section, founder of IEEE Computer & Communication Societies Joint Morocco Chapter, Founder and Chair of IEEE Antennas and Propagation Society and Microwave Theory and Techniques Society Morocco Joint Chapter and founder of IEEE Education Society Morocco Chapter. He has been also the founding Director of the Morocco Office of Arab Science and Technology Foundation, ASTF (2006–2009) and the Coordinator of ASTF RD&I Network of ElectroTechnology since 2006. He has also founded several IEEE Student Branches in different Moroccan universities and engineering schools.

He has authored and co-authored 5 books and more than 120 papers in international refereed journals and conferences in the field of Electrical, Information and Communication Technologies. He has been the Editor-in-Chief of International Journal on Information and Communication Technologies, Serial Publications, India since 2007. He is also an active member of the editorial boards of several IEEE and other indexed international journals in the field of information and communication technologies.

Prof. Essaaidi also founded and has been the General Chair of the Mediterranean Microwave Symposium since the year 2000, Information and Communication Technologies International Symposium in 2005 and 2007, the International Conference on Multimedia Systems and Computing in 2009, 2011 and 2012 and the International Conference on Complex Systems in 2012. He has also been involved in the Organizing and Scientific Committees of other several international conferences held worldwide.

He holds four patents on antennas for very high data rate UWB and multi-band wireless communication systems and high resolution medical imaging systems. Furthermore, he has co-organized/been involved in the juries of several national and international competitions aiming at fostering research, development and innovation such as Moroccan Engineers Week 2006, 2007, "Made in Morocco", Arab Science and Technology Foundation (ASTF) "Made in Arabia" Competitions in 2007 and 2009, Qatar Foundation Stars of Science 2010 and Intel Science Competition 2011.

He was also a member of the IEEE 802.16 Sponsor Ballot Pool of IEEE Standard Association that defined the technical specifications for WiMmax.

Moreover, he has supervised several Ph D and Masters Theses and has been the principal investigator and the project manager for several international research projects with US, Canada, Spain, Finland, Italy, Korea, Egypt, Tunisia and UAE.

Chapter 16
Perspectives on Engineering Education Quality in Tunisia After 50 Years of Statehood

Jennifer DeBoer

16.1 Introduction

Tunisia's 50th anniversary of statehood and its subsequent 2011 revolution inspire a reflection on how this small, diverse country has developed. Tunisia's historically liberal and secular social environment contrasts with its centralized national government, which has oversight of the large primary, secondary, and tertiary public school system. The confluence of liberal social values with a palpable, integrated government presence and a changing cultural environment has created an interesting sociological context for education, which permeates the engineering education (EE) sphere. For example, strong legal protections for gender equality guarantee women and men similar rights to participate in the labor force, including those in engineering fields; these are enacted through numerous high-stakes exams throughout training and for entry into government engineering work. However, the proportion of women and men working in public-sector jobs versus those in the engineering industry is far from equal. Added to this complex educational and social setting are Tunisia's historical connections to France, the country's newly developing private sector, and numerous other characteristics that make it a provocative case study. This case study demonstrates a unique and fluctuating situation, but it can serve to inform some of the major questions of higher education policy around the globe.

The chapter begins with background information on the Tunisian EE system. It proceeds with a description of the research design and methods and a description of the people and questions that comprise the dataset. The study then details themes that emerge from respondents' interviews, beginning with their references to historical touch points, their view of the curriculum, the theme of regulation, and the theme of tension between internal and external pressures. Respondents address internationally comparative issues such as the French inheritance in Tunisia's system,

J. DeBoer (✉)
Purdue University, West Lafayette, IN, USA
e-mail: deboerj@purdue.edu

M. Abdulwahed et al. (eds.), *Advances in Engineering Education in the Middle East and North Africa,* DOI 10.1007/978-3-319-15323-0_16

comparisons to the Anglo-American model, and international student exchange programs, and they enunciate differing opinions on how well this system delivers on its mission to its students and its country. The chapter concludes with a brief case study that illustrates the emergent themes from participant responses.

16.2 Background

This chapter highlights recent developments and stakeholder perspectives on the current structure of the engineering diploma in Tunisia. Beyond describing how the system works, the chapter focuses on how well it is working in the eyes of the constituent Tunisians who are part of it. Interviews with professors, students, administrators, and industry leaders yield a broad spectrum of opinions of what is effective and what is unproductive about the Tunisian EE system. Does this system of educating engineers answer to the needs of Tunisians? The chapter begins with a section of background information and relevant literature.

16.2.1 Higher Education and National Priorities

Education has been a topic of high importance since the inception of Tunisian sovereignty. Tunisia's first president, Habib Bourguiba, emphasized the development of the national primary, secondary, and tertiary education system as a top-priority engine for the new nation. Historically, Tunisia has devoted a large amount of its financial resources to education (approximately 7.5 % GDP (UNESCO 2003)). This means that, as the "top national priority," more than 20 % of the annual government budget goes to education (Hamdy 2007). Due to its strong institutional inheritance from and continued linkage to France, Tunisia's tertiary system closely parallels the university system of the former colonial occupier. As the chapter will discuss, one major issue in Tunisian EE that respondents identify is the number of remnants of the French colonial system that still exist today. It is doubtful whether these factors have been adapted to fit the diverging needs of this particular country, particularly in the engineering sector.

EE in Tunisia is a fertile ground for investigation. Tunisian policy-makers are conscious of national priorities for training proficient engineers to answer the Millennium Development Goals (UN 2005) as well as to increase competitiveness in the contemporary knowledge economy (World Bank 2007). The country is currently trying to define how universities themselves can serve as jumping-off points for business development. Selectivity and prestige of the field, rigor of the training, and curriculum balance between theory and practice have been major parts of this debate (e.g., World Bank 2010; Catusse et al 2010). As this chapter will demonstrate, stakeholders in Tunisia note that the country needs to decide what fields are the most vital for its own development.

16.2.2 Educational Data and Accountability

In addition to making the education system a top fiscal priority, Tunisia has demonstrated its desire to gather educational data and serve as an educational policy entrepreneur in the region. For example, Tunisia has shown itself to be a unique educational testing environment at the primary and secondary levels in the Middle East and North African (MENA) region. Tunisia is the only African country to participate in PISA, the Programme for International Student Assessment (OECD 2010). Tunisia, as a policy entrepreneur in the region, was one of the first North African countries to participate in the Trends in International Math and Science Study (TIMSS) in 1999. This international assessment now occurs every 4 years, and, two rounds later, all but Libya in the North African region participated in TIMSS, as did most of the MENA region. Tunisia's experiment may have helped to give the policy the necessary foothold in that region to allow it to spread (DeBoer 2012).

At the tertiary level, Tunisian national policies have also reflected an interest in educational data, testing, and cross-national alignment. The government put forth a set of policies under the heading of a Development Program for Higher Education and Support for Quality (*Programme de Développement de l'Enseignement Supérieur et d'Appui à la Qualité*), supported in part by the World Bank (World Bank 2013). Among the policies included in this program were efforts to increase access and efficiency through support for institutional autonomy and instructor support for improved teaching methods. Tunisia is one of numerous Maghreb countries that have advanced large-scale structural and accountability system changes in line with the Bologna process (Buckner 2011). However, the LMD structure has not been applied to engineering programs, similar to the choice France has also made; interestingly, although the Tunisian government has taken steps to align educational research and practice with broad international activities, the country's governmental policies often align more closely with the direction of French higher education than with the rest of Africa or even with other MENA countries.

16.2.3 Demographic Trends

Contemporary demographic shifts highlight the importance of this topic and this study, both at the time of data collection (2007) and at the time of publication (2014). The shifting demographics and employment status of young adults and college graduates were a large driver of Tunisia's spark of the "Arab Spring" in late 2010 and 2011; at the time of research design and data collection, the strains of a young population bubble were becoming apparent. Youth unemployment was rising from 16.9 % in 2006, the year prior to this work, up to 21.9 % in 2009 (Gassab and Ben Ouada Jamoussi 2011). While Tunisia as a socioeconomic entity continued to develop and urbanize, the universities remained firmly on their own track (Haddad 2009). On one hand, "massification," or the drastic increase in access to higher education, was seen as a major success by national policy-makers, broadening participation in tertiary educational opportunities. However, this was combined with economic strain and mismatch between graduate preparation and employer needs, meaning that graduate unemployment became widespread. Tunisian universities and the Tunisian Ministry of Higher Education, Research, Science, and Technology (MESRST, whose name and acronym recently changed to the Tunisian Ministry of Higher Education and Scientific Research (MESRS)) broadly responded by investing even more in centralized public institutions. However, the implications of increased public higher education purview as well as the outcomes of other solutions to address the oversupply of qualified graduates remain unclear (Melonio and Mezouaghi 2010). Public expenditure on higher education has steadily increased, but it may not be sufficient to buttress against threats to quality due to demographic shifts (Abdessalem 2011). Engineering graduates, in particular, are often perceived as immune to this kind of demographic bubble, but, as this chapter discusses, the oversupply of engineering and technical diplomas has affected the Tunisian EE system as well.

That oversupply, combined with other demographic and economic trends, reached a breaking point in 2011. With the recent change in political power and the Tunisian Revolution, higher education has taken on an even higher priority according to numerous groups (Ben Othman 2011). Interestingly, the revolution was both driven by students and young people and made use of new social media technologies (e.g., YouTube) when traditional media channels were stifled (Arif 2014). This chapter draws on information on Tunisian higher education directly from students who were experiencing changing demographics in 2007. In addition, the chapter incorporates data directly from the responsible ministry. This objective information is openly available from MESRST as well as from individual institution sites (Republic of Tunisia; MESRST 2005). The chapter also draws on the overview of the field provided by Ben Sedrine and Gobe (2004), which details all aspects of professional engineers—the historical politics behind their education, the job market, and the changing economy over the years—to inform the semi-structured interview protocol. Ultimately, the chapter presents shared and contrasting perspectives of diverse stakeholders. The concentration of activities in Tunis, the capital, belies the diversity of experiences of Tunisian students, teachers, and engineers across the country. The country is multilingual (with many students and EE stakeholders speaking Arabic, French, English, and some Italian) and diverse in numerous dimensions throughout its small geography. This chapter highlights the opinions of numerous varied stakeholders.

Is the system in a given country effective for its respective body of engineering students? This chapter goes directly to the source and asks participants in the field of EE to describe the effectiveness of Tunisia's system for its respective body of engineering students. The Tunisian system of educating engineers has been in constant flux for the past 15 years. It is becoming more aligned with international systems but has yet to maximize the potential of the Tunisian students abroad or the global community in Tunisia.

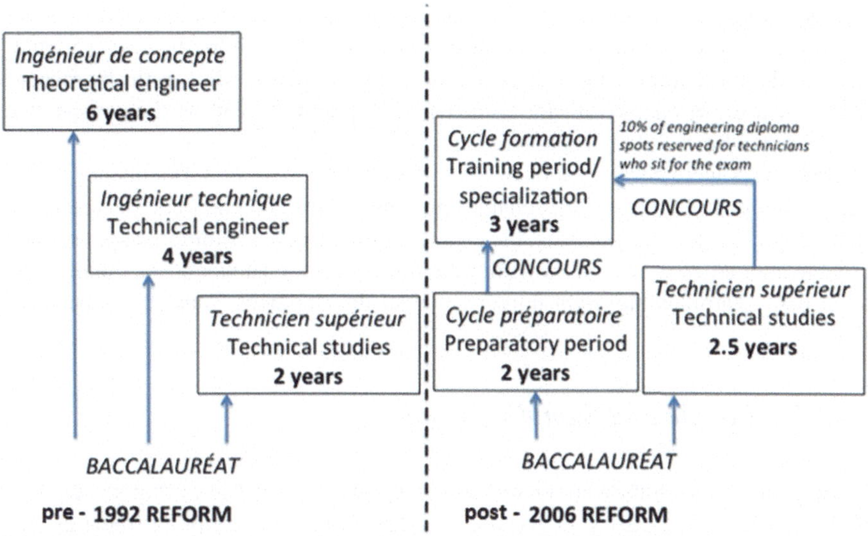

Fig. 16.1 Broad summary of diploma structure reforms (from descriptions available on MESRS 2014)

16.2.4 Recent Reforms

Figure 16.1 summarizes the recent structural reform to the Tunisian engineering diploma. In 1992, the structure of the diploma had three main tracks post *baccalauréat* ("*bac*"). One track led to a technical degree after 2 years, one to a basic engineer's qualification after 4 years, and a third to a prestigious, theoretical engineering degree after 6 years of study (Fig. 16.1). After major changes in 1992 and further adaptation towards quality improvement and the LMD system in 2006, structural reforms responded to supply as well as demand needs and combined the two engineer's degrees into a 5-year diploma while further allowing a pathway from the technician's degree to further engineering study. The upper-level engineering studies were further reformed to allow for increasing practical application. The 1992 change and subsequent World Bank support throughout the 1990s fostered the creation and connection of institutes for technical studies (Salmi 2013).

16.3 Methods

16.3.1 Research Design and Rationale

The investigation employs qualitative methods to gather the direct perspective of current participants in the Tunisian EE system. In this study, qualitative data and stakeholder narratives generate broader hypotheses about the current state and future possibilities for Tunisian EE. Participant groups were first selected to cover perspectives from important stakeholder groups and then sampling frames including key demographics. The interview protocol was created to unearth the respondents' own experiences with the Tunisian EE system as well as their respective perceptions of the system's characteristics, success, and trajectory.

Interviews were conducted over a period of approximately 3 months during the early summer of 2007, and this chapter presents varied stakeholders' perspectives. In addition to interview analysis, the chapter contextualizes common responses and notable anomalies in responses within the framework of Tunisian EE. Finally, two short case examples highlight university-level development from USA's perspective.

16.3.2 Sample and Sampling Frame

Participants in the study were identified via a purposive sample of four stakeholder groups: academia, industry, government, and students. The sampling frame of academic representatives included instructors—university faculty of varying ranks—as well as school administrators. The sampling frame of industry representatives included both public- and private-sector employers. The government sampling frame comprised officials from MESRST and other ministries that oversee engineering degrees (e.g., agriculture) and administrators. The sampling frame included any officially recognized subdisciplines of engineering study conceived broadly, which includes information technology (IT) degrees. The particular respondents in this study were first identified through convenience sampling via the contact networks of key leaders of an international engineering organization with headquarters in Tunis. Additional contacts were developed through the author's relationships with other EE leaders. Snowball sampling then helped to identify additional respondents.

The final sample comprised 26 interviewees. Interviews were primarily conducted in French by the author, though six interviews were conducted in English due to the respondents' comfort with the language and their preference during the interview. Interview quotes provided in this chapter have been translated from French where necessary (Table 16.1).

While the sampling frame was restricted to institutions that grant the *diplôme d'ingénieur* (the current 5-year engineering diploma mentioned above), respondents included representatives from classic engineering *écoles* (prestigious universities) as well as tertiary institutions under the supervision of both the Ministry of Higher Education (MESRST) and another ministry (such as the Ministry of Agriculture), and recently created private universities.

16.3.3 Data Collection

Twenty-six interviews were conducted from April through June 2007. Interviews lasted from 50 to 90 min (typically 80–90 min), and each was tape recorded. Interviews were organized as in-depth, semi-structured conversations in which the

Table 16.1 Final sample characteristics

Sampling frame	Demographics included	Rank/positions included
Academic	Male, female; currently working across the country (inside and outside of Tunis)	Tenured faculty, untenured faculty, visiting faculty from France, USA; university founders/directors; research group and lab directors; directors of studies; preparatory high school directors; faculty with and without international experience, either studying and/or working abroad (USA, France, Germany, Canada); deans
Government	Male; based in Tunis	Industry liaisons, trade association directors, national utilities directors, ministry officials
Industry	Male, female; Tunis	Industry advisors for apprenticed students; technical managers; company associate directors; CEOs; practicing engineering project directors (subdiscipline of construction)
Students	Male, female; Tunis	Fourth- and fifth-year engineering diploma students

respondent described at length his or her own experiences and perceptions. Initial interviews helped to develop the overall structure of the interviews and the semi-structured topics used for the protocol. Appendix A comprises the protocol including more detail on the broad areas that guided the semi-structured interviews. These broad areas included the following: structural perspectives, perceptions of quality, student qualifications, teacher qualifications, industry connections, cultural context, social context, and political context. The focus of the study was on the traditional 5-year *diplôme d'ingénieur*. However, questions related to the *technicien supérieur* (advanced technical postsecondary degree) were included as related to discussions about the engineering diploma.

Within each broad area listed above, the interview focused more in-depth on questions and follow-up probes pertinent to the respondents' respective profiles, demographics, and professions. Professors, on "the front lines" of the education process, offered valuable input regarding the recruitment and classroom experiences of students. Industry perspectives, especially in EE, are important drivers of the conversation around the required skill sets of engineering graduates (e.g., Hundley et al. 2013). For this project, a number of industry employers in both the private and the large Tunisian public sector and across multiple subdisciplines were interviewed about the competence and qualities of newly graduated engineers from Tunisian universities. Finally, students gave their own opinions on how well the system was serving them at a critical point in their studies, close to graduation.

16.3.4 Analysis

Respondents provided their respective personal experiences and perceptions. Their stories are aggregated using a grounded theory approach (Creswell and Plano-Clark 2007) in order to uncover emergent patterns across respondents. Using inductive methods, recurrent themes are constructed from aggregated stories, combining convergent trends into broader and broader categories and analyzing the resulting macro-perspectives to understand the story and answer the research questions at hand.

The sample includes multiple respondents in the same stakeholder group by design, both to provide additional, different perspectives and to serve as a way of triangulating the information from respondents. The analyses include participant language verbatim accounts, highlighting discrepant data (Bashir et al. 2008) to increase validity. There are, however, some limitations to this study. In selecting respondents, the researcher found participants who would be the most willing and able to participate in an interview. Such participants might provide biased perspectives that are not indicative of the experiences of Tunisian EE stakeholders in general. Further, respondents who were referred by others via snowball sampling might contain their own biases. As this study employs qualitative methods, there is no attempt made to generalize findings here; they are provided to shed light on diverse perspectives and suggest hypotheses that could be further explored in subsequent quantitative data-gathering and analysis stages.

To support the statements made here, multiple respondents have concurred with these opinions, and where their opinions were divided, their divisions are indicated, and the respondent's profile is indicated to give an idea of the genesis of different perspectives on the Tunisian system. Statistics provided by the MESRST as well as by prior quantitative research studies support some of the responses given by the study participants. By talking to numerous representatives of the different groups who contribute to the education of the country's engineers, the overall picture begins to emerge.

16.4 Results

Results are organized into six major themes that emerged from participant interviews: *history, curriculum, regulation, differentiation, external pressures,* and *perceived quality*. First, nearly all participants situate their responses within a historical perspective; the importance of a historical lens forms a major theme across interviews. The chapter proceeds to a discussion of the respondents' focus on the EE curriculum, where they point out the mismatch between skills required of graduates and the training they receive. There are ostensible opportunities for hands-on preparation, but in practice, this is not very frequent. Respondents note a continuing tension between theory and practical classes. The next theme that is discussed addresses the conundrum of regulation in Tunisian EE. Respondents see regulation as highly centralized, though they also point out the diffuse centers of power in different universities and faculties. Quality control and increasing privatization are seen as threats to the traditional mechanisms of regulation. In the subsequent section, respondents discuss the complex challenges and opportunities of the differentiated Tunisian EE pipeline. The theme of notable external pressures is cited by numerous respondents, as are perceptions of quality. The chapter closes with a brief case study comparison of two new universities and possible future directions for Tunisian EE.

16.4.1 Historical Perspective

Many interviewees reference Tunisia's first president, Habib Bourguiba, in their responses. One of Bourguiba's main priorities for the nascent country was education, and this emphasis on education as a key to Tunisia's future has continued through the subsequent administration and into the new leadership established in 2011. The historical reference to education as an elemental priority seems to ground perspectives on Tunisian EE in a deep and valued history.

Respondents frequently reference the hierarchy between the levels of the previous engineering degree, a historically trichotomous structure. The engineering diploma was previously structured to create two "types" of engineer and one "type" of technicians. Those who studied in the technical program for 2 year after the *bac* exam (BAC +2) were given the diploma of *technicien supérieur*; those who pursued a basic engineering degree for 4 years after the *bac* (BAC +4) were given the diploma of *ingénieur technique*; and those who studied for 6 years after the *bac* (BAC +6) were given the diploma of *ingénieur de concepte*. This hierarchy bred discontent (discussed further in the section on *Differentiation* and the subtheme of *Prestige*).

In 1992, a major change was made to the pedagogical structure of engineering training. Purportedly to try and combat the problems that arose from having a more elite corps of engineers, the two engineering diplomas were combined into one degree of 5 years after the *bac* (BAC +5). The formation of technicians is still entirely separate from that of engineers, except where now 10% of spots going into the *cycle de spécialisation* are reserved for those who finished their five semesters of technician studies and apply for the exam as the engineering students do. Before, according to respondents, three times as many engineers as technicians were being produced; now that statistic is reversed. The major reform years serve as important landmarks around which respondents structured their answers. Other smaller changes were tried over the past 40 years of Tunisian EE. For example, a more rigorous accreditation system was reportedly tried in the 1980s, but only lasted 2–3 years. These historical reference points, primarily the major structural reforms, frequently differentiate eras and serve as milestones in respondents' narratives, providing a longitudinal comparison and a sense that Tunisian EE is continuing to improve.

16.4.2 Curriculum

Respondents designate the rigorous EE curriculum as rigid and difficult to change. Perspectives on quality in the curriculum bring out subthemes around the skills required of engineering graduates and the training they actually receive. The tension here is consistently between theoretical training and practical skill demands in the workplace.

16.4.2.1 Professional and Interpersonal Knowledge, Skills, and Attitudes

Employers repeatedly note that engineers still require a lot of training when they enter a job. Public-sector employers convey satisfaction with the current system, while private-sector employers appear dismissive of Tunisian graduates' abilities. According to one professor, the purpose of engineering pedagogy is to make engineers capable of learning. One employer points out that they are looking for workers—whether engineers or technicians—whoever could learn. According to private-sector employers, they are not as sensitive to differences between the engineering and technical degrees; their priority is what the employees could do.

Unfortunately, local companies do not have many resources to support trainers for new workers; they need engineers with the right, targeted skills. According to numerous industry respondents, "Private enterprises look for engineers who are directly operational…" Private companies demand an engineer who is employable right away; to have this gift, they may need to invest more in training the engineers while they are in university than they currently are (according to academic respondents).

In general, employers feel that the interns they have and the engineers they do ultimately hire are unprepared. Deficiencies that are cited included hands-on or practical knowledge, a lack of "soft" skills, and a need for multilingual training (not only in French but also often in English, Italian, or German) to participate in the regional and global economies. Most laboratory funding comes from the government to the schools; according to respondents, part of the reason instruction is very theoretical is a lack of equipment. Some highly successful faculty can circumvent this obstacle by obtaining other funding: "At universities (I know of the Engineering University of XYZ), improvement is a direct result of ability to obtain outside funding [*for labs and practical tools*]" (note added by author). Given some of the limits of laboratory availability, some recent studies suggest online/blended learning and remote lab opportunities as one way to mitigate resource constraints (Naddami et al 2014). Using remote laboratories may also facilitate university networks and consolidation in the Middle East and North Africa (Salah et al. 2014).

Because of the "lack of practical experience" cited by industry employers and the general need for an engineer to undergo much training on the job, state-run public-sector employers are more willing to take on new student engineers. According to numerous respondents, they have the time and resources to invest in training them during their first year on the job. For large public enterprises, engineers have to be versatile and know multiple domains. For public enterprises, research is not a priority; the job is to give continued service. Despite more openness to new graduates and a lack of focus on research, employment in the public sector is strongly related to higher educational attainment (Pekkarinen and Pellicer 2013). This "misallocation" of talent may hamper the growth of both private and public elements of the Tunisian economy.

In the faculty of an engineering institution, respondents report that efforts are made to occasionally bring industry into the classroom to teach a module or a class. This is an efficient and effective way of incorporating industry professionals and innovative perspectives into the schools. However, it is reportedly difficult to find industry people ready to teach for more long-term assignments. Additionally, from the industry's perspective, teaching staff are "very academic and not open to collaborate or do work for industries." Although each respective university might claim that its graduates are slightly more suitable for employment, one employer sees little difference between the abilities of graduates from public and private schools.

16.4.2.2 Hands-on Opportunities for Practice

There are hands-on work opportunities built into the engineering diploma—apprentice-like opportunities for students as they finish their degrees—and respondents pointed to these as the few existing connections between academic curricula and early industry experiences. Each of the 3 years in the final cycle of the *diplome d'ingénieur* also comprises a specific kind of project for the student, which gets progressively longer and more independent in each respective year. Summer internships also contribute to the student's real-world experience. The first year, a small project is done, which normally takes about a month. The second year, the *projet de fin d'année* (end-of-year project) is slightly longer, normally 2 months. During the third year, the *projet de fin d'études* (senior project) takes up most of the final semester of study. To find these internships and subjects for their projects, students must often turn to their own resources. The MESRST demands practical experience internships and reportedly encourages the collaboration of universities and industries. However, respondents point out the disconnect between theory and reality, as these internships often end up being completed within the university rather than "in the field." Respondents feel that few companies have comprehensive internship programs. Respondents note that students had to go door to door, since industry "does not approach students." Students who are already active in different student organizations make some contacts through informal networks, but universities often end up providing the resources, lab space, and topics for internships or research projects. Here, again, respondents highlight the importance of organic and informal networks.

These internships give students the opportunity to observe practices in industry. Moreover, the third extended project during their final year challenges them to take on a real-world problem provided by an engineering company. Sometimes, students proceed from their internship to employment at that same company, though respondents say that, in reality, this is rare. The mismatch between the skills required and students' preparedness for industry seems to be a missed opportunity for realizing theoretical best practices in teaching engineering.

16.4.2.3 The Potential of Technoparks

Technopoles, or "technoparks" that physically juxtapose small and medium enterprises (SMEs) and university campuses, were set up just before the new millennium to "constitute the nerve centers of a national network of scientific and technical research…" (Republic of Tunisia, MESRST 1999). Respondents see these Tunisian *technopoles* as an attempt to bring together the three facets of engineering: industry, education, and research. However, *technopoles* are also seen as an import from the American EE ecosystem, and respondents perceive that it is more the case that foreign businesses take up residence at the parks. Professors perceive the *technopoles* as useful for internships for their students, though employers themselves do not concur; corporate respondents saw the major benefit of the *technopole* as the discussion between different stakeholder groups at the administrative level and visibility for the companies, not recruiting.

From the industry side, respondents report an "immense effort" to get "alternating" education. This method would alternate students' schedules between classroom work and industry internships. This kind of schedule reportedly "works already" with technical high schools that prepare students for practical technical jobs. However, higher education is said to want a contract that is "overly demanding" so that industry representatives have not succeeded in forming a mutually satisfactory arrangement. Some more flexible (usually private) universities experiment with alternating education for its senior projects in recent years. Students attend classes for 3 days and work at their internships for 3 days. They have a number of complaints about this arrangement, saying that the changes back and forth were too frequent for them.

16.4.2.4 Innovative Capacity and Entrepreneurship

Research and innovation are cited as potential benefits of close industry/university collaboration. However, the private sector gives no more concrete support for research and innovation than the public enterprises do. According to most respondents, little research and development (R&D) is done at the university. The university is said to end up supporting the private sector with its own resources. Former president Ben Ali and his administration are described as actively engaged in fostering the development of private enterprises, a goal made even more pressing as the Tunisian market was opened in 2004. At the time of interviews, industry "fabric" was only in its formative stages. Between 2004 and 2009, 2000 enterprises were planned for creation. The push for individual entrepreneurship was part of this same initiative. In 2003, President Ben Ali advocated for universities training "creators of enterprises and not just demanders of employment" (Mabrouk 2007). For this development to occur, the government would need to assist in bringing the private sector into the education system as well; respondents perceive this to be at a nascent stage and not yet really happening.

Entrepreneurship skills are nominally required in all tertiary degrees as integrated activities for any discipline, as required by the 2002 Educational Reform Act (Gyimah-Brempong and Ondiege 2011). However, these courses are barely mentioned, and students do not feel prepared to practice entrepreneurship. There are numerous business schools in Tunisia, and student evaluations show some satisfaction with the experience, mixed with a need for more institutional self-assessment and improvement (Belhaj et al. 2013). More recently, incubators are springing up as potential vehicles for innovation, although one study noted that such local systems of innovation were still emergent (Fayolle et al. 2010).

A large number of engineers are employed by state-run or semi-state-run public companies. Beyond their roles as employers, these companies are seen as doing little to encourage innovation. These semi-state-run companies want "studies" from apprenticed students rather than "real research projects." There is said to be not a culture of R&D—instead, it is reported to be more of a service and commercial market. Often, employers do not see the value of R&D. There is, therefore, little industry lab funding. While academic and industry respondents see different reasons for this lack of industry investment in innovative capacity, both admit this is the case. Interestingly, however, the number of researchers for the size of the country is large (UIS 2014). Respondents anecdotally note this, another example of an oversupply of highly qualified human capital.

16.4.3 Regulation

Respondents nearly universally describe facing challenges in trying to innovate in a strict system serving multiple regulatory rulers. Some academic tasks are highly centralized, while at the same time, certain organizations enjoy a great deal of autonomy. Both before and subsequent to the overthrow of the Ben Ali regime, this has led to perceptions of corruption and integrity challenges throughout primary and secondary education and into tertiary education (Milovanovitch 2014). Teacher quality is a major concern for administrators and students alike. The perception of increasing privatization and autonomy provokes uncertainty from respondents as to how this will ultimately change the EE system. Respondents are not sure if regulatory changes will indeed persist.

16.4.3.1 Centralization Versus Autonomy of Some Tasks

Respondents describe the education system as highly federal and concentrated, with a council made up of representatives from all of the universities overseeing individual institutions from within the MESRST itself. Despite extensive centralized standards, implementation is rather autonomous, and university faculties are bonded together rather loosely. Acceptance and instruction for the engineering diploma are done on the part of the numerous school faculties within each university, or *établissements* (approximately analogous to a college within an American university). Each of the *établissements* is governed by a faculty counsel from within the school. The university system serves an almost purely organizational purpose, as opposed to the classic university campus system seen in North America. Classes almost never overlap, and there is no notion of a liberal arts curriculum. Often, different faculties are not even in the same city. For this study, respondents rarely referenced their *école* as their institution, more frequently describing their *établissement,* with some also discussing their specific departments.

Tunisia's main professional organization for engineers is the *Conseil de l'Ordre des Ingénieurs*. This group is a loose affiliation, in which membership is often required for employment. There are smaller organizations, but they are not as active. As one respondent says, it is easy to form such societies but difficult to have them play a large role. Besides professional societies, there are scientific societies, which are older and more numerous. These scientific societies organize conferences and communicate with international peer organizations. Both the professional societies and individual academic departments operate on their own, but their members (faculty and students) still feel dominated by the national authority body.

16.4.3.2 Quality Control for Faculty

Respondents see "quality control" as a new concept for the Tunisian system, recognizing that de facto self-policing was the accepted norm. As long as national regulations were met, schools could operate independently. Teacher quality is seen by nearly every respondent as a key obstacle for the Tunisian EE system. The idea of student surveys is foreign, and numerous instructors themselves even cite difficulties to train the reticence out of their students to respond to questionnaires. The stimulus for these kinds of measures exists, but it has not yet been institutionalized. Although respondents could cite very specific metrics for tenure and promotion, they then relate these specific benchmarks to issues of corruption, lack of actual quality assessed, and continuous improvement. For example, the number of theses that a professor is supervising lends prestige to their reputation. In practice, professors may take on projects for students that they will supervise but will then repeatedly assign them to their own doctoral students. Further, the highest quality engineering graduates are seen as universally heading to industry, leaving less qualified graduates to be "relegated" to academia.

Massification has only increased the perceived threat to teacher quality. The side effects of *massification* are perceived as being felt by strained human resources, and, in this area, EE is not immune. Visiting professors are seen as an outlet through which the government is trying to alleviate the problem of a growing student body and stable faculty numbers. In practice, teachers from one institution often teach at another institution, and this untenable solution does not have the capacity to support the whole system. The problem of teacher quality is even further exacerbated by *massification*; schools do not have enough money to ensure that all of their teaching staff members are BAC +12 professors. This qualification (BAC +12) comprises a 12-year process of learning to teach, beyond just normal graduate research work, which takes more time than is available to administrators who desperately need to staff their classrooms. Sometimes, younger doctoral students are taken as visiting professors as well. The problem with having current doctoral students teach is not only that they are not as qualified as those who have gained their teaching qualification, but, as respondents pointed out, it also slows down the doctoral candidates in the process of completing their degree.

16.4.3.3 Privatization and Growing Autonomy

The landscape of Tunisian EE is overwhelmingly made up of public state schools. A new law passed in 2000 grants private universities the authority to issue ministry-approved degrees. The respective ministry usually oversees these schools, gives authorization to give out official diplomas, and gives equivalence of diplomas, as it does with foreign diplomas. In 2003, the government publicly recognized that it would look to private universities to help accommodate the burdensome numbers of students who made up the large demographic shift (La Presse 2003). There are currently ten private universities that offer the *diplome d'ingénieur* as opposed to 16 public *établissements* within the larger 13 universities (Republic of Tunisia 2006). Private schools are marketed as an avenue through which motivated students, who may not have scored as highly on the *bac* as their high school classmates, can enter the engineering world. The general opinion of private institutions, however, has been that they are "diploma factories" and offer little educational quality. The opinion of respondents is: "Why should one have to pay for a good education?"

Even with private universities, the central Tunisian government still controls the diploma, dictating the number of hours required, the average score necessary to receive a diploma, and the number of professors at the school (see MESRST references for more details). On the other hand, private institutions are still perceived as having more freedom than their public counterparts to innovate in terms of the structure of the diploma and the curriculum. These universities can court business connections that both public and private US universities already have, and that public Tunisian universities reportedly cannot or do not care to nourish, according to respondents. While private universities do have to report to the MESRST, they have more freedom in determining their curriculum. Respondents do not perceive any degrees of freedom at state schools. New classes or programs can be rapidly integrated into private schools, and the private schools do not have to follow such a strict plan. Private school respondents had a few complaints about their unique situation, but these were infrequent. Because there are so many visiting professors and similar classes are required, it is as if the private universities have the same classes as public ones, but more time from the professors is allotted to the students. A student satisfaction survey found that students at a newer private university were moderately satisfied with their education (Chabchoub 2013).

Besides private universities, the government has recently supported other efforts towards privatization, increased autonomy, and support for market reforms as ways of alleviating demographic and quality pressures. The government launched the *Université Virtuelle Tunisien*, which now partners with some residential *établissements* in Tunisia as well as numerous universities abroad (see also, e.g., *Université Libre de Tunis*). However, it does not offer *diplomes d'ingénieurs*. It has begun, however, to offer courses in conjunction with other technical degrees, so it could further support the purpose of training a broader base of engineers (reportedly potentially relieving about 20 % of residential university courses). Another example of government reform efforts comes in the form of the *double tutelle* programs. While all degree-granting institutions fall under the supervision, or *tutelle*, of

MESRST, there are some that have a *double tutelle* because of their subject focus. For example, the telecommunication school for technicians in Tunis, also known as ISET'COMM, also falls under the *tutelle* of the Ministry of Technology. This allows the schools a bit more freedom than their single *tutelle* peers, which respondents perceive as valuable for students and perhaps desirable for industry and opportunities to get hands-on practice.

At the same time, market-oriented solutions are seen as only slightly increasing industry opportunities for students and private-sector employment opportunities. Respondents from the academic side point out that companies are not accommodating to students; while the addition of market-based initiatives is growing, respondents feel it is sometimes awkward. Even when students provide real solutions for companies to implement, there is no remuneration. Also, many companies vacation during the summer months when students need to work or intern. Private-sector employment still makes up the minority of spots for engineering students despite its growing presence and the government's encouragement of its development. Respondents mention that most private enterprises are SMEs and are usually family run, so personal contacts are the main method of entrance into private employment. According to respondents, the private sector is a more desirable, albeit more difficult, goal for graduates.

The focus of private (as well as semi-state-run) businesses is telecom. There is less diversity in job types in the private sector, since the private industry caters to the demands of Tunisia's large telecom business. Diversifying disciplines within the university is also difficult (e.g., for robotics in young mechanical engineering schools), with few schools offering certain majors and even fewer offering new specialties (Romdhane and Mlika 2011). Such flexibility is seen as key by respondents for supporting long-term curriculum reform success. Other studies have found some growth in disciplines such as biomedical engineering, at a level competitive with and adherent to multiple sets of international standards (Abu-Faraj 2014). Respondents noted that Tunisian policy-makers need to decide which disciplines and subspecialties are key to development and do more to support them.

16.4.3.4 Continuing or Alternating Education

Around 25–30 % of graduates go on to do postgraduate work. Some students choose to continue after the engineering degree to graduate school for a master's (1–2 years of study) or a doctorate (3–4 years of independent research). In general, respondents feel that younger generations of Tunisian society always want to go further with studies to get another degree; part of it is prestige, and students "continue just to continue." A perceived added benefit for those who continue to study is to be able to avoid the difficult job market, but the primary reason for continuing is the hierarchy in which regard is given to those with more degrees.

Few Tunisians return to school after work to continue their studies. This is changing, however, as new continuing education programs are beginning to cater to those with degrees who find themselves stopped in their careers. In engineering, a new program has been started to aid older BAC +4 engineers who are currently blocked in the development of their careers. The advancement offered by this program is often paid for by employers (from semi-state-run companies) and allows recipients to gain an equivalency for a full engineering degree (formerly BAC +6, now BAC +5) to advance their careers. Administrators perceive this new offering as desirable, but few other stakeholder groups in this study made note of it.

The flexibility of continuing education or the more recent proposals for alternate education are seen as possible ways of addressing gaps in the standard curriculum. Overall, the emergent theme of *curriculum* highlights the mismatch between the skills demanded by industry and the skills and experiences supported by both industry and academia. Constraints such as time, the increasing student–teacher ratio, and few private business positions only exacerbate the challenges of a curriculum delivering all the demands made on engineering graduates.

16.4.4 Differentiation

The restrictions—perceived as cumbersome—of a highly tracked education system and the growth in the supply of certain groups of engineers emerge as a theme for nearly all respondents. They provide perspectives around the role of the myriad tests in Tunisian EE as well as the limits these tests place on engineering students' autonomy. They note that the hierarchical system, including the frequent tests, inhibits any gains in efficient sorting that the differentiated system might have. Respondents point to the outcomes for female engineers as a striking example of a system that supports their meritocratic advancement in some dimensions but hampers it in others.

16.4.4.1 Testing Culture

Respondents frequently make note of the numerous tests required of students to advance to each subsequent stage of the engineering degree and the high level of determinism attached to test results. After their first year of *lycée* (high school), students select a track of study, either geared towards humanities or technical skills (Republic of Tunisia, MESRST 2006). More and more, these secondary schools incorporate information and communication technology (ICT) into their teaching practice (Abdelwahed 2012). After 3 years of high school, they must sit for the one of the most important determinant exams of their academic careers—the *bac*. Respondents very frequently refer to the *bac* as a key milestone in their interviews.

Students' *bac* score helps to determine their placement in university. Those who score high enough can pass into the first phase of an engineering diploma, which is 2 years of intense theoretical instruction. Everyone who passes the *bac* is guaranteed a placement in tertiary schooling, though, as some academic respondents note, sometimes with this placement system, students do not even get one of their top ten choices. At the end of this 2-year phase, another exam determines where they can be placed for the second part of the *diplome,* courses oriented more towards practical experience. After three large projects and summer internships, engineers can chose to search for a private company for employment or pass yet another exam, which will give them a placement in one of the state-run or semi-state-run enterprises, such as the gas and electric company ("STEG"). This exam, even more so than those prior, is comprehensive and demands a whole portfolio from the applicant. The idea behind this abundance of exams at every step of the engineering pathway is to ensure equality, according to respondents. Respondents note that the preponderance of examinations continues for those doctoral students who stay on the academic track; there is even a sequence of exams for tenure tracking professors' advancement, including an oral component.

16.4.4.2 Efficiency

The *bac* and subsequent exams are state run, so *établissements* actually figure little into the decision-making process. The higher the student's score, the more likely he or she will get the first choice for university. The principle behind the rigorous exam is seen as giving all students an objective and egalitarian method of getting into university. These students, having graduated high school, are guaranteed a place in a university. According to government projections given by respondents in 2007, the enormous influx of students will fall off somewhat after reaching its peak of approximately 500,000 students around 2011. Levels will stabilize around 340,000 in 2015, still a 55 % growth from 2000 and approximately equal to the 2005 level. In about 5 years, *massification* will have begun to decline, so leaders are described as being content to wait and make only temporary fixes.

According to respondents, *massification* has made use of the "efficient" sorting system and strained it. This problem has been differentially felt in some very particular engineering disciplines where demand is decreasing and supply has traditionally been large (e.g., agricultural engineering). However, according to all respondents, *massification* in Tunisia (in 2007) is more the problem of other areas of study. Humanities disciplines—for example, philosophy, law, and language—are feeling the crunch, they say. In a response foreboding the events that sparked the Arab Spring, one participant notes: "There are vegetable sellers with MBAs, but it is not yet an engineering problem." All parties acknowledge that engineers are luckier than their peers in other disciplines. Engineering classes may, however, be reaching their saturation point, as noted by an adamant handful of respondents.

One of the reasons engineers have been lucky is the limit on the number of diplomas awarded; the exclusivity of the engineering diploma ensures that few engineers were actually formed. The study of engineering holds a high amount of prestige. And, after university, engineers from many different backgrounds report feeling the unemployment crunch much less than their fellow students. Because of the constant high demand for engineering expertise, engineers have less trouble finding employment than other overcrowded disciplines, they note. Even so, some of the student respondents continuing their education feel that having more and more degrees had become obligatory. The general attitude for students is to "continue as long as possible" in school, and sometimes this attitude is promoted by the appeal of avoiding the job market. The job market for engineers may have become more affected by *massification* shortly after data collection; the year 2007 was to be the final stage in the opening of Tunisia's market.

16.4.4.3 Women

Since the birth of the independent country, women in Tunisia have enjoyed complete equality under the law. First President Habib Bourguiba immediately granted women equal rights. Their presence in the higher education sphere is strong, sometimes the majority—57.2 % of university students are female, for example (Republic of Tunisia, MHE 2005)—in universities and public service jobs. However, engineering is an exception, and both male and female respondents corroborate this.

Despite forward-thinking attitudes since Tunisia's birth, traditional attitudes towards women in engineering persist. Two respondents from very different demographics say, "Engineering is not convenient for women from the point of view of family," and "It is against nature." In her book, Riadh Zghal looks closely at the situation and perspective of women engineers in Tunisia—before and after university, in the workforce, and without employment (Zghal 2006). This can be framed in the larger conflict between conservative and liberal ideologies and the government's reform policy response (Megahed and Lack 2011).

Even with these attitudes, respondents deny that any prejudice existed in the academic world or in public-sector employment, thanks to the objective evaluation process. "Exams give the same chance to all," is a common response. This may be true; most female respondents report the same thing, saying the most visible discrimination happens outside of the public sector and in certain subdisciplines of engineering. These types of attitudes could only be harmful where the strict government rules do not guarantee equal treatment. For example, even in public employment, selection later in the career is not done with exams, so the opportunity for discrimination exists. "After the exams, selection is more personal…assumptions are made about women's needs," describes one female respondent.

Industry reportedly will only hire women over 40 or very young because of their assumed duties as mothers. Instead of dealing with these kinds of prejudices, most women choose to follow the academic track in engineering. Some respondents point out that women engineers choose teaching since it is difficult culturally and logistically to be on an engineering project site. Nearly all respondents note that women go into teaching in the engineering field rather than in other engineering sectors.

16.4.5 Internal and External Pressures

Respondents note that there is a tense push and pull between external pressures from France, aid agencies, and macro trends in accreditation on the one hand and internal pressures from Tunisian politics, society, and culture on the other. This is experienced by participants who moved between the internal and external domains as exchange students or faculty as well as by those who stayed local. The influence of France is never far away from respondent's narratives, and often the feeling of external pressure is tied to funding. Internal pressures relate both to strong cultural norms in higher education and to national economic fears.

16.4.5.1 External Pressures

A major theme through participants' personal experiences and their perceptions of the Tunisian system's success is the burden and opportunity of the historical alignment with France. As a French protectorate, Tunisia had close institutional ties with France. Before statehood, and even up until the formation of the Tunisian National University of Engineering (ENIT) in 1968, Tunisia's first engineering university, many engineers studied in France. From its official beginnings, the national Tunisian system was modeled after the French *grandes écoles*, a rigid and centralized, hierarchical structure. As a thought-provoking corollary, this historical artifact connecting French and Tunisian universities may have supported transnational student activism and provided Paris as a locus for Tunisian student groups' operations after state repression in Tunisia in the 1960s (Hendrickson 2014).

As previously noted, though, respondents cite the emphasis on theoretical training as an evident "burdensome" remnant of the French presence. Industry respondents as well as some academic respondents describe the curriculum's lack of emphasis on practical knowledge and experience (and the resultant preparedness of engineers) as stemming from the system's French heritage. The French language itself may be seen as an impediment to progress; although science and engineering classes are largely taught in French, students are increasingly using Arabic in their quotidian study and practice. An investigation in Libya found that students perceived study in Arabic as easier and more accessible for them, even though they saw English as the gateway to global opportunity (Tamtam et al. 2013). Many respondents note that even today, whenever changes are made to the educational system in France, Tunisia follows suit. However, in creating and implementing the changes, Tunisia "lags behind," they say. This lead-and-follow relationship prompts one professor to describe the attitude in Tunisia as a persistent feeling of colonization.

Since statehood, respondents remark that the Tunisian model of education has always been based on the French system, which itself has recently been moving towards pan-European settings. However, respondents characterize the creation of the Bologna Process (Reinalda and Kulesza 2005) and the subsequent development of Euro-level standards (European Network for the Accreditation of Engineering Education; Augusti et al. 2007) as a move more towards the American model.

As Europe changed to a three-cycle system (License–Masters–Doctorate or LMD), Tunisia was preparing itself to do the same. It moved all of its degrees to the LMD system except for medicine and engineering. At the time of data collection, the Ministry of Higher Education (MESRST) was in the process of planning the changeover for engineering, projected for the next few years. According to respondents, "The change to LMD is being made too quickly; the infrastructure is unprepared." It is unclear if this change will continue as scheduled, especially given French reluctance to transition its engineering schools. Further, regional obstacles to the LMD adoption in the Maghreb have emerged, such as the lack of mobilization of business in supporting LMD and the lack of R&D and technology transfer as part of LMD (Melyani 2014). The institution of INSAT, an intermediate step towards the LMD system, is a small success for the Tunisian government. However, with the French still reticent to change their *grands écoles* for engineers over to the European LMD system, respondents doubt Tunisia will enthusiastically do so without them.

Pressures for an emphasis on quality and decentralization are seen as coming strongly from the World Bank. Respondents took note that the World Bank was investing heavily in quality assurance through a program called *Programme d'appui à la qualité* (PAQ; for more information, see worldbank.org for project summaries). Respondents focus their discussions on the teacher evaluation component of the program and its focus on changing the culture around quality before enforcing it. Their comments reflect frustration with another attempt to put a reform with theoretical potential into practice without major buy-in, administrative support, or incentives.

Top Tunisian students do often study abroad, so the Tunisian EE structure must deal with systems for international exchanges. Problems exist for the engineers who studied abroad when (or if) they return to Tunisia, despite the regard given to studying abroad. The strict mirroring of the French system makes the integration of even other European models difficult. One respondent who had studied in Germany outlines the differences with the system there. She sees that in Germany, engineers had chosen their path for themselves and, therefore, more readily assumed life responsibilities. Thesis students are not paid in Tunisia, so she perceives that they did not assume the grown-up attitude of students studying abroad. The same professor who described persistent feelings of colonization notes that France sometimes even questions its own exchange programs as possibly neo-colonialist.Fifty years after independence, Tunisia retains a number of logical and philosophical connections with France—the education system being one of the strongest reflections of the French. These connections are seen as historically significant and persistent, albeit fraught with numerous tensions.

16.4.5.2 Specter of Brain Drain

The internal Tunisian requirements often do not match with the impulse to send engineers abroad with an eye to reintegrate them (sometimes referred to as "brain circulation"; Saxenian 2005). For students and workers who studied and worked abroad before returning to teach in Tunisia, numerous difficulties are described. Because of the necessary personal contacts to get recommendations, even initial steps are difficult and good research positions are hard to come by. Although the top students are seen as leaving for France, they are also seen as having many difficulties upon return. For professors to advance, they need to have an advisor: "Those who study abroad are orphaned without a supervisor to help them start to get research funding."

Students who come back from studying abroad have a different host of difficulties than graduates from Tunisian schools as they attempt to reintegrate themselves into the Tunisian workforce. International companies, however, are more willing to hire these individuals. Often, the students' training has been more focused on practical experience—for example, if they studied in Germany—or at a more prestigious university in France. If the opportunity exists, students who studied in Tunisia are seen as likely to travel abroad for employment.

On the other hand, professor exchanges bring a number of teachers from France to Tunisian universities. They are placed at universities as *professeurs vacataires,* or visiting professors. By law, they cannot be employed as full professors unless they are Tunisian or are married to a Tunisian. They are seen as useful and highly qualified additions to a strained workforce. Besides professors and undergraduates, France retains close ties with Tunisia through graduate students studying at each others' institutions and a growing number of joint degrees. Tunisia has recently begun to actively capitalize on its international intellectual pool by starting a network for Tunisians living and working abroad. Both the respondents and recent studies of emigration/immigration patterns note that the demographic trends in Tunisia have, in particular, pushed highly qualified sectors of the labor market abroad without domestic opportunities (Fourati 2010).

16.4.5.3 Prestige Versus Practice

The internal pressure of capital in the form of perceived prestige of an institution is incredibly strong, according to the descriptions of participants. In the older EE diploma system (BAC +2, +4, and +6), prestige drove both individual and institutional decision-making. Incoming students generally applied to this highest level of study, for reasons of reputation and postgraduation salary projections. There was also little market for the BAC +4 engineers over the BAC +6. Those whose qualifications did not meet the standard for BAC +6 studies were relegated to *ingénieur technique* formation, and those who did not make the cut for the BAC +4 degree had to go into *technicien* studies instead. This bred discontent among the ranks of the shorter degree recipients, whose diplomas were held in lower regard, but whose skills were essential. Furthermore, the high desirability of the more advanced degree caused an inverted pyramid of human capital. The problem was a "too many generals, not enough army" problem (according to two respondents). The 1992 change was intended to address this tension. The duration of the technician training was extended to 2.5 years; this change coupled with the consolidation of the engineering diplomas was intended to elevate the status of the technician and encourage enrollment. According to respondents, though, this problem persists.

Respondents observe a predictable hierarchy of choices for technical baccalaureates when they select their tertiary institutions. There are no indicators available about schools' qualities to aid in these decisions, but based on reputations and traditional tendencies, preferences normally fall in a certain pattern. The very top students will leave immediately to start their engineering studies in France and Germany. The next, approximately, 150 will enroll in the Preparatory Institute for Scientific and Technical Studies (IPEST), located near Tunis, which prepares students to apply for the French engineering exams. This is because, according to one respondent, "Tunisia doesn't have anything to offer its best students." After these "most desirable" options, the next highest scoring students will go into medicine or pharmaceutical studies. When spots for these pursuits fill up, students then prefer the traditional Tunisian preparatory schools (with the one located in Tunis preferred the highest), followed by the National Institute for Applied Sciences and Technology (INSAT), and last the technical colleges (ISETs) and other science or technical pursuits.

Engineering students have a certain percentage of spots reserved to study in France or Germany, and almost certainly, the students with the best scores will take this opportunity. The best engineers are leaving for Europe, and in more cases than not, they are not coming back. According to one professor, brain drain is not actually a problem; "for every engineer that leaves, ten more are formed in Tunisia." After the *bac*, about 20 students are sent to France to study engineering and about 20 to Germany. (About 1.4 % of first-year engineering students are sent abroad.) A small number of students depart with parents who can afford to send them to Europe or North America. The few foreign students who come to Tunisia are usually from francophone sub-Saharan Africa.

France is certainly the foreign country benefiting the most from Tunisia's production of rigorously trained high school graduates (UIS 2014). Germany as a receiving country follows France for the number of Tunisian engineering students who study abroad. The students who go to France receive the same 2 years of preparatory training followed by a stage of practical instruction. However, students who choose to study in Germany undergo a formation markedly different from the francophone/Tunisian system. Therefore, they report difficulties reintegrating into the Tunisian system if they come back. Besides numerous specific problems, at a broad level, the state provides money to send students to Germany, but then does not recognize the diploma. The respondents who had experienced these mismatches readily point out their difficulties and perceive them as another example of a rigid system that misses out on talent because of its inelasticity.

16.4.6 Comparative Case Examples

This chapter highlights the example of one particular private university (ESPRIT) to illustrate recent developments in Tunisian EE. This is juxtaposed with a nascent graduate institution (Masdar) developing in the Middle East and more explicitly modeling itself on an American exemplar. These two short case examples inform the Tunisian context more broadly.

16.4.6.1 ESPRIT

In 2003, three experienced leaders from university administration and ministry management formed a private university. These leaders already had the traditional system experience and the social capital and connections to make such a move. ESPRIT (Private University of Engineering and Technology) has tried to emphasize communication and language skills more. According to one description, "The government does not yet have confidence in private universities…There is little encouragement so few private universities are formed." This is the case despite the perceived potential of private education to offset the rapid expansion of student enrollment in state higher education (Abdessalem 2011). The founders experienced a number of difficulties in starting up this new concept and new school. With ESPRIT, the government provided 25 % of the university's start-up costs.

Student respondents choose private school and choose to go to this particular school for different reasons; some students are at ESPRIT because they failed *préparatoire* (preparatory school) and could not continue otherwise, and often, technicians who fail the engineering exam end up at ESPRIT. There is a "much greater diversity" in age and background in this student body. Private universities like ESPRIT are more amenable to students coming back to school, even to move past their technician degree to an engineering degree. Continuing education and such flexible opportunities in public schools could be expanded further, note some private school respondents.

Another benefit of ESPRIT reported by students and faculty is the focus on "soft skills." All of the exams involve an oral defense in front of the entire class and the opportunity to receive critiques on nontechnical skills. Some modules are in English. Nearly all students at ESPRIT have internships outside of the school. However, there are some complaints registered. For the ESPRIT students, alternating every three days in school with three days at their senior project is very difficult—too short a frequency for each part—and they might have preferred the structure in public engineering schools.

Students appreciate the "better, respectful atmosphere" and that professors act more like "informal equals." Students think it is a good mix of soft skills and practical experience. Because they choose to be at the school and make the choice to pay, some students also have chosen to work of their own volition. Now, some respondents said that more and more students choose to go to private universities as their top priority, while 10 years before, they had to do so and had no other alternative. According to numerous private and public academic respondents, in private school, students can demand quality, while public schools do not have the time to devote to students. Smaller classes make it easier to do things by hand and receive more practical experience. From the earliest days of the university, the student–teacher ratio at ESPRIT was kept at an extreme low. This is changing in recent years, though, as the university has grown (Chabchoub 2013). Teaching evaluation satisfaction is around 84 %, although student satisfaction varies according to the student's year in school (Chabchoub 2013). One ESPRIT student is even completing her final project with a company in Marseille by e-mail. Most ESPRIT graduates go into industry. Students and faculty alike see these indicators of flexibility as important positive effects of being part of a new, small, private university.

16.4.6.2 Masdar

ESPRIT, a novel private school, could be compared and contrasted to the Masdar Institute of Science and Technology, a graduate institution focused on alternative energy and sustainable technology in Abu Dhabi. Its vision is targeted at creating graduates for the knowledge economy and placing them directly into the laboratory environment of Masdar City (Masdar Institute 2006). Beyond merely weaving the environments of theory and practice together in the curriculum, the institute is affiliating with and modeling its structure on MIT, envisioning the development of a highly qualified science and technology workforce learning from visiting MIT scholars and implementing some of MIT's characteristic programs (MIT 2013).

Within Tunisia, schools formed in partnership with American or other international universities are not unique. For example, in partnership with an American university, the South Mediterranean University was formed and now boasts a rank as one of the top 1000 business schools in the world. It very explicitly builds on American higher education practices (Systèmes d'information n. d.). While its focus is not on engineering, the advent of an American-accredited university operating under a partnership with a US institution (University of Maryland) is seen as having the potential to drastically change the landscape of the Tunisian system if more similar projects were developed.

16.4.7 Perceptions of Developing Quality and Ambiguity About Future Directions

Respondents for this study were drawn from a wide variety of backgrounds, a successful sampling frame, and their opinions reflect the diversity of their experiences with the Tunisian education system. However sufficient this system might be, all respondents recognize that, given the chance, students would choose to study outside of the country. Most interviewees point out or accede that students lack the kind of practical experience they should have as an engineer. Many stages of advancement depend heavily on knowing the right people. Personal connections, while sometimes utilitarian, can be an obstacle to merit-based advancement and, further, go against Tunisia's egalitarian mission. Quality control is only now entering the policy-making debate, and, according to numerous respondents, it has happened only as a result of exterior aid. Self-regulation may be desirable for the bureaucracy, but independent evaluation will undoubtedly improve the quality of teaching. Some teachers have started student evaluations on their own, but they are few and have reported difficulties. Reform policies and practices are happening, but they are not institutionalized norms. With the recent revolution and political turmoil in Tunisia, there again seem to be opportunities for change but uncertainty about the path that developments will take.

16.5 Conclusions and Implications

The recent political revolution dramatically altered the tenor of higher education discussions in Tunisia. Indeed, the outcry and protest of young, qualified, and unemployed citizens were the catalyst for the subsequent chain of events and administrative overthrow. The uprising grew out of longstanding tensions between conservative and liberal forces, including a highly educated populace with fewer and fewer job opportunities and a restrictive government, as well as a government-mandated secularism despite calls for a revival of Islam in the public sphere (Hart 2014). As the first free-and-fair presidential elections approach in the autumn and winter of 2014, thus far, a national balance has prevailed between new political powers and rights for women, between moderate Islamist parties and liberal progressives (Klaas and Dirsus 2014). The themes identified by participants here continue to ring true, though the ambiguity expressed by participants about how the tension between centralized and external pressures with internal demographic and social needs would be resolved also persists. Perhaps, future development after the revolution will continue towards increased autonomy and decentralization. On the other hand, the need for stability and creating jobs might lead to the system leaning more heavily on its traditional educational structures.

One of the obstinate themes in respondent narratives is the tension among autonomous, ambiguous, and flexible control and centralized standards and regulation. Tunisia's selection system is merit based; the best technical students in high school will be engineers. But is Tunisia maximizing its selection of good engineers? The *bac* as a single test may not be a good predictor of a successful—creative, industrious, and motivated—engineer. The material covered in engineering courses spans a wide array of difficult, profound topics. Tunisia's engineers are rigorously trained and are prepared to learn. They are strong in theoretical subjects, and to receive their diploma, they are required to complete internships in industry. They have been selected from the highest achievers in the student population.

However, the system of selecting courses after the *bac* does not lend itself to picking the most motivated engineers. It strictly limits the number of engineers available to Tunisia, and continues to relegate the rest of the massive amounts of high school graduates to disciplines already teeming with unemployed university graduates. Because of the traditional preferences, the best students will become engineers, not necessarily those who want to practice engineering. Prestige is seen to play a large part in deciding a number of things for Tunisian engineers—continuing their education, what school to pick, even their choice of engineering at all. Students want to go to the best schools—engineering schools or medical/pharmaceutical institutions—and get the highest degree possible. Prolonging the elitist system of highly exclusive *grands écoles* is not productive for Tunisia, especially if these elite students leave the country at the first opportunity. Initial efforts reported to provide better indicators of university quality are promising, but they face a number of organizational and cultural barriers before these metrics can supersede the public's assumptions about a university's reputation.

Although the centralized system of tests and accreditation have ensured that a small, elite cadre of engineers emerges from Tunisia, and although the traditional reliance on universities' hierarchical prestige is difficult to forget, the social pressures of *massification*, increasingly diverse engineers, and autonomous private industry demands may ultimately make further decentralization necessary. Theoretical education reforms have not been successfully implemented in practice, as cited by every respondent in one or more domains. Without policy implementation, respondents do not see Tunisia realizing its full potential for knowledge economy competitiveness and engineering employability. Even restrictions in the banking sector are reportedly hampering the growth of small technology-focused start-ups (Shahani 2014). Tenuous labor market reforms thus far have not alleviated unemployment; higher education's focus on relevant skills preparation must be drastically increased to address the needs of current students (Haouas et al. 2012).

There are areas of great potential. The development of professional and scientific societies as active parts of the puzzle could provide a model for facilitating interaction between universities and industry. The government here could act more as a facilitator of interaction between the two parties rather than a bureaucratic organization requiring a specific number of credit hours or internships. The Tunisian system is advanced in actually requiring three industry internships before graduation, but could be made more robust by bringing this practical component into classroom instruction itself.

Tunisian EE wants to provide a fertile environment matched with industry needs and open to compete on a global stage. The system is starting to focus on skills and experiences like global competence, but it does not execute when it comes to incorporating exchange students and faculty back home. The current situation is that the best technical minds in the country leave every year and normally do not come back. These talents are ever more necessary to Tunisia as it positions itself globally, especially if it wants to be competitive in the European sphere. Tunisian respondents within universities and leading university and corporate administrations want to see their universities as world class. One of Tunisia's biggest challenges now is to make their universities competitive with engineering schools abroad in the eyes of their own students.

Tunisia's relationship with France is the system's largest exterior support, but it may be one of its biggest impedances as well. Such close ties with France have not completely limited the international perspective of Tunisian students and schools, but their focus is severely narrowed. Reliance on France for direction is hampering Tunisia's ability to adapt its system to suit its specific needs. For now, the system is functioning well, as respondents report that the country retains its close cultural relationship with France. However, it appears that Tunisia is reluctant to try something different. As an expanding and developing country, Tunisia must take advantage of the changing university structure as well as the privatization of companies and universities to address the needs of national infrastructure and economic development. Opportunities with other institutions in the Euro–Mediterranean region may provide immediate areas for growth (Ezzine 2014). Indeed, current indicators of exports and Tunisian multinational companies hint that the Euro–Mediterranean region is alive and growing as a regional partner and that Tunisia's engagement with countries within the region is diversifying (Gana and Richard 2014).

The tenuous struggle between theory and reality of opportunity in Tunisian EE will not persist if, as these respondents indicated, all stakeholders recognize the need for some kind of change. The recent political changes in the country may themselves be the necessary catalyst for Tunisian EE to make a leap forward in achieving the education reforms it has already set out for itself. University students have taken to the polls, with the aspiration for Tunisia to be a model for its own local geographical region. This chapter's Tunisian case study broadly informs major questions in EE and higher education around the world. The countervailing forces of mobility, equity, cost structures, and individual needs are resounding themes for the Tunisian respondents.

Acknowledgments The author would like to acknowledge the thoughtful feedback from anonymous reviewers in the preparation of this chapter. Data collection was facilitated by gracious hosts of the authortion shaped thace at the office of the World Federation of Engineering Organizations in Tunis. Finally, this chapter would not have been possible without the kind investment of time and openness provided by the interview respondents who shared their general perceptions and personal experiences.

Appendices

Appendix A—Semi-structured interview protocol

Interviews covered the following topics broadly. Greater emphasis was placed on questions in domains pertinent to the respondent's individual profile (e.g., faculty member, student). The following interview protocol served as a semi-structured framework for the in-depth interview. Follow-up, probe questions built upon responses to these general questions.

Structural Perspective:

Describe to me your understanding of the development of the current diploma over the past 50 years? What have been some of the key milestones or change points in the structure of the EE system?

What are the essential components of the current Tunisian engineering diploma system?

Perceptions of Quality:

What is your perception of the success of the Tunisian diploma system?

What components of the Tunisian diploma system are the most successful? Which components still need to be revised?

Student Qualifications:

Tell me about your experience as a student (current or former) in the Tunisian EE system. What experiences stick out in your mind?

What knowledge, skills, attitudes, and experiences do Tunisian graduates have? What tools do they still lack upon graduation? How are these skills assessed before matriculation and after graduation?

Teacher Qualifications:

Tell me about your experience as a professor (or in working with professors) in the Tunisian EE system. What are your tasks (research, teaching, mentoring, etc.), and how are you supported?

What are the skills and experiences expected of professors? What is the conventional training experience, and what was your own experience becoming a professor?

Industry Connections:

What is your experience as an entrepreneurial entity in Tunisia?

What experiences does industry require of engineering graduates? What experiences are seen as lacking?

What institutional connections exist between industry and academia in Tunisia?

Cultural Context:

What cultural factors do you see interacting with the structure of the EE diploma in Tunisia?

How do you perceive cultural factors to have interacted with the development of the engineering diploma requirements and curriculum?

Social Context:

What demographic characteristics of Tunisian engineering graduates can be observed in recent graduating classes?

How do individual- and society-level demographics interact with the environment of EE in Tunisia?

Political Context:

How have historical political factors such as connections to France, the structure of examinations, the prestige of engineering diplomas over technical diplomas, and the financial situation shaped the EE system?

How have the major recent political changes shaped the current state of EE or its foreseeable future?

References

Abdelwahed, S. (2012). Integration of ICT in teaching practices: Why and how successful development of use of ICT in Tunisia? *The Journal of Quality in Education, 3,* 40.

Abdessalem, T. (2011). Scope, relevance and challenges of financing higher education: The case of Tunisia. *Prospects, 41*(1), 135–155. doi:10.1007/s11125-011-9177-7.

Abu-Faraj, Z. O. (2014). Biomedical engineering education in the Middle East and North Africa. Biomedical engineering (MECBME), 2014 Middle East conference on 17–20 Feb. 2014, Doha Qatar. IEEE.

Arif, R. (2014). *Social movements, YouTube and political activism in authoritarian countries: A comparative analysis of political change in Pakistan, Tunisia & Egypt.* Theses and Dissertations. http://ir.uiowa.edu/etd/4564.

Augusti, G., Borri, C., & Guberti, E. (2007). EUR-ACE and ENAEE: A proposal and a tool for a European system for accreditation of engineering education. International conference on engineering education—ICEE 2007.

Belhaj, F., Ben-Ayed, O., & Kammoun, R. (2013). Student perception of higher education quality in Tunisian business schools: Case of the University of Sfax. *Journal of Applied Research in Higher Education, 5*(1), 5–16.
Ben Othman, R. (2011). Quality reform and academic freedom. Arab network for quality assurance in higher education, Abu Dhabi Conference, 2011.
Ben Sedrine, S., & Gobe, E. (2004). *Les Ingénieurs Tunisiens: Dynamiques récentes d'un groupe professionel*. Paris: L'Harmattan.
Buckner, E. (2011). The role of higher education in the Arab state and society: Historical legacies and recent reform patterns. *Comparative & International Higher Education, 3,* 21–26.
Catusse, M., Destremau, B., & Verdier, E. (2010). *L'Etat face aux débordements du social au Maghreb. Formation, travail et protection sociale*. Paris: KARTHALA Editions.
Chabchoub, A. (2013). Satisfaction surveys of students, an approach to measure the quality of education: The case of engineering schools in Tunisia. *Journal of Quality in Education, 4,* 13.
Creswell, J. W., & Plano Clark, V. L. (2007). *Designing and conducting mixed methods research*. Thousand Oaks: Sage.
DeBoer, J. (2012). *Why the fireworks?: Theoretical perspectives on the explosion in international assessments and implications for the Gulf.* Presentation to the Gulf research meeting, July, 2012, University of Cambridge, UK.
European Commission. (2007). Bologna declaration. http://ec.europa.eu/education/policies/educ/bologna/bologna.pdf. Accessed 1 June 2007.
Fayolle, A., Ben Salah, A., Ben Salah, I., & Belkacem, L. (2010). Efficacité des pépinières dans la création d'entreprise innovante: cas de la Tunisie. *Innovations, 33*(3), 157–179. doi:10.3917/inno.033.0157.
Fourati, H. (2010). La migration hautement qualifiée depuis et vers la Tunisie. http://cadmus.eui.eu/handle/1814/13455.
Gana, A., & Richard, Y. (2014). *La régionalisation du monde: Construction territoriale et articulation*. Tunis: IRMC Institut de recherche sur le Maghreb contemporain, Paris.
Gassab, M., & Ben Ouada Jamoussi, H. (2011). *Determinants of Graduate Unemployment in Tunisia*. Working Papers, 16, AlmaLaurea Inter-University Consortium.
Gyimah-Brempong, K., & Ondiege, P. (2011). *Reforming higher education: Access, equity, and financing in Botswana, Ethiopia, Kenya, South Africa and Tunisia*. Geneva: World Economic Forum, the World Bank and the African Development Bank.
Haddad, I. (2009). *Les étudiants, l'université de masse et la ville durable. Le cas de la Tunisie. THÈSE NO 4542 (2009). Présentée Le 3 Décembre 2009, À La Faculté Environnement Naturel, Architectural Et Construit, Laboratoire De Construction Et Conservation, Programme Doctoral En Architecture Et Sciences De La Ville*. Lausanne: École Polytechnique Fédérale De Lausanne.
Haouas, I., Sayre, E., & Yagoubi, M. (2012). Youth unemployment in Tunisia: Characteristics and policy responses. *Topics in Middle Eastern and African economies* Vol. 14.
Hart, S. (2014). Revolution in Tunisia: A historical analysis of the uprisings that began the Arab spring. 13th Annual celebration for undergraduate research and creative performance (2014). http://digitalcommons.hope.edu/curcp_13/85.
Hendrickson, B. (2014). *Imperial fragments and transnational activism: 1968(s) in Tunisia, France, and Senegal*. History dissertations. http://iris.lib.neu.edu/history_diss/15.
Hundley, S., Brown, L. G., Fox, P., Didion, C., & Sayre, D. R. (2013). *The attributes of a global engineer: Purpose, perspectives, and progress*. American society for engineering education, Paper ID # 6198.
Klaas, B., & Dirsus, M. (19 September 2014). Bridging the two Tunisias. *Foreign Policy*.
La Presse de Tunisie. (22 October 2003) 30.000 étudiants a l'horizon 2006. 5.
Mabrouk, S. (25 March 2007). L'union fait la force. *Jeune Afrique,* pp. 62.
Masdar Institute. (2006).Press release: Classes start for first students at Masdar Institute of Science and Technology. Technology and development progam, MIT.

Megahed, N., & Lack, S. (2011). Colonial legacy, women's rights and gender-educational inequality in the Arab world with particular reference to Egypt and Tunisia. *International Review of Education, 57*(3–4), 397–418. doi:10.1007/s11159-011-9215-y.

Melonio, T., & Mezouaghi, M. (2010). *Financing higher education in the Mediterranean region: The case of Egypt, Lebanon and Tunisia*. Agence Francaise Developpement.

Melyani, M. (2014). *Interstices N°3: Enseignement supérieur et nouvelle gouvernance au Maghreb*. Paris: Revue Semestrelle, L'Harmattan.

Milovanovitch, M. (2014). *Trust and institutional corruption: The case of education in Tunisia* (SSRN Scholarly Paper No. ID 2439311). Rochester, NY: Social Science Research Network. http://papers.ssrn.com/abstract=2439311.

MIT. (2013). MIT & MIT: About us. http://web.mit.edu/mit-mi-cp/about/index.html. Accessed 1 Jan 2013.

Naddami, A., Fahli, A., Gourmaj, M., Pester, A., & Oros, R. (2014). Importance of a network of online labs in Magrebian countries. In remote engineering and virtual instrumentation (REV), 2014 11th international conference on (pp. 77–78). Porto: IEEE. doi:10.1109/REV.2014.6784227.

Organisation for Economic Co-operation and Development. (2013). Programme for international student assessment. http://www.oecd.org/pisa/. Accessed 1 Jan 2013.

Pekkarinen, T., & Pellicer, M. (2013). Education and allocation of skills in Tunisia: Evidence from an education reform. *IZA Journal of Labor & Development, 2*(1), 1–21. doi:10.1186/2193-9020-2-14.

Reinalda, B., & Kulesza, E. (2005). *The Bologna Process—harmonizing Europe's higher education: Including the essential original texts*. Opladen: Barbara Budrich.

Republic of Tunisia, Ministry of Higher Education, Scientific Research, and Technology (MESRST). (1999). Technoparks and innovation. Republic of Tunisia, Ministry of Higher Education, Scientific Research and Technology.

Republic of Tunisia, Ministry of Higher Education, Scientific Research, and Technology (MESRST). (2005). Les Indicateurs de l'Enseignement Supérieur. Ministry of Higher Education. Republic of Tunisia.

Republic of Tunisia, Ministry of Higher Education, Scientific Research, and Technology (MESRST). (2006). Baccalaureate orientation guide. Republic of Tunisia, Ministry of Higher Education, Scientific Research and Technology.

Republic of Tunisia, Ministry of Higher Education, Scientific Research, and Technology (MESRST). (2007). Websites of higher education and scientific research institutions and students and university services. Republic of Tunisia, Ministry of Higher Education, Scientific Research and Technology. http://www.universites.tn/anglais/divers/site_etab_ang.htm.

Romdhane, L., & Mlika, A. (2011). Teaching and research in mechanism theory and robotics in Tunisia. In M. Ceccarelli (Ed.), *Technology developments: The role of mechanism and machine science and IFToMM* (pp. 451–459). Netherlands: Springer. http://link.springer.com/chapter/10.1007/978-94-007-1300-038.

Salah, R. M., Alves, G. R., & Guerreiro, P. (2014). Reshaping higher education systems in the MENA region: The contribution of remote and virtual labs. In remote engineering and virtual instrumentation (REV), 2014 11th International Conference on. Porto: IEEE. doi:10.1109/REV.2014.6784265.

Salmi, J. (2013). Constructing a differentiated tertiary education system. Policy note prepared for AusAID, May, 2013.

Shahani, A. (20 October 2014). Tunisia's emerging tech sector hampered by old policies. http://www.npr.org/blogs/alltechconsidered/2014/10/20/357590073/in-tunisia-democracy-doesn-t-create-jobs. Accessed 30 Nov 2014.

Systèmes, d'information: Un master en partenariat Tunisie-USA. (n.d.). http://www.tunisia-today.com/archives/10702.

Tamtam, A., Gallagher, F., Naher, S., & Olabi, A. (2013). The impact of language of instruction on quality of science and engineering education in Libya: Qualitative study of faculty members. *European Scientific Journal, 9*(31). http://eujournal.org/index.php/esj/article/view/2045.

UNESCO Institute of Statistics. (2014). UIS. http://www.uis.unesco.org. Accessed 1 Jan 2014.
United Nations. (2005). The UN Millennium Development Goals. http://www.un.org/millenniumgoals/. Accessed 1 June 2007.
World Bank. (2010). Tunisia country brief. http://www.worldbank.org/. Accessed 1 Jan 2011.
World Bank. (2013). Tunisia country program evaluation. Retrieved from the World Bank Independent Evaluation Group, ieg.worldbankgroup.org. Accessed 1 Dec 2013.
Zghal, R. (2006). Les Femmes Ingénieurs en Tunisie: Représentations, Roles et Contribution. Credif.

Jennifer DeBoer is currently an assistant professor of engineering education at Purdue University. Her research interests include the use of technology in education and the structure of engineering training for local capacity building. She employs both quantitative and qualitative methods. Currently, she is analyzing information collected from massive open online courses (MOOCs) to understand their utility for different groups of students as well as studying methods of assessment of global competence.

She completed her postdoctoral work at MIT's Teaching and Learning Laboratory, and she completed her doctoral work at Vanderbilt University in international education policy, where she received a National Science Foundation/American Educational Research Association (AERA) dissertation award for one of her papers. Prior to her doctoral studies, she worked with the Hole-in-the-Wall project, conducted independent research on engineering education in Tunisia, and interned in international education organizations in Washington, DC. She completed her bachelor's degrees in mechanical engineering and foreign languages and literatures at MIT in 2005.

She served as the president and forum chair for the Student Platform for Engineering Education Development (SPEED), which works to empower students around the globe to make improvements in engineering education systems themselves. She is currently on the executive committee of the International Federation of Engineering Education Societies (IFEES) and the International Studies SIG in AERA.

Chapter 17
Social Justice and the Engineering Profession: Challenging Engineering Education to Move Beyond the Technical

Ramzi N. Nasser and Michael H. Romanowski

17.1 Introduction

Historically, the engineering profession has celebrated itself as a technical profession serving the status quo, remaining relatively unresponsive to public concern (Riley 2008). Recently, "some leaders in the engineering community in the United States and other countries have been seeking to cast engineering as a profession in service to humanity" (Riley 2008, p. v). Consequently, ethics in engineering has become a standard of compliance in engineering education. Nevertheless, without the pressure for change from within the community of scholars and educators (Riley 2008), practice in the engineering profession has not seen significant change toward matters of human concerns. Current implications suggest that engineers have a very narrow technical focus empty of deontological responsibilities and related consequences (Pantazidou and Nair 1999), and generally, engineering education centers on problem-solving technical analytical approaches (Riley 2008). Thus, when knowledge about the world is addressed, it is generally an "objective" version of reality limiting ethical concerns to the environment and ecosystem. Still, however, social justice remains elusive in the engineering field.

The growth of the engineering sector specifically in the Middle East throughout the 1960s and well into the 1990s had drawn large cooperation, companies and contractors often employing engineers at all ranks. It is often the case that engineers are positioned as leaders in multinational or even single companies that provide the livelihood for more than 100,000 laborers toiling in unbearable working conditions. Prepared or not, the engineer is thrown in the multinational business world of contracting, drawn into dubious contracting practices, managing human resources

R. N. Nasser (✉)
Dhofar University, Salalah, Oman
e-mail: rnasser@du.edu.om

M. H. Romanowski
Qatar University College of Education, Qatar University, Doha, Qatar

M. Abdulwahed et al. (eds.), *Advances in Engineering Education in the Middle East and North Africa,* DOI 10.1007/978-3-319-15323-0_17

and project overestimation, project kickbacks, and clientelism (Green 2005). The engineer becomes the keeper, the manager, and businessman who controls the life of laborers often protected by laws and regulations of local national constituents. Such blaring malpractices are rarely confronted, analyzed, and reflected upon in schools of engineering. Rather, most of the ethical issues surrounding the engineering profession are transformed into regulations and standards for safety precautions tempered with the myth of objectivity.

Still we argue that the engineering profession is missing the fundamental point—that engineers fail to identify and, more importantly, address issues of injustice compounded by the fact that education and awareness is inadequate (Shuman et al. 2004; McGinn 2003). For example, according to Sax (2000), engineering students express significantly lower commitment to social action than their peers. Furthermore, nearly 80% of engineering graduates still are not required to take ethics-related courses (Herkert 2011). Not only there is a lack of teaching ethics of the engineering profession but there are serious implications for engineers beyond the clients they serve but for society as a whole requiring engineers to embrace ethical behavior as an integral part of the profession (Riley 2008; Catalano 2006; Catalano and Baillie 2006; Kabo et al. 2009).

While impetus of change is profoundly outlined in a number of groundbreaking perspectives and visions emphasizing the transformation of the engineer to a heroic and good engineer (Broome and Peirce 1997), a new language and discourse have surfaced for the engineering profession that are slowly being embraced in higher education. This is evident in the more recently evolving discussion about expanding the ethical education of engineers (Tucker and Ferguson 2007). In 2000, the Accreditation Board for Engineering and Technology (ABET) addressed their outcome standards by including a number of nontechnical learning outcomes engineering students must develop, including "an understanding of professional and ethical responsibility; an ability to design a system component of process to the desired needs within realistic constraints such as economic, environmental, social, political, ethical, health and safety manufacturability and sustainability; and a knowledge of contemporary issues" (ABET 2008, p. 3).

However, such a movement will require a paradigmatic shift that provides oppositional perspectives to mainstream thinking, specifically in the context of the Middle East. This paradigm shift must spark and develop discussion connecting to larger themes of social justice, including struggles against racism, militarism, environmental devastation, human rights, natural depletion, corruption, and poverty. Suggesting a paradigm shift toward embodying a transformative vision of engineering education and aspiring to change the educational and field practice of the engineer—the project manager, designer, or problem-solver—demands a deep and complex understanding of issues dealing with human power, social forces, justice, hegemony, human rights, and equality, all of which are seldom found not only in engineering programs or courses but are initiated in many disciplines such as education, business, and many of the social sciences.

Teaching social justice will require colleges of engineering to resist and move away from "tacked on" approaches in their undergraduate programs; engineers must be able to internalize social justice as a perspective on engineering using it as a critical lens when viewing the world, as there is a need to develop commitments to social justice during the undergraduate education process. Engineering students need to develop a complex understanding of social justice and skills enabling them to critically reflect and identify social injustices in human relations, thus providing rational arguments of social justice as the basis for plans for social action that eliminates injustices. It is vital that engineering educators determine the most effective curricula and pedagogical approaches to accomplish this task.

As professors of education, we understand that education is not a neutral enterprise and by the very nature of the institution, the educator is involved, whether he or she is conscious of it or not, in a political act (Apple 1993). Therefore, it is pertinent that all educators give up their retreat into the myth of political or ethical neutrality based on the "science" of their work and confront their responsibility to not only embrace but also teach students about social justice and encourage them to enter the struggle for a qualitatively better life for all. In this context, we argue that engineers must work to lessen the most pressing issues facing humankind without being blinded by a belief in the infallibility of engineering. The fundamental question is: How can social justice become an integrated element that shapes how engineers view their profession?

One way of accomplishing this objective is to provide engineering students a new paradigm for engineering that centers on issues of social justice rather than profit or technical wizardry (Catalano and Baillie 2006). Engineering students need to learn to utilize social justice as a critical conceptual lens that allows them to move their thinking to the integration level, uncovering patterns and connections between their practice, profession, and social justice.

In this chapter, we engage in defining what social justice is. We then compare and contrast this definition with engineering professors and students at a college of engineering in the Middle East. Building upon these findings in light of ABET standards and the curriculum at this study's site, we offer recommendations regarding how engineers can develop social justice as a critical lens that informs their profession and practice.

17.1.1 Social Justice and Education

This examination of social justice calls upon various traditions (religious and secular) and schools of thought. The term social justice is often used in different contexts that are grounded in the individual's experience and restricted by time, place, and social location (Riley 2008). Teasley and Rice (1996) argue that social justice draws upon the normative elements of what should be and what we see as unjust, and these normative aspects are highly impacted by influences such as race, ethnicity, class and class culture, gender, religion, culture, experiences, and access to power among other factors. More importantly, these influences can also serve as the source of social justice; that is, ethics, feminism, and religion.

Context and tradition play a significant role in education. These traditions can play a significant role in defining social justice. For this chapter, we will evoke several traditions to tease out intricate aspects of social justice. Faith, not traditions, provides insight into social justice because religious traditions are linked to essential human values such as peace, social justice, and equity and can contribute to a better understanding of the term and the areas of social injustice. For example, Qutb (1953) argues that Islamic values include freedom of conscience, economic and racial equality, and social interdependence. Islam includes mandatory alms-giving (Zakah); charity, kindness, and voluntary contribution with moral motives toward a just distribution of the wealth, thus pointing to the need to consider the oppressed, marginalized, and powerless. Thomas Aquinas developed the concept of "common good" as an ethical and moral imperative based on Christian theology (Blackburn 1996). The common good is central to the tenets of many religious faiths such as liberation theology.

This chapter examines understanding of social justice in the context of engineering education, but more importantly, we want to study if the teaching of social justice is truly embedded in engineering programs or "tacked on" to satisfy program accreditation. Furthermore, we desire to examine the sociocultural factors and pedagogical challenges that might either hinder or assist the development of engineers as professionals and intellectuals that stand against oppression and social injustice, not only in awareness but also as an agent of change that strives for human rights and social justice. This requires a space where professors have opportunities to awaken their consciousness and discover ways to develop this kind of consciousness with their students (Duncan-Andrade 2005; Freire 2000).

However, we understand the best sources of information regarding issues of social justice and its role in engineering are those directly involved in engineering education—students and professors. These voices are invaluable in understanding the role social justice plays and should play in engineering education and the profession. Through interviews, these voices will be heard and used to develop an understanding of the current role social justice plays in one particular college of engineering and to offer recommendations for engineering education to empower and engage professors and students in the process of learning and action for social justice. Although the geographical parameters implied in this study center on the Middle East, findings and educational initiatives could reach wider global interests.

17.2 Research Methodology and Design

In this study, we used qualitative approaches by triangulating the data collection procedure. Our philosophical position was to approach issues without a priori notion to the extent that we could identify emerging issues. Therefore, we used grounded theory methodology that allows the researcher to study a phenomenon without a set framework but instead lets us see that specific issues emerge (Kezar

2005). The aim was to inquire on issues that are grounded in the accounts and experiences of the participants.

The triangulation method collected data through interviews with engineering faculty, followed by data from students, and then we identified the courses that included ethical components. These course syllabi were content analyzed, looking for ABET-accreditated student outcomes (c and f) and two additional policy documents.

17.2.1 Sample

The sample in this study included $n=9$ faculty members and $n=20$ students from the college of engineering. The student sample was selected conveniently from different departments of a fairly large college of engineering in the Middle East. No specific criteria led to the selection of the sample of faculty and student participants, but inclusion was based on students having completed the required ethics course. The distribution of students is shown in Table 17.1. Students came from diverse backgrounds and programs grouped by gender. Faculty also came from diverse backgrounds and nationalities; there were three electrical engineering professors, mechanical and industrial systems engineering ($n=1$), industrial engineering ($n=1$), computer science ($n=1$), civil engineering ($n=1$), computer engineering ($n=1$), and chemical engineering ($n=1$). The experiences of the faculty ranged from 8 to 25 years.

Table 17.1 Student characteristics grouped by gender

Variable		Female	Male
Major	Computer engineering	5	
	Electrical engineering	3	
	Industrial engineering	2	
	Chemical engineering	2	
	Computer science	1	
	Civil engineering		7
Level	3rd year (junior)	3	
	4th year (senior)	10	7
Courses	Ethics course	13	7
	Capstone course	10	5
Nationality	Egyptian	2	
	Indian	1	
	Iraqi		2
	Jordan	2	2
	Palestinian		1
	Sudanese	1	
	Syrian	1	
	Qatari	6	2

17.2.2 Content Analysis

Content analysis was conducted on selected course syllabi against the ABET standards. The two key student outcomes identified by ABET as Criterion 3 based on student outcomes are: (c) an understanding of professional and ethical responsibility and (f) an ability to design a system, component, or process to meet desired needs within realistic constraints such as economic, environmental, social, political, ethical, health and safety, manufacturability, and sustainability. The process was to identify these two standards as mapped in the outcomes of each course syllabus and identify what activities or objectives were evident. The courses that address the ABET student outcomes (c and f listed above). The content analysis of the syllabi was more of the enumeration and identification of whether courses covered aspects, notions, or case studies addressing social justice. The process was to locate social justice objectives, learning outcomes, and any assignments that demonstrate how the college of engineering is addressing ethics and/or social justice. In addition, two documents that could influence the fields of engineering were examined. *Achieving Vision for Civil Engineering for 2022* and *The Engineer of 2020: Visions of Engineering in the New Century* (National Academy of Engineering 2004) were also reviewed to assess the extent to which social justice was covered in these two documents.

Procedure, Focus Groups, and Interview Questions

The uniqueness of this study is that it addresses social justice as perceived by students and faculty. We developed open-ended guide questions that were used with students and faculty to start discussions related to social justice in the engineering program profession. Participants were probed to bring further understanding and to provide examples. The interviews were conducted individually with faculty members and in groups or individually with students. The theoretical lens was the extent that social justice was integrated in the engineering program. The following seven questions guided our discussions with faculty and students.

1. Please write a short definition regarding how you would define social justice.
2. What is the relationship between ethics and social justice?
3. What are some of the issues facing engineers regarding social justice?
4. How and where do you address social justice in your courses? In the program? Please provide examples.
5. Do you think it is important to integrate social justice in the engineering program? Why or why not?
6. If you have to integrate social justice into the engineering program, how would you do this?
7. Do you think issues of war, peace, justice, equality, equity, and gender are things that an engineer must be aware of in his engineering education? Provide example please.

Qualitative Data Analysis

Bogdan and Biklen (2002) define qualitative data analysis as "working with data, organizing it, breaking it into manageable units, synthesizing it, searching for patterns, discovering what is important and what is to be learned, and deciding what you will tell others" (p. 145). The interview data was examined in the context of the research questions and involved selecting, focusing, condensing, and transforming data guided by thinking about which data best answer the research questions. During the analysis, data was organized identifying themes, patterns, and connections (Denzin and Lincoln 2007; Taylor-Powell and Renner 2003; Miles and Huberman 1994). As several themes emerged, they were content analyzed and relevant quotes were integrated into various themes in order to support or refute particular findings. In addition, we also started to make conclusions, verifications, and further recommendations. The rationale was to provide an accurate account of the professors' and students' understanding of social justice in the engineering programs where they teach and study. We also sought to determine the extent, if at all, of issues and themes of social justice that emerged in the classroom, the curriculum, and the teaching in this particular college of engineering.

17.3 Findings

17.3.1 ABET Standards, Reports, and College of Engineering Syllabi

Engineering professors who wish to incorporate social, ethical, and global issues into their classes face the dilemmas of covering the required content and training or resources to integrate this knowledge (Tucker and Ferguson 2007). An important point regarding this study is that the analysis and discussion centers only on the courses in the college of engineering. There are other courses in the university as in core requirement (courses that generally require all students would have to take) such as human rights, law, or social work that students could opt to enroll in. These are electives and not all students chose these courses. With this in mind, there are several sources that are looked at when considering ways to include ethics and ethical responsibility into engineering programs. First, ABET developed several achievable program outcomes. Among the learning outcomes specified as essential by ABET accreditation criteria and mentioned above: "an understanding of professional and ethical responsibility; an ability to design a system component of process to the desired needs within realistic constraints such as economic, environmental, social, political, ethical, health and safety, manufacturability and sustainability; and a knowledge of contemporary issues" (ABET 2008, p. 3).

In addition to examining the course syllabi, two additional documents, *Achieving Vision for Civil Engineering for 2022* (American Society of Civil Engineers 2009) and *The Engineer of 2020: Visions of Engineering in the New Century* (National Academy of Engineering 2004), were examined. *Visions of Engineering in the New Century* provides some discussion regarding the ethical components of the engineering profession. Since these documents may have some influence on engineering programs, they were read and analyzed. The first report states that ethics education should be promoted as part of the civil engineering curricula and should encourage the development of codes of ethics and provide ethical guidance. The document positions engineers as being trusted by society to ethically act as "planners, designers, constructors, and operators of society's economic and social engine," and be the stewards of the environment, able to manage risk and uncertainty caused by natural events, accidents, and other threats. A second document known as *Engineer of 2020: Visions of Engineering in the New Century* argued for educated engineers who are ethically grounded, adapting solutions in an ethical way, and calling for green engineering. It is stated in *Engineer 2020* that engineers "must work within the constraints provided by technical, economic, business, political, social, and ethical issues" (p. 7)… "and to ethically assist the world in creating a balance in the standard of living for developing and developed countries alike" (p. 51).

The sample and study were carried out in a college of engineering that has been shaped and molded by accreditation standards and requirements. There is significant evidence that the curriculum meets ABET standards for ethics and engineering. Through the formal, explicit, or written curriculum (the planned program of objective, learning outcomes, assessment, and resources), it is clear that the college of engineering addresses ethical issues in the profession. A content analysis of syllabi illustrates this aspect of the formal curriculum with evidence of terms such as professional ethics, ethics homework, and professional conduct which includes the alignment with the code of ethics, engineering ethics (Codes of Ethics in Engineering Disciplines and Steps in Resolving Ethical Dilemmas) and makes them aware of engineering ethics. Other areas addressed in the formal curriculum include philosophical ethical theory and morality; codes ethics and professional practice; use of moral principles or codes of ethics to make decisions regarding conflicts of an ethical nature; and an overview of computing ethics and practice and one particular resource, *Ethics in Information Technology.*

The engineering program offered a course in ethics, and ethics are integrated throughout the curriculum through micro-insertions where relevant. However, examining the ABET Criteria for Accrediting Engineering Programs, other reports mentioned above and the formal curriculum from the engineering college at the Gulf University quickly illustrate that social justice is not a fundamental concern for engineering programs. The terms social justice or justice itself are absent from the discourse. If mentioned, social justice is couched in ethics as a moral and professional responsibility that centers on environmental codes and omitting a wide range of social injustices that could have been addressed in the context of what engineers do in the field.

17.3.2 Student Perceptions

The data gained from student interviews and focus groups are quite complex and highly influenced by the type of engineering, gender, and nationality. To begin with, it was evident that students clearly understood the engineering code of ethics and other ethical codes needed for their respective engineering program. Regarding students' definitions of social justice, comments such as "to treat everyone equally, no racism or such issues," "equal rights for everyone," "equal rights and the same rules apply for everyone," "equality between individuals in the workplace," "no wasta" (an Arabic word that is translated in "clout" or who you know), and "universal criterion for everyone" illustrate the understanding of social justice for these students.

These students argued that ethics and social justice are dependent upon each other and that ethics serve as a prerequisite for social justice. Students indicated that ethics should be applied in the practice of social justice. One student summed up the majority of these students' thinking when he stated, "If a society or people are doing things ethically there will be social justice." However, when asked what issues engineers face regarding social justice and to list examples, this is where the particular field, gender, and nationality played a significant role in shaping their responses. For example, three expatriate females in electrical and computer engineering seemed to limit their responses to unequal pay for engineers from various nationalities and they seemed to indicate that they were not prepared to discuss issues of social justice. One Qatari female student stated that a social injustice is that engineers work for many hours.

On the other hand, a focus group of five expatriate male civil engineers provided extensive discussion and numerous examples of social injustices. When asked where they gained this knowledge, they were quick to cite their experiences "on the site…and never in the classroom" and interacting with workers and the project. Several of these students visited workers camps and cited the poor working conditions, low wages, the harsh living conditions, and how companies add glucose to the workers' water "only to keep worker working and the machine running." They stated that when the safety officer comes to the site, workers are told to make it look like they are doing well. These five students pointed out the unethical practice of how change orders are used to make profit and this is a common and accepted practice. These valuable lessons and realities of the workplace were learned "during summer training experience…never in the classroom." More importantly, the students did not document this in their summer report that required students to address technical requirements and were not required to address "ethical or social issues." All five students agreed that female students would not have this opportunity and experience raising issues of gender inequality in education.

Many of the students interviewed seem to be limited by their own understanding of the field of engineering. The following quotes, followed by discussion, illustrate this claim. Students stated, "our studies are technology centered and we should not think about the implications about the technology. Companies have specific department for ethical issues," "these are just ethical issues and it has nothing to do with our studies," "It is the firm's role and I am not looking for justice," and "my study is engineering and calculations." For some students, this thinking seems to create a bifurcation between engineers and ethics beyond professional codes of ethics. Hence, this view of engineering prevents the development of social justice.

The majority of the students seemed to couch social justice within the various engineering code of ethics. For example one student stated, "engineers have a code of ethics and once they follow the code of ethics there should not be any issues of social justice. Laws should regulate social justice and code of ethics." It seems that students limit their responsibility for social justice to the various ethical codes in their fields, "we all must know and follow the code of ethics…we lose our license if engineers do not follow this."

Regarding the sources of knowledge of ethics, students listed their ethics course as one source of learning about ethics but not social justice and did not report that the course "did not really change my thinking." Several students could not recall the curriculum of the ethics course but some did recall one case study used in the class, the Challenger explosion. One student stated there is no instruction on "social justice in the engineering program. The ethics course is about professional codes but no application to work and no application of ethics. Nothing in the courses about social justice." Another student stated, "the ethics course taught us to be a professional and problem solver. Having ethics is part of being a professional engineer." Several students stated that they learned about ethics "through the ethics course, right and responsibilities, protection of copyrights and plagiarism" while another male student stated "professor refer to it every now and then." Another student provided valuable insight when she stated, "ABET standards cannot touch the reality working in the field." Finally one student commented, "in every profession people should know about [social justice]. The engineering program is not providing this insight to social justice—you learn it from life."

Finally, the question of how social justice should be added to the engineering programs was asked and students offered valuable insight. One student stated, "engineers must have a role in promoting social justice" while another student added, "social justice is something that is very important." However, not all students communicated this position. Five students participating in a focus group agreed with this statement "social justice is nice to talk about at the table but in the field is not important."

Nevertheless, those students who thought social justice should be a factor in engineering education offered several suggestions. Students who indicate that it is important argued that social justice could be integrated into various aspects of the programs such as the internships, ethics courses, and projects. One student stated, "Yes it should be integrated across because only then the values will be embraced." Another student thought that social justice was "very important, it could be integrated in the project management course" while one student believed that the "professors must enlighten students" on this issue and another student thought that "professors should open students' mind—these ideas should be injected throughout the teaching process—professors should share experiences." Several students told of a professor who was previously a professional engineer in the field and taught for one semester; students learned a lot about field-based experience, suggesting that social justice issues should really come from the field rather than from theories and academic exercises. Some students thought that since workshops on safety are offered, workshops that center on social justice would be effective. Another student suggests field trips to sites and workers' camps to open the eyes of students.

However, not all students felt this way. One female student when asked about addressing social justice in her program responded negatively and stated, "no it would be redundant if it was always mentioned in classes…one class and then move on" and another stated, "should be taught in some other subject." Several students painted a gloomy although possibly a realistic picture of the current situation with the following quotes, "the engineering program is not providing this insight to social justice—you learn it from life" and "profit over social justice…the engineer will get fired if he thinks about social justice." This may be the case, but as one student stated, "this value [social justice] must come from the inside in order to change things."

17.3.3 Professor Interviews

Similar questions were asked to faculty that were asked to students, on whether faculty definitions of social justice meant equality. One faculty member told researchers that respecting diversity, respecting cultural differences, treating people equally independent of gender, ethnicity, and religion was the key to social justice. Other faculty members suggested that social justice does not only imply equality but also provides equal opportunity or equity.

Regarding the relation between ethics and social justice, the large majority of the faculty felt that ethics was the overarching frame that social justice operated under. One faculty member said, "if social justice is not applied then there is something wrong with ethics." A large majority felt if there were no ethics, social justice would not exist. Other faculties saw ethics and social justice in a discrete sense where ethics is a subset of social justice.

In terms of what faculty thought about social justice in the engineering profession and issues of social justice that engineers faced, there were several themes that emerged. Ethnicity was a major cause of injustice and that engineers are treated differently whether in pay or treatment based on their national identity. Likewise, faculty felt that issues of gender, discrimination, and what women can and cannot do are prevalent in the field. For example, it is taken for granted that a woman cannot be in the field and generally is given an office job even though she willingly approves of being in the field.

Other faculty members touched on the treatment of workers as a key issue facing engineers and that the engineering field has not addressed this issue of human rights. One faculty member went as far as saying that workers in this Gulf nation are more like slaves and that engineers should be aware of worker rights and quality of life. Otherwise, all of it will backfire, on safety, productivity, and efficiency. He further noted that cheap labor means expensive costs later. An industrial engineer addressed social justice being embedded at the heart of the industrial engineering profession:

> It is quite obvious of the impact of engineers on society and the environment but not sure how it is linked to social justice. Industrial engineering tries to develop ways for companies to produce things faster and make higher profits. If companies do better then they make more profits and then they can hire more people. We design human ergonomics and help the workers. When working conditions are improved then workers are more productive. We design to improve the production system and the main purpose is society because we must produce a product of high quality, low cost and fast.

This professor points out that as industrial engineers try to improve the whole system, social justice is subsumed in what industrial engineers do.

In response to addressing social justice in courses or the engineering program, a large majority of the faculty members interviewed stated that the engineering program did not address social justice. In fact, most believed that ethics was covered tactically rather than being adopted wholeheartedly. One faculty member stated:

> In my course I am only focused on the technical—in the whole program we address this in ABET evaluation—we do the ABET evaluation in a technical way to please ABET standards. Not in any deep way. Even in the senior design project we do not touch deeply on this issue.

A substantial number felt that social justice being very important to engineers is not really covered in the sense that students can assimilate the values that are embedded in ethics or even social justice. One professor stated, "the worse way to integrate social justice is in an ethics course…it will be technical" and professors generally assess students by either having that skill or not. One professor told us that "in the senior project, ethics is a main part and students must prove they did not violate anything." Faculty believed that there should be an intentional engagement across the curriculum rather than a one-shot ethics course.

However, there were several examples given by professors that illustrate how social justice was evident and integrated in the program through engineering projects. For example, one professor told how engineering faculty worked with students "who won prizes in Germany and Britain for designing a bullet proof vest for ordinary people rather than soldiers. The vest is unique and new and available on the market at low cost targeting poor people who could suffer from war." The key aspect of this exercise was to get students to think design not for profit but to protect those needing the technology. Another way students could learn about social issues was have students work in teams and treat others fairly while working on projects. This begins the process of caring for others and being foundational for social justice.

Several faculty members felt the ethics course that students took in the first year was not useful and generally not practice-based. Instead, ethics should be inserted across the curriculum when it is appropriate. In addition, one faculty member argued that an ethics course in the first year is not effective since students are novices about the engineering profession. He suggested that a course on ethics and society should be offered at the senior level after the students have been exposed to industry practices and projects. Finally, across the board, faculty felt the importance of social justice and ideally that it should be integrated within, not with what engineers do but in what they should do and how they should think.

17.4 Discussion

From our experiences as professors, we would argue that these findings would be reflected in most colleges of engineering and similar findings in colleges of education, business, and law, and one could make this argument for many other fields of study at the university level. With this insight, it is clear from these findings that there are times when some faculty and students understand the role of social justice in engineering while others seem to not fully grasp this dimension. While students do not fully embrace the idea of social justice, some are willing to bifurcate the roles of engineers being socially responsible and the other as engineers in the field. The professors, on the other hand, clearly see the importance of social justice and as we do, struggle with the conception, application significance, and implications of the power interactions in a world that seems to push social justice to the margins.

As professors, we struggle with issues of social justice, but to limit this discussion to the engineering field is quite discriminatory and fails to reflect the power teaching and learning in every classroom. It is unavoidable, but every classroom can perpetuate the social relations and orders existing within society or can challenge the status quo. Thus, by bringing social justice to the engineering field, we are bringing the awareness or the missing link between the technical and the intellectual between theory and praxis.

Our intention is not to be labeled as liberals trying to impose social issues on the engineering profession but rather as professors who understand that all education shapes students and society, and that in every program and level of the university it is important to raise issues of privilege and, more importantly, to engage the privileged in learning about systematic and hidden injustices. Even if one disagrees that these issues should be raised in the university, one must keep in mind that the university should serve as a marketplace of ideas and there should be a multitude of narratives presented in order to expose the structures of power and privilege to both those who reap the benefits of privilege and those who are left to struggle.

17.5 Recommendations: Good Sense and Critical Theory

Our overall argument is that it is important to understand that higher education, or all education or learning for that matter, cannot occur in a vacuum. Rather, universities and schools are entities embedded within a sociocultural structure that impacts the knowledge transmitted and produced in these places of learning. We believe that the engineering profession, just as *all* professions, whether education, law, or business, must resist the trivialization of education. That is "the evasion or neglect of larger, more critical topics and the stress put on technical rather than on social political and moral issues" (Purpel 1989, p. 2). Rather, the discourse of all education should include the task of challenging and transforming many of our basic institutions and belief systems in order to develop a more just society. In addition, this research and discussion addresses the formal curriculum but cannot account for the hidden curriculum (the unintended outcomes of learning) or the curriculum-in-use (the actual curriculum that is delivered and presented by each professor). There is an abundant amount of opportunities in every class for professors to raise ethical issues and social justice. With that in mind, we offer the following recommendations.

Although social justice is defined in various ways, there are common elements of social justice that can be found across these traditions and schools of thought mentioned above. First, social justice requires an awareness that people in every time and place continue to struggle for social justice and it is not something to be achieved and done, but rather it is a continuous process and ongoing struggle (Riley 2008). Thus, faculty must move beyond the technical knowledge by integrating designs, problem solving, and case studies as a way to address social justice. Second, social justice requires an openness to change that requires those in power to give up privilege and serve others and develop a more equal distribution of resources whether food, finances, shelter, or health care. Engineering education must be aware of the social issues and must draw on practices in the field, thus bringing issues of injustice much closer. Third, social justice requires naming: identifying and defining social and economic relationships, particularly the inequitable distribution of power and resources (Fine 1992). Engineering programs must become interdisciplinary opening spaces for discussions that go beyond the code of ethics and begin to deconstruct the notion of a project, companies, hierarchy of labor, and profession, and reconstruct the meaning of engineering in the context of social justice. Finally, social justice is the struggle to end different kinds of oppression, to seek economic equality, to uphold the rights of all humans, and acknowledge the dignity and worth of every being to restore right relationships among all people and the environment (Ford Foundation 2007).

Furthermore, integrating social justice into accreditation and educational programs opens the space to address the unexpressed narratives or systems that operate in a covert manner and on those who perpetuate them. It allows to challenge Gramsci's notion of the structure of common sense that includes the taken for granted, unexamined, hardly noticed beliefs and assumptions characterizing the conformist thinking of the mass of people in a given social order (Coben 2002). Gramsci makes it clear that common sense is established by consent to the dominant class attitudes and interests and accepted by society as being in its own general interests. Common sense is naturalized to the point of being taken for granted. For example, an engineer could be hired on a mega project and completely fulfill the ethics code—a common sense practice. However, the engineer never really challenges the common sense of the way the project was awarded, the clientelism, possible kickbacks, and other corruptive actions. This is where the common sense must be critiqued for the complicity of the system and corruptive behaviors and injustices. The good sense must replace these constant injustices questioning and replacing the system that perpetuates the inequalities, demeaning human equality and the eradication of the well-being of society.

That engineers must be able to use the rationale of good sense and be able to reflect on social and economic conditions rationally. Good sense is "inherently coherent and critical" (Coben 2002, p. 269) and good sense develops from a critique of common sense. It is good sense that seeks to delegitimize common sense and challenge the interests of the dominant culture. Although common sense is deeply embedded in the thinking of society, it can be changed by education that must establish sound criticism of common sense in order to develop alternative views and understanding of the world. This awareness allows individuals to reflect upon what is and why things are the way they are and develop new ways of thinking outside common sense. Instead of the common sense of engineers considering the impact of projects on the environment and the economic bottom line, the good sense would question the common sense and consider alternative viewpoints regarding the impact of engineering on workers and their families.

In education, critical theory has provided valuable insight and skills required not only to be aware of injustices in the educational systems but also the power struggles in the classroom and also to act to alleviate the oppression by focusing on critique and reform. Critical theory draws on Marxist theory and suggests an unequal social stratification in our society based on the socially constructed notion of class, race, and gender. Those in power control society and create and maintain unequal conditions that marginalize, oppress, and disenfranchise those who lack access to resources and power. Those who maintain the inequality receive the economic and social benefits, creating injustices that are embedded within social institutions and structures. Because they want to retain privilege, they continue to oppress others by the powers and hegemonic structures in place. That is the "maintenance of domination not by force but primarily through consensual social practices, social forms, and social structures, produced in specific sites such as the public institutions, the school and mass media, the political system and the family" (McLaren 2007, p. 203). The key aspect of hegemony is that it is a struggle in which the powerful win the consent of those who are oppressed, unknowingly participating in their own oppression.

McLaren (2007) argues that social justice educators must offer students a language of critique and a language of possibility "so they can conceptualize, analyze, theorize and critically reflect upon their experience" (p. 51). Giroux (1983, 1988) makes a clear distinction between a "language of critique" and a "language of possibility." Giroux points out that a common failure of radical critics is that they offer the former but not the latter. He argues that the aim of critical educators is not just to offer a critique but "to raise ambitions, desires, and real hope for those who wish to take seriously the issue of educational struggle and social justice" (Giroux 1988, p. 177). Individuals must raise the level of consciousness about the social injustices and about the way they perceive the world. For social justice to prevail, individuals "must learn to apply critical analysis and utopian thinking" (McLaren 2007).

The engineering profession demands critical thinking to recognize the best courses of action. However, engineers must develop critical thinking that is not limited to a positivist technical framework. We argue elsewhere that this type of critical thinking moves beyond mere logic or problem solving and raises issues of social justice that inevitability invoke questions about value-laden issues embedded in culture, power, and politics, such as workers' rights (Romanowski and Nasser 2012) and other issues that are relevant to engineering professors.

This requires an understanding and embracing of critical theory, opening the way for a language of critique and possibilities where students are active, skeptical, and aware of their own understanding of the world and the ideologies and the moral knowledge that govern life experiences. Engineers must have both the desire and ability "to disengage themselves from the tacit assumptions of discursive practices and power relations in order to exert more conscious control of their everyday lives" (Kincheloe 2000, p. 24). Riley (2008) argues that by developing critical thinking skills drenched with critical theory, students better

> understand the institutional nature of prejudice, the connectedness of oppression, and the notion of intersectionality. It also allows for understanding sociotechnical systems and the co-construction of science and society. It helps us understand the behavior of individuals within organizations and organizations in larger political and cultural landscapes (p. 113).

Critical theory can assist students by allowing them to "name" the injustice and understand whether injustice is maintained or dismantled. It provides them with what Freire (2000) termed a critical consciousness, that is, the ability to analyze and critically examine social, political, and economic oppression and to take action against the oppressive elements of society. This awareness begins the process of questioning, developing into an active struggle against social justice and leads to transformative action. This means that engineering education must be inserting directly into the political sphere so education represents not securing technical knowledge and employment, but rather a struggle for meaning and a place where "critical reflection and action become part of a fundamental social project to help students develop a deep and abiding faith in the struggle to overcome economic, political and social injustices, and to further humanize themselves as part of this struggle" (Giroux 2012). Central to this type of education is the task of making the pedagogical more political and the political more pedagogical (Aronowitz and Giroux 1985).

17.5.1 Social Justice and Ethic of Caring

Pantazidou and Nair (1999) suggest that there should be an integration of an ethic of care into engineering education and practice. Gilligan (1982) and Noddings (1984), who are feminist philosophers of moral education and closely identified with the ethics of care, make the argument that caring should be the foundation for ethical decision-making. Gilligan (1982) provided an alternative to what she views as a masculine concept of rights or justice. Gilligan argued that men discuss ethical questions and issues in terms of rights and language, and women speak about their issues in terms of empathy, concern, and sympathy for others. Noddings (2002) argues

> The key, central to care theory, is this: caring-about (or, perhaps a sense of justice) must be seen as instrumental in establishing the conditions under which caring-for can flourish. Although the preferred form of caring is cared-for, caring-about can help in establishing, maintaining, and enhancing it. Those who care about others in the justice sense must keep in mind that the objective is to ensure that caring actually occurs. Caring-about is empty if it does not culminate in caring relations. (pp. 23–24)

How can an ethic of care be integrated within engineering programs? Noddings (2002) argues that education from the care perspective has several key components. *Modeling* requires students move beyond simple moral reasoning and professors demonstrate in their thought and actions what it means to care. Noddings (1995) points out in order to teach students to care "we do not merely tell them to care and give them texts to read on the subject, we demonstrate our caring in our relations with them" (p. 190). This idea supports the professor's strategy to get his students to work together and be fair with each other as earlier discussed. *Dialogue* requires that professors engage students in dialogue about caring and not simply treat it as a learning outcome but rather that caring must be fully embodied in the engineering discourse. It is then vital to discuss and explore how caring can be manifested in different ways. Students should be given opportunities to critique and evaluate their own and others' attempts to engage in caring. Caring requires *practice* and if we want to produce students who will care for others, then opportunities must be provided for students to practice in caring and reflection on that practice.

17.5.2 Accrediting Agencies

Almost a decade ago, the National Council for Accreditation of Teacher Education (NCATE) had adopted social justice as an example of what teaching programs might consider as "when evaluating prospective students, 'dispositions' helps schools of education measure how these prospective teachers would respond in a classroom setting" (Applebaum 2009, pp. 377–378). Under tremendous pressure from various "conservative" forces that equated "social justice" as a code word for a liberal agenda and thus an ideological annoyance, NCATE removed social justice awareness and understanding and simplified social justice by linking it to Bush's No Child Left Behind, stating "social justice demands that we take appropriate action to fulfill these promises by assuring high quality education for all children" (NCATE 2008, p. 6). However, any mention or development of social justice in teacher education or dispositions is absent. Although ethics is spotted in NCATE standards, this might be suggested as a code word for considerations of social justice. ABET, like other professional associations, has been receptive to the idea of ethics as possibly an overarching term to social justice. The main underlying issue in this study is even though ABET addresses these in terms of ethical behaviors and even NCATE's notion of diversity as a celebration of difference in attempts to abrogate racism and discrimination but still far removed from social justice education.

17.6 Conclusions

In closing, it is clear that it is difficult to integrate social justice in not only engineering but also many other fields of study. It is also evident that the development of social justice in students is not limited to the field of engineering. Based on our own experiences with colleges of education and accrediting programs, similar arguments could be made for the lack of teaching social justice and the movement toward a technical trivialized education. However, we would argue that all of these approaches led us to a widening of ethical thinking in engineering that might conflict with the current direction of industry. Although a large number of faculty members saw the importance of the integration of social justice in the profession and engineering education, the leadership still needs to move forward on issues of equity and equality and the larger well-being of society. It is then the role of the university to not simply technically train professionals for employment opportunities but rather the university has a societal obligation to develop good citizens who in turn improve society by powerfully accelerating social justice, and we would argue this demands teaching social justice.

References

ABET. (2008). Criteria for accrediting engineering programs. http://www.abet.org/Linked Documents-UPDATE/Criteria and PP/E001 07-08 EAC Criteria 11-15-06.pdf.

American Society of Civil Engineers. (2009). *The vision for civil engineering for 2025: A roadmap for the profession*. Reston: American Society of Civil Engineers.

Apple, M. W. (1993). *Official knowledge: Democratic education in a conservative age*. New York: Routledge.

Applebaum, B. (2009). Is teaching for social justice a "Liberal Bias"? *Teachers College Record, 111*(2), 376–408.

Aronowitz, S., & Giroux, H. A. (1985). *Education under Siege*. Massachusetts: Bergin and Garvey Publishers.

Blackburn, S. (1996). "Common good" and "difference principle" The Oxford Dictionary of Philosophy. New York: Oxford University Press. Oxford Reference Online. http://www.oxfordreference.com/.

Bogdan, R. C., & Biklen, S. K. (2002). *Qualitative research for education: An introduction to theory and methods* (4th ed.). Needham Heights: Allyn and Bacon.

Broome, T. H., & Peirce, J. (1997). The heroic engineer. *Journal of Engineering Education, 86*(1), 51–56.

Catalano, G. D. (2006). *Engineering ethics: Peace, justice and the earth. Synthesis lectures on engineers, technology and society*. San Rafael: Morgan & Claypool Publishers.

Catalano, G. D., & Baillie, C. (2006). *Engineering, social justice and peace: A revolution of the heart*. Paper presented at ASEE Annual Conference & Exposition, Chicago, Illinois, June 18–21, 2006.

Coben, D. C. (2002). Metaphors for an educative politics: "Common sense," "good sense," and educating adults. In C. Borg, J. Buttigieg, & P. Mayo (Eds.), *Gramsci and education* (pp. 262–290). New York: Rowman & Littlefield Publishers, Inc.

Denzin, N. L., & Lincoln, Y. (2007). *The landscape of qualitative research* (3rd ed.). New York: Sage Publications, Inc.

Duncan-Andrade, J. M. R. (2005). Developing social justice educators. *Educational Leadership, 62*(6), 70–73.

Fine, M. (1992). *Disruptive voices: The possibilities of feminist research*. Ann Arbor: University of Michigan Press.

Ford Foundation (2007). Peace and social justice. http://www.fordfound. org/program/peace_main.cfm.

Freire, P. (2000). *Pedagogy of the oppressed*. (30th anniversary ed.). New York: Continuum Publishing Company.

Gilligan, C. (1982). *In a different voice: Psychological theory and women's development*. Cambridge: Harvard University Press.

Giroux, H. A. (1983). *Theory and resistance in education*. South Hadley: Bergin Garvey.

Giroux, H. A. (1988). *Teachers as intellectuals: Toward a critical pedagogy of learning*. South Hadley: Bergin Garvey.

Giroux, H. A. (2012). The war against teachers as public intellectuals in dark times. *Truthout*. http://truth-out.org/opinion/item/13367-the-corporate-war-against-teachers-as-public-intellectuals-in-dark-times.

Green, P. (2005). Disaster by design: Corruption, construction and catastrophe. *British Journal of Criminology, 45*(4), 528–546.

Herkert, J. R. (2011). Engineering ethics education in the USA: Content, pedagogy and curriculum. *European Journal of Engineering Education, 25*(4), 303–313.

Kabo, J., Day, R. J. F., & Baillie, C. (2009). Engineering and social justice: How to help students cross the threshold. *Practice and Evidence of the Scholarship of Teaching and Learning in Higher Education, 4*(2), 126–146.

Kezar, A. J. (2005). Consequences of radical change in governance: A grounded theory approach. *The Journal of Higher Education, 76*(6), 634–668.

Kincheloe, J. L. (2000). Making critical thinking critical. In D. Weil & H. K. Anderson (Eds.), *Perspectives in critical thinking: Essays by teachers in theory and practice*. New York: Peter Lang.

McGinn, R. E. (2003). Mind the gaps: An empirical approach to engineering ethics, 1997–2001. *Science and Engineering Ethics, 9*(4), 517–542.

McLaren, P. (2007). *Life in schools: An introduction to critical pedagogy in the foundations of education* (5th ed.). Boston: Allyn and Bacon.

Miles, M. B., & Huberman, A. M. (1994). *Qualitative data analysis* (2nd ed., pp. 10–12). Newbury Park: Sage.

National Academy of Engineering. (2004). *The engineer of 2020: Visions of engineering in the new century*. Washington, DC: The National Academies Press.

NCATE. (2008). *National professional standards accreditation of teacher preparation institutions*. National Council for Accreditation of Teacher Education, Washington, DC.

Noddings, N. (1984). *Caring, a feminine approach to ethics and moral education*. Berkeley: University of California Press.

Noddings, N. (1995). *Philosophy of education*. Colorado: Harper Collins.

Noddings, N. (2002). *Starting at home: Caring and social policy*. Berkeley: University of California Press.

Pantazidou, M., & Nair, I. (1999). Ethic of care: Guiding principles for engineering teaching and practice. *Journal of Engineering Education, 88*(2), 205–212.

Purpel, D. E. (1989). *The moral and spiritual crisis in education*. Massachusetts: Bergin & Garvey Publishers, Inc.

Qutb, S. (1953). *Social justice in Islam.* (trans: J. B. Hardie). Washington: American Council of Learned Societies.
Riley, D. (2008). *Engineering and social justice.* San Francisco: Morgan Claypool Publishers.
Romanowski, M. H., & Nasser, R. (2012). Critical thinking and Qatar's education for a new era: Negotiating possibilities. *International Journal of Critical Pedagogy, 4*(1), 118–134.
Sax, L. J. (2000). Citizenship development and the American college student. In T. Ehrlich (Ed.), *Civic responsibility and higher education* (pp. 3–18). Phoenix: Oryx Press.
Shuman, L. J., Sindelar, M. F., Besterfield-Sacre, M., Wolfe, H., Pinkus, R. L., Miller, R. L., Olds, B. M., & Mitcham, C. (2004). *Can our students recognize and resolve ethical dilemmas?* Paper presented at American Society for Engineering Education Annual Conference and Exposition, Salt Lake City, UT. http://www.asee.org/conferences/annual/2004highlights.cfm, Session Number 1526. Accessed 19 Feb 2008.
Taylor-Powell, E., & Renner, M. (2003). *Analyzing qualitative data.* Madison: Board of Regents of the University of Wisconsin.
Teasley, M., & Rice, J. (1996). *What is social justice? Perspectives on multiculturalism and cultural diversity, 6*(2), 1–3. http://www.students.vcu.edu/counsel/MC/justice.html.
Tucker, J., & Ferguson, D. (2007). *Incorporating Ethics and Social Responsibility in Undergraduate Engineering Education*, International Conference on Engineering Education—ICEE 2007, Coimbra, Portugal.

Ramzi Nasser is Professor and Dean of the College of Arts and Applied Sciences at Dhofar University in Oman. Ramzi Nasser has taught math and education courses in a career spanning 23 years in Academia. He has contributed to international journals and publications in the area of institutional research, psycho-social behavior, educational research, mathematics education, teacher professional development and continues to research and work with school teachers, practitioners and school leaders in Qatar and more recently in Oman. He has contributed to the development of educational research agenda, enjoys writing, researching in multidiscipline areas in the social sciences. Ramzi Nasser is also an Editor-in-Chief of the Near and Middle Eastern Journal of Research in Education, the only English language Journal addressing research in Education in the Arab World and the Middle East.

Michael H. Romanowski is Professor of Education and Coordinator of the Masters of Education in Educational Leadership at Qatar University. He earned his PhD from Miami University. A former high school teacher in the American Public School System, Professor Romanowski brings to the classroom diverse educational experiences including academic positions in the United States, China and currently Qatar. He has published his research and scholarship in various academic books and international journals and has managed external and internal research grants examining various important educational issues, recently addressing the national education reform in Qatar. In 2011, he was awarded the Qatar University Outstanding Teaching Award 2011 and the College of Education Quality and Excellence Award. He continues to research and write on various educational and culture issues.

Part VI
Industry Perspectives, Future Insights, and Conclusions

Chapter 18
Engineering Education in the Middle East and North Africa: An Industry Perspective

Mabrouk Ouederni

18.1 Introduction

With a total population of almost 300 million, vast reserves of natural resources, and a strategic location, the Middle East and North Africa (MENA) region has played an important role in world economy. Experts gathered at the World Economic Forum in 2013 agreed that this region is poised to play an even bigger role on the world stage due to several factors. Fueled by the discovery of oil and gas in the mid-twentieth century, several industries have developed throughout the region taking advantage of revenues from energy and a need to grow national economies. Industrial development was not limited to oil and gas exploration and processing but included chemicals and petrochemicals manufacturing, fertilizers, agriculture, and infrastructure.

At the dawn of the twenty-first century, the role of industrial development will be critical to the growth and diversification of economies and well-being of societies in the MENA region (National Academy of Engineering 2004, 2005; NRC 2011). There is no doubt that the region faces several challenges to diversify its economy and sustain its industrial development (Khair et al. 2012). One of the most important challenges in this regard is the development of human capital (Abdulwahed et al. 2013). The engineering profession has been essential in the development of the various industrial sectors of the region. The role of engineers will be even more critical to help the industry face renewed challenges and find sustainable solutions.

Recent studies in engineering education emphasize that engineering graduates should possess—in addition to strong fundamentals in math, science, and engineering theories—a number of soft skills and competences needed in their professional career, such as teamwork, communication, inter/multidisciplinary knowledge, analytical thinking, ingenuity, creativity, technological innovation, business and management skills, leadership, ethics, and professionalism (Somerville et al. 2005;

M. Ouederni (✉)
Qatar Petrochemical Company (QAPCO), Doha Qatar
e-mail: MOuederni@qapco.com.qa

M. Abdulwahed et al. (eds.), *Advances in Engineering Education in the Middle East and North Africa,* DOI 10.1007/978-3-319-15323-0_18

Sheppard et al. 2004; Swearengen and Barnes 2002; Abdulwahed et al. 2013; Shuman et al. 2005).

Several engineering schools have been opened to support the industrial growth in the region and compensate for the heavy dependence on expatriate engineers. In this chapter, we will examine engineering education in the region from the perspective of a large industrial organization. We will list the essential skills needed of engineering graduates after they join industry based on experience that spans over several years. A case will be made for the critical role engineering programs can and should play in helping the industry in the MENA region succeed and thrive. Finally, some recommendations will be made on how industry views the partnership with educational institutions in order to establish the most effective collaboration in the engineering and scientific fields.

18.2 Petrochemical Industry in the State of Qatar

From nomadic herding, fishing, and pearl diving to oil and liquefied natural gas (LNG) production, Qatar's economy has traveled a long way on an ever-adapting and fascinating road. Today, Qatar's economic path is opening up to a number of ever-increasing diverse and lucrative industries. Qatar's energy sector has grown and changed tremendously since the discovery of oil and gas. Between 2000 and 2009, Qatar has tripled its production of natural gas and has become the world leader in LNG.[1] Petrochemicals play a large part in the diversification of Qatar's energy-based economy. In fact, petrochemical products are serving the fundamental needs of society in areas such as housing, food, health care, and transportation. The largest petrochemical industries exist in the USA and Western Europe, but the largest growth capacity is in the Middle East and Asia. With its manufacturing capabilities evolving and expanding, Qatar is constantly adopting improvements in technology and state-of-the-art facilities. Petroleum-manufactured materials other than oil and gas account for a large percentage of the economy, and this sector is a strong force behind the diversification of different sources of national income. Since the 1940s, petrochemicals have played a major role in improving the quality of our lives in health care, transportation, and agriculture. Petrochemicals such as plastics will also play a major role in helping us meet new challenges and shaping our future.

18.3 The Need for Engineers

The link between industry and the engineering profession is as old as industry itself. The various waves of industrial revolution that have taken place since the mid-eighteenth century have all been fueled by the power of engineering. From the textile

[1] Official IMF data 2012.

industry, railways, electricity, oil and gas, automobiles, and mass production to the more recent advances in telecommunications and smart manufacturing, the role of engineers is at center stage. Indeed, industrial revolutions in the eighteenth and nineteenth centuries have led to the establishment of formal engineering education schools.

The industrial growth of the region requires an adequate supply of engineers to sustain it. It is beyond the scope of this chapter to analyze the number of graduates and how many the industry needs. It is, however, clear that in a region such as the Gulf Co-operation Council (GCC), due to the tremendous growth in industries such as energy and construction, the current number of graduates does not satisfy the demand for engineering students. Since the supply and demand for engineers may vary from one MENA country to the other, it is difficult to discuss specific numbers. But throughout the region, it is extremely important that engineering schools and industry should work closely to identify the needs and assess future requirements so that schools can make better plans to meet the demand for specific engineering disciplines.

Deviating from quantitative needs of engineers toward the qualities industry requires in engineers, it is safe to state that problem solving, design and development, and communications competencies are at the core of industry-required competencies in graduate engineers. Since the region is becoming a hub for chemical and petrochemical production, it is important to illustrate the dependence on talented engineers needed by the industry to maintain its competitiveness and sustain its growth. The global trend shows that most developments are carried out internally within industrial organizations, as shown in the figure below.[2]

Industrial and academic collaboration comes in the second place, highlighting the importance of the ability of engineers in the industry to work with their counterparts in industry and academia to solve problems and develop new products and processes.

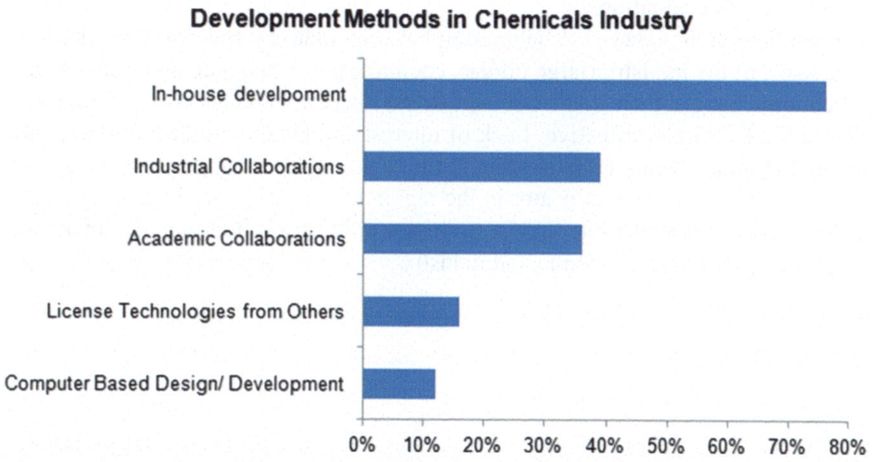

[2] Battelle R&D Magazine, 2014.

18.4 Engineering Education: An Industry Perspective

The MENA region is not homogeneous and the overall state of both its industrial base and engineering schools varies from one country to another. However, for the purpose of this chapter, the region is treated as a whole where the main qualities and skills required by industry from engineering education are the same.

As a leading petrochemical company in the MENA region, Qatar Petrochemical Company (QAPCO) has been depending on engineers, training them, and developing them for the last 30 years or so.[3] The following sections do not represent a comprehensive list of all the requirements by industry from an engineering education program; rather, they provide a main set of qualities that make a successful engineer from industry perspective.

18.4.1 Align Education and Research Programs with Industry Needs

As engineering schools develop program curricula and look for the best ways to prepare their students for the real world, it is extremely important that industry is involved at an early stage of this process. This is important in order to ensure alignment between academic programs and industrial needs. A number of engineering schools in the region have already obtained Accreditation Board for Engineering and Technology (ABET) accreditation, and others are working toward it. The involvement of industry experts and experienced engineers is also very important in this regard. Our experience shows that the input of industry experts into industrial advisory boards of engineering departments in Qatar is very valuable. Industry can assist on both ends of the education spectrum: curriculum development and program outcomes evaluation.

Even though the level of collaboration between industry and academia depends on the size of the industry (large oil/gas, chemical/petrochemical, and utilities tend to be more engaged than small-size industries), there are always areas of improvement to make it more effective. Lack of interest and clear cultural differences are among the issues facing collaboration in the region, according to an assessment of the situation by a veteran educator in the region (Akili 2012). In the next two subsections, we propose two interventions which would help in building alignment and collaboration between academia and industry.

[3] Qatar Petrochemical Company (QAPCO) was established in 1974 and started production of polyethylene in 1981 as the first plant of its kind in the region.

18.4.1.1 Engineering Faculty Residency Programs in Industries

Another possible way to improve coordination and industry input into engineering education programs is for industry to host engineering faculty members to spend a period of time on the industrial site. This exchange model proved effective in other parts of the world in industries such as aerospace (McMasters 2006). In a report titled "The Green Report: Engineering Education for a Changing World," the American Society for Engineering Education strongly recommends that engineering faculty members spend time in industry.[4] Even though circumstances in both industry and engineering schools may be different, we have no reason to believe that it will not work in the oil and gas or petrochemical industry in the MENA region. Such an exchange is expected to familiarize engineering faculty with industrial operations and facilitate transfer of technology from academia to industry. QAPCO has proposed this idea in various settings to the academic community; engineering education leaders and industry leaders should explore ways to formalize this exchange model.

18.4.1.2 Provisioning of Internships in Engineering Education

An internship is a student's first encounter with the industrial world and is critical to his/her development. The role of the internship program is to offer learning experiences that go beyond the theory taught in the classroom and can help the student gain a good understanding of the industry. It will also enhance the student's academic and career goals connecting academic experience with the professional world. Developing the technical knowledge and skills of students also provides a significant competitive advantage to the student and strengthens the collaboration between industry and academia.

The current QAPCO internship program is scheduled for 8 weeks during the months of July and August for the majority of students. In the first week, a 3-day inclusion course is organized to introduce the student to QAPCO. This course also includes health and safety training, issuing of personal protection equipment, and orientation to the various plants. Special training workshops are also organized in order to educate the students on subject matter and topics related to the organization and its operations. Lectures on leadership and motivation are also a part of this training.

At the end of the program, all technical students deliver a presentation detailing what they have learned during their time at QAPCO.

[4] American Society for Engineering Education (ASEE), 2010 report.

18.4.2 Industry Perspectives on Most Important Engineering Competencies

If we were to come up with a wish list of qualities (skills/competencies) desired in an engineer from an industry perspective, the following list would be representative of most industrial organizations in the region:

- Strong understanding of basic math and science
- Strong understanding of engineering fundamentals
- Good knowledge of design principles and manufacturing processes
- Safety awareness
- Strong communication skills
- Strong critical thinking and problem-solving skills
- Collaboration and teamwork
- High engineering ethics standards
- Awareness about diversity and working in a multicultural environment
- Practical competencies
- Research and innovation competencies

The role of a sound foundation in math, science, and engineering principles is well established. It is very important, however, to emphasize the role of "soft skills" in making a successful engineer as will be detailed in subsequent sections of this chapter. Our view of essential competencies of engineers has also been reported in several studies (Abdulwahed et al. 2013). Even though all the skills/competencies listed above are important, this chapter is not intended to treat each one of them in detail; rather, we will focus on some aspects that are extremely important from our perspective, namely critical thinking and problem solving.

18.4.2.1 Critical Thinking and Problem Solving

Soft skills are extremely important in the success of engineering graduates in the real world. From the perspective of an industrial organization, according to a Boeing Company executive, "Rather, engineering majors who fail in industry are those who have all the right technical competencies but not the soft or people skills to be successful. Specifically, they tend to lack the ability to work well in teams, communicate effectively, define problems and consider alternative and creative solutions" (Stephens 2013).

From the perspective of an industrial organization, engineering schools need to strive to make the "professional" engineer with the right "soft skills" as opposed to the "traditional" engineer who is limited to a strong technical background. Skills such as communication, leadership, and teamwork are essential to industrial organizations. But our experience and monitoring of graduating engineers who join the workforce shows that the ability to think critically and solve problems stands out as the skill that needs more attention in the region due to its vitality to successful industry operations in the MENA region.

Whether the engineer's task at hand is to design a process for a plant production line, or to find a solution to a global environmental issue, the innovative mind is his/her best tool. A United Nations Educational, Scientific and Cultural Organization (UNESCO) report on the role of engineering in development states that "resolving these issues will require tremendous innovation and ingenuity by engineers, working alongside other technical and non-technical disciplines. It requires the engineer's ability to synthesize solutions and not simply their ability to analyze problems" (Jowitt 2010).

Critical thinking is the use of those cognitive skills or strategies that increase the probability of a desirable outcome (Halpern 1996). Fresh engineers who join industrial companies will face issues and problems to resolve on a regular basis; it is a part of the job. A critical step in solving the problem is to identify it in the first place. Then he/she should be able to define the barriers (obstacles), explore the available options, decide on an agreed course of action and execute it, and, finally, evaluate (test) if it is working and be ready to take another course of action. It is extremely important that engineering programs provide students with the right set of tools to master this process.

From an industry perspective, problem solving is not the sole responsibility of senior engineers or managers. As problems faced by industry in the MENA region become more challenging and multifaceted, individuals at all levels (including engineers) are expected to contribute to problem solving. "Successful problem solving translates into enhanced productivity and increased profit" (Marone and Blauth 2004).

In the year 2013, QAPCO has registered 41,480 training hours for its workforce.[5] Through its intensive training programs, QAPCO Learning and Development tracks trainees individually in order to help them improve their skills and maximize their adjustment to the professional life. Feedback from trainers consistently stresses the importance of soft skills and the overall need for improvement in critical thinking and problem-solving skills.

18.4.2.2 Safety Awareness

Safety awareness and training are high on the priority of large industrial companies in the region. Whether in hydrocarbon processing, which represents a large portion of industrial activities in the region, or other industries, more focus will be devoted in the future to safety performance and overall record. Incidents such as the Bhopal disaster in India and the Piper Alpha accident in the North Sea put even more attention on process safety in particular. An examination of modern industrial plants reveals that operations are becoming more and more interconnected. The modern engineer is exposed to various aspects of plant operations and not just his/her specific work environment. Process safety, therefore, does not concern only chemical

[5] QAPCO Sustainability Report, 2013.

engineers; rather, all engineering disciplines should take into account the vital importance of teaching safety, process safety in particular.

Large oil and gas companies in the region, as an example, have extensive safety training for their engineers, but these programs would be more effective if new engineers come into a plant or a research lab with previous exposure to safety fundamentals. Safety should be integrated into design projects and lab environment, and engineering curricula should encourage a culture of safety.

18.4.2.3 Research and Innovation Competencies

The waves of industrial revolution mentioned in section 18.3 are sometimes referred to as waves of innovation and industrial development. They are also called Kondratiev waves, or K-waves. What matters for the purpose of this chapter is not the theory of cycles and waves in industrial development, but rather the word "Innovation," which is common to all those historical industrial developments. Research and innovation are directly related to and dependent on the soft skills mentioned in the previous section. Critical thinking, problem solving, and collaboration are key to successful R&D programs in industrial laboratories.

As the MENA region, and the GCC in particular, increases its industrial development activities, the role of research and innovation becomes critical for sustaining this growth. In the case of the State of Qatar, according to official government planning, the petrochemical sector is expected to significantly increase its production capacity by the year 2020.[6] The State of Qatar also set ambitious goals for itself in order to fulfill the Qatar National Vision 2013 (QNV 2030).[7] The industry is expected to align its activities and R&D goals with the national vision and contribute to the knowledge-based economy of the future.

If we take the example of the chemical industry, which is heavily dependent on new product and process innovation, it is clear that the region's spending on R&D is not up to the level of other regions of the world (see table below).[8] It is not our intent to analyze the R&D spending situation in the region, but one factor that would stimulate research and innovation programs in the industry is the availability of talented and well-prepared researchers from engineering schools in the MENA region.

Our present assessment is that there is not enough talent produced by the local engineering schools to meet the demand for researchers with engineering background. It is not only the number of graduate programs that matters but also the inclusion of research- and innovation-oriented activities in undergraduate programs that is needed.

[6] This ambitious plan to diversify the economy and grow the petrochemical sector is led by the Ministry of Energy and Industry of the State of Qatar.

[7] Information on QNV 2030 can be found at the following link:

[8] Note: Chemicals excluding pharmaceuticals, Source: CEFIC, 2013.

Global R&D spending in chemicals industry ($US million)	2012
EU-27	12,140.9
USA	10,763.5
China	10,229.0
Japan	9427.2
Korea, Republic	2177.0
India	1326.9
Switzerland	463.8
Australia	397.4
GCC	*380.0*
Taiwan	325.9
Rest of world	1320.7
World	*48,952.4*

18.5 Conclusions

Industrial development has always been linked to the ingenuity and innovation of engineering minds. As the MENA region invests in diversifying its economy and growing its industrial sector, it will critically need an adequate supply of locally educated engineers. Engineering education in the MENA region was examined from the perspective of a large industrial organization that has been depending on engineers, training them, and developing them for more than 30 years. The chapter discussed the critical engineering skills required by industry and the role of engineering education in helping industry in the region succeed and grow. Close collaboration between industry and engineering schools is very important in order to ensure that students are well prepared to join the workforce and contribute to solving industry's and society's problems. This collaboration should focus not only on technical abilities but also on developing soft skills. Critical thinking and problem solving represent must-have skills that are highly required for engineering graduates to succeed in industry. Based on our experience, these skills need more attention from engineering programs in the region in order to prepare the professional engineers needed by industry. Other important aspects of engineering education in the region include the role of internships, research and innovation, and the critical importance of integrating safety into engineering education.

Acknowledgments The author is grateful to QAPCO management for their support in completing this work. Special thanks to Dr. Mohamed Yousuf Al-Mulla (VC and CEO), Mr. Yousuf Rebeeh (COO), Mr. Abdulla Naji (learning and development manager), and Mr. Ahmad Yaseen (PR manager) for their input and support.

References

Abdulwahed, M., Balid, W., Hasna, M. O., & Pokharel, S. (2013). *Skills of engineers in knowledge based economies: A comprehensive literature review, and model development*. Teaching, Assessment and Learning for Engineering (TALE), 2013 IEEE International Conference on (pp. 759–765). IEEE.

Akili, W. (2012). Academia-industry collaboration, American Society for Engineering Education (ASEE) 2012 Annual Conference, St. Antonio Texas, June 2012, AC2012-3248.

Halpern, D. (1996). Thought and knowledge: An introduction to critical thinking. Lawrence Erlbaum Associates: Mahwah.

Jowitt, P. (2010). *Engineering: Issues, challenges and opportunities for development*. UNESCO Report, p 39.

Khair, R., Hasna, M. O., & Abdulwahed, M. (2012). *Transition into a knowledge based economy in Qatar: The role of engineering and engineering education*. Proceedings of the World Engineering Education Forum WEEF, 15th to 18th of October, Buenos Aires, Reno: Argentina.

Marone, M., & Blauth D. (2004). Creating a problem solving culture. Reno: AchieveGlobal Inc.

McMasters, J. (2006). Influencing Student Learning: An Industry Perspective. *Int Journal of Engineering Education, 22*(3), 447–459.

National Academy of Engineering. (2004). *The engineer of 2020: Visions of engineering in the new century*. Washington, DC: National Academies Press.

National Academy of Engineering. (2005). *Educating the engineer of 2020: Adapting engineering education to the new century*. Washington, DC: National Academies Press.

National Research Council. (Ed.). (2011). *Assessing 21st century skills: Summary of a workshop*. Washington, DC: National Academies Press.

Sheppard, K., Dominick, P., & Aronson, Z. (2004). Preparing engineering students for the new business paradigm of international teamwork and global orientation. *International Journal of Engineering Education, 20*(3), 475–483.

Shuman, L. J., Besterfield-Sacre, M., & McGourty, J. (2005). The ABET 'Professional Skills'—Can they be taught? Can they be assessed? *Journal of Engineering Education, 49*(1), 41–55.

Somerville, M., Anderson, D., Berbeco, H., Bourne, J. R., Crisman, J., Dabby, D., Donis-Keller, H., et al. (2005). The Olin Curriculum: Thinking Toward the Future. *IEEE Transactions on Education, 48*(1), 198–205.

Stephens, R. (2013). Aligning Engineering Education and Experience to Meet the Needs of Industry and Society. *The Bridge, National Academy of Engineering, 43*(2), 31–34.

Swearengen, J., & Barnes, S. (2002). Globalization and the undergraduate manufacturing engineering curriculum. *Journal of Engineering Education, 49*(1), 255–261.

Dr. Mabrouk Ouederni is the head of Research & Development at Qatar Petrochemical Company (QAPCO), a leading producer of polyethylene and other petrochemicals. Prior to joining QAPCO, he has spent nearly 15 years in the US polymer industry where he held responsibilities as a senior R&D engineer and a technical R&D manager with leading chemical companies such as Hoechst Celanese, Invista (KoSa), and Wacker Chemical Corporation. Dr. Mabrouk has also spent several years in academia with full-time faculty positions in polymer science and engineering at Sultan Qaboos University in Oman and Jubail University College in Saudi Arabia. He also interacts extensively with engineering students at Qatar University and Texas A&M University through lectures and research projects.

He is a member of the Research and Innovation Committee of the Gulf Petrochemical and Chemical Association (GPCA) and serves on several academic engineering advisory committees. He is the author of several papers and patents in the field of polymer science and engineering. He graduated with a BS/MS in materials science and engineering from Tunisia in 1989 and a PhD in polymer engineering from the University of Tennessee in the USA in 1994.

Chapter 19
An Insight into the Future of Engineering Education, Recommendations for Implementations, and Conclusions

Mahmoud Abdulwahed, Mazen O. Hasna and Jeffrey E. Froyd

19.1 Insights into the Future of Engineering Education and Research in the Middle East and North Africa

The Middle East and North Africa (MENA) region in general, and the Gulf Cooperation Council (GCC) region in particular, is a significantly dynamic and fast-changing part of the world, and when coupled with an ever-dynamic field (engineering), prediction becomes a real challenge. However, understanding current challenges and practices helps a lot in predicting what might help in avoiding such challenges in the future.

Most, if not all, engineering curricula in the GCC region follow the North American model of education, and hence will follow to a large extent the changes that will happen globally in that regard. Nevertheless, the specificity of the region will make it even more essential in the future to contextualize the curricula to the local needs (more on this is presented in Sect. 19.3), especially to focus on the roles of engineering disciplines in transforming and maintaining a knowledge-based economy (KBE), and to drive innovations. On the delivery part, an increasing focus is expected to be given to student motivation and leadership skills.

The other dimension has to do with the growth of the population in the region. The role of private engineering schools is expected to grow (as the engineering profession is still looked at as a high-standard career, and there is a high demand for various engineering disciplines), and quality assurance on the national and regional levels should play an important role to keep the standards that have traditionally been maintained by governmental schools.

Another important dimension is the need to enhance student mobility among engineering schools in the GCC region. Despite the similarity in the curricula, mo-

M. Abdulwahed (✉) · M. O. Hasna
Qatar University, Doha, Qatar
e-mail: m.abdulwahed@qu.edu.qa

J. E. Froyd
Engineering Academic and Student Affairs, Texas A&M University,
College Station, TX, USA

M. Abdulwahed et al. (eds.), *Advances in Engineering Education in the Middle East and North Africa*, DOI 10.1007/978-3-319-15323-0_19

bility of students is weak and there will be a need to develop a regional framework for the recognition of mutual certificates and articulations for credit transfer among engineering schools.

Finally, the growth of graduate studies and research in the GCC region has been led so far by engineering colleges. Given that the focus on research and graduate studies is expected to grow (in line with the KBE focus), engineering schools should be ready to partner with other disciplines (especially in the medical field) to create new research directions and, possibly, new multidisciplinary programs. The focus on topics such as energy, water, and cyber security is becoming key in almost all GCC countries, and alignment of efforts is becoming essential in the future for more impact.

In conclusion, the economy of the GCC region will continue to be carbon-based for years to come. In the transformation to KBE, the role of engineering is central (Abdulwahed et al. 2013), and the responsibility of engineering schools will be to make sure that they are in the driving seat and not in the passenger one. The first boom was missed by engineers back in the 1970s and the 1980s when they could have played a bigger role in utilizing the production of oil into more added-value industries. The chance to contribute and lead is available again, and it is for the engineers to decide where to sit.

19.2 Insights into the Future of Engineering Education and Research in North America

Predicting the future of engineering education, like predicting the evolution of any field, is more likely wrong than right. Therefore, authors venturing into this territory are more likely to offer a vision for what they hope will occur than definitive forecasts for what will happen. At the beginning of the third millennium, four prominent engineering educators offered their six-part vision for engineering education: (i) vision, (ii) effective teaching methods, (iii) approaches for developing critical skills in engineering graduates, (iv) approaches for supporting engineering educations as they learned to teach, (v) evaluating teaching, and (vi) promoting reform (Felder et al. 2000a, b, c; Rugarcia et al. 2000; Stice et al. 2000; Woods et al. 2000). In the USA, the National Academy of Engineering took its turn in a program entitled "The Engineer of 2020." The first volume (National Academy of Engineering 2004) described desired attributes for engineering graduates who could address complex socio-technological challenges of the twenty-first century. The second volume (National Academy of Engineering 2005) laid pathways for transforming engineering education programs so that engineering graduates in 2020 would possess the desired attributes. At this point, it appears that "The Engineer of 2020" will not arrive within the initially envisioned time frame. A third attempt at envisioning the future of engineering education and taking steps toward realizing the vision was made by the American Society of Engineering Education, again in a series of two reports. The first report asserted that "innovation in engineering education depends on a vibrant community of practitioners and researchers working in collaboration to advance the frontiers of knowledge and practice…based on a continual cycle of educational practice and research" (Jamieson and Lohmann 2009, p. 1). Furthermore, it noted that the primary "responsibility for the quality of the engineering educational experience rests with the engineering faculty and administrators" (Jamieson and Lohmann 2009, p. 1). Consequently, the need is to "strengthen career-long professional development in teaching and learning, starting with the doctoral programs that produce most engineering faculty" (Jamieson and Lohmann 2009, p. 1). The report concluded that expansion of the needed community of practitioners and researchers, as well as professional development programs, should be based on "a scientifically credible and shared knowledge base" (Jamieson and Lohmann 2009, p. 1) for engineering learning and teaching. The second report did not describe pathways for achieving the vision; instead, it "set out to understand the current 'state of the culture' by conducting a survey of faculty committees, chairs, and deans" (Jamieson and Lohmann 2012, p. 6). It contained several observations about the current culture and offered a set of recommendations for transforming the people and institutions from which the culture is derived. Readers interested in the observations and recommendations are encouraged to obtain a copy of either the executive summary or full report. Recently, the Curricula and Pedagogy Committee of the Institute of Electrical and Electronics Engineers (IEEE) generated four scenarios for possible futures of engineering education (Froyd et al. 2014). Based on its scenario generation exercise, the committee concluded that the two most important uncertainties in forecasting the future of engineering education were the "degree to which engineering programs adapt and innovate" (Froyd et al. 2014, p. 3) and "values and competencies of faculty members" (Froyd et al. 2014, p. 3). For the first uncertainty, the extremes of programmatic adaptation were ineffective or very effective adaptation. For the second uncertainty, the extremes were maintaining the current priority on disciplinary research or valuing student learning. Combining the two extremes for each of the two uncertainties led to the generation of four scenarios. To gather information to estimate the likelihood of the four scenarios, the committee, with the support of the IEEE, conducted a survey of engineering educators across the world (Ohland et al. 2014). Results of the survey are still being analyzed, but preliminary results suggest that adoption of new assessment and instructional approaches will occur slowly because 2176 survey respondents tended to prefer instructional approaches, assessment approaches, and instructional technologies that were similar to their current practices.

19.3 Future Recommendations and Concluding Remarks of the Book

Recent advances in engineering education practice, scholarship, and research have brought ample evidence on the importance of various recommendations for enhancement of engineering education. While the majority of engineering institutions in the MENA region follow classical approaches for engineering education, some engineering colleges started to adopt further innovations and novel practices. In this section, we provide a set of recommendations for implementation in engineering colleges. These recommendations are generally applicable for MENA institutions and elsewhere. The proposed set of recommendations spans four dimensions: (1) context and industry, (2) innovation and venturing, (3) curriculum and structure, and (4) pedagogy and scholarship:

- Context and industry
  - Contextualize engineering education to the local needs while keeping a global flavor
  - Foster inherent linkage with industry
- Innovation and venturing
  - Foster multidisciplinary design
  - Focus on technology and engineering entrepreneurship in engineering education
  - Focus on technology and engineering leadership in engineering education
- Curriculum and structure
  - Deploy an integrated engineering education curriculum with focus on theoretical knowledge and technical and soft competencies development
  - Develop and deploy structured first-year engineering experience programs
  - Develop and deploy structured pathways for engineering faculty development in engineering innovation and education
  - Develop and deploy driving organizational structures in the form of a center or department for engineering innovation and education
- Pedagogy and scholarship
  - Foster constructivist pedagogy and experiential approaches in engineering education
  - Utilize recent innovations in technology-enhanced and digital learning in engineering education
  - Foster a culture of conducting scholarship and research in engineering innovation and education among engineering faculty

The book provided a synopsis of recent innovations in the practice of engineering education in the MENA region. While the culture of conducting scholarship in engineering education is still in its infancy in the MENA region, a noticeable interest in deploying novel practices in engineering education has been observed. The book is one of the early documented scholarly works on recent innovations in engineering education in the MENA region, and it is hoped that it will contribute to the establishment of a further systematic approach to sustainable innovations in engineering education in the region.

References

Abdulwahed, M., Balid, W., Hasna, M. O., & Pokharel, S. (2013). *Skills of engineers in knowledge based economies: A comprehensive literature review, and model development.* Teaching, Assessment and Learning for Engineering (TALE), 2013 IEEE International Conference on (pp. 759–765). IEEE.

Felder, R. M., Rugarcia, A., & Stice, J. E. (2000a). The future of engineering education V. Assessing teaching effectiveness and educational scholarship. *Chemical Engineering Education, 34*(3), 198–207.

Felder, R. M., Stice, J. E., & Rugarcia, A. (2000b). The future of engineering education VI. Making reform happen. *Chemical Engineering Education, 34*(3), 208–215.

Felder, R. M., Woods, D. R., Stice, J. E., & Rugarcia, A. (2000c). The future of engineering education II. Teaching methods that work. *Chemical Engineering Education, 34*(1), 26–39.

Froyd, J. E., Lord, S. M., Ohland, M. W., Prahallad, K., Lindsay, E. D., & Dicht, B. (2014). *Scenario planning to envision potential futures for engineering education.* Paper presented at the Frontiers in Education Conference, Madrid, Spain.

Jamieson, L. H., & Lohmann, J. R. (2009). *Creating a culture for scholarly and systematic innovation in engineering education, Phase 1.* Washington, DC: American Society for Engineering Education.

Jamieson, L. H., & Lohmann, J. R. (2012). *Innovation with impact: Creating a culture for scholarly and systematic innovation in engineering education.* Washington, DC: American Society for Engineering Education.

National Academy of Engineering. (2004). *The engineer of 2020: Visions of engineering in the new century.* Washington, DC: The National Academies Press.

National Academy of Engineering. (2005). *Educating the engineer of 2020: Adapting engineering education to the new century.* Washington, DC: The National Academies Press.

Ohland, M. W., Dicht, B., Froyd, J. E., Lindsay, E. D., Lord, S. M., & Prahallad, K. (2014). *Using an international survey to inform scenario of the future of engineering education.* Paper presented at the Australasian Association for Engineering Education Conference, Wellington, New Zealand.

Rugarcia, A., Felder, R. M., Woods, D. R., & Stice, J. E. (2000). The future of engineering education I. A vision for a new century. *Chemical Engineering Education, 34*(1), 16–25.

Stice, J. E., Felder, R. M., Woods, D. R., & Rugarcia, A. (2000). The future of engineering education IV. Learning how to teach. *Chemical Engineering Education, 34*(2), 118–127.

Woods, D. R., Felder, R. M., Rugarcia, A., & Stice, J. E. (2000). The future of engineering education III. Developing critical skills. *Chemical Engineering Education, 34*(2), 108–117.

Mahmoud Abdulwahed completed his BSc, MSc, and PhD in electrical, control, and systems engineering. He continued his postgraduate studies in Germany, Sweden, France, and the UK where he worked before joining the College of Engineering at Qatar University in fall 2011. His main expertise is in innovation and education. He has published in areas of technology-enhanced and e-learning, engineering and STEM education, entrepreneurship, and leadership. Mahmoud has written more than 60 peer-reviewed conference and journal articles, and attained a number of recognitions, awards, and fellowships alongside his studies and career. He is currently adviser of strategic initiatives for the vice president and chief academic officer at Qatar University, where his role focuses on inception, conceptualization, and/or management of large-scale strategic initiatives; he is also a faculty member at the College of Engineering.

Mazen O. Hasna received the BS degree from Qatar University, Doha, Qatar, in 1994; the MS degree from the University of Southern California (USC), Los Angeles, in 1998; and the PhD degree from the University of Minnesota, Twin Cities, in 2003—all in electrical engineering.

In 2003, Dr. Hasna joined the Department of Electrical Engineering at Qatar University as an assistant professor. Currently, he serves as the vice president and chief academic officer of Qatar University. His research interests span the general area of digital communication theory and its application to performance evaluation of wireless communication systems over fading channels. His current specific research interests include cooperative communications, ad hoc networks, cognitive radio, and network coding.

Jeffrey E. Froyd is a Texas Engineering Experiment Station (TEES) research professor in the Office of Engineering Academic and Student Affairs at Texas A&M University, College Station. He received the BS degree in mathematics from the Rose-Hulman Institute of Technology and the MS and PhD degrees in electrical engineering from the University of Minnesota, Minneapolis. He was an assistant professor, associate professor, and professor of electrical and computer engineering at the Rose-Hulman Institute of Technology. At Rose-Hulman, he cocreated the Integrated, First-Year Curriculum in Science, Engineering and Mathematics, which was recognized in 1997 with a Hesburgh Award Certificate of Excellence. He served as a project director with a National Science Foundation (NSF) Engineering Education Coalition in which six institutions systematically renewed, assessed, and institutionalized innovative undergraduate engineering curricula. He has authored more than 70 journal articles and conference papers and offered more than 30 workshops on faculty development, curricular change processes, curriculum redesign, and assessment. He has served as a program cochair for three Frontiers in Education conferences and the general chair for the 2009 conference. He also serves on the IEEE Curricula and Pedagogy Committee, which is part of the University Resources Committee, which is part of the Educational Activities Board. Prof. Froyd is a fellow of the IEEE, a fellow of the American Society for Engineering Education (ASEE), an Accreditation Board for Engineering and Technology Accreditation (ABET) program evaluator, the editor in chief for the *IEEE Transactions on Education*, a senior associate editor for the *Journal of Engineering Education*, and an associate editor for the *International Journal of STEM Education*.

Index

M. Abdulwahed et al. (eds.), *Advances in Engineering Education in the Middle East and North Africa,* DOI 10.1007/978-3-319-15323-0

The manufacturer's authorised representative in the EU is Springer Nature Customer Service Centre GmbH, Europaplatz 3, 69115 Heidelberg, Germany. If you have any concerns regarding our products, please contact ProductSafety@springernature.com

Printed and bound by CPI Group (UK) Ltd, Croydon, CR0 4YY
30/09/2026
02225572-0017